OXFORD PAPERBACK REFERENCE

A Dictionary of
Biology

Oxford
Paperback
Reference

The most authoritative and up-to-date reference books for
both students and the general reader.

A Dictionary of

Biology

THIRD EDITION

Oxford   New York

OXFORD UNIVERSITY PRESS

Oxford University Press, Walton Street, Oxford OX2 6DP

Oxford New York
Athens Auckland Bangkok Bombay
Calcutta Cape Town Dar es Salaam Delhi
Florence Hong Kong Istanbul Karachi
Kuala Lumpur Madras Madrid Melbourne
Mexico City Nairobi Paris Singapore
Taipei Tokyo Toronto

and associated companies in
Berlin Ibadan

Oxford is a trade mark of Oxford University Press

First published 1985 as A Concise Dictionary of Biology
Second edition 1990
Third edition 1996

British Library Cataloguing in Publication Data
Data available
ISBN 0-19-280032-9

10 9 8 7 6 5 4 3 2

Printed in Great Britain by
Mackays PLC
Chatham, Kent.

Preface

This dictionary is derived from the *Concise Science Dictionary*, first published by Oxford University Press in 1984 (third edition, 1996). It consists of all the entries relating to biology and biochemistry in this dictionary, together with those entries relating to geology that are required for an understanding of palaeontology and soil science and a few entries relating to physics and chemistry that are required for an understanding of the physical and chemical aspects of biology (including laboratory techniques for analysing biological material). It also includes a selection of the words used in medicine and palaeoanthropology. For this new edition over 1500 new entries have been added, including terms in molecular genetics and biotechnology (including genetic engineering) to reflect recent advances in these fields. In addition, the coverage of human biology, environmental science, and food technology and production has been expanded in accordance with the requirements of the GCSE and A-level biology syllabuses. Taxonomic entries have been updated to reflect the five-kingdom system of classification, now widely accepted and taught, and many phyla have been included as new entries. Over 60 new illustrations have been added for this third edition.

An asterisk placed before a word used in an entry indicates that this word can be looked up in the dictionary and will provide further explanation or clarification. However, not every word that appears in the dictionary has an asterisk placed before it. Some entries simply refer the reader to another entry, indicating either that they are synonyms or abbreviations or that they are most conveniently explained in one of the dictionary's longer articles. Synonyms and abbreviations are usually placed within brackets immediately after the headword. Terms that are explained within an entry are highlighted by being printed in italic type.

The more chemical aspects of biochemistry and the chemistry itself will be found in *A Dictionary of Chemistry*; this and *A Dictionary of Physics* are companion volumes to this dictionary.

SI units are used throughout this book and its companion volumes.

1996   E.A.M.

Editors

Elizabeth Martin MA
Michael Ruse BSc, PhD
Elaine Holmes BSc, PhD

Advisers

B. S. Beckett BSc, BPhil, MA (Ed)
R. A. Hands BSc
Michael Lewis MA
W. D. Phillips PhD

Contributors

Tim Beardsley BA
Lionel Bender BSc
Belinda Cupid MSc, PhD
E. K. Daintith BSc
Malcolm Hart BSc, MIBiol
Robert S. Hine BSc, MSc

Anne Lockwood BSc
J. Valerie Neal BSc, PhD
R. A. Prince MA
Brian Stratton BSc, MSc
Elizabeth Tootill BSc, MSc

A band The region of a striated muscle fibre that contains both thick (myosin) and thin (actin) filaments. It is visible as a dark band with a lighter central zone (*see* **H zone**) in the middle of a *sarcomere.

abdomen The posterior region of the body trunk of animals. In vertebrates it contains the stomach and intestines and the organs of excretion and reproduction. It is particularly well defined in mammals, being separated from the *thorax by the *diaphragm. In many arthropods, such as insects and spiders, it may be segmented.

abductor (levator) A type of muscle whose function is to move a limb away from the body. Abductors work antagonistically with *adductors.

abiogenesis The origin of living from nonliving matter, as by *biopoiesis. *See also* **spontaneous generation**.

abiotic factor Any of the nonliving factors that make up the *abiotic environment* in which living organisms occur. They include *edaphic factors and all the aspects of *climate, geology, and atmosphere that may affect the biotic environment. *Compare* **biotic factor**.

abomasum One of four chambers that form the stomach of ruminants. The abomasum is referred to as the 'true stomach' as it is in this chamber that protein digestion occurs, in acidic conditions. *See* **Ruminantia**.

ABO system One of the most important human *blood group systems. The system is based on the presence or absence of *antigens A and B on the surface of red blood cells and *antibodies against these in blood serum. A person whose blood contains either or both these antibodies cannot receive a transfusion of blood containing the corresponding antigens as this would cause the red cells to clump (*see* **agglutination**). The table illustrates the

Group	Antigens on red cell surface	Antibodies in serum	Blood group of people donor can receive blood from	Blood group of people donor can give blood to
A	A	anti-B	A,O	A, AB
B	B	anti-A	B,O	B, AB
AB	A and B	none	A,B, AB, O	AB
O	neither A nor B	anti-A and anti-B	O	A, B, AB, O

The ABO blood group system

basis of the system: people of blood group O are described as 'universal donors' as they can give blood to those of any of the other groups. *See also* **immune response**.

abscisic acid A naturally occurring plant *growth substance that promotes leaf ageing, leaf fall (*see* **abscission**), and *apical dominance and induces *dormancy in seeds and buds.

abscission The separation of a leaf, fruit, or other part from the body of a plant. The process is controlled by growth substances, notably *abscisic acid; it involves the formation of an *abscission zone*, at the base of the part, within which a layer of cells (*abscission layer*) breaks down.

absolute refractory period *See* **refractory period**.

absorption The movement of fluid or a dissolved substance across a cell membrane. In animals, for example, soluble food material is absorbed into the circulatory system through cells lining the alimentary canal. In plants, water and mineral salts are absorbed from the soil by the *roots. *See* **osmosis**.

absorption spectrum *See* **spectrum**.

abyssal zone The lower depths of the ocean (below approximately 2000 metres), where there is effectively no light penetration. Abyssal organisms are adapted for living under high pressures in cold dark conditions.

acceptor (in biochemistry) A compound, molecule, ion, etc., to which electrons are donated in the formation of a coordinate bond.

accessory bud A bud that is situated at the side of or above an axillary bud (*see* **axil**).

accessory chromosome Any nonessential chromosome that is found in addition to the regular *karyotype of a species; in plants such chromosomes are known as *B-chromosomes*. Accessory chromosomes are heterochromatic (*see* **chromatin**). The frequency of accessory chromosomes in a species varies among populations but their presence does not appear to alter the appearance (phenotype) of the organisms.

accessory pigment A *photosynthetic pigment that traps light energy and channels it to chlorophyll *a*, the primary pigment, which initiates the reactions of photosynthesis. Accessory pigments include the *carotenoids, *phycobiliproteins, and chlorophylls *b*, *c*, and *d*.

acclimatization **1.** The progressive adaptation of an organism to any change in the environment that subjects it to physiological stress. **2.** The overall sum of processes by which an organism attempts to compensate for conditions that would substantially reduce the amount of oxygen delivered to its cells.

accommodation **1.** *(in animal physiology)* Focusing: the process by which the focal length of the *lens of the eye is changed so that clear images of objects at a range of distances are displayed on the retina. In man and some other mammals accommodation is achieved by reflex adjustments in the shape of the lens brought about by relaxation and contraction of muscles within the *ciliary body. **2. (in botany)** The ability of a plant to adapt itself to gradually changing environmental conditions. **3. (in animal behaviour)** Adjustments made by an animal's nervous or sensory systems in response to continuously changing environmental conditions.

acellular Describing tissues or organisms that are not made up of separate cells but often have more than one nucleus (see **syncytium**). Examples of acellular structures are muscle fibres. *Compare* **unicellular**.

acentric Describing an aberrant chromosome fragment that lacks a centromere. Such fragments are normally lost because they are unable to orientate properly during cell division.

acetabulum (cotyloid cavity) The semicircular cavity in the *pelvic girdle that houses the ball-shaped head of the *femur.

acetate (ethanoate) A salt or ester of acetic acid.

acetic acid (ethanoic acid) A carboxylic acid, CH_3COOH, that is used as a carbon source by certain green algae. Combined with coenzyme A (see **acetyl coenzyme A**), it plays a crucial role in energy metabolism.

acetone See **ketone; ketone body**.

acetylation The introduction of an acetyl group (CH_3CO-) into a compound. The acetylation of *coenzyme A to acetyl coenzyme A is an important stage in the *Krebs cycle: the acetyl group is derived from pyruvate after the removal of a molecule of carbon dioxide and two hydrogen atoms.

acetylcholine (ACh) A substance that is released at some (*cholinergic*) nerve endings (see **neurotransmitter**). Its function is to pass on a nerve *impulse to the next nerve (i.e. at a *synapse) or to initiate muscular contraction. Once acetylcholine has been released it has only a transitory effect because it is rapidly broken down by the enzyme *cholinesterase.

acetylcholinesterase See **cholinesterase**.

acetyl coenzyme A (acetyl CoA) A compound formed in the mitochondria when an acetyl group (CH_3CO-), derived from the breakdown of fats, proteins, or carbohydrates (via *glycolysis), combines with the thiol group ($-SH$) of *coenzyme A. Acetyl CoA feeds into the energy generating *Kreb's cycle and also plays a role in the synthesis and oxidation of fatty acids.

achene A dry indehiscent fruit formed from a single carpel and containing a single seed. An example is the feathery achene of clematis. Variants of the achene include the *caryopsis, *cypsela, *nut, and *samara. *See also* **etaerio**.

acid A type of compound that contains hydrogen and dissociates in water to produce positive hydrogen ions. The reaction, for an acid HX, is commonly written:

$$HX \leftrightarrows H^+ + X^-$$

In fact, the hydrogen ion (the proton) is hydrated, and the complete reaction is:

$$HX + H_2O \leftrightarrows H_3O^+ + X^-$$

The ion H_3O^+ is the *hydroxonium ion*. The strength of an acid depends on the extent to which it dissociates: *strong acids* (e.g. sulphuric acid and hydrochloric acid) are almost completely dissociated in water; *weak acids* (e.g. carbonic acid) are only partially dissociated. *See also* **buffer; pH scale**. *Compare* **base**.

acid–base balance The regulation of the concentrations of acids and bases in blood and other body fluids so that the *pH remains within a physiologically acceptable range. This is achieved by the presence of natural *buffer systems, such as the haemoglobin, bicarbonate ions, and carbonic acid in mammalian blood. By acting in conjunction, these effectively mop up excess acids and bases and therefore prevent any large shifts in blood pH. The acid–base balance is also influenced by the selective removal of certain ions by the kidneys and the rate of removal of carbon dioxide from the lungs.

acidic stains *See* **staining**.

acidosis A condition in which the body fluids become more acidic, i.e. the pH is less than 7.4, and the capacity of the body to *buffer hydrogen ions is diminished. A decrease in the elimination of carbon dioxide from the body gives rise to *respiratory acidosis*, while a deficiency of hydrogencarbonate results in *metabolic acidosis*.

acid protease A protein-digesting enzyme (*see* **protease**) that exhibits maximum activity and stability in acid conditions (pH 2.0–5.0) and is inactivated at pH values above 6.0. Acid proteases have a low *isoelectric point and are low in basic amino acids. Two types are widely used in the food and beverage industries: those from *Aspergillus*, which resemble pepsin; and those from *Mucor*, which resemble rennin.

acid rain Precipitation having a pH value of less than about 5.0, which has adverse effects on the fauna and flora on which it falls. Rainwater typically has a pH value of 5.6, due to the presence of dissolved carbon dioxide (forming carbonic acid). Acid rain results from the emission into the atmosphere of various pollutant gases, in particular sulphur dioxide and various oxides of nitrogen, which originate from the burning of fossil fuels and from car exhaust fumes, respectively. These gases dissolve in atmospheric water to form sulphuric and nitric acids in rain, snow, or hail (*wet deposition*). Alternatively, the pollutants are deposited as gases or minute particles (*dry deposition*). Both types of acid deposition affect plant growth – by damaging the leaves and impairing photosynthesis and by increasing the acidity of the soil, which results in the leaching of essential nutrients. This acid pollution of the soil also leads to acidification of water draining from the soil into lakes and rivers, which become unable to support fish life. Lichens are particularly sensitive to changes in pH and can be used as indicators of acid pollution (*see* **indicator species**).

acinus (*pl.* **acini**) The smallest unit of a multilobular gland, such as the pancreas. Each acinus in the pancreas is made up of a hollow cluster of *acinar cells*, which produce the digestive enzymes secreted in pancreatic juice. Minute ducts from the pancreatic acini eventually drain into the pancreatic duct.

acoelomate Describing any bilaterally symmetrical animal of the subkingdom Metazoa that does not possess a *coelom (*see also* **body cavity**). Examples of acoelomate animals are the platyhelminths.

acquired characteristics Features that are developed during the lifetime of an individual, e.g. the enlarged arm muscles of a tennis player. Such characteristics are not genetically controlled and cannot be passed on to the next generation. *See also* **Lamarckism; neo-Lamarckism**.

acquired immune deficiency syndrome *See* **AIDS**.

acquired immunity *See* **immunity**.

Acridine

acridine A chemical (see formula) that is capable of causing *frameshift mutations in the DNA sequence. Several derivatives of acridine (such as acridine orange) are used as dyes or biological stains.

acrocentric *See* **centromere**.

acromegaly A chronic condition developing in adulthood due to overproduction of (or oversensitivity to) *growth hormone, usually caused by a tumour in the pituitary gland. This leads to a gradual enlargement of the bones, causing characteristic coarsening of the facial features and large hands and feet.

acrosome *See* **spermatozoon**.

ACTH (adrenocorticotrophic hormone, corticotrophin) A hormone produced by the anterior *pituitary gland in response to stress that controls secretion of certain hormones (the *corticosteroids) by the adrenal glands. It can be administered by injection to treat such disorders as rheumatic diseases and asthma, but it only relieves symptoms and is not a cure.

actin A contractile protein found in muscle tissue, in which it occurs in the form of filaments (called thin filaments). Each thin filament consists of two chains of actin molecules, around which is twisted a strand of *tropomyosin and interspersed *troponin. Units of muscle fibre (*see* **sarcomere**) consist of actin and *myosin filaments, which interact to bring about muscle contraction (*see also* **sliding filament theory**). Actin is also found in the *microfilaments* that form part of the *cytoskeleton of all cells.

Actinobacteria (Actinomycetes; Actinomycota) A phylum of Gram-positive mostly anaerobic nonmotile bacteria. All species are fungus-like, with filamentous cells producing reproductive spores on aerial branches similar to the spores of certain moulds. The phylum includes bacteria of the genera *Actinomyces*, some species of which cause disease in animals and man; and *Streptomyces*, which are a source of many important antibiotics (including streptomycin).

actinomorphy *See* **radial symmetry.**

Actinomycetes *See* **Actinobacteria.**

action potential The change in electrical potential that occurs across a cell membrane during the passage of a nerve *impulse. As an impulse travels in a wavelike manner along the *axon of a nerve, it causes a localized and transient switch in electric potential across the cell membrane from −60 mV (millivolts; the *resting potential) to +45 mV. The change in electric potential is caused by an influx of sodium ions. Nervous stimulation of a muscle fibre has a similar effect.

action spectrum A graphical plot of the efficiency of electromagnetic radiation in producing a photochemical reaction against the wavelength of the radiation used. For example, the action spectrum for photosynthesis using light shows a peak in the region 670–700 nm. This corresponds to a maximum absorption in the absorption *spectrum of chlorophylls in this region.

activated sludge process A *sewage and waste-water treatment. The sludge produced after primary treatment is pumped into aeration tanks, where it is continuously stirred and aerated, resulting in the formation of small aggregates of suspended colloidal organic matter called *floc*. Floc contains numerous slime-forming and nitrifying bacteria, as well as protozoans, which decompose organic substances in the sludge. Agitation or air injection maintains high levels of dissolved oxygen, which helps to reduce the *biochemical oxygen demand. Roughly half the sewage in Britain is treated using this method.

activation energy Symbol E_a. The minimum energy required for a chemical reaction to take place. In a reaction, the reactant molecules come together and chemical bonds are stretched, broken, and formed in producing the products. During this process the energy of the system increases to a maximum, then decreases to the energy of the products. The activation energy is the difference between the maximum energy and the energy of the reactants; i.e. it is the energy barrier that has to be overcome for the reaction to proceed. The activation energy determines the way in which the rate of the reaction varies with temperature. It is usual to express activation energies in joules per mole of reactants.

activator 1. A compound (usually a protein) that is capable of initiating the expression of a specific gene (*see* **gene expression**). *Compare* **repressor**. **2.** A substance that − by binding to an *allosteric site on an enzyme − enables the active site of the enzyme to bind to the substrate. **3.** Any compound that potentiates the activity of a drug or other foreign substance in the body.

active immunity *Immunity acquired due to the body's response to a foreign antigen.

active site (**active centre**) The site on the surface of an *enzyme molecule that binds the substrate molecule. The properties of an active site are determined by the three-dimensional arrangement of the polypeptide chains of the enzyme and their constituent amino acids. These govern the nature of the interaction that takes place and hence the degree of substrate specificity and susceptibility to *inhibition.

active transport The movement of substances through membranes in living cells, often against a *concentration gradient: a process requiring metabolic energy. Organic molecules and inorganic ions are transported into and out of both cells and their organelles. The substance binds to a *carrier molecule embedded in the membrane, which carries it through the membrane and releases it on the opposite side. Active transport serves chiefly to maintain the normal balance of ions in cells, especially the concentration gradients of sodium and potassium ions crucial to the activity of nerve and muscle cells. *Compare* **facilitated diffusion**.

actomyosin The complex formed from the interaction of the proteins *actin and *myosin during the process of muscle contraction. *See also* **sliding filament theory**.

adaptation 1. *(in evolution)* Any change in the structure or functioning of an organism that makes it better suited to its environment. *Natural selection of inheritable adaptations ultimately leads to the development of new species. Increasing adaptation of a species to a particular environment tends to diminish its ability to adapt to any sudden change in that environment. **2.** *(in physiology)* The alteration in the degree of sensitivity (either an increase or a decrease) of a sense organ to suit conditions more extreme than normally encountered. An example is the adjustment of the eye to vision in very bright or very dim light.

adaptive radiation (**divergent evolution**) The evolution from one species of animals or plants of a number of different forms. As the original population increases in size it spreads out from its centre of origin to exploit new habitats and food sources. In time this results in a number of populations each adapted to its particular habitat: eventually these populations will differ from each other sufficiently to become new species.

A good example of this process is the evolution of the Australian marsupials into species adapted as carnivores, herbivores, burrowers, fliers, etc. On a smaller scale, the adaptive radiation of the Galapagos finches provided Darwin with crucial evidence for his theory of evolution (*see* **Darwin's finches**).

adductor (depressor) A type of muscle whose function is to pull a limb inwards, towards the body of an animal. *Compare* **abductor**.

adenine A *purine derivative. It is one of the major component bases of *nucleotides and the nucleic acids *DNA and *RNA.

adenohypophysis *See* **pituitary gland**.

adenosine A nucleoside comprising one adenine molecule linked to a D-ribose sugar molecule. The phosphate-ester derivatives of adenosine, AMP, ADP, and *ATP, are of fundamental biological importance as carriers of chemical energy.

adenosine diphosphate (ADP) *See* ATP.

adenosine monophosphate (AMP) *See* ATP.

adenosine triphosphate *See* ATP.

adenovirus One of a group of DNA-containing viruses found in rodents, fowl, cattle, monkeys, and man. In man they produce acute respiratory-tract infections with symptoms resembling those of the common cold. They are also implicated in the formation of tumours (*see* **oncogenic**).

adenylate cyclase The enzyme that catalyses the formation of *cyclic AMP. It is bound to the inner surface of the plasma membrane. Many hormones and other chemical messengers exert their physiological effects by increased synthesis of cyclic AMP through the activation of adenylate cyclase. The hormone binds to a receptor on the outer surface of the plasma membrane, which then activates adenylate cyclase on the inner surface via *G-protein.

ADH *See* **antidiuretic hormone**.

adipocyte *See* **fat cell**.

adipose tissue A body tissue comprising cells containing *fat and oil. It is found chiefly below the skin (*see* **subcutaneous tissue**) and around major organs (such as the kidneys and heart), acting as an energy reserve and also providing insulation and protection.

adjuvant A nonantigenic substance (such as aluminium hydroxide) that, in combination with an antigen, enhances antibody production by inducing an inflammatory response, which leads to a local influx of antibody-forming cells. Adjuvants are used therapeutically in the preparation of vaccines, since they increase the production of antibodies against small quantities of antigen and lengthen the period of antibody production.

adolescence The period in human development that occurs during the teenage years, between the end of childhood and the start of adulthood, and is characterized by various physical and emotional changes associated with development of the reproductive system. It starts at *puberty*, when the reproductive organs begin to function, and is marked by the start of menstruation (*see* **menstrual cycle**) in females and the appearance of the *secondary sexual characteristics in both sexes. In males the secondary sexual characteristics are controlled by the hormone testosterone and include deepening of the voice due to larynx enlargement, the appearance of facial and pubic hair, rapid growth of the skeleton and muscle, and an increase in *sebaceous gland secretions. In females the secondary sexual characteristics are controlled by oestrogens and include growth of the breasts, broadening of the pelvis, redistribution of fat in the body, and appearance of pubic hair.

ADP *See* ATP.

adrenal cortex The outer layer of the *adrenal gland, in which several steroid hormones, the *corticosteroids, are produced.

adrenal glands A pair of endocrine glands situated immediately above the kidneys (hence they are also known as the *suprarenal glands*). The inner portion of the adrenals, the *medulla*, contains *chromaffin tissue and secretes the hormones *adrenaline and *noradrenaline; the outer *cortex* secretes small amounts of sex hormones (*androgens and *oestrogens) and various *corticosteroids, which have a wide range of effects on the body. *See also* **ACTH**.

Adrenaline

adrenaline (epinephrine) A hormone (see formula), produced by the medulla of the *adrenal glands, that increases heart activity, improves the power and prolongs the action of muscles, and increases the rate and depth of breathing to prepare the body for 'fright, flight, or fight' (*see* **alarm response**). At the same time it inhibits digestion and excretion. Similar effects are produced by stimulation of the *sympathetic nervous system. Adrenaline can be administered by injection to relieve bronchial asthma and reduce blood loss during surgery by constricting blood vessels.

adrenal medulla The inner part of the *adrenal gland, in which *adrenaline is produced.

adrenergic Describing a nerve fibre that either releases *adrenaline or *noradrenaline when stimulated or is itself stimulated by these substances. *Compare* **cholinergic**.

adrenocorticotrophic hormone *See* ACTH.

adsorption The formation of a layer of solid, liquid, or gas on the surface of a solid or, less frequently, of a liquid. There are two types depending on the nature of the forces involved. In *chemisorption* a single layer of molecules, atoms, or ions is attached to the adsorbent surface by chemical bonds. In *physisorption* adsorbed molecules are held by the weaker physical forces. The property is utilized in adsorption *chromatography.

adventitious Describing organs or other structures that arise in unusual positions. For example, ivy has adventitious roots growing from its stems.

aeciospore (aecidiospore) An asexual spore formed in the rust fungi (*see* **rusts**) from the fusion of cells. The nuclei from the two cells do not fuse and the spore is binucleate. Aeciospores develop in a sorus known as an *aecidium*.

aeolian soil A type of soil that is transported from one place to another by the wind.

aerobe *See* **aerobic respiration**.

aerobic respiration A type of *respiration in which foodstuffs (usually carbohydrates) are completely oxidized to carbon dioxide and water, with the release of chemical energy, in a process requiring atmospheric oxygen. The reaction can be summarized by the equation:

$$C_6H_{12}O_6 + 6O_2 \rightarrow 6CO_2 + 6H_2O + energy$$

Most organisms have aerobic respiration (i.e. they are *aerobes*); exceptions include certain bacteria and yeasts. *Compare* **anaerobic respiration**.

aerotaxis *See* **taxis**.

aestivation 1. *(in zoology)* A state of inactivity occurring in some animals, notably lungfish, during prolonged periods of drought or heat. Feeding, respiration, movement, and other bodily activities are considerably slowed down. *See also* **dormancy**. *Compare* **hibernation**. 2. *(in botany)* The arrangement of the parts of a flower bud, especially of the sepals and petals.

aetiology The study of causes, especially the causes of medical conditions.

afferent Carrying (nerve impulses, blood, etc.) from the outer regions of a body or organ towards its centre. The term is usually applied to types of nerve fibres or blood vessels. *Compare* **efferent**.

affinity chromatography A type of *chromatography that is dependent on the affinity between specific molecules. The matrix, which can be packed into a column, is made from a material that is able to specifically bind to the molecule under investigation. Antibody purification can be achieved by this method when the matrix is composed of an appropriate antigen.

aflatoxin Any of four related toxic compounds produced by the mould *Aspergillus flavus*. Aflatoxins bind to DNA and prevent replication and transcription. They can cause acute liver damage and cancers: man may be poisoned by eating stored peanuts and cereals contaminated with the mould.

afterbirth The *placenta, *umbilical cord, and *extraembryonic membranes, which are expelled from the womb after a mammalian fetus is born. In most nonhuman mammals the afterbirth, which contains nutrients and might otherwise attract predators, is eaten by the female.

agamospermy *See* apomixis.

agar An extract of certain species of red seaweeds that is used as a gelling agent in microbiological *culture media, foodstuffs, medicines, and cosmetic creams and jellies. *Nutrient agar* consists of a broth made from beef extract or blood that is gelled with agar and used for the cultivation of bacteria, fungi, and some algae.

agglutination The clumping together by antibodies of microscopic foreign particles, such as red blood cells or bacteria, so that they form a visible pellet-like precipitate. Agglutination is a specific reaction, i.e. a particular antigen will only clump in the presence of its specific antibody; it therefore provides a means of identifying unknown bacteria and determining *blood group. When blood of incompatible blood groups (e.g. group A and group B – *see* **ABO system**) is mixed together agglutination of the red cells occurs (*haemagglutination*). This is due to the reaction between antibodies in the plasma (*agglutinins*) and *agglutinogens (antigens) on the surface of the red cells.

agglutinogen Any of the antigens that are present on the outer surface of red blood cells (erythrocytes). There are more than 100 different agglutinogens and they form the basis for identifying the different *blood groups. Antibodies in the plasma, known as *agglutinins*, react with the agglutinogens in blood of an incompatible blood group (*see* **agglutination**).

aggressin A toxic substance that is secreted by certain parasitic microorganisms and inhibits the natural defence mechanisms of a host organism.

aggression Behaviour aimed at intimidating or injuring another animal of the same or a competing species. Aggression between individuals of the same species often starts with a series of ritualized displays or contests that can end at any stage if one of the combatants withdraws, leaving the victor with access to a disputed resource (e.g. food, a mate, or *territory) or with increased social dominance (*see* **dominant**). It is also often seen in *courtship. Aggression or threat displays usually appear to exaggerate the performer's size or strength; for example, many fish erect their fins and mammals and birds may erect hairs or feathers. Special markings may be prominently exhibited, and *intention movements* may be made: dogs bare their teeth, for example. Some animals have evolved special structures for use in aggressive interactions (e.g. antlers in deer) but these are seldom used to cause actual injury; the opponent usually flees first or adopts *appeasement postures. Fights 'to the death' are comparatively rare. *See* **agonistic behaviour; display behaviour; ritualization**.

agitator A bladelike instrument used in fermenters and *bioreactors to mix the medium continuously in order to maintain the rate of oxygen transfer and to help keep the cells in suspension.

Agnatha A class or subphylum of marine and freshwater vertebrate chordates (*see* **Chordata**) that lack jaws. They are fishlike animals with cartilaginous skeletons and well-developed sucking mouthparts with horny teeth. The only living agnathans are lampreys and hagfishes (order Cyclostomata), which are parasites or scavengers. Fossil agnathans, covered in an armour of bony plates, are the oldest known fossil vertebrates. They have been dated from the Silurian and Devonian periods, 440–345 million years ago. *Compare* **Gnathostomata**.

agonist A drug, hormone, or transmitter substance that forms a complex with a receptor site that is capable of triggering an active response from a cell. *Compare* **antagonist**.

agonistic behaviour Any form of behaviour associated with *aggression, including threat, attack, *appeasement, or flight. It is often associated with defence of a territory; for example, a *threat display* by the defending individual is often met with an appeasement display from the intruder, thus avoiding harmful conflict.

agranulocyte Any white blood cell (*see* **leucocyte**) with a nongranular cytoplasm and a large spherical nucleus; *lymphocytes and *monocytes are examples. Agranulocytes are produced either in the lymphatic system or in the bone marrow and account for 30% of all leucocytes. *Compare* **granulocyte**.

agriculture The study and practice of cultivating land for the growing of crops and the rearing of livestock. The increasing demands for food production during the second half of the 20th century have seen many developments in agricultural technology and practices that have greatly increased crop and livestock production. However, these advances in modern *intensive farming* techniques have had their impact on the environment, particularly with increased use of *fertilizers and *pesticides. The now widespread practice of crop *monoculture* (in which one crop is grown densely over an extensive area) has required an increase in the use of *pesticides, as monoculture provides an ideal opportunity for crop pests. Monoculture also requires vast areas of land, which has meant that natural habitats have been destroyed. *Deforestation has resulted from the clearing of forests for crop production and cattle rearing. Advances in technology have included ploughing machines with hydraulic devices that can control the depth to which the soil is ploughed, and seed drills that automatically implant seeds in the soil so that ploughing is not necessary. Food supply in many less-developed countries relies on *subsistence farming*, in which the crops and livestock produced are used solely to feed the farmer and his family. In such countries a system known as *slash and burn* is common, in which the vegetation in an area is cut down and then burnt, thus returning the minerals to the soil. The area can then be used for crop cultivation until the soil fertility drops, at which point it is then abandoned for a number of years and another site is cultivated.

The selective *breeding of crop plants and farm animals has had an enormous impact on productivity in agriculture. Modern varieties of crop plants have increased nutritional value and greater resistance to disease, while animals have been selectively bred to enhance their yields of milk, meat, and other products. Recent developments in genetic engineering have enabled the potential use of *transgenic organisms in agriculture to be explored.

AI *See* **artificial insemination**.

AIDS (acquired immune deficiency syndrome) A disease of humans characterized by defective cell-mediated *immunity and increased susceptibility to infections. It is caused by the retrovirus *HIV (human immunodeficiency virus). This infects and destroys helper *T cells, which are essential for combating infections. HIV is transmitted in blood, semen, and vaginal fluid; the major routes of infection are unprotected vaginal and anal intercourse, intravenous drug abuse, and the administration of contaminated blood and blood products. A person infected with HIV is described as *HIV-positive*; after the initial infection the virus can remain dormant for up to ten years before AIDS develops.

air bladder *See* **swim bladder**.

air pollution (atmospheric pollution) The release into the atmosphere of substances that cause a variety of harmful effects to the natural environment. Most air pollutants are gases that are released into the troposphere, which extends about 8 km above the surface of the earth. The burning of fossil fuels, for example in power stations, is a major source of air pollution as this process produces such gases as sulphur dioxide and carbon dioxide. Released into the atmosphere, both these gases are thought to contribute to the *greenhouse effect. Sulphur dioxide and nitrogen oxides, released in car exhaust fumes, are air pollutants that are responsible for the formation of *acid rain; nitrogen oxides also contribute to the formation of *photochemical smog. *See also* **ozone layer; pollution**.

air sac 1. Any one of a series of thin-walled sacs in birds that are connected to the lungs and increase the efficiency of ventilation. Some of the air sacs penetrate the internal cavities of bones. **2.** A structural extension to the *trachea in insects, which increases the surface area available for the exchange of oxygen and carbon dioxide in respiration.

akinete A nonmotile reproductive cell of certain filamentous blue-green bacteria (Cyanobacteria). An akinete is an enlarged resting cell with a thick wall and large amounts of food reserves and DNA. After cell division has occurred within the akinete the cell wall ruptures, releasing a filament of cells.

alanine *See* amino acid.

alarm response An immediate response to any stimulus that potentially threatens the wellbeing of an organism. It involves the release of *adrenaline and noradrenaline from the adrenal glands, triggered by increased sympathetic nervous activity. These hormones enhance the effects of the sympathetic nervous system (e.g. by increasing heart and breathing rates) and promote glycogen breakdown, which supplies large amounts of glucose for increased respiration and energy release. *Compare* **resistance response**.

alarm signal A warning signal given by an animal to other members of its population in response to perceived danger, usually the approach of a predator. This is a form of *altruism, since the animal that has perceived the danger may waste valuable time in giving the signal (or attract the attention of the predator by doing this) and thus reduce its own chances of survival. For example, the alarm signal of a rabbit to a threatening situation involves thumping the ground and then flashing the white of its tail as it runs, which alerts the rabbits nearby.

albinism Hereditary lack of pigmentation (*see* **melanin**) in an organism. Albino animals and human beings have no colour in their skin, hair, or eyes (the irises appear pink from underlying blood vessels). The *allele responsible is *recessive to the allele for normal pigmentation.

albumen *See* albumin.

albumin One of a group of globular proteins that are soluble in water but form insoluble coagulates when heated. Albumins occur in egg white (the protein component of which is known as *albumen*), blood, milk, and plants. Serum albumins, which constitute about 55% of blood plasma protein, help regulate the osmotic pressure and hence plasma volume. They also bind and transport fatty acids. α-lactalbumin is one of the proteins in milk.

albuminous cell *See* companion cell.

alburnum *See* sapwood.

alcohol An organic compound that contains the −OH group bound to a carbon atom. In systematic chemical nomenclature alcohol names end in the suffix *-ol*. Examples are methanol, CH_3OH, and *ethanol, C_2H_5OH. Alcohols that have two −OH groups in their molecules are *diols* (or *dihydric alcohols*), those with three are *triols* (or *trihydric alcohols*), etc.

alcoholic fermentation *See* fermentation.

aldohexose *See* monosaccharide.

aldose *See* monosaccharide.

aldosterone A hormone produced by the adrenal glands (*see* **corticosteroid**) that controls excretion of sodium by the kidneys and thereby maintains the balance of salt and water in the body fluids. *See also* **angiotensin**.

aleurone layer The outer layer of living cells of the endosperm of wheat and other grain species. It is a single layer of cells that synthesizes the enzyme α-amylase, which is secreted during germination into the starch-filled endosperm and breaks down the starch into maltose and glucose. Studies on the barley grain have suggested that *gibberellins exert a degree of control over the synthesis of the enzyme, possibly by switching on the genes for the synthesis of the specific RNA that codes for the α-amylase protein. However, the mode of action is unclear.

algae (*sing.* **alga**) A group of unrelated simple organisms that contain chlorophyll (and can therefore carry out photosynthesis) and live in aquatic habitats and in moist situations on land. The algal body may be unicellular or multicellular (filamentous, ribbon-like, or platelike). Formerly regarded as plants, algae are now classified as members of the kingdom *Protoctista; they are assigned to separate phyla based primarily on the composition of the cell wall, the nature of the stored food reserves, and the other photosynthetic pigments present. *See* **Bacillariophyta; Chlorophyta; Chrysophyta; Phaeophyta; Rhodophyta**.

The organisms formerly known as blue-green algae are now classified as bacteria (*see* **Cyanobacteria**).

algal bloom The rapid increase in populations of algae and other phytoplankton, in particular the blue-green bacteria (cyanobacteria), that occurs in inland water systems, such as lakes. The density of the organisms may be such that it may prevent light from passing to lower depths in the water system. Algal blooms are caused by an increase in levels of nitrate, a mineral ion essential for algal and bacterial growth. The source of increased nitrate may be from agricultural *fertilizers, which are leached into water systems from the land, or *sewage effluent. Algal blooms contribute to the eutrophication of water systems. *See also* **eutrophic**.

algin (**alginic acid**) A complex polysaccharide occurring in the cell walls of the brown algae (Phaeophyta). Algin strongly absorbs water to form a viscous gel. It is produced commercially from a variety of species of *Laminaria* and from *Macrocystis pyrifera* in the form of *alginates*, which are used mainly as stabilizers and texturing agents in the food industry.

alimentary canal (**digestive tract; gut**) A tubular organ in animals that is divided into a series of zones specialized for the ingestion, *digestion, and *absorption of food and for the elimination of indigestible material (see illustration). In most animals the canal has two openings, the mouth (for the intake of food) and the *anus (for the elimination of waste). Simple animals, such as coelenterates (e.g. *Hydra* and jellyfish) and flatworms, have only one opening to their alimentary canal, which must serve both functions.

The human alimentary canal

alkali A *base that dissolves in water to give hydroxide ions (OH⁻).

alkaline phosphatase An enzyme that catalyses the hydrolysis of phosphoric acid esters under conditions of alkaline pH. In humans the blood levels of alkaline phosphatase are measured as part of the assessment of liver function; in the blood the enzyme also breaks down phosphates required for mineralization of bone. There are two distinct forms of the enzyme (*see* **isozyme**), liver alkaline phosphatase and bone alkaline phosphatase.

alkaloid One of a group of nitrogenous organic compounds derived from plants and having diverse pharmacological properties. Alkaloids include morphine, cocaine, atropine, quinine, and caffeine, most of which are used in medicine as *analgesics or anaesthetics. Some alkaloids

are poisonous, e.g. strychnine and coniine, and *colchicine inhibits
cell division.

alkalosis A condition in which the body fluids become more alkaline, i.e.
the pH is more than 7.4.

alkaptonuria (alcaptonuria) An inherited metabolic disorder that results
from a deficiency of the enzyme homogentisic acid oxidase, which is
required for the complete breakdown of the amino acids tyrosine and
phenylalanine. The accumulation of the intermediate product,
homogentisic acid, which imparts a dark colour to the urine, damages
connective tissue and causes joint disease.

allantois One of the membranes that develops in embryonic reptiles,
birds, and mammals as a growth from the hindgut. It acts as a urinary
bladder for the storage of waste excretory products in the egg (in reptiles
and birds) and as a means of providing the embryo with oxygen (in reptiles,
birds, and mammals) and food (in mammals; *see* **placenta**). *See also*
extraembryonic membranes.

allele (allelomorph) One of the alternative forms of a gene. In a diploid
cell there are usually two alleles of any one gene (one from each parent),
which occupy the same relative position (*locus) on *homologous
chromosomes. One allele is often *dominant to the other (known as the
*recessive), i.e. it determines which aspects of a particular characteristic
the organism will display. Within a population there may be several alleles
of a gene; each has a unique nucleotide sequence.

allele frequency (gene frequency) The occurrence of an *allele in a
population in relation to other alleles of a gene at the same *locus.

allelomorph *See* allele.

allelopathy The secretion by plants of chemicals, such as phenolic and
terpenoid compounds, that inhibit the growth or reproduction of other
plant species, with which they are competing. For example, the aromatic
oils released by certain shrubs of the Californian chaparral pass into the
soil and inhibit the growth of herbaceous species nearby. Some plants
produce chemicals that are toxic to grazing herbivorous animals.

allergen An antigen that provokes an *immune response. Common
allergens include pollen and dust (*see* **allergy**).

allergy A condition in which the body produces an abnormal *immune
response to certain *antigens (called *allergens*), which include dust, pollen,
certain foods and drugs, or fur. In allergic individuals these substances,
which in a normal person would be destroyed by antibodies, stimulate the
release of *histamine, leading to inflammation and other characteristic
symptoms of the allergy (e.g. asthma or hay fever). This response is a type
of *anaphylaxis.

allochthonous Describing an organism that originates from a place other than that in which it is found. The organism is usually a transient member of a community. *Compare* **autochthonous**.

allogamy Cross-fertilization in plants. *See* **fertilization**.

allogenic 1. Relating to or caused by a change in the environment or an individual organism brought about by some external factor. For example, the increased predation in a habitat caused by an immigrant predator would be described as allogenic. *Compare* **autogenic**. **2.** (*or* **allogeneic**) Describing the variation that exists in the genotypes of different individuals, usually when these belong to the same species.

allograft *See* graft.

allometric growth The regular and systematic pattern of growth such that the mass or size of any organ or part of a body can be expressed in relation to the total mass or size of the entire organism according to the allometric equation: $Y = bx^\alpha$, where Y = mass of the organ, x = mass of the organism, α = growth coefficient of the organ, and b = a constant.

allopatric Describing or relating to groups of similar organisms that could interbreed but do not because they are geographically separated. *Compare* **sympatric**. *See* **speciation**.

allopolyploid A *polyploid organism, usually a plant, that contains multiple sets of chromosomes derived from different species. Hybrids are usually sterile, because they do not have sets of *homologous chromosomes and therefore *pairing cannot take place. However, if doubling of the chromosome number occurs in a hybrid derived from two diploid ($2n$) species, the resulting tetraploid ($4n$) is a fertile plant. This type of tetraploid is known as an *allotetraploid*; as it contains two sets of homologous chromosomes, pairing and crossing over are now possible. Allopolyploids are of great importance to plant breeders as advantages possessed by different species can be combined. The species of wheat, *Triticum aestivum*, used to make bread is an *allohexaploid* ($6n$), possessing 42 chromosomes, which is six times the original haploid number (n) of 7. *Compare* **autopolyploid**.

all-or-none response A type of response that may be either complete and of full intensity or totally absent, depending on the strength of the stimulus; there is no partial response. For example, a nerve cell is either stimulated to transmit a complete nervous impulse or else it remains in its resting state; a stinging *thread cell of a coelenterate is either completely discharged or it is not.

allosteric enzyme An enzyme that has two structurally distinct forms, one of which is active and the other inactive. In the active form, the quaternary structure (*see* **protein**) of the enzyme is such that a substrate can interact with the enzyme at the active site (*see* **enzyme–substrate complex**). The conformation of the substrate-binding site becomes altered in the inactive form and interaction with the substrate is not possible.

Allosteric enzymes tend to catalyse the initial step in a pathway leading to the synthesis of molecules. The end product of this synthesis can act as a feedback inhibitor (*see* **inhibition**) and the enzyme is converted to the inactive form, thereby controlling the amount of product synthesized.

allosteric site A binding site on the surface of an enzyme other than the *active site. In noncompetitive *inhibition, binding of the inhibitor to an allosteric site inhibits the activity of the enzyme. In an *allosteric enzyme, the binding of a regulatory molecule to the allosteric site changes the overall shape of the enzyme, either enabling the substrate to bind to the active site or preventing the binding of the substrate.

allozyme Any one of a number of different forms of the same enzyme that are coded by different alleles at the same locus.

••• hydrogen bond
Ⓡ = amino-acid side chain

Alpha helix

alpha helix The most common form of secondary structure in *proteins, in which the polypeptide chain is coiled into a helix. The helical structure is held in place by weak hydrogen bonds between the N–H and C=O groups in successive turns of the helix (see illustration). *Compare* **beta sheet**.

alpha-naphthol test A biochemical test to detect the presence of carbohydrates in solution, also known as *Molisch's test* (after the Austrian chemist H. Molisch (1856–1937), who devised it). A small amount of alcoholic alpha-naphthol is mixed with the test solution and concentrated sulphuric acid is poured slowly down the side of the test tube. A positive reaction is indicated by the formation of a violet ring at the junction of the two liquids.

alternation of generations The occurrence within the *life cycle of an organism of two or more distinct forms (generations), which differ from each other in appearance, habit, and method of reproduction. The phenomenon occurs in parasitic protozoans, certain lower animals (e.g. coelenterates and parasitic flatworms), and in plants. The malaria parasite (*Plasmodium*), for example, has a complex life cycle involving the alternation of sexually and asexually reproducing generations. In plants the generation with sexual reproduction is called the *gametophyte and the asexual generation is the *sporophyte, either of which may dominate the life cycle, and there is also alternation of the haploid and diploid states. Thus in ferns the dominant plant is the diploid sporophyte; it produces spores that germinate into small haploid gametophytes bearing sex organs. In mosses the gametophyte is the dominant plant and the sporophyte is the spore-bearing capsule.

altricial species Any species of bird in which hatching occurs at an early stage of development. The offspring of altricial species tend to be born with no feathers and remain in the nest for a comparatively long period of time; they are described as *nidicolous*. *Compare* **precocial species**.

altruism Behaviour by an animal that decreases its chances of survival or reproduction while increasing those of another member of the same species. For example, a lapwing puts itself at risk by luring a predator away from the nest through feigning injury, but by so doing saves its offspring. Altruism in its biological sense does not imply any conscious benevolence on the part of the performer. Altruism can evolve through *kin selection, if the recipients of altruistic acts tend on average to be more closely related to the altruist than the population as a whole. *See also* **alarm signal; inclusive fitness**.

alveolus 1. The tiny air sac in the *lung of mammals and reptiles at the end of each *bronchiole. It is lined by a delicate moist membrane, has many blood capillaries, and is the site of exchange of respiratory gases (carbon dioxide and oxygen). **2.** The socket in the jawbone in which a tooth is rooted by means of the *periodontal membrane.

Alzheimer's disease A neurological disease characterized by progressive loss of intellectual ability. The disease, which is named after German physician A. Alzheimer (1864–1915), is associated with general shrinkage of the brain tissue, with *amyloid deposits, and changes in the neurotransmitter systems within the brain that include a loss in the activity of *cholinergic neurones.

amacrine cell A type of nerve cell found in the *retina of the eye. Amacrine cells receive sensory information from the rod and cone receptor cells in the retina and are able to integrate sensory information before sending this to the brain.

amber A yellow or reddish-brown fossil resin. The resin was exuded by certain trees and other plants and often contains preserved insects, flowers, or leaves that were trapped by its sticky surface before the resin hardened. Amber is used for jewellery and ornaments. It also has the property of acquiring an electrical charge when rubbed (the term electricity is derived from *electron*, the Greek name for amber). It occurs throughout the world in rock strata from the Cretaceous to the Pleistocene, but most commonly in Cretaceous and Tertiary rocks.

ambient Describing the conditions and factors that are present in the immediate environment. For example, the ambient temperature is the local temperature in a specific environment.

ameloblast Any of the epithelial cells that secrete *enamel during tooth formation. They die before the tooth erupts, therefore damaged enamel in a tooth cannot be replaced. *Compare* **odontoblast**.

amensalism An association between two species that is detrimental to one of the species but has no effect on the other. A common example of amensalism is the release of chemical toxins by plants that can inhibit the growth of other plant species (*see* **allelopathy**).

Ames test A method of assessing the extent to which a drug or other *xenobiotic is capable of causing a mutation. The drug is added to a container of agar containing a special strain of *Salmonella* bacteria that require the amino acid histidine for growth. If the xenobiotic is mutagenic and interacts with the DNA of the bacteria, some mutant strains able to grow without histidine will arise and colonies of bacteria appear on the agar.

amine Any one of a group of organic compounds derived by replacing one or more of the hydrogen atoms in ammonia by organic groups. *Primary amines* have one hydrogen replaced, e.g. methylamine, CH_3NH_2. They contain the functional group $-NH_2$ (the *amino group*). *Secondary amines* have two hydrogens replaced, e.g. methylethylamine, $CH_3(C_2H_5)NH$. *Tertiary amines* have all three hydrogens replaced, e.g. *trimethylamine. Amines are produced by the decomposition of organic matter.

amino acid Any of a group of water-soluble organic compounds that possess both a carboxyl (–COOH) and an amino (–NH$_2$) group attached to the same carbon atom, called the α-carbon atom. Amino acids can be represented by the general formula R–CH(NH$_2$)COOH. R may be hydrogen or an organic group, which may be nonpolar, basic, acidic, or polar; the nature of the R group determines the properties of any particular amino acid. Through the formation of peptide bonds, amino acids join together to form short chains (*peptides) or much longer chains (*polypeptides). Proteins are composed of various proportions of about 20 commonly occurring amino acids (see table). The sequence of these amino acids in the protein polypeptides determines the shape, properties, and hence biological role of the protein. Some amino acids that never occur in proteins are nevertheless important, e.g. *ornithine and citrulline, which are intermediates in the urea cycle.

Plants and many microorganisms can synthesize amino acids from simple inorganic compounds, but animals rely on adequate supplies in their diet.

alanine (Ala)

$$CH_3 - \overset{\overset{\displaystyle H}{|}}{\underset{\underset{\displaystyle NH_2}{|}}{C}} - COOH$$

*arginine (Arg)

$$H_2N - \overset{}{\underset{\underset{\displaystyle NH}{\|}}{C}} - NH - CH_2 - CH_2 - CH_2 - \overset{\overset{\displaystyle H}{|}}{\underset{\underset{\displaystyle NH_2}{|}}{C}} - COOH$$

asparagine (Asn)

$$H_2N - \overset{}{\underset{\underset{\displaystyle O}{\|}}{C}} - CH_2 - \overset{\overset{\displaystyle H}{|}}{\underset{\underset{\displaystyle NH_2}{|}}{C}} - COOH$$

aspartic acid (Asp)

$$HOOC - CH_2 - \overset{\overset{\displaystyle H}{|}}{\underset{\underset{\displaystyle NH_2}{|}}{C}} - COOH$$

cysteine (Cys)

$$HS - CH_2 - \overset{\overset{\displaystyle H}{|}}{\underset{\underset{\displaystyle NH_2}{|}}{C}} - COOH$$

glutamic acid (Glu)

$$HOOC - CH_2 - CH_2 - \overset{\overset{\displaystyle H}{|}}{\underset{\underset{\displaystyle NH_2}{|}}{C}} - COOH$$

glutamine (Gln)

$$\overset{\displaystyle H_2N}{\underset{\displaystyle O}{>}}C - CH_2 - CH_2 - \overset{\overset{\displaystyle H}{|}}{\underset{\underset{\displaystyle NH_2}{|}}{C}} - COOH$$

glycine (Gly)

$$H - \overset{\overset{\displaystyle H}{|}}{\underset{\underset{\displaystyle NH_2}{|}}{C}} - COOH$$

*histidine (His)

$$HC = \overset{}{\underset{\underset{\displaystyle N}{}}{C}} - CH_2 - \overset{\overset{\displaystyle H}{|}}{\underset{\underset{\displaystyle NH_2}{|}}{C}} - COOH$$

*isoleucine (Ile)

$$CH_3 - CH_2 - \overset{}{\underset{\underset{\displaystyle CH_3}{|}}{CH}} - \overset{\overset{\displaystyle H}{|}}{\underset{\underset{\displaystyle NH_2}{|}}{C}} - COOH$$

*leucine (Leu)

$$\overset{\displaystyle H_3C}{\underset{\displaystyle H_3C}{>}}CH - CH_2 - \overset{\overset{\displaystyle H}{|}}{\underset{\underset{\displaystyle NH_2}{|}}{C}} - COOH$$

* an essential amino acid

*lysine (Lys)

$$H_2N - CH_2 - CH_2 - CH_2 - CH_2 - \overset{\overset{\displaystyle H}{|}}{\underset{\underset{\displaystyle NH_2}{|}}{C}} - COOH$$

*methionine (Met)

$$CH_3 - S - CH_2 - CH_2 - \overset{\overset{\displaystyle H}{|}}{\underset{\underset{\displaystyle NH_2}{|}}{C}} - COOH$$

*phenylalanine (Phe)

$$\bigcirc - CH_2 - \overset{\overset{\displaystyle H}{|}}{\underset{\underset{\displaystyle NH_2}{|}}{C}} - COOH$$

proline (Pro)

$$\begin{matrix} H_2C - CH_2 \\ H_2C \quad CH - COOH \\ \backslash \ N \ / \\ H \end{matrix} \longrightarrow \begin{matrix} H - C - CH_2 \\ OH \quad \quad \quad \\ H_2C \quad CH - COOH \\ \backslash \ N \ / \\ H \end{matrix}$$

4-hydroxyproline

serine (Ser)

$$HO - CH_2 - \overset{\overset{\displaystyle H}{|}}{\underset{\underset{\displaystyle NH_2}{|}}{C}} - COOH$$

*threonine (Thr)

$$CH_3 - \overset{}{\underset{\underset{\displaystyle OH}{|}}{CH}} - \overset{\overset{\displaystyle H}{|}}{\underset{\underset{\displaystyle NH_2}{|}}{C}} - COOH$$

*tryptophan (Trp)

$$\begin{matrix} & C - CH_2 - \overset{\overset{\displaystyle H}{|}}{\underset{\underset{\displaystyle NH_2}{|}}{C}} - COOH \\ & \| \\ & CH \\ & N \\ & H \end{matrix}$$

tyrosine (Tyr)

$$HO - \bigcirc - CH_2 - \overset{\overset{\displaystyle H}{|}}{\underset{\underset{\displaystyle NH_2}{|}}{C}} - COOH$$

*valine (Val)

$$\overset{\displaystyle H_3C}{\underset{\displaystyle H_3C}{>}}CH - \overset{\overset{\displaystyle H}{|}}{\underset{\underset{\displaystyle NH_2}{|}}{C}} - COOH$$

The amino acids occurring in proteins

The *essential amino acids must be present in the diet whereas others can be manufactured from them.

aminopeptidase A membrane-bound enzyme (an *exopeptidase) in the small intestine that breaks down peptides and dipeptides into amino acids.

amino sugar Any sugar containing an amino group in place of a hydroxyl group. The *hexosamines* are amino derivatives of hexose sugars and include *glucosamine* (based on glucose) and *galactosamine* (based on galactose). The former is a constituent of *chitin and the latter occurs in cartilage.

ammonia A colourless gas, NH_3, with a strong pungent odour. Ammonia is produced by the *deamination of excess amino acids in the liver. Industrially it is made from its constituent elements by the Haber process for use in the manufacture of nitric acid, ammonium nitrate, ammonium phosphate, and urea (the last three as fertilizers), explosives, dyestuffs, and resins. The participation of ammonia in the *nitrogen cycle is a most important natural process. By means of *nitrogenase enzymes, nitrogen-fixing bacteria are able to achieve similar reactions to those of the Haber process, but under normal conditions of temperature and pressure. The reactions release ammonium ions, which are converted by nitrifying bacteria into nitrite and nitrate ions.

ammonite An extinct aquatic mollusc of the class *Cephalopoda. Ammonites were abundant in the Mesozoic era (225–65 million years ago) and are commonly found as fossils in rock strata of that time, being used as *index fossils for the Jurassic period. They were characterized by a coiled shell divided into many chambers, which acted as a buoyancy aid. The external suture lines on these shells increased in complexity with the advance of the group.

ammonotelic Describing animals that excrete nitrogenous waste in the form of *ammonia. Most aquatic animals are ammonotelic. *Compare* **ureotelic; uricotelic**.

amniocentesis The taking of a sample of amniotic fluid from a pregnant woman to determine the condition of an unborn baby. A hollow needle is inserted through the woman's abdomen and wall of the uterus and the fluid drawn off. Chemical and microscopical examination of cells shed from the embryo's skin into the fluid are used to detect spina bifida, *Down's syndrome, or other serious biochemical or chromosomal abnormalities.

amnion A membrane that encloses the embryo of reptiles, birds, and mammals within the *amniotic cavity*. This cavity is filled with *amniotic fluid*, in which the embryo is protected from desiccation and from external pressure. *See also* **extraembryonic membranes**.

amniote A vertebrate whose embryos are totally enclosed in a fluid-filled sac – the *amnion. The evolution of the amnion provided the necessary fluid environment for the developing embryo and therefore allowed animals to breed away from water. Amniotes comprise the reptiles, birds, and mammals. *Compare* **anamniote**.

amniotic egg The type of egg produced by reptiles, birds, and prototherian (egg-laying) mammals (*amniotes), in which the embryo develops inside an *amnion. The shell of the egg is either calcium-based or leathery.

Amoeba A genus of *protozoa of the phylum *Rhizopoda, members of which have temporary body projections called *pseudopodia. These are used for locomotion and feeding and result in a constantly changing body shape (*see* **amoeboid movement**). Most species are free-living in soil, mud, or water, where they feed on smaller protozoans and other single-celled organisms, but a few are parasitic. The best known species is the much studied *A. proteus*.

amoebocyte An animal cell whose location is not fixed and is therefore able to wander through the body tissues. Amoebocytes are named after their resemblance, especially in their movement, to *Amoeba* (*see* **amoeboid movement**) and they feed on foreign particles (including invading bacteria). They occur, for example, in sponges and mammalian blood (e.g. some *leucocytes).

amoeboid movement The mechanism of movement demonstrated by *Amoeba* and other cells that are capable of changing their shape (e.g. *phagocytes). The cytoplasm of *Amoeba* consists of a central fluid endoplasm surrounded by a more viscous ectoplasm. The fluid endoplasm slides towards the front of the cell, forming a *pseudopodium and propelling the cell forward. On reaching the tip of the pseudopodium, the endoplasm is converted into ectoplasm; at the same time the ectoplasm at the rear of the cell is converted into endoplasm and streams forward, thus maintaining continuous movement.

amount of substance Symbol n. A measure of the number of entities present in a substance. The specified entity may be an atom, molecule, ion, electron, photon, etc., or any specified group of such entities. The amount of substance of an element, for example, is proportional to the number of atoms present. The SI unit of amount of substance is the *mole.

AMP *See* **ATP; cyclic AMP**.

amphetamine A drug, 1-phenyl-2-aminopropane (or a derivative of this compound), that stimulates the central nervous system by causing the release of the transmitters noradrenaline and dopamine from nerve endings. It inhibits sleep, suppresses the appetite, and has variable effects on mood; prolonged use can lead to addiction.

Amphibia The class of vertebrate chordates (*see* **Chordata**) that contains the frogs, toads, newts, and salamanders. The amphibians evolved in the Devonian period (about 370 million years ago) as the first vertebrates to occupy the land, and many of their characteristics are adaptations to terrestrial life. All adult amphibians have a passage linking the roof of the mouth with the nostrils so they may breathe air and keep the mouth closed. The moist scaleless skin is used to supplement the lungs in gas exchange. They have no diaphragm, and therefore the muscles of the mouth and pharynx provide the pumping action for breathing. Fertilization is usually external and the eggs are soft and prone to desiccation, therefore reproduction commonly occurs in water. Amphibian larvae are aquatic, having gills for respiration; they undergo metamorphosis to the adult form.

amphibolic pathway A biochemical pathway that serves both anabolic and catabolic processes. An important example of an amphibolic pathway is the *Krebs cycle, which involves both the catabolism of carbohydrates and fatty acids and the synthesis of anabolic precursors for amino-acid synthesis (e.g. α-ketogluturate and oxaloacetate).

amphimixis True sexual reproduction, involving the fusion of male and female gametes and the formation of a zygote. *Compare* **apomixis**.

amphioxus Another name for the lancelet: *see* **Cephalochordata**.

Amphistylic jaw suspension

amphistylic jaw suspension A type of jaw suspension seen in certain sharks, in which the upper jaw is braced against the cranium and is also supported by the hyomandibular (*see* **hyoid arch**). See illustration. *Compare* **autostylic jaw suspension; hyostylic jaw suspension**.

amphoteric Describing a compound that can act as both an acid and a base. Amino acids, which contain both acidic and basic groups in their molecules, can be described as amphoteric. Solvents, such as water, that can both donate and accept protons are usually described as *amphiprotic* (*see* **solvent**).

ampulla 1. An enlargement at one end of each of the *semicircular canals of the inner ear. Each ampulla contains a group of receptors – sensory hair cells – embedded in a gelatinous cap (*cupula*), which detects movement in one particular dimension, corresponding to the plane of the canal. Movement of the head causes the cupula (and the hairs within it) to bend in a direction opposite to that of the head movement (see illustration); this stimulates nerve impulses in the receptors, which are interpreted by the brain as movement in a particular dimension. **2.** Any small vesicle or saclike process.

cupula

movement of cupula (and hairs)

endolymph

sensory hair cells

movement of head

nerve fibres to brain

Ampulla

amylase Any of a group of closely related enzymes that degrade starch, glycogen, and other polysaccharides. Plants contain both α- and β-amylases; the name *diastase* is given to the component of malt containing β-amylase, important in the brewing industry. Animals possess only α-amylases, found in pancreatic juice (as *pancreatic amylase*) and also (in humans and some other species) in saliva (as *salivary amylase* or *ptyalin*). Amylases cleave the *glycosidic bonds of the long polysaccharide chains, producing a mixture of glucose and maltose.

amyloid Fibrous tissue that may accumulate between cells in various animal tissues. Amyloid deposits are insoluble and can exert pressure on various vital organs. These deposits are generally detected by staining with the dye Congo red. A build-up of amyloid tissue in the brain is a feature of *Alzheimer's disease and *bovine spongiform encephalopathy.

amylopectin A *polysaccharide comprising highly branched chains of glucose molecules. It is one of the constituents (the other being amylose) of *starch.

amyloplast An organelle in plants that stores starch. Amyloplasts are often found in nonphotosynthetic tissue, such as roots and storage tubers.

amylose A *polysaccharide consisting of linear chains of between 100 and 1000 linked glucose molecules. Amylose is a constituent of *starch. In water, amylose reacts with iodine to give a characteristic blue colour.

anabolic steroid Any steroid compound that promotes tissue growth, especially of muscles. Naturally occurring anabolic steroids include the male sex hormones (*androgens). Synthetic forms of these are used medically to help weight gain after debilitating diseases; their use by athletes to build up body muscles can cause liver damage and is banned by most athletic authorities.

anabolism The metabolic synthesis of proteins, fats, and other constituents of living organisms from molecules or simple precursors. This process requires energy in the form of ATP. *See* **metabolism**. *Compare* **catabolism**.

anaemia A condition that arises when either there are too few erythrocytes (red blood cells), the erythrocytes do not contain sufficient amounts of haemoglobin, or the erythrocytes are abnormal in other respects. Anaemia often results from loss of blood or from a deficiency in the factors necessary to synthesize haemoglobin (e.g. iron) or erythrocytes (e.g. folic acid and vitamin B_{12}). Increased destruction of erythrocytes may be induced by certain drugs or severe infection, and an abnormal form of haemoglobin results in sickle-cell anaemia (*see* **polymorphism**).

anaerobe *See* anaerobic respiration.

anaerobic respiration A type of *respiration in which foodstuffs (usually carbohydrates) are partially oxidized, with the release of chemical energy, in a process not involving atmospheric oxygen. Since the substrate is never completely oxidized the energy yield of this type of respiration is lower than that of *aerobic respiration. It occurs in some yeasts and bacteria and in muscle tissue when oxygen is absent (*see* **oxygen debt**). *Obligate anaerobes are organisms that cannot use free oxygen for respiration; *facultative anaerobes are normally aerobic but can respire anaerobically during periods of oxygen shortage. Alcoholic *fermentation is a type of anaerobic respiration in which one of the end products is ethanol.

anaesthetic Any compound that can render an animal unconscious of painful stimuli. Anaesthetics can be used either at a local level, when sensation is removed from a specific area of the body; or at a general level, when a state of complete unconsciousness is induced.

analgesic A substance that reduces pain without causing unconsciousness, either by reducing the pain threshold or by increasing pain tolerance. There are several categories of analgesic drugs, including morphine and its derivatives (*see* **opiate**), which produce analgesia by acting on the central nervous system; nonsteroidal anti-inflammatory drugs (e.g. *aspirin); and local anaesthetics.

analogous Describing features of organisms that are superficially similar but have evolved in different ways. The wings of butterflies and birds are analogous organs.

anamniote A vertebrate that lacks an *amnion and whose embryos and larvae must develop in water. Anamniotes comprise the agnathans, fishes, and amphibians. *Compare* **amniote**.

anaphase One of several stages of cell division. In *mitosis the chromatids of each chromosome move apart to opposite ends of the spindle. In the first anaphase of *meiosis, the paired homologous chromosomes separate and move to opposite ends; in the second anaphase the chromatids move apart, as in mitosis.

anaphylaxis An abnormal *immune response that occurs when an individual previously exposed to a particular *antigen is re-exposed to the same antigen. Anaphylaxis may follow an insect bite or the injection of a drug (such as penicillin). It is caused by the release of *histamine and similar substances and may produce a localized reaction or a more generalized and severe one, with difficulty in breathing, pallor, or drop in blood pressure, unconsciousness, and possibly heart failure and death. *See also* **allergy**.

anatomy The study of the structure of living organisms, especially of their internal parts by means of dissection and microscopical examination. *Compare* **morphology**.

androecium The male sex organs (*stamens) of a flower. *Compare* **gynoecium**.

androgen One of a group of male sex hormones that stimulate development of the testes and of male *secondary sexual characteristics (such as growth of facial and pubic hair in men). *Testosterone* is the most important. Androgens are produced principally by the testes when stimulated with *luteinizing hormone but they are also secreted in smaller amounts by the adrenal glands and the ovaries. Injections of natural or synthetic androgens are used to treat hormonal disorders of the testes and breast cancer and to build up body tissue (*see* **anabolic steroid**).

anemophily Pollination of a flower in which the pollen is carried by the wind. Examples of anemophilous flowers are those of grasses and conifers. *Compare* **entomophily; hydrophily**.

aneuploid Describing a nucleus, cell, or organism in which one or more chromosomes have been added to or deleted from the complete set, so that the total number of chromosomes is not an exact multiple of the haploid number (n); for example, $2n + 1$ (*see* **trisomy**) or $2n - 1$ (*monosomy*). *Compare* **euploid**.

Angiospermophyta A phylum comprising the flowering plants. Angiosperm gametes are produced within *flowers and the ovules (and the seeds into which they develop) are enclosed in a carpel (*compare* **Coniferophyta**). The angiosperms are the dominant plant forms of the present day. They show the most advanced structural organization in the plant kingdom, enabling them to inhabit a very diverse range of habitats. There are two classes within this group: the *Monocotyledonae with one seed leaf (cotyledon) in the seed, and the *Dicotyledonae with two seed leaves.

angiotensin Either of two related peptide hormones that raise blood pressure. Angiotensin I is derived, by the action of the enzyme *renin, from a protein (α-globulin) secreted by the liver into the bloodstream. As blood passes through the lungs, another enzyme splits angiotensin I, forming angiotensin II. This causes constriction of blood vessels and stimulates the release of *antidiuretic hormone and *aldosterone, which increase blood pressure.

angstrom Symbol Å. A unit of length equal to 10^{-10} metre. It was formerly used to measure wavelengths and intermolecular distances but has now been replaced by the nanometre. 1 Å = 0.1 nanometre. The unit is named after the Swedish pioneer of spectroscopy A. J. Ångstrom (1814–74).

animal Any living organism of the kingdom Animalia, characterized by an inability to manufacture its own food, so that it feeds on other organisms or organic matter (*see* **heterotrophic nutrition**). Animals are therefore typically mobile (to search for food) and have evolved specialized sense organs for detecting changes in the environment; a *nervous system coordinates information received by the sense organs and enables rapid responses to environmental stimuli. Animal *cells lack the cellulose cells walls of *plant cells. For a classification of the animal kingdom, see Appendix.

animal behaviour The activities that constitute an animal's response to its external environment. Certain categories of behaviour are seen in all animals (e.g. feeding, reproduction) but these activities involve different movements in different species and develop in different ways. Some movements are highly characteristic of a species (*see* **instinct**) whereas others are more variable and depend on the interaction between innate tendencies and *learning during the individual's lifetime. Physiologists study how changes in the body (e.g. hormone levels) affect behaviour, psychologists study the mechanisms of learning, and ethologists study the behaviour of the whole animal: how this develops during the individual's lifetime and how it evolved through natural selection (*see* **ethology**).

animal starch *See* glycogen.

anion A negatively charged *ion, such as the chloride ion (Cl⁻). *Compare* **cation**.

anisogamy Sexual reproduction involving the fusion of gametes that differ in size and sometimes also in form. *See also* **oogamy**. *Compare* **isogamy**.

Annelida A phylum of invertebrates comprising the segmented worms (e.g. the earthworm). Annelids have cylindrical soft bodies showing *metameric segmentation, obvious externally as a series of rings separating the segments. Each segment is internally separated from the next by a membrane and bears stiff bristles (*see* **chaeta**). Between the gut and other body organs there is a fluid-filled cavity called the *coelom, which acts as a hydrostatic skeleton. Movement is by alternate contraction of circular and longitudinal muscles in the body wall. The phylum contains three classes: *Polychaeta, *Oligochaeta, and *Hirudinea.

annual A plant that completes its life cycle in one year, during which time it germinates, flowers, produces seeds, and dies. Examples are the sunflower and marigold. *Compare* **biennial; ephemeral; perennial**.

annual rhythm The occurrence of a process or a function in a living organism on a yearly basis. Events that display an annual rhythm can include life cycles, such as those of *annual plants; mating behaviour; some kinds of movement, such as *migration; or growth patterns, such as the *growth rings of woody plant stems. *See also* **biorhythm**.

annual ring *See* growth ring.

annulus 1. *(in botany)* **a.** A ragged ring of tissue that remains on the stalk of a mushroom or toadstool. Also called a *velum*, it is formed from the ruptured membrane that originally covered the lower surface of the cap. **b.** The region of the wall of a fern sporangium that is specialized for spore dispersal. It consists of cells that are thickened except on their outer walls. On drying out, the cells contract and the sporangium ruptures, releasing the spores. The annulus springs back into position when the residual water in the cells vaporizes and any remaining spores are dispersed. **2.** *(in zoology)* Any of various ring-shaped structures in animals, such as any of the segments of an earthworm or other annelid.

Anoplura *See* **Siphunculata**.

anoxic Lacking or not involving or requiring oxygen. For example, a culture of anaerobic microorganisms is called an *anoxic culture*.

anoxic reactor A *bioreactor in which the organisms being cultured are anaerobes or in which the reaction being exploited does not require oxygen.

ANS *See* **autonomic nervous system**.

antagonism 1. The interaction of two substances (e.g. drugs, hormones, or enzymes) having opposing effects in a system in such a way that the action of one partially or completely inhibits the effects of the other. For example, one group of anti-cancer drugs acts by antagonizing the effects of certain enzymes controlling the activities of the cancer cells. *See also* **antagonist. 2.** An interaction between two muscles, known as *antagonistic muscles*, in which contraction of one prevents that of the other. For example, the *biceps and triceps are an antagonistic pair. *See* **voluntary muscle. 3.** An interaction between two organisms (e.g. moulds or bacteria) in which the growth of one is inhibited by the other. *Compare* **synergism**.

antagonist A drug that inhibits the effect of an *agonist in such a way that the combined biological effect of the two substances becomes smaller than the sum of their individual effects. *Competitive antagonists* act by binding to agonist receptors, while *noncompetitive antagonists* do not bind to the same receptor sites as the agonist.

antenna A long whiplike jointed mobile paired appendage on the head of many arthropods, usually concerned with the senses of smell, touch, etc. In insects, millipedes, and centipedes they are the first pair of head appendages and are specialized and modified in many insects. In crustaceans they are the second pair of head appendages, the first pair (the *antennules*) having the sensory function, while the antennae are modified for swimming and for attachment.

antennule *See* antenna.

anterior 1. Designating the part of an animal that faces to the front, i.e. that leads when the animal is moving. In man and bipedal animals the anterior surface corresponds to the *ventral surface. **2.** Designating the side of a flower or axillary bud that faces away from the flower stalk or main stem, respectively. *Compare* **posterior**.

anther The upper two-lobed part of a plant *stamen, usually yellow in colour. Each lobe contains two pollen sacs within which are numerous pollen grains, which are released when the anther ruptures.

antheridium (*pl*. **antheridia**) The male sex organ of algae, fungi, bryophytes, clubmosses, horsetails, and ferns. It produces the male gametes (*antherozoids*). It may consist of a single cell or it may have a wall that is made up of one or several layers forming a sterile jacket around the developing gametes. *Compare* **archegonium**.

antherozoid (spermatozoid) The motile male gamete of algae, fungi, bryophytes, clubmosses, horsetails, ferns, and certain gymnosperms. Antherozoids usually develop in an *antheridium but in certain gymnosperms, such as *Ginkgo* and *Cycas*, they develop from a cell in the pollen tube.

anthocyanin One of a group of *flavonoid pigments. Anthocyanins occur in the cell vacuoles of various plant organs and are responsible for many of the blue, red, and purple colours in plants (particularly in flowers).

Anthozoa *See* Cnidaria; coral.

antibiotics Substances obtained from microorganisms, especially moulds, that destroy or inhibit the growth of other microorganisms, particularly disease-producing bacteria and fungi. Common antibiotics include *penicillin, *streptomycin*, and *tetracyclines*. They are used to treat various infections but tend to weaken the body's natural defence mechanisms and can cause allergies. Overuse of antibiotics can lead to the development of resistant strains of microorganisms.

antibody A protein (*see* **immunoglobulin**) produced by certain white blood cells (*lymphocytes) in response to entry into the body of a foreign substance (*antigen) in order to render it harmless. An antibody–antigen

reaction is highly specific. Antibody production is one aspect of the
*immune response and is stimulated by such antigens as invading bacteria,
foreign red blood cells (*see* **ABO system**), inhaled pollen grains or dust, and
foreign tissue *grafts. *See also* **immunity; monoclonal antibody**.

anticlinal *(in botany)* At right angles to the surface of an organ or part.
In anticlinal cell division the plane of division is at right angles to the
surface of the plant body. *Compare* **periclinal**.

anticoagulant A substance that prevents the formation of blood clots.
*Heparin is a natural anticoagulant, which is extracted to treat such
conditions as thrombosis and embolism. Synthetic anticoagulants include
*warfarin.

anticodon A sequence of three nucleotides (trinucleotide) on a strand of
transfer *RNA that can form base pairs (*see* **base pairing**) with a specific
trinucleotide sequence (*see* **codon**) on a strand of messenger RNA during
*translation. *See also* **protein synthesis**.

antidiuretic hormone (ADH; vasopressin) A hormone, secreted by the
posterior *pituitary gland, that stimulates absorption of water by the
kidneys and thus controls the concentration of body fluids. ADH is
produced by specialized nerve cells in the hypothalamus of the brain and is
transported to the posterior pituitary in the bloodstream. Deficiency of
ADH results in a disorder known as *diabetes insipidus*, in which large
volumes of urine are excreted; it is treated by administration of natural or
synthetic hormone.

antigen Any substance that the body regards as foreign and that
therefore elicits an *immune response. Antigens may be formed in, or
introduced into, the body. They are usually proteins. *Histocompatibility
antigens* are associated with the tissues and are involved in the rejection of
tissue or organ *grafts (*see* **histocompatibility**); an example is the group
of antigens encoded by the *HLA system. A graft will be rejected if the
recipient's body regards such antigens on the donor's tissues as foreign. *See
also* **antibody**.

antigenic variation The ability of certain pathogenic microorganisms,
particularly viruses, to alter the antigens on their outer surface. This
prevents the pathogen from being easily recognized and destroyed by the
immune system of the host.

antigen-presenting cell A cell, such as a macrophage, that 'presents'
foreign antigens to helper *T cells as part of the *immune response. A
direct reaction between these lymphocytes and foreign antigens is not
common. The foreign antigen forms a complex with class II
*histocompatibility antigens on the surface of the antigen-presenting cell.

antihistamine Any drug that inhibits the effects of *histamine in the
body and is therefore used to relieve and prevent the symptoms associated
with allergic reactions, such as hay fever. Since one of the side-effects

produced by antihistamines is sleepiness, some are used to prevent motion sickness and induce sleep.

antioxidants Substances that slow the rate of oxidation reactions. Various antioxidants are used to preserve foodstuffs and to prevent the deterioration of rubber, synthetic plastics, and many other materials. Some antioxidants inhibit the oxidation reaction by removing oxygen *free radicals. Naturally occurring antioxidants with this ability include *vitamin E and β-*carotene; they limit the cell and tissue damage caused by foreign substances, such as toxins and pollutants, in the body.

antipodal cells The three haploid cells in the mature *embryo sac of flowering plants that are situated at the opposite end to the micropyle.

antiport A membrane protein that effects the *active transport of a solute across a biological membrane but only when it also transports a second solute (usually of a different type) in the opposite direction. *Compare* **symport**; **uniport**.

antisense DNA DNA that prevents the expression of certain genes. It consists of a stretch of single-stranded DNA that is complementary to a messenger RNA (mRNA) molecule; by binding to the mRNA it prevents the synthesis of a specific protein. Antisense DNA has been produced by techniques of genetic engineering and is undergoing clinical trials for its ability to inhibit the action of a gene that causes uncontrolled division of cancer cells.

antiseptic Any substance that kills or inhibits the growth of disease-causing microorganisms but is essentially nontoxic to cells of the body. Common antiseptics include hydrogen peroxide, the detergent cetrimide, and ethanol. They are used to treat minor wounds. *Compare* **disinfectant**.

antitoxin An antibody produced in response to a bacterial *toxin.

anucleate Describing any cell that does not contain a nucleus. For example, mature *erythrocytes are anucleate.

anus The terminal opening of the *alimentary canal in most animals, through which indigestible material (*faeces) is expelled.

aorta The major blood vessel in higher vertebrates through which oxygenated blood leaves the *heart from the left ventricle. The aorta branches to form many smaller arteries, which in turn branch many times to supply oxygen and essential nutrients to all living cells in the body. *See also* **dorsal aorta**; **ventral aorta**.

aortic arches (arterial arches) Six pairs of arteries in vertebrate embryos, which connect the *ventral aorta to the *dorsal aorta by running between the gill slits. The arches are numbered I to VI from the anterior end. In adult tetrapods arches I and II are lost, III gives rise to the *carotid arteries, IV (one side of which is lost in birds and mammals) becomes the *systemic arch supplying the trunk and limbs, V is lost (*see* **ductus**

arteriosus), and VI gives rise to the pulmonary arch supplying the lungs (*see* **pulmonary artery**). In adult fish four to six arches persist as branchial arteries supplying the gills.

aortic body *See* **ventilation centre**.

aortic valve *See* **semilunar valve**.

apatite A complex mineral form of the salt calcium phosphate, $Ca_5(PO_4)_3(OH,F,Cl)$, that is the main constituent of the enamel of teeth.

aphotic Not exposed to light. The aphotic zone of the ocean comprises the lowest depths where no light penetrates. *See* **abyssal zone**.

apical dominance Inhibition of the growth of lateral buds in a plant by the presence of a growing apical bud. It is brought about by the action of auxins (produced by the apical bud) and abscisic acid.

apical meristem A region at the tip of each shoot and root of a plant in which cell divisions are continually occurring to produce new stem and root tissue, respectively (*see* **meristem**). The new tissues produced are known collectively as the *primary tissues* of the plant. *See* **ground meristem; procambium; protoderm**. *Compare* **cambium**.

Apicomplexa (Sporozoa) A phylum of the Protoctista containing parasitic *protozoa that may have a number of different animal hosts. Their complex life cycle involves the alternation of asexual reproduction (multiple fission) and sexual reproduction and the production of resistant spores (see illustration). The phylum includes the agents causing malaria (*Plasmodium*) and toxoplasmosis (*Toxoplasma*).

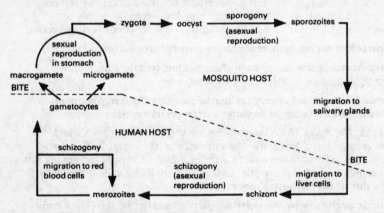

Life cycle of Plasmodium vivax *(malaria parasite)*

apocarpy The condition in which the female reproductive organs (*carpels) of a flower are not joined to each other. It occurs, for example, in the buttercup. *Compare* **syncarpy**.

apocrine secretion *See* secretion.

apodeme An inward projection from the exoskeleton of an arthropod, to which muscles that enable movement of the limbs are attached.

apoenzyme An inactive enzyme that must associate with a specific *cofactor molecule or ion in order to function. *Compare* **holoenzyme**.

apomixis (agamospermy) A reproductive process in plants that superficially resembles normal sexual reproduction but in which there is no fusion of gametes. In apomictic flowering plants there is no fertilization by pollen, and the embryos develop simply by division of a *diploid cell of the ovule. *See also* **parthenocarpy; parthenogenesis**.

apoplast An interconnected system in plants that consists of all the cell walls and the water that exists in them (the cell wall is composed of cellulose fibres, between which are spaces filled with water). The movement of water (and dissolved ions and solutes) through the cell walls is known as the *apoplast pathway*. This is the main route by which water taken up by a plant travels across the root cortex to the *endodermis (*see also* **Casparian strip**). *Compare* **symplast**.

apoptosis The process of cell death, which occurs naturally as part of the normal development, maintenance, and renewal of tissues within an organism. During embryonic development it plays a vital role in determining the final size and form of tissues and organs. For example, the fingers are 'sculpted' on the spadelike embryonic hand by apoptosis of the cells between them; the tubules of the embryonic kidney are hollowed out by a similar process. Cancer is associated with the suppression of apoptosis (*see also* **growth factor**), which also occurs when viruses infect cells – in order to inhibit the activity of *killer cells. Apoptosis differs from cell *necrosis, in which cell death may be stimulated by a toxic substance.

aposematic coloration *See* warning coloration.

appeasement Behaviour that inhibits aggression from another animal of the same species, frequently taking the form of a special posture or *display emphasizing the weakness of the performer. Threatening structures (e.g. antlers) and markings are covered or turned away, and vulnerable parts of the body may be exposed. Appeasement is seen in *courtship, in greeting ceremonies, and often (from the loser) after a fight.

appendicular skeleton The components, collectively, of the vertebrate skeleton that are attached to the main supporting, or *axial, skeleton. The appendicular skeleton is made up of paired appendages (e.g. legs, wings, arms) together with the *pelvic girdle and *pectoral girdle.

appendix (vermiform appendix) An outgrowth of the *caecum in the alimentary canal. In humans it is a *vestigial organ containing lymphatic tissue and serves no function in normal digestive processes. Appendicitis is caused by inflammation of the appendix.

apposition The addition of layers of cellulose to the inner surface of a plant cell wall, at its junction with the plasmalemma. This type of growth results in thickening and strengthening the cell wall and usually occurs when elongation of the cell is complete. *Compare* **intussusception**.

Apterygota A subclass of small primitive wingless insects in which metamorphosis is slight or absent. It includes the orders Thysanura and Diplura (bristletails), *Collembola (springtails), and Protura. *Compare* **Pterygota**. *See* **Insecta**.

aqueous humour The fluid that fills the space between the cornea and the lens of the vertebrate eye. In addition to supplying the cornea and lens with nutrients, the aqueous humour helps to maintain the shape of the eye. It is produced and renewed every four hours by the *ciliary body.

Arabidopsis A genus of flowering plants of the family Cruciferae. The species *A. thaliana* (Thale cress) is widely used as a research tool in molecular genetics and developmental biology because it has a small and simple genome (five pairs of chromosomes), over half of which codes for protein, and it can be easily cultured, having a life cycle of only 6–8 weeks.

arachidonic acid A polyunsaturated fatty acid, $CH_3(CH_2)_3(CH_2CH:CH)_4(CH_2)_3COOH$, that is essential for growth in mammals. It can be synthesized from *linoleic acid. Arachidonic acid acts as a precursor to several biologically active compounds, including *prostaglandins, and plays an important role in membrane production and fat metabolism. The release of arachidonic acid from membrane *phospholipids is triggered by certain hormones. *See* **essential fatty acid**.

Arachnida A class of terrestrial arthropods of the subphylum *Chelicerata, comprising about 65 000 species and including spiders, scorpions, harvestmen, ticks, and mites. An arachnid's body is divided into an anterior *prosoma and a posterior *opisthosoma. The prosoma bears *chelicerae, *pedipalps, and four pairs of walking legs. The opisthosoma may bear various sensory or silk-spinning appendages (*see* **spinneret**). Arachnids are generally carnivorous, feeding on the body fluids of their prey or secreting enzymes to digest prey externally. Spiders immobilize their prey with poison injected by the fanglike chelicerae, while scorpions grasp their prey in large clawed pedipalps and may poison it using the posterior stinging organ. Ticks and some mites are parasitic but most arachnids are free-living. They breathe either via *tracheae (like insects) or by means of thin highly folded regions of the body wall called *lung books.

arachnoid membrane One of the three membranes (*meninges) that surround the brain and spinal cord of vertebrates. It lies between the *pia mater and the *dura mater. The arachnoid membrane is very delicate and carries *cerebrospinal fluid, which sustains and cushions the nervous tissue.

arbovirus Obsolete name for any RNA-containing virus that is transmitted from animals to man through the bite of mosquitoes and ticks (i.e. arthropods, hence *arthropod-borne* viruses). They cause various forms of encephalitis (inflammation of the brain) and serious fevers, such as dengue and yellow fever.

Archaebacteria A subkingdom of bacteria containing the *methanogens, which produce methane; the thermoacidophilic bacteria, which live in extremely hot and acidic environments (such as hot springs); and the halophilic bacteria, which can only function at high salt concentrations and are abundant in the world's oceans. Archaebacteria are distinguished from the *Eubacteria in that their membrane lipids are ether-linked as opposed to being ester-linked, their RNA and ribosomes have a characteristic structure and shape, and they lack *peptidoglycan in their cell walls. They are possibly descendants of the earliest forms of life.

archegonium (*pl.* **archegonia**) The multicellular flask-shaped female sex organ of bryophytes, clubmosses, horsetails, ferns, and many gymnosperms. Such plants are described as *archegoniate* to distinguish them from algae, which do not possess archegonia. The dilated base, the *venter*, contains the oosphere (female gamete). The cells of the narrow neck liquefy to allow the male gametes to swim towards the oosphere. The archegonium is thus an adaptation to the terrestrial environment as it provides a means for the male gametes to reach the female gamete. *Compare* **antheridium**.

archenteron (gastrocoel) A cavity within an animal embryo at the *gastrula stage of development. All or part of the archenteron eventually forms the cavity of the gut. It is connected to the outside by an opening (the *blastopore*), which becomes either the mouth, the mouth and anus, or the anal opening of the animal.

areolar connective tissue A type of *connective tissue consisting of a gel-like matrix incorporating strands of protein fibres (*collagen and *elastin) and such cells as *fibroblasts, *mast cells, *macrophages, and fat cells. This tissue is found throughout the body under the skin and linking organs and other tissues.

arginine *See* **amino acid**.

aril An outgrowth that grows around and may completely enclose the testa (seed coat) of a seed. It develops from the placenta, funicle, or micropyle of an ovule. The aril surrounding the nutmeg seed forms the spice mace. *See also* **caruncle**.

arousal A level of physiological and behavioural responsiveness in an animal, which tends to vary between sleep and full alertness. It is controlled by a particular part of the brain (the *reticular activating system*) and can be detected by changes in brain electrical activity, heart rate and muscle tone, responsiveness to new stimuli, and general activity.

arrector pili A small muscle in the dermis of the skin that is attached to the base of a hair follicle. Contraction of the arrector pili, in response to cold or fear, pulls the hair into a vertical position, thus trapping an insulating layer of air around the body and causing 'goose flesh' in humans.

arrhythmia A disturbance in the normal rhythm of the heart, usually caused by reduced blood flow within the heart or a defect in the *pacemaker (the sinoatrial node) of the heart.

arteriole A small muscular blood vessel that receives blood from the arteries and carries it to the capillaries.

artery A blood vessel that carries blood away from the heart towards the other body tissues. Most arteries carry oxygenated blood (the *pulmonary artery is an exception). The large arteries branch to form smaller ones, which in turn branch into *arterioles. All arteries have muscular walls, whose contraction aids in pumping blood around the body. The accumulation of fatty deposits in the walls of the arteries leads to *atherosclerosis*, which limits and may eventually block the flow of blood. *Compare* **vein**.

Arthropoda A phylum of invertebrate animals comprising over one million species – the largest in the animal kingdom. Arthropods inhabit marine, freshwater, and terrestrial habitats worldwide. Characteristically, they possess an outer body layer – the *cuticle – that functions as a rigid protective exoskeleton; growth is thus possible only by periodic moults (*see* **ecdysis**). The arthropod body is composed of segments (*see* **metameric segmentation**) usually forming distinct specialized body regions, e.g. head, thorax, and abdomen. These segments may possess hardened jointed appendages, modified variously as *mouthparts, limbs, wings, reproductive organs, or sense organs. The main body cavity, containing the internal organs, is a blood-filled *haemocoel, within which lies the heart. The origins and relationships of the various groups of arthropods remain uncertain, but they have been assigned to three subphyla or superclasses according to the basic structure of their appendages: *Crustacea (shrimps, barnacles, crabs, etc.); *Uniramia, including the classes *Insecta (insects), *Chilopoda (centipedes), and *Diplopoda (millipedes); and *Chelicerata, including the *Arachnida (spiders, scorpions, mites, and ticks).

articulation The attachment of two bones, usually by means of a *joint. The thigh bone (femur), for instance, articulates with the pelvic girdle.

artificial insemination (AI) The deposition of semen, using a syringe, at the mouth of the uterus to make conception possible. It is used in the selective *breeding of domestic animals and also in humans in some cases of impotence and infertility. It is timed to coincide with ovulation in the female.

artificial selection The modification of species by selective *breeding. Animals or plants with desirable characteristics are interbred with the aim of altering the *genotype and producing a new strain of the organism for a specific purpose. For example, sheep are bred by means of artificial selection in order to improve wool quality.

Artiodactyla An order of hooved mammals comprising the even-toed ungulates, in which the third and fourth digits are equally developed and bear the weight of the body. The order includes cattle and other ruminants (*see* **Ruminantia**), camels, hippopotamuses, and pigs. All except the latter are herbivorous, having an elongated gut and teeth with enamel ridges for grinding tough grasses. *Compare* **Perissodactyla**.

ascending tracts *See* spinal cord.

ascocarp The reproductive body of fungi of the phylum *Ascomycota, which contains *ascus cells. An ascocarp may be a closed sphere (*cleistothecium*), a flask-shaped structure (*perithecium*) with a small opening (*ostiole*), or an open cup-shaped structure (*apothecium*). The ascocarp consists of both ascus-bearing and vegetative *hyphae.

ascogenous hyphae The hyphae, in fungi of the Ascomycota, that grow from the *ascogonium after it has fused with the *antheridium. The ascogenous hyphae are made up of binucleate cells containing one nucleus derived from the male antheridium and the other from the female ascogonium. This condition is represented as $n + n$, rather than $2n$, as the cells are not true diploid cells. Asci (*see* **ascus**) develop from the ascogenous hyphae.

ascogonium The female gametangium of certain fungi of the Ascomycota, from which *ascogenous hyphae develop.

Ascomycota A phylum of fungi, formerly classified as a class (Ascomycetes) or a subdivision (Ascomycotina). It includes the *yeasts, some species of edible fungi, and *Claviceps purpurea*, which causes *ergot in rye. Sexual reproduction is by means of ascospores produced within an *ascus. The asci are usually grouped together in an *ascocarp.

ascorbic acid *See* vitamin C.

ascospore The spore of fungi belonging to the phylum Ascomycota. *See* ascus.

ascus (*pl.* **asci**) A specialized cell in fungi of the phylum *Ascomycota, which contains two haploid nuclei that fuse during sexual reproduction and then undergo meiosis, giving rise to eight *ascospores* contained within the ascus.

asepsis The condition of an environment that is free of pathogens. This is achieved by *sterilization techniques.

asexual reproduction Reproduction in which new individuals are produced from a single parent without the formation of gametes. It occurs chiefly in lower animals, microorganisms, and plants. In microorganisms and lower animals the chief methods are *fission (e.g. in protozoans), *fragmentation (e.g. in some aquatic annelid worms), and *budding (e.g. in coelenterates and yeasts). The principal methods of asexual reproduction in plants are by *vegetative propagation (e.g. bulbs, corms, tubers) and by the formation of sexual *spores. Spore formation occurs in mosses, ferns, and other plants showing alternation of generations, as a dormant stage between sporophyte and gametophyte, and in some algae and fungi, to produce replicas of the plant. *Compare* **sexual reproduction**.

asparagine *See* **amino acid**.

aspartic acid *See* **amino acid**.

aspirin (acetylsalicylic acid) A drug that reduces inflammation, combats fever, and alleviates pain. Aspirin works by inhibiting the formation of *prostaglandins, which are major factors in the inflammation process. It also reduces the aggregation of blood platelets, hence its use in maintaining blood flow following heart and circulatory disorders.

assimilation The utilization by a living organism of absorbed food materials in the processes of growth, reproduction, or repair.

association An ecological unit in which two or more species occur in closer proximity to one another than would be expected on the basis of chance. Early plant ecologists recognized associations of fixed composition on the basis of the *dominant species present (e.g. a coniferous forest association). Associations now tend to be detected by using more objective statistical sampling methods. *See also* **consociation**.

association centre The part of the brain that links a primary sensory area (the part of the cerebral cortex that receives primary sensory impulses) with other parts of the brain, such as memory and motor areas, and deals with the interpretation and meaning of the primary sensory imput. For example, the auditory association area interprets a 'moo' sound as that coming from a cow.

aster A starlike arrangement of small fibres radiating from a *centriole. Asters become conspicuous in animal cells at the ends of the spindle when cell division starts.

astigmatism A lens defect in which when rays in one plane are in focus those in another plane are not. The eye can suffer from astigmatism, usually when the cornea is not spherical. It is corrected by using an anastigmatic lens, which has different radii of curvature in the vertical and horizontal planes.

astrocyte *See* **glia**.

atherosclerosis Obstruction of the arteries by localized deposits of fatty material (including *cholesterol) on their inner walls. Atherosclerosis is associated with high blood levels of cholesterol, particularly in the form of low-density *lipoprotein; it can result in heart failure if it affects the coronary arteries (*see* **coronary vessels**).

atlas The first *cervical vertebra, a ringlike bone that joins the skull to the vertebral column in terrestrial vertebrates. In advanced vertebrates articulation between the skull and atlas permits nodding movements of the head. *See also* **axis**.

ATP

ATP (adenosine triphosphate) A nucleotide that is of fundamental importance as a carrier of chemical energy in all living organisms. It consists of adenine linked to D-ribose (i.e. adenosine); the D-ribose component bears three phosphate groups, linearly linked together by covalent bonds (see formula). These bonds can undergo hydrolysis to yield either a molecule of *ADP* (*adenosine diphosphate*) and inorganic phosphate or a molecule of *AMP* (*adenosine monophosphate*) and pyrophosphate (*see* **ATPase**). Both these reactions yield a large amount of energy (about 30.6 kJ mol^{-1}) that is used to bring about such biological processes as muscle *contraction, the *active transport of ions and molecules across cell membranes, and the synthesis of biomolecules. The reactions bringing about these processes often involve the enzyme-catalysed transfer of the phosphate group to intermediate substrates. Most ATP-mediated reactions require Mg^{2+} ions as *cofactors.

ATP is regenerated by the rephosphorylation of AMP and ADP using the chemical energy obtained from the oxidation of food. This takes place during *glycolysis and the *Krebs cycle but, most significantly, is also a result of the reduction–oxidation reactions of the mitochondrial *electron transport chain, which ultimately reduces molecular oxygen to water (*oxidative phosphorylation).

ATPase Any of a group of enzymes that bring about the hydrolysis of ATP. This results in the cleavage of either one phosphate group, with the formation of ADP and inorganic phosphate (P_i), or of two phosphate groups, with the formation of AMP and pyrophosphate (PP_i); the second reaction yields twice as much energy as the first. ATPase activity is associated with many energy-consuming processes; for example, in muscle contraction it is associated with *myosin when activated by actin. One form of ATPase, *ATP synthetase*, can catalyse the synthesis of ATP, for example in the mitochondrial *electron transport chain (*see also* **chemiosmotic theory**).

atrioventricular node (AVN) A specialized group of *cardiac muscle fibres situated in the fibrous ring between the right atrium and ventricle of the heart. The AVN is the only pathway between the atria and the ventricles through which electrical impulses can pass. Thus, following the contraction of the atria, the AVN initiates a wave of contraction in the ventricles via the *bundle of His.

atrioventricular valve *See* bicuspid valve; tricuspid valve.

atrium **1.** (*or* **auricle**) A chamber of the *heart that receives blood from the veins and forces it by powerful muscular contraction into the *ventricle(s). Fish have a single atrium but all other vertebrates have two. **2.** Any of various cavities or chambers in animals, such as the chamber surrounding the gill slits of the lancelet and other invertebrate chordates.

atrophy The degeneration or withering of an organ or part of the body.

atropine A poisonous crystalline alkaloid, $C_{17}H_{23}NO_3$. It can be extracted from deadly nightshade and other solanaceous plants and is used in medicine to treat colic, to reduce secretions, and to dilate the pupil of the eye.

attenuation **1.** *(in medicine)* A process of reducing the disease-producing ability of a microorganism. It can be achieved by chemical treatment, heating, drying, irradiation, by growing the organism under adverse conditions, or by serial passage through another organism. Attenuated bacteria or viruses are used for some *vaccines. **2.** *(in mycology)* The conversion by yeasts of carbohydrates to alcohol, as in brewing and wine and spirit production.

attenuator *See* operon.

atto- Symbol a. A prefix used in the metric system to denote 10^{-18}. For example, 10^{-18} second = 1 attosecond (as).

audibility The limits of audibility of the human ear are between about 20 hertz (a low rumble) and 20 000 hertz (a shrill whistle). With increased age the upper limit falls quite considerably.

audiometer An instrument that generates a sound of known frequency and intensity in order to measure an individual's hearing ability.

auditory Of or relating to the *ear. For example, the *auditory meatus* is the canal leading from the pinna to the tympanum (eardrum).

auditory nerve The nerve that transmits sensory information from the ear to the brain. *See also* organ of Corti.

auricle **1.** *See* atrium. **2.** *See* pinna.

Australopithecus A genus of fossil primates that lived 1.3–4 million years ago, coexisting for some of this time with early forms of men (*see* **Homo**). They walked erect and had teeth resembling those of modern man, but the brain capacity was less than half that of a human. Various finds have been made, chiefly in East and South Africa (hence the name, which means 'southern ape'). *Australopithecus* and related genera are known as *australopithecines*.

autecology The study of ecology at the level of the species. An autecological study aims to investigate the ecology of *populations or individuals of a particular species, including habitat, distribution, life cycle, etc. This should enable a full description of the *ecological niche of the organism to be made. *Compare* **synecology**.

autochthonous Describing an organism that is native to the place in which it is found. *Compare* **allochthonous**.

autoclave A strong steel vessel used for carrying out chemical reactions, sterilizations, etc., at high temperature and pressure.

autoecious Describing rust fungi (*see* **rusts**) that pass their life cycle in association with only one host plant. *Compare* **heteroecious**.

autogamy 1. A type of reproduction that occurs in single isolated individuals of ciliate protozoans of the genus *Paramecium*. The nucleus divides into two genetically identical haploid nuclei, which then fuse to form a diploid zygote. The onset of autogamy is associated with changing environmental conditions and may be necessary to maintain cell vitality. **2.** Self-fertilization in plants. *See* **fertilization**.

autogenic Relating to or caused by a change in the environment or an individual organism due to some *endogenous factor, i.e. one that comes from within the environment or organism. *Compare* **allogenic**.

autograft *See* **graft**.

autoimmunity A disorder of the body's defence mechanism in which an *immune response is elicited against its own tissues. Rheumatoid arthritis is basically an autoimmune disease.

autolysis The process of self-destruction of a cell, cell organelle, or tissue. It occurs by the action of enzymes within or released by *lysosomes. *See also* **lysis**.

autonomic nervous system (ANS) The part of the vertebrate *peripheral nervous system that supplies stimulation via motor nerves to the smooth and cardiac muscles (the involuntary muscles) and to the glands of the body. It is divided into the *parasympathetic and the *sympathetic nervous systems, which tend to work antagonistically on the same organs (see illustration). The activity of the ANS is controlled principally by the *medulla oblongata and *hypothalamus of the brain.

autopolyploid A *polyploid organism in which the multiple sets of chromosomes are all derived from the same species. For example, doubling of the chromosome number during mitotic cell division, possibly induced by *colchicine, gives rise to a tetraploid known as an *autotetraploid*. *Compare* **allopolyploid**.

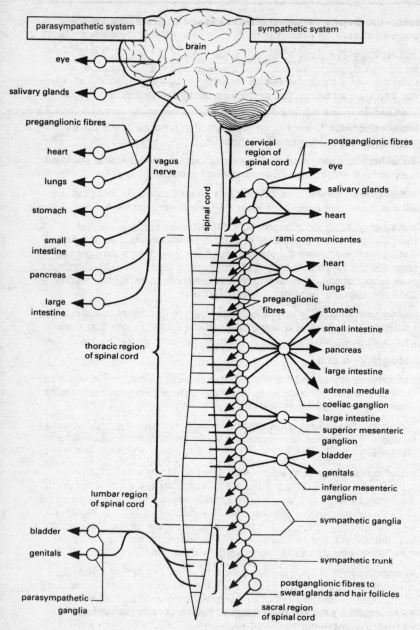

The autonomic nervous system (showing one side of each system of the ANS)

autoradiography An experimental technique in which a radioactive specimen is placed in contact with (or close to) a photographic plate, so as to produce a record of the distribution of radioactivity in the specimen. The film is darkened by the ionizing radiation from radioactive parts of the sample. Autoradiography is used to study the distribution of particular substances in living tissues and cells. A radioactive isotope of the substance is introduced into the organism or tissue, which is killed, sectioned, and examined after enough time has elapsed for the isotope to be incorporated into the substance.

autosome Any of the chromosomes in a cell other than the *sex chromosomes.

cranium

upper jaw

lower jaw

Autostylic jaw suspension

autostylic jaw suspension A type of jaw suspension seen in the lungfish (*see* **Dipnoi**) and tetrapods, in which the upper jaw is connected directly to the cranium (see illustration). *Compare* **amphistylic jaw suspension; hyostylic jaw suspension**.

autotomy The shedding by an animal of part of its body followed by the regeneration of the lost part. Autotomy is achieved by the contraction of muscles at specialized regions in the body. It serves as a protective mechanism if the animal is damaged or attacked (e.g. tail loss in certain reptiles) and is common as a method of asexual reproduction in polychaete worms, in which both new head and tail regions may be regenerated.

autotrophic nutrition A type of nutrition in which organisms synthesize the organic materials they require from inorganic sources. Chief sources of carbon and nitrogen are carbon dioxide and nitrates, respectively. All green plants are autotrophic and use light as a source of energy for the synthesis, i.e. they are *photoautotrophic* (*see* **photosynthesis**). Some bacteria are also photoautotrophic; others are *chemoautotrophic*, using energy derived from chemical processes (*see* **chemosynthesis**). *Compare* **heterotrophic nutrition**.

auxanometer Any mechanical instrument or measuring device used to study the growth or movement of plant organs. One type of auxanometer consists of a recording device that translates any increase in stem height into movement of a needle across a scale.

auxin Any of a group of plant *growth substances responsible for such processes as the promotion of growth by cell enlargement, the maintenance of *apical dominance, and the initiation of root formation in cuttings. Naturally occurring auxins, such as *indoleacetic acid (IAA)*, are synthesized in the shoot tips. Synthetic auxins, such as *2,4-D* and *2,4,5-T*, are used as weedkillers.

Aves The birds: a class of bipedal vertebrate chordates (*see* **Chordata**) with *feathers, wings, and a beak. They evolved from reptilian ancestors in the Jurassic period (190–136 million years ago) and modern birds still have scaly legs, like reptiles. Birds are warm-blooded (*see* **homoiothermy**). The skin is dry and loose and has no sweat glands, so cooling is effected by panting. Their efficient lungs and four-chambered heart (which completely separates oxygenated and deoxygenated blood) ensure a good supply of oxygen to the tissues. Birds can therefore sustain a high body temperature and level of activity necessary for *flight. The breastbone bears a keel for the attachment of flight muscles. The skeleton is very light; many of the bones are tubular, having internal struts to provide strength and *air sacs to reduce weight and provide extra oxygen in flight. Their feathers are vital for flight, streamlining the body, and insulation against heat loss.

Many birds show a high degree of social behaviour in forming large flocks and pair bonding for nesting, egg incubation, and rearing young. Fertilization is internal and the female lays hard-shelled eggs. *See also* **Ratitae**.

avidin A glycoprotein component of egg white that binds strongly to the vitamin *biotin. Proteins and nucleic acids can be linked to biotin (*biotinylated*), and the avidin–biotin reaction can then be used in a number of assay methods, such as antigen–antibody reactions or *DNA hybridization. For example, enzymes conjugated with avidin can be used to bind to biotinylated antibodies.

AVN *See* **atrioventricular node**.

awn A stiff bristle that projects from the tip of a plant part or organ. The bracts in a grass inflorescence commonly bear awns (*see* **spike**).

axenic culture A *culture medium in which only one type of microorganism is growing. Such cultures are widely used in microbiology to determine the basic growth requirements or degree of inhibition by antibiotics or other chemicals of a particular species.

axial skeleton The main longitudinal section of the vertebrate *skeleton, including the *skull, the *vertebral column, and the rib cage. *Compare* **appendicular skeleton**.

axil The angle between a branch or leaf and the stem it grows from. *Axillary* (or *lateral*) *buds* develop in the axil of a leaf. The presence of axillary buds distinguishes a leaf from a leaflet.

axillary bud *See* **axil**.

axis The second *cervical vertebra, which articulates with the *atlas (the first cervical vertebra, which articulates with the skull). The articulation between the axis and atlas in reptiles, birds, and mammals permits side-to-side movement of the head. The body of the axis is elongated to form a peg (the *odontoid process*), which extends into the ring of the atlas and acts as a pivot on which the atlas (and skull) can turn.

axon The long threadlike part of a nerve cell (*neurone). It carries the nerve impulse (in the form of an *action potential) away from the *cell body of a neurone towards either an effector organ or the brain. *See also* **nerve fibre**.

Axoneme structure in cross section of a cilium

axoneme The core of a *cilium or *flagellum, consisting of two central *microtubules surrounded by nine other pairs of microtubules (see illustration). The outer microtubules are associated with the protein *dynein, which is responsible for the movement of the organelle.

B

Bacillariophyta A phylum of *algae comprising the diatoms. These marine or freshwater unicellular organisms have cell walls (*frustules*) composed of pectin impregnated with silica and consisting of two halves, one overlapping the other. Diatoms are found in huge numbers in plankton and are important in the food chains of seas and rivers. Past deposition has resulted in diatomaceous earths (kieselguhr) and the oil reserves of these species have contributed to oil deposits.

bacillus Any rod-shaped bacterium. Generally, bacilli are large, Gram-positive, spore-bearing, and have a tendency to form chains and produce a *capsule. Some are motile, bearing flagella. They are ubiquitous in soil and air and many are responsible for food spoilage. The group also includes *Bacillus anthracis*, which causes anthrax.

backbone *See* **vertebral column**.

back cross A mating between individuals of the parental generation (P) and the first generation of offspring (F_1) in order to identify hidden *recessive alleles. If an organism displays a *dominant characteristic, it may possess two dominant alleles (i.e. it is homozygous) or a dominant and a recessive allele for that characteristic (i.e. it is heterozygous). To find out which is the case, the organism is crossed with one displaying the recessive characteristic. If all the offspring show the dominant characteristic then the organism is homozygous, but if half show the recessive characteristic, then the organism is heterozygous. *See also* **test cross**.

background radiation Low-intensity *ionizing radiation present on the surface of the earth and in the atmosphere as a result of cosmic radiation and the presence of radioisotopes in the earth's rocks, soil, and atmosphere. The radioisotopes are either natural or the result of nuclear fallout or waste gas from power stations.

bacteria A diverse group of ubiquitous microorganisms all of which consist of only a single *cell that lacks a distinct nuclear membrane and has a *cell wall of a unique composition (see illustration). Bacteria comprise the kingdom Prokaryotae (*see* **prokaryote**) and most of them are identified by means of *Gram's stain; they are usually classified on the basis of their metabolic requirements (e.g. whether or not they require oxygen: *see* **aerobic respiration; anaerobic respiration**) and shape. A bacterial cell may be spherical (*see* **coccus**), rodlike (*see* **bacillus**), spiral (*see* **spirillum**), comma-shaped (*see* **vibrio**), corkscrew-shaped (*see* **spirochaete**), or filamentous, resembling a fungal cell. The majority of bacteria range in size from 0.5 to 5 µm. Many are motile, bearing flagella, possess an outer slimy *capsule, and produce resistant spores (*see* **endospore**). In general bacteria reproduce only asexually, by simple division of cells, but a few groups undergo a form of sexual reproduction (*see* **conjugation**). Bacteria are largely responsible for decay and decomposition of organic matter, producing a cycling of such chemicals as carbon (*see* **carbon cycle**), oxygen, nitrogen (*see* **nitrogen cycle**), and sulphur (*see* **sulphur cycle**). A few bacteria obtain their food by means of *photosynthesis, including the *Cyanobacteria (blue-green bacteria); some are saprotrophs; and others are parasites, causing disease. The symptoms of bacterial infections are produced by *toxins.

Bacteria are divided into two subkingdoms. The *Archaebacteria comprise the *methanogens and other primitive forms; the Eubacteria comprise the remaining (and vast majority of) species.

A generalized bacterial cell

bacterial growth curve A curve on a graph that shows the changes in size of a bacterial population over time in a culture. The bacteria are cultured in sterile nutrient medium and incubated at the optimum temperature for growth. Samples are removed at intervals and the number of viable bacteria is counted. A logarithmic growth curve is plotted, which shows various phases (see graph).

In the *lag* (or *latent*) *phase* there is only a small increase in numbers as the bacteria imbibe water, and synthesize ribosomal RNA and subsequently enzymes, in adjusting to the new conditions. The length of this phase depends on which medium was used to culture the bacteria before the investigation and which phase the cells are already in. As the life span (generation time) of the cells decreases, they enter the *log* (or *exponential*) *phase*, in which the cells reach a maximum rate of reproduction and the number of bacteria increases directly with time, giving a straight slope on a logarithmic scale (*see* **exponential growth**). For example, the fastest generation time for *E. coli* is 21 minutes. Growth rate can be estimated in this phase. With time, as the population grows, it enters the *stationary phase*, when the nutrients and electron acceptors are depleted and the pH drops as carbon dioxide and other waste poisons accumulate. As the cell's energy stores are depleted the rate of cell division decreases. The *death* (or *final*) *phase* occurs when the rate at which the bacteria die exceeds the rate at which they are produced; the population declines as the levels of nutrients fall and toxin levels increase. *See also* **population growth**.

Bacterial growth curve

bactericidal Capable of killing bacteria. Common bactericides are some *antibiotics, *antiseptics, and *disinfectants. *Compare* **bacteriostatic**.

bacteriochlorophyll A form of chlorophyll found in photosynthetic bacteria, such as the purple sulphur bacteria. There are two types, bacteriochlorophyll *a* and bacteriochlorophyll *b*, which are structurally similar to the chlorophyll *a* and chlorophyll *b* found in plants; the purple bacteria contain either of these two types of bacteriochlorophyll, depending on the species. Bacteriochlorophyll is located in membrane-bound *chromatophores*.

bacteriophage (phage) A virus that is parasitic within a bacterium. Each phage is specific for only one type of bacterium. Most phages (*virulent phages*) infect, quickly multiply within, and destroy (lyse) their host cells. However, some (*temperate phages*) remain dormant in their hosts after initial infection: their nucleic acid becomes integrated into that of the host and multiplies with it, producing infected daughter cells (*see* **lysogeny**). Lysis may eventually be triggered by environmental factors Phages are used experimentally to identify bacteria, to control manufacturing processes (such as cheese production) that depend on bacteria, and, because they can alter the genetic make-up of bacterial cells, they are important tools in genetic engineering as cloning *vectors.

bacteriorhodopsin A membrane-bound protein of the halophilic (salt-resistant) bacterium *Halobacterium halobium*. When activated by light, it pumps protons out of the cell; this creates a concentration gradient, which enables ATP to be synthesized. Bacteriorhodopsin is composed of seven α-helix segments, which span the membrane and are joined together by short amino-acid chains. It contains the prosthetic group retinal, which is also found in the pigment *rhodopsin in the rod cells of vertebrates.

bacteriostatic Capable of inhibiting or slowing down the growth and reproduction of bacteria. Some *antibiotics are bacteriostatic. *Compare* **bactericidal**.

baker's yeast Strains of the yeast *Saccharomyces cerevisiae* that are used in bread-making to enable the dough to rise. Water is added to flour, which activates the *amylase enzymes that hydrolyse the starch in flour to glucose. Baker's yeast is then added, which uses the glucose as a substrate for *aerobic respiration. The carbon dioxide produced from yeast respiration causes bubbles to form in the dough; these become larger during heating in an oven, giving bread its typical texture.

balance **1.** *(in animal physiology)* Equilibrium in the posture of the body. In vertebrates balance is sensed and maintained by the *vestibular apparatus of the inner ear. **2.** *(in nutrition) See* **diet**.

balanced polymorphism *See* **polymorphism**.

baleen *See* **whalebone**.

Interlocking barbs of a contour feather

barb **1.** *(in zoology)* Any one of the stiff filaments forming a row on each side of the longitudinal shaft of a feather (see illustration). Together the barbs form the expanded part (*vane*) of the feather. *See* **barbule**. **2.** *(in botany)* A hooked hair.

barbiturate Any one of a group of drugs derived from barbituric acid, which have a depressant effect on the central nervous system. Barbiturates were originally used as sedatives and sleeping pills but their clinical use is now limited due to their toxic side-effects; prolonged use can lead to addiction. Specific barbiturates in clinical use include phenobarbitone, for treating epilepsy, and methohexitane sodium, used as an anaesthetic.

barbule Any of the minute filaments forming a row on each side of the *barb of a feather. In a *contour feather adjacent barbules interlock by means of hooks (*barbicels*) and grooves, forming a firm vane. Down feathers have no barbicels.

Barfoed's test A biochemical test to detect monosaccharide (reducing) sugars in solution, devised by the Swedish physician C. T. Barfoed (1815–99). *Barfoed's reagent*, a mixture of ethanoic (acetic) acid and copper(II) acetate, is added to the test solution and boiled. If any reducing sugars are present a red precipitate of copper(II) oxide is formed. The reaction will be negative in the presence of disaccharide sugars as they are weaker reducing agents.

bark The protective layer of mostly dead cells that covers the outside of woody stems and roots. It includes the living and dead tissues external to the xylem, including the phloem and periderm. The term can be used more specifically to describe the periderm together with other tissues isolated by the activity of the *cork cambium. In some species, such as birch, there is one persistent cork cambium but in the older stems of certain other species a second cork cambium becomes active beneath the periderm and further periderm layers are formed every few years. The result is a composite tissue called *rhytidome*, composed of cork, dead cortex, and dead phloem cells.

baroreceptor A *receptor that responds to changes in pressure. The *carotid sinus in the carotid artery contains baroreceptors that respond to changes in arterial pressure and are therefore involved in the regulation of blood pressure and heart beat.

Barr body A structure consisting of a condensed X chromosome (*see* **sex chromosome**) that is found in nondividing nuclei of female mammals. The presence of a Barr body is used to confirm the sex of athletes in sex determination tests. It is named after the Canadian anatomist M. L. Barr (1908–), who identified it in 1949.

basal ganglia Small masses of nervous tissue within the brain that connect the *cerebrum with other parts of the nervous system. They are involved with the subconscious regulation of voluntary movements. The largest of the basal ganglia is the *corpus striatum*.

basal metabolic rate (BMR) The rate of energy metabolism required to maintain an animal at rest. BMR is measured in terms of heat production per unit time and is usually expressed in kilojoules of heat released per square metre of body surface per hour ($kJ\,m^{-2}\,h^{-1}$). It indicates the energy consumed in order to sustain such vital functions as heartbeat, breathing, nervous activity, active transport, and secretion. Different tissues have different metabolic rates (e.g. the BMR of brain tissue is much greater than that of bone tissue) and therefore the tissue composition of an animal determines its overall BMR. For any comparable group of animals (such as mammals) BMR is proportional to body weight according to the allometric equation (*see* **allometric growth**); small animals tend to have a higher metabolic rate per unit weight than large ones.

base A compound that reacts with an *acid to give water (and a salt). A base that dissolves in water to produce hydroxide ions is called an *alkali*. For example, ammonia reacts as follows:

$$NH_3 + H_2O \leftrightarrows NH_4^+ + {}^-OH$$

Similar reactions occur with organic *amines (see also **nitrogenous base**).

basement membrane A thin sheet of collagen in animals that underlies and supports the cells of an *epithelium, separating this from the underlying tissue. The basement membrane is made up of *glycosaminoglycan (mucopolysaccharide) and fibrous material. In some tissues, such as the kidney *glomerulus and the lung *alveoli, the basement membrane acts as a filter (see **ultrafiltration**).

base pairing The chemical linking of two complementary nitrogenous bases in *DNA and in certain types of *RNA molecules. Of the four such bases in DNA, adenine pairs with thymine and cytosine with guanine. In RNA, thymine is replaced by uracil. Base pairing is responsible for holding together the two strands of a DNA molecule to form a double helix and for faithful reproduction and reading of the *genetic code. The links between bases take the form of *hydrogen bonds.

basic stains See **staining**.

Basidiomycota A phylum of fungi, formerly classified as a class (Basidiomycetes) or a subdivision (Basidiomycotina). Sexual reproduction is by means of basidiospores produced externally on a club-shaped or cylindrical *basidium. Basidia are often grouped together on fruiting structures, such as mushrooms, puffballs, and bracket fungi. Exceptions are the *rusts and *smuts, which do not produce obvious fruiting bodies.

basidiospore The characteristic spore of fungi belonging to the phylum *Basidiomycota. See **basidium**.

basidium (pl. **basidia**) A specialized cell in fungi of the phylum *Basidiomycota, in which nuclear fission and meiosis occur. This results in the formation of four *basidiospores*, which are attached to the basidium by means of *sterigmata* (short stalks).

basilar membrane See **organ of Corti**.

basket cell A type of cell found in the *cerebral cortex. Basket cells form a layer next to the *Purkyne cells, from which they receive stimuli.

basophil A type of white blood cell (*leucocyte) that has a lobed nucleus surrounded by granular cytoplasm (see **granulocyte**). Basophils are produced continually by stem cells in the red bone marrow; like *mast cells, they produce histamine and heparin as part of the body's defences at the site of an infection or injury (see **inflammation**).

bast An old name for *phloem.

batch culture A technique used to grow microorganisms or cells. A limited supply of nutrients for growth is provided; when these are used up, or some other factor becomes limiting, the culture declines. Cells, or products that the organisms have made, can then be harvested from the culture.

Batesian mimicry *See* mimicry.

bats *See* Chiroptera.

B cell (B lymphocyte) A *lymphocyte that is derived from stem cells in the bone marrow but does not mature in the thymus (*compare* **T cell**); in birds it matures in the bursa of the cloaca (hence *B* cell). Each B cell is programmed to recognize a specific antigen. In the presence of this antigen the B cell divides repeatedly to form a clone of cells, which mature into *plasma cells; these produce large amounts of specific antibody, which circulates in the blood and lymph and binds to the corresponding antigen. After a few days of antibody production the plasma cells die. However, some cells from the clone remain in the form of *memory cells*, which initiate a more rapid immune response on subsequent exposure to the same antigen.

B-chromosome *See* accessory chromosome.

becquerel Symbol Bq. The SI unit of activity (*see* **radiation units**). The unit is named after the discoverer of radioactivity, A. H. Becquerel (1852–1908).

bees *See* Hymenoptera.

beetles *See* Coleoptera.

beet sugar *See* sucrose.

behaviour The sum of the responses of an organism to internal or external stimuli. The behaviour of an animal can be either instinctive (*see* **instinct**) or learned. *See* **animal behaviour**.

behavioural genetics The branch of genetics concerned with determining the relative importance of the genetic constitution of animals as compared to environmental factors in influencing animal behaviour.

Benedict's test A biochemical test to detect *reducing sugars in solution, devised by the US chemist S. R. Benedict (1884–1936). *Benedict's reagent* – a mixture of copper(II) sulphate and a filtered mixture of hydrated sodium citrate and hydrated sodium carbonate – is added to the test solution and boiled. A high concentration of reducing sugars induces the formation of a red precipitate; a lower concentration produces a yellow precipitate. Benedict's test is a more sensitive alternative to *Fehling's test.

benthos Flora and fauna occurring on the bottom of a sea or lake. Benthic organisms may crawl, burrow, or remain attached to a substrate. *Compare* **nekton; neuston; plankton**.

beriberi A disease caused by a low intake of vitamin B_1 (thiamine; *see* **vitamin B complex**), resulting in damage to peripheral nerves and heart failure. Beriberi is most common in regions of the Far East where the diet is based on polished white rice, which lacks the thiamine-rich seed coat.

berry A fleshy fruit formed from either one carpel or from several fused together and containing many seeds. The fruit wall may have two or three layers but the inner layer is never hard and stony (as in some drupes). Examples of berries are grapes and tomatoes. A berry, such as a cucumber, that develops a hard outer rind is called a *pepo*. One that is segmented and has a leathery rind, such as a citrus fruit, is called a *hesperidium*. The rind contains oil glands and is lined by the white mesocarp, commonly called *pith*.

beta cells (β cells) *See* **islets of Langerhans.**

Beta sheet

beta sheet (β-pleated sheet) A form of secondary structure in *proteins in which extended polypeptide chains lie parallel to each other and are linked by hydrogen bonds between the N–H and C=O groups (see illustration). Beta sheets occur in many globular proteins and link polypeptides of the same type in certain fibrous proteins, including fibroin (the protein of silk). *Compare* **alpha helix.**

bicarbonate *See* **hydrogencarbonate.**

biceps A muscle that runs along the large bone of the upper arm (*humerus) and is connected to the *radius at one end and the shoulder bone (*scapula) at the other. Contraction of the biceps causes the arm to flex at the elbow joint (*see* **flexor**). It works antagonistically with the triceps, which contracts to extend the arm (*see* **antagonism**). *See also* **voluntary muscle.**

bicuspid valve (left atrioventricular valve; mitral valve) A valve, consisting of two flaps, situated between the left atrium and the left ventricle of the heart of birds and mammals. When the left ventricle contracts, forcing blood into the aorta, the bicuspid valve closes the aperture to the left atrium, thereby preventing any backflow of blood. The valve reopens to allow blood to flow from the atrium into the ventricle. *Compare* **tricuspid valve**.

biennial A plant that requires two growing seasons to complete its life cycle. During the first year it builds up food reserves, which are used during the second year in the production of flowers and seeds. Examples are carrot and parsnip.

bilateral symmetry A type of arrangement of the parts and organs of an animal in which the body can be divided into two halves that are mirror images of each other along one plane only (usually passing through the midline at right angles to the dorsal and ventral surfaces). Bilaterally symmetrical animals are characterized by a type of movement in which one end of the body always leads. In botany this type of symmetry is usually called *zygomorphy* when applicable to flowers (e.g. foxglove and antirrhinum flowers are zygomorphic). *Compare* **radial symmetry**.

bile (gall) A bitter-tasting greenish-yellow alkaline fluid produced by the *liver, stored in the *gall bladder, and secreted into the *duodenum of vertebrates. It assists the digestion and absorption of fats by the action of *bile salts*, which chemically reduce fatty substances and decrease the surface tension of fat droplets so that they are broken down and emulsified. Bile may also stimulate gut muscle contraction (*peristalsis). Bile also contains the *bile pigments*, *bilirubin* and *biliverdin*, which are produced by the breakdown of the blood pigment *haemoglobin.

bile duct The tube through which bile passes from the *liver or (when present) the *gall bladder to the duodenum.

bilirubin *See* **bile**.

biliverdin *See* **bile**.

binary fission *See* **fission**.

binding site An area on the surface of a molecule that combines with another molecule. Binding sites on enzymes can be *active sites or *allosteric sites.

binocular vision The ability, found only in animals with forward-facing eyes, to produce a focused image of the same object simultaneously on the retinas of both eyes. This permits three-dimensional vision and contributes to distance judgment.

binomial nomenclature The system of naming organisms using a two-part Latinized (or scientific) name that was devised by the Swedish botanist Linnaeus (Carl Linné; 1707–78); it is also known as the *Linnaean system*. The first part is the generic name (*see* **genus**), the second is the specific epithet or name (*see* **species**). The Latin name is usually printed in italics, starting with a capital letter. For example, in the scientific name of the common frog, *Rana temporaria*, *Rana* is the generic name and *temporaria* the specific name. The name of the species may be followed by an abbreviated form of the name of its discoverer; for example, the common daisy is *Bellis perennis* L. (for Linnaeus). There are several International Codes of Taxonomic Nomenclature that lay down the rules for naming organisms. *See also* **classification; taxonomy**.

bioaccumulation An increase in the concentration of chemicals, such as *pesticides, in organisms that live in environments contaminated by a wide variety of organic compounds. These compounds are not usually decomposed in the environment (i.e. they are not biodegradable) or metabolized by the organisms, so that their rate of absorption and storage is greater than their rate of excretion. The chemicals are normally stored in fatty tissues. *DDT is known as a *persistent pesticide*, as it is not easily broken down and bioaccumulates along *food chains, so that increasing concentrations occur in individual organisms at each trophic level.

bioactivation A metabolic process in which a product that is chemically reactive is produced from a relatively inactive precursor. *See also* **prodrug**.

bioassay (biological assay) A controlled experiment for the quantitative estimation of a substance by measuring its effect in a living organism. For example, the amount of the plant hormone auxin can be estimated by observing its effect on the curvature of oat coleoptiles – the concentration of the hormone is proportional to the curvature of the coleoptile.

biochemical evolution (molecular evolution) The gradual changes that occur at the molecular level in organisms over a period of time. Brought about by *mutation, it results in the evolution of novel genes and proteins.

biochemical fuel cell A system that exploits biological reactions for the conversion of biomass (chemical energy) to electricity (electrical energy). One potential application is the generation of electricity from industrial waste and *sewage. *Methyltrophic* organisms (i.e. organisms that use methane or methanol as their sole carbon sources) are being investigated for their potential use in biochemical fuel cells.

biochemical oxygen demand (BOD) The amount of oxygen taken up by microorganisms that decompose organic waste matter in water. It is therefore used as a measure of the amount of certain types of organic pollutant in water. BOD is calculated by keeping a sample of water containing a known amount of oxygen for five days at 20°C. The oxygen content is measured again after this time. A high BOD indicates the presence of a large number of microorganisms, which suggests a high level of pollution.

biochemical taxonomy *See* **taxonomy**.

biochemistry The study of the chemistry of living organisms, especially the structure and function of their chemical components (principally proteins, carbohydrates, lipids, and nucleic acids). Biochemistry has advanced rapidly with the development, from the mid-20th century, of such techniques as chromatography, X-ray diffraction, radioisotopic labelling, and electron microscopy. Using these techniques to separate and analyse biologically important molecules, the steps of the metabolic pathways in which they are involved (e.g. *glycolysis and the *Krebs cycle) have been determined. This has provided some knowledge of how organisms obtain and store energy, how they manufacture and degrade their biomolecules, how they sense and respond to their environment, and how all this information is carried and expressed by their genetic material. Biochemistry forms an important part of many other disciplines, especially physiology, nutrition, and genetics, and its discoveries have made a profound impact in medicine, agriculture, industry, and many other areas of human activity.

biodegradable *See* **pollution**.

biodiversity (biological diversity) The existence of a wide variety of species (*species diversity*) or other taxa of plants, animals, and microorganisms in a natural community or habitat, or of communities within a particular environment (*ecological diversity*), or of genetic variation within a species (*genetic diversity; see* **genetic variability**). The maintenance of a high level of biodiversity is important for the stability of ecosystems. Certain habitats, especially rainforests, have a rich species diversity, which is threatened by the continued destruction of habitats (*see* **deforestation; desertification**). Such ecosystems typically support large numbers of rare species, and population sizes of individual species tend to be small; they are therefore especially vulnerable to habitat destruction. Biodiversity in natural habitats also represents an important pool of species and genetic material of potential use to human societies. For example, wild plants continue to be used as a source of new drugs and other products, and the development of new strains and varieties of crop plants with increased disease resistance usually depends on incorporating genetic material from wild plants.

bioelement Any chemical element that is found in the molecules and compounds that make up a living organism. In the human body the most common bioelements (in decreasing order of occurrence) are oxygen, carbon, hydrogen, nitrogen, calcium, and phosphorus. Other bioelements include sodium, potassium, magnesium, and copper. *See* **essential element**.

bioenergetics The study of the flow and the transformations of energy that occur in living organisms. Typically, the amount of energy that an organism takes in (from food or sunlight) is measured and divided into the amount used for growth of new tissues; that lost through death, wastes, and (in plants) transpiration; and that lost to the environment as heat (through respiration).

bioengineering 1. The use of artificial tissues, organs, and organ components to replace parts of the body that are damaged, lost, or malfunctioning, e.g. artificial limbs, heart valves, and heart pacemakers. **2.** The application of engineering knowledge to medicine and zoology.

biofeedback The technique whereby a subject can learn to control certain body functions, such as heart rate or blood pressure, that are usually unconsciously regulated by the autonomic nervous system. It is facilitated by the use of monitoring devices, such as pulse monitors, electroencephalographs, and electromyographs, and can be useful in treating high blood pressure, migraine, epilepsy, and other disorders.

biofuel A gaseous, liquid, or solid fuel that contains an energy content derived from a biological source. The organic matter that makes up living organisms provides a potential source of trapped energy that is beginning to be exploited to supply the ever-increasing energy demand around the world. An example of a biofuel is rapeseed oil, which can be used in place of diesel fuel in modified engines. The methyl ester of this oil, *rapeseed methyl ester (RME)*, can be used in unmodified diesel engines and is sometimes known as *biodiesel*. Other biofuels include *biogas and *gasohol.

biogas A mixture of methane and carbon dioxide resulting from the anaerobic decomposition of such waste materials as domestic, industrial, and agricultural sewage. The decomposition is carried out by methanogenic bacteria (*see* **methanogen**); these obligate anaerobes produce methane, the main component of biogas, which can be collected and used as an energy source for domestic processes, such as heating, cooking, and lighting. The production of biogas is carried out in special *digesters*, which are widely used in China and India. As well as providing a source of fuel, these systems also enable *sewage, which contains pathogenic bacteria, to be digested, thereby removing the danger to humans that could otherwise result from untreated domestic and agricultural waste.

biogenesis The principle that a living organism can only arise from other living organisms similar to itself (i.e. that like gives rise to like) and can never originate from nonliving material. *Compare* **spontaneous generation**.

biogeochemical cycle (nutrient cycle) The cyclical movement of elements between living organisms (the biotic phase) and their nonliving (abiotic) surroundings (e.g. rocks, water, air). Examples of biogeochemical cycles are the *carbon cycle, *nitrogen cycle, *oxygen cycle, *phosphorus cycle, and *sulphur cycle.

biogeography The branch of biology that deals with the geographical distribution of plants and animals. *See* **plant geography; zoogeography**.

biological clock The mechanism, presumed to exist within many animals and plants, that produces regular periodic changes in behaviour or physiology. Biological clocks underlie many of the *circadian, tidal, and *annual rhythms seen in organisms (e.g. hibernation in animals). They continue to run even when conditions are kept artificially constant, but eventually drift out of step with the natural environment without the specific signals that normally keep them synchronized.

biological control The control of *pests by biological (rather than chemical) means. This may be achieved, for example, by breeding disease-resistant crops or by introducing a natural enemy of the pest, such as a predator or a parasite. This technique, which may offer substantial advantages over the use of pesticides or herbicides, has been employed successfully on a number of occasions. Examples include the control of the prickly pear cactus (*Opuntia*) in Australia by introducing the cactus moth (*Cactoblastis cactorum*), whose caterpillars feed on the plant's growing shoots, and the use of the ladybird to prey upon the scale insect (*Icerya*), which kills citrus fruit trees. Insect pests have also been subjected to genetic control, by releasing large numbers of males of the pest species that have been sterilized by radiation: infertile matings subsequently cause a decline in the pest population. This method has been used to control the screw worm fly (*Cochliomyia hominivora*), which lays its eggs in the open wounds of domestic cattle. Biological control is considered to reduce a number of the problems associated with chemical control using *pesticides, but care should be taken to avoid upsetting the natural ecological balance; for example, a particular predator may also destroy harmless or beneficial species.

biological rhythm *See* biorhythm.

biological warfare The military use of microorganisms, such as bacteria, viruses, fungi, and other microorganisms, including the agents for anthrax and botulism, to induce disease or death among humans, livestock, and crop plants. Though officially banned in most countries, research continues with the aim of developing virulent strains of existing microorganisms, using genetic engineering and other techniques.

biology The study of living organisms, which includes their structure (gross and microscopical), functioning, origin and evolution, classification, interrelationships, and distribution.

bioluminescence The emission of light without heat by living organisms. The phenomenon occurs in glow-worms and fireflies, bacteria and fungi, and in many deep-sea fish (among others); in animals it may serve as a means of protection (e.g. by disguising the shape of a fish) or species recognition or it may provide mating signals. The light is produced during the oxidation of a compound called *luciferin* (the composition of which varies according to the species), the reaction being catalysed by an enzyme, *luciferase*. Bioluminescence may be continuous (e.g. in bacteria) or intermittent (e.g. in fireflies).

biomarker A normal metabolite that, when present in abnormal concentrations in certain body fluids, can indicate the presence of a particular disease or toxicological condition. For example, abnormal concentrations of glucose in the blood can be indicative of diabetes mellitus (*see* **insulin**).

biomass The total mass of all the organisms of a given type and/or in a given area; for example, the world biomass of trees, or the biomass of elephants in the Serengeti National Park. It is normally measured in terms of grams of *dry mass per square metre. *See also* **pyramid of biomass**.

biome A major ecological community or complex of communities that extends over a large geographical area characterized by a dominant type of vegetation. The organisms of a biome are adapted to the climate conditions associated with the region. There are no distinct boundaries between adjacent biomes, which merge gradually with each other. Examples of biomes are *tundra, tropical *rainforest, *taiga, *chaparral, *grassland (temperate and tropical), and *desert.

biomechanics The application of the principles of mechanics to living systems, particularly those living systems that have coordinated movements. Biomechanics also deals with the properties of biological materials, such as blood and bone. For example, biomechanics would be used to analyse the stresses on bones in animals, both when the animals are static and when they are moving. Other types of problems in biomechanics include the fluid mechanics associated with swimming in fish and the aerodynamics of birds flying. It is sometimes difficult to perform realistic calculations in biomechanics because of complexity in the shape of animals or the large number of parts involved (for example, the large number of muscles involved in the movement of a human leg).

biomolecule Any molecule that is involved in the maintenance and metabolic processes of living organisms (*see* **metabolism**). Biomolecules include carbohydrate, lipid, protein, nucleic acid, and water molecules; some biomolecules are *macromolecules.

biophysics The study of the physical aspects of biology, including the application of physical laws and the techniques of physics to study biological phenomena.

biopoiesis The development of living matter from complex organic molecules that are themselves nonliving but self-replicating. It is the process by which life is assumed to have begun. *See* **origin of life**.

biopsy The removal of a small section of tissue from a potentially diseased organ or tissue in a living organism. The biopsied tissue is usually analysed by microscopic techniques in order to identify the nature of the disease.

bioreactor (**industrial fermentor**) A large stainless steel tank used to grow producer microorganisms in the industrial production of enzymes and other chemicals. After the tank is steam-sterilized, an inoculum of the producer cells is introduced into a medium that is maintained by probes at optimum conditions of temperature, pressure, pH, and oxygen levels for enzyme production. An *agitator mixes the medium, which is constantly aerated. It is essential that the culture medium is sterile and contains the appropriate nutritional requirements for the microorganism. When the nutrients have been utilized the product is separated; if the product is an extracellular compound the medium can be removed during the growth phase of the microorganisms, but an intracellular product must be harvested when the batch culture growth stops. Some bioreactors are designed for *continuous culture.

biorhythm (biological rhythm) A roughly periodic change in the behaviour or physiology of an organism that is generated and maintained by a *biological clock. Well-known examples are the *annual and *circadian rhythms occurring in many animals and plants. The term biorhythm has unfortunately also become associated with a pseudoscientific cult postulating that man's physical, emotional, and intellectual behaviour is influenced by three biological cycles with periodicities of exactly 23 days, 28 days, and 33 days respectively. There is no reliable evidence to support this theory.

biosensor A device that uses an immobilized agent to detect or measure a chemical compound. The agents include enzymes, antibiotics, organelles, or whole cells. A reaction between the immobilized agent and the molecule being analysed is transduced into an electronic signal. This signal may be produced in response to the presence of a reaction product, the movement of electrons, or the appearance of some other factor (e.g. light). Biosensors are being used increasingly in diagnostic tests: these allow quick, sensitive, and specific analysis of a wide range of biological products, including antibiotics, vitamins, and other important biomolecules (such as glucose), as well as the determination of certain *xenobiotics, such as synthetic organic compounds.

biosphere The whole of the region of the earth's surface, the sea, and the air that is inhabited by living organisms.

biosynthesis The production of molecules by a living cell, which is the essential feature of *anabolism.

biosystematics *See* systematics.

biotechnology The development of techniques for the application of biological processes to the production of materials of use in medicine and industry. For example, the production of antibiotics, cheese, and wine rely on the activity of various fungi and bacteria. *Genetic engineering can modify bacterial cells to synthesize completely new substances, e.g. hormones, vaccines, *monoclonal antibodies, etc.

biotic factor Any of the factors of an organism's environment that consist of other living organisms and together make up the *biotic environment*. These factors may affect an organism in many ways; for example, as competitors, predators, parasites, prey, or symbionts. In time, the distribution and abundance of the organism will be affected by its interrelationships with the biotic environment. *Compare* **abiotic factor**.

biotin A vitamin in the *vitamin B complex. It is the *coenzyme for various enzymes that catalyse the incorporation of carbon dioxide into various compounds. Adequate amounts are normally produced by the intestinal bacteria in animals although deficiency can be induced by consuming large amounts of raw egg white. This contains a protein, *avidin, that specifically binds biotin, preventing its absorption from the gut. Other sources of biotin include cereals, vegetables, milk, and liver.

bipedalism The ability to maintain a standing position, and to move about, using only two legs. Birds and humans are bipedal. The development of bipedalism was an important factor in hominid evolution.

bipolar cell *See* retina.

bipolar neurone A *neurone that has two processes, an axon and a dendron, extending in different directions from its cell body. Many sensory neurones are bipolar neurones. *Compare* **multipolar neurone; unipolar neurone.**

A generalized biramous appendage

biramous appendage A type of appendage that is characteristic of arthropods of the subphylum *Crustacea. It forks from the basal *protopodite* to form two branches, the inner *endopodite* and the outer *exopodite* (see illustration). Each of these branches can be composed of either one or more segments. There are many variations on this generalized structure; the branches often possess highly specialized extensions – the *epipodite* (on the coxopodite), the *exite* (on the exopodite), and the *endite* (on the endopodite). *Compare* **uniramous appendage**.

birds *See* Aves.

birth *See* parturition.

birth control (contraception) The intentional avoidance of pregnancy by methods that do not normally hinder sexual activity. The methods used can be 'natural' or 'artificial'. Natural methods, often used because of religious or moral objections to artificial methods, include the *rhythm method*, in which sexual intercourse is avoided during times when ovulation occurs; and *coitus interruptus*, an unreliable method in which the penis is withdrawn from the vagina before ejaculation. The rhythm method requires a monitoring of the woman's menstrual cycle and may be unsuitable in those women with irregular cycles. Artificial methods use devices or other agents (*contraceptives*) to prevent pregnancy. They include the *condom*, a rubber sheath placed over the penis to trap the sperm; and the *diaphragm*, a rubber cap placed over the cervix. Contraceptives that prevent *implantation include the *intrauterine device* (*IUD*), a metal or plastic coil placed in the uterus by a doctor (which may cause unacceptable bleeding in some women); and the 'morning-after pill', taken within three days after sexual intercourse. Other *oral contraceptives prevent ovulation. *Sterilization is usually considered to be irreversible but attempts at reversing the process are possible. For casual relationships, or relationships involving a partner whose sexual history is not known, health workers advise the use of condoms with all other forms of contraception as this provides the safest means of reducing the risk of infection by *sexually transmitted diseases.

birth rate (natality) The rate at which a particular species or population produces offspring. The birth rate of a species is used to measure its fecundity (reproductive capability). It is also an important factor in controlling the size of a population. *Compare* **death rate**.

bisexual *See* **hermaphrodite**.

biuret test A biochemical test to detect proteins in solution, named after the substance *biuret* ($H_2NCONHCONH_2$), which is formed when urea is heated. Sodium hydroxide is mixed with the test solution and drops of 1% copper(II) sulphate solution are then added slowly. A positive result is indicated by a violet ring, caused by the reaction of *peptide bonds in the proteins or peptides. Such a result will not occur in the presence of free amino acids.

bivalent *(in genetics) See* **pairing**.

bivalves *See* **Pelecypoda**.

bladder 1. *(in anatomy)* **a.** A hollow muscular organ in most vertebrates, also known as the *urinary bladder*, in which urine is stored before being discharged. In mammals urine is conveyed from the *kidneys to the bladder by the *ureters and is discharged to the outside through the *urethra. **b.** Any of various other saclike organs in animals for the storage of liquid or gas. *See* **gall bladder; swim bladder. 2.** *(in botany)* **a.** A modified submerged leaf of certain aquatic insectivorous plants, such as the bladderwort (*Utricularia*). It forms a hollow with a single opening that is sealed by a valve to trap small aquatic invertebrates after they have been sucked in. **b.** An air-filled cavity in the thallus of certain seaweeds, such as the bladderwrack (*Fucus vesiculosus*).

bladderworm (cysticercus) A larval stage of some tapeworms (*see* Cestoda), consisting of a fluid-filled sac containing an inverted scolex. It develops in the muscle of the secondary host and matures into an adult worm in a primary host that eats this infected tissue.

blastocoel *See* blastula.

blastocyst *See* blastula; implantation.

blastoderm The layer of cells that surrounds the central cavity (blastocoel) of a *blastula. In yolky eggs, such as those of insects, the blastoderm forms a layer covering the yolk. *See also* **imaginal disc**.

blastopore *See* archenteron.

blastula The stage of *development of an animal embryo that results from *cleavage of a fertilized egg. This stage generally resembles a hollow ball with the dividing cells (*blastomeres*) of the embryo forming a layer (*blastoderm*) around a central cavity (*blastocoel*). In vertebrates the blastula forms a disc (*blastodisc*) on the surface of the yolk. In mammals the blastula stage is known as a *blastocyst*. *See also* **gastrula**.

blending inheritance The early theory that assumed that hereditary substances from parents merge together in their offspring. Mendel showed that this does not occur (*see* **Mendel's laws**). In breeding experiments an appearance of blending may result from co-dominant alleles (*see* **co-dominance**) and *polygenes but close study shows that the alleles retain their identity through successive generations. *Compare* **particulate inheritance**.

blind spot The portion of the retina at which blood vessels and nerve fibres enter the optic nerve. There are no rods or cones in this area, so no visual image can be transmitted from it.

Compositon of mammalian blood

blood A fluid body tissue that acts as a transport medium within an animal. It is contained within a blood *vascular system and in vertebrates is circulated by means of contractions of the *heart. Oxygen and food are carried to tissues, and carbon dioxide and chemical (nitrogenous) waste are transported from tissues to excretory organs for disposal (*excretion). In addition blood carries *hormones and also acts as a defence system. Blood consists of a liquid (*see* **blood plasma**) containing blood cells (*see* **erythrocyte; leucocyte**) and *platelets (see illustration).

blood–brain barrier The mechanism that controls the passage of substances from the blood to the cerebrospinal fluid bathing the brain and spinal cord. It takes the form of a semipermeable lipid membrane permitting the passage of solutions but excluding particles and large molecules. This barrier provides the central nervous system with a constant environment, while not interfering with the transport of essential metabolites.

blood capillary *See* **capillary**.

blood cell (**blood corpuscle**) Any of the cells that are normally found in the blood plasma. These include red cells (*see* **erythrocyte**) and white cells (*see* **leucocyte**).

blood clotting (**blood coagulation**) The production of a mass of semisolid material at the site of an injury that closes the wound, helping to prevent further blood loss and bacterial invasion. The clot is formed by the action of *clotting factors and *platelets. Calcium ions and platelets trigger the release of a lipoprotein, *thromboplastin*, from injured tissues. This converts *prothrombin in the blood to its enzymically active form *thrombin*. Thrombin catalyses the formation of the insoluble protein *fibrin* from soluble *fibrinogen*; the fibrin forms a fibrous network in which blood cells become enmeshed, producing a clot.

blood groups The many types into which an individual's blood may be classified, based on the presence or absence of certain antigenic proteins (*agglutinogens) on the surface of the red blood cells. Blood of one group contains *antibodies in the serum that react against the agglutinogens on the cells of other groups. *Incompatibility* between groups results in clumping of cells (*agglutination), so knowledge of blood groups is important for blood transfusions. In man, the two most important blood group systems are the *ABO system and the system involving the *rhesus factor.

blood pigment Any one of a group of metal-containing coloured protein compounds whose function is to increase the oxygen-carrying capacity of blood. *See* **respiratory pigment**.

blood plasma The liquid part of the *blood (i.e. excluding blood cells). It consists of water containing a large number of dissolved substances, including proteins, salts (especially sodium and potassium chlorides and bicarbonates), food materials (glucose, amino acids, fats), hormones, vitamins, and excretory materials. *See also* **blood serum; lymph**.

blood platelet *See* **platelet**.

blood pressure The pressure exerted by the flow of blood through the major arteries of the body. This pressure is greatest during the contraction of the ventricles of the heart (*systolic pressure*; *see* **systole**), which forces blood into the arterial system. Pressure falls to its lowest level when the heart is filling with blood (*diastolic pressure*; *see* **diastole**). Blood pressure is measured in millimetres of mercury using an instrument called a *sphygmomanometer*. Normal blood pressure for a young average adult human is in the region of 120/80 mmHg (the higher number is the systolic blood pressure; the lower number the diastolic blood pressure), but individual variations are common. Abnormally high blood pressure (*hypertension*) may be associated with disease or it may occur without an apparent cause.

blood serum Blood plasma from which the fibrin and clotting factors have been removed by centrifugation or vigorous stirring, so that it cannot clot. Serum containing a specific antibody or antitoxin may be used in the treatment or prevention of certain infections. Such serum is generally derived from a nonhuman mammal (e.g. a horse).

blood vascular system The tissues and organs of an animal that transport blood through the body. In vertebrates it consists of the heart and blood vessels. *See* **vascular system**.

blood vessel A tubular structure through which the blood of an animal flows. *See* **artery**; **arteriole**; **capillary**; **venule**; **vein**.

blue-green bacteria (blue-green algae) *See* **Cyanobacteria**.

BOD *See* **biochemical oxygen demand**.

body cavity The internal cavity of the body of an animal, which is present in most invertebrates and all vertebrates and contains the major organs. The body cavity of vertebrates and many invertebrates is the *coelom. In vertebrates the body cavity is divided by a transverse septum just posterior to the heart into the abdominal and thoracic cavities (*see* **abdomen**; **thorax**). In mammals the septum is the *diaphragm.

body fluid Any of the fluids found within animals, including blood, lymph, tissue fluid, urine, bile, sweat, and synovial fluid. Body fluids are generally involved with the processes of transport, excretion, or lubrication. They allow the distribution of oxygen and nutrients to the tissues and organs and the transport of waste products from the tissues, enabling their elimination from the body.

Bohr effect The effect of pH on the dissociation of oxygen from haemoglobin, first discovered by the Danish physiologist Christian Bohr (1855–1911). An increase in carbon dioxide concentration makes the blood more acidic and decreases the efficiency of the uptake of oxygen by haemoglobin molecules. This shifts the *oxygen dissociation curve to the right and increases the tendency of haemoglobin to release oxygen (*see* **haemoglobinic acid**). Thus in actively respiring tissues, where the concentration of carbon dioxide in the blood is high, haemoglobin readily releases its oxygen, while in the lungs, where blood carbon dioxide is low (due to its continual diffusion into the alveoli), haemoglobin readily binds oxygen.

bolus The ball of chewed food bound together with saliva that is formed in the mouth by the action of the tongue. The bolus is shaped to a size that allows it to pass into the oesophagus after being swallowed (*see* **deglutition**).

bomb calorimeter An apparatus used for measuring heats of combustion. It consists of a strong container in which the sample is sealed with excess oxygen and ignited electrically. The heat of combustion at constant volume can be calculated from the resulting rise in temperature. The apparatus is used to calculate the *calorific values of foods and the energy content of a sample of biomass, required for constructing a *pyramid of energy.

Haversian canal

concentric lamellae

calcified matrix

bone cell

single Haversian system

Structure of compact bone

bone The hard connective tissue of which the *skeleton of most vertebrates is formed. It comprises a matrix of *collagen fibres (30%) impregnated with bone salts (70%), mostly calcium phosphate (hydroxyapatite, $Ca_{10}(PO_4)_6(OH)_2$), in which are embedded bone cells: *osteoblasts (which secrete the matrix) and *osteocytes. Bone generally replaces embryonic *cartilage and is of two sorts – compact bone and spongy bone. The outer *compact bone* is formed as concentric layers (*lamellae*) that surround small holes (*Haversian canals): see illustration. The inner *spongy bone* is chemically similar but forms a network of bony bars. The spaces between the bars may contain bone marrow or (in birds) air for lightness. *See also* **cartilage bone; membrane bone; periosteum**.

bone marrow A soft tissue contained within the central cavity and internal spaces of a bone. At birth and in young animals the marrow of all bones is concerned with the formation of blood cells: it contains *haemopoietic tissue and is known as *red marrow*. In mature animals the marrow of the long bones ceases producing blood cells and is replaced by fat, being known as *yellow marrow*.

bony fishes *See* Osteichthyes.

bony labyrinth *See* labyrinth.

borax carmine A red dye, used in optical microscopy, that stains nuclei and cytoplasm pink. It is frequently used to stain large pieces of animal tissue.

botany The scientific study of plants, including their anatomy, morphology, physiology, biochemistry, taxonomy, cytology, genetics, ecology, evolution, and geographical distribution.

botulinum toxin A toxin produced by the bacterium *Clostridium botulinum*, which can cause fatal *food poisoning. It is the most toxic substance known to man.

bovine spongiform encephalopathy (BSE) A degenerative disease of the brain that affects cattle and is caused by an abnormal form of a cellular protein (*see* **prion**). Known colloquially as 'mad cow disease', it results in a build-up of *amyloid tissue in the brain.

Bowman's capsule (renal capsule) The cup-shaped end of a kidney *nephron. Its epithelium contains *podocytes, which facilitate the passage of glomerular filtrate from the blood into the nephron. It is named after its discoverer, the British physician Sir William Bowman (1816–92).

Brachiopoda A phylum of marine invertebrates – the lamp shells. They live in shallow waters, attached to a firm substratum by means of a flexible stalk (*peduncle*), and are protected by a bivalved shell consisting of dorsal and ventral valves. A food-gathering *lophophore protrudes from the shell. Brachiopods thrived in Palaeozoic times but are now much less numerous; living brachiopods include *Terebratella*, with an articulated shell; and *Lingula*, in which the shell valves are held together by muscles only.

bract A modified leaf with a flower or inflorescence in its axil. Bracts are often brightly coloured and may be mistaken for the petals of a flower. For example the showy 'flowers' of poinsettia and *Bougainvillea* are composed of bracts; the true flowers are comparatively inconspicuous. *See also* **involucre**.

bracteole A reduced leaf that arises from the stalk of an individual flower.

bradycardia A decrease in heart rate. *See* **tachycardia**.

bradykinin *See* kinin.

brain 1. The enlarged anterior part of the vertebrate central nervous system, which is encased within the cranium of the skull. Continuous with the spinal cord, the brain is surrounded by three membranes (*see*

meninges) and bathed in cerebrospinal fluid, which fills internal cavities (*ventricles). It functions as the main coordinating centre for nervous activity, receiving information (in the form of nerve impulses) from sense organs, interpreting it, and transmitting 'instructions' to muscles and other *effectors. It is also the seat of intelligence and memory. The embryonic vertebrate brain is in three sections (*see* **forebrain; hindbrain; midbrain**), which become further differentiated during development into specialized regions. The main parts of the adult human brain are a highly developed *cerebrum in the form of two cerebral hemispheres, a *cerebellum, *medulla oblongata, and *hypothalamus (see illustration). **2.** A concentration of nerve *ganglia at the anterior end of an invertebrate animal.

The human brain

brain death The permanent absence of vital functions of the brain, which is marked by cessation of breathing and other reflexes (including the *pupillary reflex) controlled by the *brainstem and by a zero reading on an *electroencephalogram. Organs may be removed for transplantation when brain death is established, which may not necessarily be associated with permanent absence of heart beat.

brainstem The part of the brain comprising the *medulla oblongata, the *midbrain, and the *pons. It resembles and is continuous with the spinal cord. The midbrain controls and integrates reflex activities (such as respiration) that originate in higher centres of the brain via a network of nerve pathways (the *reticular formation*).

branchial Of or relating to the gills. Most fish possess five pairs of *branchial arches*, skeletal supports for the gills, which are contained in a cavity known as the *branchial chamber*. Water is drawn into the branchial chamber through narrow *gill slits (or *branchial clefts*) in the wall of the pharynx and passes over the gills, where gaseous exchange takes place. Blood is transported to and from the gills in afferent and efferent *branchial arteries*, respectively.

breathing *See* **expiration; inspiration; respiratory movement.**

breed A domesticated *variety of an animal or, rarely, a cultivated variety of plant. Cultivated plants are more often simply called varieties or, more correctly, *cultivars. Examples of animal breeds are Friesian cattle and Shetland sheepdogs.

breeding The process of sexual *reproduction and bearing offspring. *Selective breeding* of both plants and animals is used in *agriculture to produce offspring that possess the beneficial characters of both parents (*see also* **artificial insemination**). *Inbreeding is the production of *homozygous phenotypically uniform offspring by mating between close relatives. Plants that self-fertilize, such as wheat and tomatoes, are inbreeders. *Outbreeding is the production of *heterozygous phenotypically variable offspring by mating between unrelated organisms.

breeding season (mating season) A specific season of the year in which many animals, including mammals and birds, mate, which ensures that offspring are produced only at a certain time of the year. This timing is important as it enables animals to give birth at a time of the year when environmental conditions and food supply are at their optimum. The breeding season of most animals is in the spring or summer. The stimulus to mate is the result of a photoperiodic response (*see* **photoperiodism**), which is thought to be controlled by day length.

brewing The process by which beer is made. *Fermentation of sugars from barley grain by the yeasts *Saccharomyces cerevisiae* and *S. uvarum* (or *S. carlsbergenesis*) produces alcohol (ethanol). In the first stage the barley grain is soaked in water, a process known as *malting*. The grain is then allowed to germinate and the natural enzymes of the grain (the amylases and the maltases) convert the starch to maltose and then to glucose. The next stage is *kilning* or *roasting*, in which the grains are dried and crushed. The colour of a beer depends on the temperature used for this process: the higher the temperature, the darker the beer. In the next stage, *mashing*, the crushed grain is added to water at a specific temperature and any remaining starch is converted to sugar; the resultant liquid is the raw material of brewing, called *wort*. The yeast is then added to the wort to convert the sugar to alcohol, followed by hops, which give beer its characteristic flavour. Hops are the female flowers of the vine *Humulus lupulus*; they contain resins (humulones, cohumulones, and adhumulones) that give beer its distinctive bitter taste.

bronchiole A fine respiratory tube in the lungs of reptiles, birds, and mammals. It is formed by the subdivision of a *bronchus and in reptiles and mammals it terminates in a number of *alveoli.

bronchus (bronchial tube) (*pl.* **bronchi**) One of the major air tubes in the *lung. The *trachea divides into two main bronchi, one for each lung, which split into smaller bronchi and then into *bronchioles. The walls of the bronchi are stiffened by rings of cartilage.

brown algae *See* Phaeophyta.

brown fat A darker coloured region of *adipose tissue found in newborn and hibernating animals (in which it may also be called the *hibernating gland*). Compared to normal white *fat, deposits of brown fat are more richly supplied with blood vessels and have a higher proportion of unsaturated fatty acids. They can also be more rapidly converted to heat energy, especially during arousal from hibernation and during cold stress in young animals. Since the deposits are strategically placed near major blood vessels, the heat they generate warms the blood returning to the heart.

It has been said that some types of obesity in man may be linked to a lack of brown fat in affected individuals: in normal individuals, an intake of excess food is converted into brown fat and rapidly metabolized.

Brownian movement The continuous random movement of microscopic solid particles (of about 1 micrometre in diameter) when suspended in a fluid medium. First observed by the botanist Robert Brown (1773–1858) in 1827 when studying pollen particles, it was originally thought to be the manifestation of some vital force. It was later recognized to be a consequence of bombardment of the particles by the continually moving molecules of the liquid. The smaller the particles the more extensive is the motion. It can be observed in the particles of a colloidal solution and in the protoplasm of dead cells.

Brunner's glands (duodenal glands) Glands in the submucosa of the duodenum that secrete mucus and an alkaline fluid that neutralizes the acidic *chyme leaving the stomach. They are named after Swiss anatomist J. C. von Brunner (1653–1727).

brush border A region of surface epithelium that possesses densely packed microvilli (*see* **microvillus**), rather like the bristles of a brush. This greatly increases the surface area of the epithelium and facilitates the absorption of materials. Brush borders are found in the convoluted tubules of the kidney and in the lining of the small intestine.

Bryophyta A phylum of simple plants possessing no vascular tissue and rudimentary rootlike organs (rhizoids). It includes the classes *Musci (mosses) and *Hepaticae (liverworts). Bryophytes grow in a variety of damp habitats, from fresh water to rock surfaces. Some use other plants for support. Bryophytes show a marked *alternation of generations between gamete-bearing forms (gametophytes) and spore-bearing forms (sporophytes), the latter being dependent on the former for water and nutrients. The leaves, stems, and roots of the gametophyte generation are not equivalent to those of higher (vascular) plants since the whole structure is *haploid.

Bryozoa (Polyzoa) *See* **Ectoprocta**.

BSE *See* **bovine spongiform encephalopathy**.

buccal cavity (oral cavity) The mouth cavity: the beginning of the *alimentary canal, which leads to the pharynx and (in vertebrates) to the oesophagus. In vertebrates it is separated from the nasal cavity by the *palate. In mammals it contains the tongue and teeth, which assist in the mechanical breakdown of food, and the openings of the *salivary gland ducts.

bud **1.** *(in botany)* A condensed immature shoot with a short stem bearing small folded or rolled leaves. The outer leaves of a bud are often scalelike and protect the delicate inner leaves. A *terminal* (or *apical*) *bud* exists at the tip of a stem or branch while *axillary* (or *lateral*) *buds* develop in the *axils of leaves. However, in certain circumstances buds can be produced anywhere on the surface of a plant. Some buds remain dormant, but may become active if the terminal bud is removed. It is common gardening practice to remove the terminal buds of some shoots to induce the development of lateral shoots from axillary buds. *See also* **apical dominance**. **2.** *(in biology)* An outgrowth from a parent organism that breaks away and develops into a new individual in the process of *budding.

budding **1.** *(in biology)* A method of asexual reproduction in which a new individual is derived from an outgrowth (*bud*) that becomes detached from the body of the parent. In animals the process is also called *gemmation*; it is common in coelenterates (e.g. *Hydra*) and also occurs in some sponges and other invertebrates. Among fungi, budding is characteristic of the yeasts. **2.** *(in horticulture)* A method of grafting in which a bud of the scion is inserted onto the stock, usually beneath the bark.

buffer A solution that resists change in pH when an acid or alkali is added or when the solution is diluted. Acidic buffers consist of a weak acid with a salt of the acid. The salt provides the negative ion A^-, which is the conjugate base of the acid HA. An example is carbonic acid and sodium hydrogencarbonate, in which molecules H_2CO_3 and ions HCO_3^- are present. When acid is added most of the extra protons are removed by the base:

$$HCO_3^- + H^+ \rightarrow H_2CO_3$$

When base is added, most of the extra hydroxide ions are removed by reaction with undissociated acid:

$$OH^- + H_2CO_3 \rightarrow HCO_3^- + H_2O$$

Thus, the addition of acid or base changes the pH very little. Basic buffers have a weak base and a salt of the base (to provide the conjugate acid). An example is ammonia solution with ammonium chloride.

Natural buffers occur in living organisms, where the biochemical reactions are very sensitive to change in pH (*see* **acid–base balance**). The main natural buffers are H_2CO_3/HCO_3^- and $H_2PO_4^-/HPO_4^{2-}$ (*see also* **haemoglobinic acid**). Buffer solutions are also used in the laboratory (e.g. to keep microscopical preparations at their original pHs in order to prevent the formation of artefacts), in medicine (e.g. in intravenous injections), in agriculture, and in many industrial processes (e.g. dyeing, fermentation processes, and the food industry).

bugs *See* **Hemiptera**.

Development of a bulb

bulb An underground plant organ that enables a plant to survive from one growing season to the next. It is a modified shoot with a short flattened stem. A terminal bud develops at the centre of its upper surface, surrounded by swollen leaf bases that contain food stored from the previous growing season. Papery brown scale leaves cover the outside of the bulb. The stored food is used in the growing season when the terminal bud produces foliage leaves and flowers. The new leaves photosynthesize and some of the manufactured food passes into the leaf bases forming a new bulb (see illustration). If more than one bud develops, then additional bulbs form, resulting in vegetative propagation. Examples of bulb-forming plants are daffodil, onion, and tulip. *Compare* **corm**.

bulbil A small bulblike organ that may develop in place of a flower, from an axillary bud, or at the base of a stem in certain plants. If it becomes detached it develops into a new plant.

bulbourethral glands *See* **Cowper's glands**.

bulla The rounded hollow projection of bone from the skull that encloses the *middle ear in mammals.

bundle of His The specialized cardiac muscle fibres in the mammalian heart that receive electrical stimuli from the *atrioventricular node and transmit them throughout the network of *Purkyne fibres. This allows the excitation to reach all parts of the ventricles rapidly and initiates a wave of contraction to expel blood into the aorta and pulmonary artery. The fibres are named after Swiss anatomist Wilhelm His (1831–1904).

bundle sheath cells A layer of cells in plant leaves and stems that forms a sheath surrounding the vascular bundles. In C_4 plants (see **C_4 pathway**) the bundle sheath cells contain chloroplasts and are the site of the *Calvin cycle. The initial fixation of carbon dioxide to form malic acid takes place in the palisade mesophyll cells, which in C_4 plants form a circle around the bundle sheath. This arrangement, known as *Kranz anatomy* or *structure* (after the German *Kranz*, wreath), ensures that the palisade cells are in close contact with the bundle sheath cells so that the malic acid can easily pass to the bundle sheath. It also means that the products of photosynthesis can be quickly transferred from the bundle sheath to the adjacent phloem tissue for transport to other parts of the plant.

bursa (*pl.* **bursae**) A sac of fibrous connective tissue lined with synovial membrane and filled with synovial fluid. Bursae are found between bones and other tissues, such as skin, ligaments, tendons, and muscles, where they reduce friction when one part moves over another.

butanedioic acid *See* **succinate**.

butterflies *See* **Lepidoptera**.

buttress root *See* **prop root**.

C

caecum A pouch in the alimentary canal of vertebrates between the *small intestine and *colon. The caecum (and its *appendix) is large and highly developed in herbivorous animals (e.g. rabbits and cows), in which it contains a large population of bacteria essential for the breakdown of cellulose. In humans the caecum is a *vestigial organ and is poorly developed.

Cainozoic *See* **Cenozoic**.

calciferol *See* **vitamin D**.

calcitonin (thyrocalcitonin) A hormone, produced by the mammalian *thyroid gland, that lowers the concentration of calcium (and phosphate) in the blood. It operates in opposition to *parathyroid hormone.

calcium Symbol Ca. A soft grey metallic element that is an *essential element for living organisms, being required for normal growth and development. In animals it is an important constituent of bones and teeth and is present in the blood, being required for muscle contraction and other metabolic processes. In plants it is a constituent (in the form of calcium pectate) of the *middle lamella.

calcium cyclamate An organic calcium salt, $(C_6H_{11}NHSO_3)Ca.2H_2O$, that has a very sweet taste and was formerly used as a sweetening agent. Its use was banned in the UK in 1969 as it was suspected of causing cancer.

calliclone A colony derived from a single plant cell that has been removed from a suspended culture and plated out in a semisolid medium.

callus 1. (*in botany*) A protective tissue, consisting of parenchyma cells, that develops over a cut or damaged plant surface. Callus tissue can also be induced to form in cell cultures by hormone treatment. **2. (*in pathology*)** A thick hard area of skin that commonly forms on the palms of the hands and soles of the feet as a result of continuous pressure or friction. **3. (*in physiology*)** Hard tissue formed round bone ends following a fracture, which is gradually converted to new bone.

calmodulin A protein, consisting of 148 amino-acid residues, that acts as a receptor for calcium ions in many calcium-regulated processes in both animal and plant cells. Calmodulin mediates reactions catalysed by many enzymes, including *adenylate cyclase, phosphorylase kinase, and cyclic nucleotide phosphodiesterase. During the contraction of smooth muscle calmodulin binds calcium ions and subsequently activates the enzyme *myosin light chain kinase*. This enzyme phosphorylates the head of the *myosin molecule, enabling it to bind to *actin. Calmodulin may also regulate the functioning of the spindle observed during mitosis.

calorie The quantity of heat required to raise the temperature of 1 gram of water by 1°C (1 K). The calorie, a c.g.s. unit, is now largely replaced by the *joule, an *SI unit. 1 calorie = 4.186 8 joules.

Calorie (kilogram calorie; kilocalorie) 1000 calories. This unit is still in limited use in estimating the energy value of foods, but is obsolescent.

calorific value The heat per unit mass produced by complete combustion of a given substance. Calorific values are used to express the energy values of fuels; usually these are expressed in megajoules per kilogram ($MJ kg^{-1}$). They are also used to measure the energy content of foodstuffs; i.e. the energy produced when the food is oxidized in the body. The units here are kilojoules per gram ($kJ g^{-1}$), although Calories (kilocalories) are often still used in nontechnical contexts.

calorimeter Any of various devices used to measure thermal properties, such as *calorific value. *See* **bomb calorimeter**.

Calvin cycle The metabolic pathway of the dark stage of *photosynthesis, which occurs in the stroma of the chloroplasts. The pathway was elucidated by Melvin Calvin (1911–   ) and his co-workers and involves the fixation of carbon dioxide and its subsequent reduction to carbohydrate. During the cycle, carbon dioxide combines with *ribulose bisphosphate, through the mediation of the enzyme ribulose bisphosphate carboxylase, to form an unstable six-carbon compound that breaks down to form two molecules of the three-carbon compound glycerate 3-phosphate. This is converted to glyceraldehyde 3-phosphate, which is used to regenerate ribulose bisphosphate and to produce glucose and fructose.

calyptra **1.** A layer of cells that covers the developing sporophyte of bryophytes, clubmosses, horsetails, and ferns. In mosses it forms a hood over the *capsule and in liverworts it forms a sheath at the base of the capsule stalk. **2.** *See* **root cap**.

calyptrogen The region within the root *apical meristem that divides to produce the *root cap (calyptra).

calyx The *sepals of a flower, collectively, forming the outer whorl of the *perianth. It encloses the petals, stamens, and carpels and protects the flower in bud. *See also* **pappus**.

cambium (lateral meristem) A plant tissue consisting of actively dividing cells (*see* **meristem**) that is responsible for increasing the girth of the plant, i.e. it causes secondary growth. The two most important cambia are the *vascular* (or *fascicular*) *cambium* and the *cork cambium. The vascular cambium occurs in the stem and root; it divides to produce secondary *xylem and secondary *phloem (new food- and water-conducting tissues). In mature stems the vascular cambium is extended laterally to form a complete ring: the sections of this ring between the vascular bundles comprises the *interfascicular cambium*. *Compare* **apical meristem**.

Cambrian The earliest geological period of the Palaeozoic era. It is estimated to have begun about 570 million years ago and lasted for some 100 million years. During this period marine animals with mineralized shells made their first appearance and Cambrian rocks are the first to contain an abundance of fossils. Cambrian fossils are all of marine animals; they include *trilobites, which dominated the Cambrian seas, echinoderms, brachiopods, molluscs, and primitive *graptolites (from the mid-Cambrian). Trace *fossils also provide evidence for a variety of worms.

camouflage A high degree of similarity between an animal and its visual environment, which enables it to be disguised or concealed. By blending into the background the animal can elude predators or remain invisible to potential prey. *See also* **cryptic coloration; mimicry.** *Compare* **warning coloration.**

Canada balsam A yellow-tinted resin used for mounting specimens in optical microscopy. It has similar optical properties to glass.

canaliculus (*pl.* **canaliculi**) A very small channel that occurs between the cells of the liver and bone. In the liver the bile canaliculi carry bile to the bile ducts; in bone, canaliculi connect lacunae, the cavities containing bone cells.

cancer Any disorder of cell growth that results in invasion and destruction of surrounding healthy tissue by abnormal cells. Cancer cells arise from normal cells whose nature is permanently changed. They multiply more rapidly than healthy body cells and do not seem subject to normal control by nerves and hormones. They may spread via the bloodstream or lymphatic system to other parts of the body, where they produce further tissue damage (*metastases*). *Malignant tumour* is another name for cancer; *leukaemia* is cancer of white blood cells. Causative agents (*carcinogens*) include various chemicals (including those in tobacco smoke), ionizing radiation, silica and asbestos particles, and *oncogenic viruses. Hereditary factors and stress may also play a role.

cane sugar *See* **sucrose.**

canine tooth A sharp conical *tooth in mammals that is large and highly developed in carnivores (e.g. dogs) for tearing meat. There are two canines in each jaw, each situated between the second *incisor and the first *premolar. In some animals (e.g. herbivores, such as giraffes and rabbits) canine teeth are absent.

capacitation The final stage in the maturation process of a spermatozoon. This takes place inside the genital tract as the sperm penetrates the ovum.

capillarity *See* **surface tension.**

capillary (blood capillary) The narrowest type of blood vessel in the vertebrate circulatory system. Capillaries conduct blood from *arterioles to all living cells: their walls are only one cell layer thick, so that oxygen and nutrients can pass through them into the surrounding tissues. Capillaries also transport waste material (e.g. urea and carbon dioxide) to venules for ultimate excretion. Capillaries can be constricted or dilated, according to local tissue requirements.

capitulum A type of flowering shoot (*see* **racemose inflorescence**) characteristic of plants of the family Compositae, e.g. daisy and dandelion. The tip of the shoot is flattened and bears many small stalkless flowers (*florets*) surrounded by an involucre (ring) of bracts. This arrangement gives the appearance of a single flower.

capsid The protein coat of a *virus, which is made up of units called *capsomeres. The chemical nature of the capsid is important in stimulating the body's *immune response against the invading virus.

capsomere Any of the protein units that make up the regularly organized outer coat (*capsid) of the viruses. The capsid of most viruses contains more than one type of protein molecule.

capsule 1. (*in botany*) a. A dry fruit that releases its seeds when ripe; it is formed from several fused carpels and contains many seeds. The seeds may be dispersed through pores (as in the poppy), through a lid (as in plantain), or by the splitting and separation of the individual carpels (as in the crocus). Various other forms of capsules include the *silicula and *siliqua. **b.** The part of the sporophyte of mosses and liverworts in which the haploid spores are produced. It is borne on a long stalk (*seta*) and sheds its spores when mature (*see* **peristome**). **2. (*in microbiology*)** A thick gelatinous layer completely surrounding the cell wall of certain bacteria. It appears to have a protective function, making ingestion of the bacterial cell by *phagocytes more difficult and preventing desiccation. **3. (*in animal anatomy*) a.** The membranous or fibrous envelope that surrounds certain organs, e.g. the kidneys, spleen, and lymph nodes. **b.** The ligamentous sheath of connective tissue that surrounds various skeletal joints.

carapace 1. The dorsal part of the *exoskeleton of some crustaceans (e.g. crabs), which spreads like a shield over several segments of the head and thorax. **2.** The domed dorsal part of the shell of tortoises and turtles, formed of bony plates fused with the ribs and vertebrae and covered by a horny epidermal layer. The ventral part of the shell (*plastron*) is similar but flatter.

carbamide *See* **urea**.

carbohydrate One of a group of organic compounds based on the general formula $C_x(H_2O)_y$. The simplest carbohydrates are the *sugars (saccharides), including glucose and sucrose. *Polysaccharides are carbohydrates of much greater molecular weight and complexity; examples are starch, glycogen, and cellulose. Carbohydrates perform many vital roles in living organisms. Sugars, notably glucose, and their derivatives are essential intermediates in the conversion of food to energy. Starch and other polysaccharides serve as energy stores in plants, particularly in seeds, tubers, etc., which provide a major energy source for animals, including man. Cellulose, lignin, and others form the supporting cell walls and woody tissue of plants. Chitin is a structural polysaccharide found in the body shells of many invertebrate animals. Carbohydrates also occur in the surface coat of animal cells and in bacterial cell walls.

carbon Symbol C. A nonmetallic element that occurs in all organic compounds and is therefore fundamental to the structure of all living organisms. It is an *essential element for plants and animals, being ultimately derived from atmospheric carbon dioxide assimilated by plants during photosynthesis (*see* **carbon cycle**). The ubiquitous nature of carbon in living organisms is due to its unique ability to form stable covalent bonds with other carbon atoms and also with hydrogen, oxygen, nitrogen, and sulphur atoms, resulting in the formation of a variety of compounds containing chains and rings of carbon atoms.

There are two stable isotopes of carbon (proton numbers 12 and 13) and four radioactive ones (10, 11, 14, 15). Carbon–14 is used in *carbon dating.

carbon assimilation The incorporation of carbon from atmospheric carbon dioxide into organic molecules. This process occurs during *photosynthesis. *See* **carbon cycle**.

The carbon cycle in nature

carbon cycle One of the major cycles of chemical elements in the environment (*see* **biogeochemical cycle**). Carbon (as carbon dioxide) is taken up from the atmosphere and incorporated into the tissues of plants in *photosynthesis. It may then pass into the bodies of animals as the plants are eaten (*see* **food chain**). During the respiration of plants, animals, and organisms that bring about decomposition, carbon dioxide is returned to the atmosphere. The combustion of fossil fuels (e.g. coal and peat) also releases carbon dioxide into the atmosphere. See illustration.

carbon dating (radiocarbon dating) A method of estimating the ages of archaeological specimens of biological origin. As a result of cosmic radiation a small number of atmospheric nitrogen nuclei are continuously being transformed by neutron bombardment into radioactive nuclei of carbon–14. Some of these radiocarbon atoms find their way into living trees and other plants in the form of carbon dioxide, as a result of *photosynthesis. When the tree is cut down photosynthesis stops and the ratio of radiocarbon atoms to stable carbon atoms begins to fall as the radiocarbon decays. The ratio $^{14}C/^{12}C$ in the specimen can be measured and enables the time that has elapsed since the tree was cut down to be calculated. The method has been shown to give consistent results for specimens up to some 40 000 years old, though its accuracy depends upon assumptions concerning the past intensity of the cosmic radiation. The technique was developed by Willard F. Libby (1908–80) and his coworkers in 1946–47.

carbon dioxide A colourless odourless gas, CO_2, which dissolves in water to give *carbonic acid. It occurs in the atmosphere (0.04% by volume) but has a short residence time in this phase as it is both consumed by plants during *photosynthesis and produced by *respiration and by combustion.

The level of carbon dioxide in the atmosphere has increased by some 12% in the last 100 years, mainly because of extensive burning of fossil fuels and the destruction of large areas of *rainforest. This has been postulated as the main cause of the average increase of 0.5°C in global temperatures over the same period, through the *greenhouse effect; environmentalists urge that measures be taken to prevent further increases in atmospheric CO_2 concentration and subsequent global warming.

carbonic acid A weak acid, H_2CO_3, formed in solution when carbon dioxide is dissolved in water:

$$CO_2 + H_2O \leftrightarrows H_2CO_3$$

The acid is in equilibrium with dissolved carbon dioxide, and also dissociates into hydrogencarbonate and hydrogen ions. These reactions, catalysed by *carbonic anhydrase, take place in the red blood cells when carbon dioxide diffuses into them from the surrounding tissue cells. *See also* **hydrogencarbonate**.

carbonic anhydrase An enzyme, present in red blood cells and kidney cells, that catalyses the reaction between carbon dioxide and water to form carbonic acid, which subsequently dissociates:

$$CO_2 + H_2O \leftrightarrows H_2CO_3$$

$$H_2CO_3 \leftrightarrows H^+ + HCO_3^-$$

This reaction is one of the fastest known and controls the elimination of carbon dioxide from the body and the pH of urine. It also facilitates the transfer of carbon dioxide from the tissues to the blood and from the blood to the alveoli (air sacs) of the lungs. *See also* **chloride shift; haemoglobinic acid**.

Carboniferous A geological period in the Palaeozoic era. It began about 345 million years ago, following the Devonian period, and extended until the beginning of the Permian period, about 280 million years ago. In Europe the period is divided into the Lower and Upper Carboniferous, which roughly correspond to the Mississippian and Pennsylvanian periods, respectively, of North America. During the Lower Carboniferous a marine transgression occurred and the characteristic rock of this division – the Carboniferous limestone – was laid down in the shallow seas. Fauna included foraminiferans, corals, ectoprocts, brachiopods, blastoids, and other invertebrates. The Upper Carboniferous saw the deposition of the millstone grit, a mixture of shale and sandstone formed in deltaic conditions, followed by the coal measures, alternating beds of coal, sandstone, shale, and clay. The coal was formed from the vast swamp forests composed of seed ferns, lycopsids, and other plants. During the period fishes continued to diversify and amphibians became more common.

carbon monoxide A colourless odourless gas, CO. It is formed by the incomplete combustion of carbon and is present in car-exhaust gases. Carbon monoxide is able to bond with metals; this accounts for its toxicity, which is due to the binding of the CO to the iron in haemoglobin, thereby blocking the uptake of oxygen. *See* **carboxyhaemoglobin**.

carboxyhaemoglobin The highly stable product formed when *haemoglobin combines with *carbon monoxide. Carbon monoxide competes with oxygen for haemoglobin, with which it binds strongly: the affinity of haemoglobin for carbon monoxide is 250 times greater than that for oxygen. This reduces the availability of haemoglobin for combination with (and transport of) oxygen and accounts for the toxic effects of carbon monoxide on the respiratory system.

carboxylase Any enzyme involved in the transfer of carbon dioxide or carboxyl groups.

carboxyl group The organic group –CO.OH, present in *carboxylic acids.

Carboxylic acid structure

carboxylic acids Organic compounds containing the group –CO.OH (the *carboxyl group*; i.e. a carbonyl group attached to a hydroxyl group). Many long-chain carboxylic acids occur naturally as esters in fats and oils and are therefore also known as *fatty acids. *See also* **glycerides**.

carboxypeptidase An *exopeptidase enzyme in pancreatic juice that is secreted into the duodenum. The enzyme is secreted as an inactive precursor, *procarboxypeptidase*, which is activated by another pancreatic protease, *trypsin. *See also* **chymotrypsin**.

carboxysome An organelle found in certain autotrophic bacteria that contains the carbon-fixing enzyme ribulose bisphosphate carboxylase (*see* **ribulose bisphosphate**).

carcerulus A dry fruit that is a type of *schizocarp. It consists of a number of one-seeded fragments (*mericarps*) that adhere to a central axis. It is characteristic of mallow.

carcinogen Any agent that produces *cancer, e.g. tobacco smoke, certain industrial chemicals, and *ionizing radiation (such as X-rays and ultraviolet rays).

cardiac 1. Relating to the heart. **2.** Relating to the part of the stomach nearest to the oesophagus.

cardiac cycle The sequence of events that occurs in the heart during one full heartbeat. These events comprise contraction (*see* **systole**) and relaxation (*see* **diastole**) of the chambers of the heart, associated with opening and closing of the heart valves. When both the atria and the ventricles are relaxed, pressure in the heart is low and blood flows from the vena cava and pulmonary vein into the atria and through to the ventricles. The aortic and pulmonary *semilunar valves, at the junction between the left ventricle and aorta and the right ventricle and pulmonary artery, respectively, are closed; therefore, blood can enter but not leave the heart, which increases the pressure in the chambers. As the pressure in the heart increases, the atria begin to contract, forcing the blood into the ventricles and closing the *tricuspid valve and the *bicuspid valve. A wave of ventricular contraction follows, expelling the blood into the aorta and pulmonary artery to complete the cardiac cycle. At a resting heart rate, the human cardiac cycle lasts approximately 0.85 second.

cardiac muscle A specialized form of *muscle that is peculiar to the vertebrate heart. The muscle fibres are branched and interlock (*see* **Purkyne fibres**) and the muscle itself shows spontaneous contraction and does not need nervous stimulation (*see* **pacemaker**). The vagus nerve to the heart can, however, affect the rate of contraction (*see* **tachycardia**).

cardiac output The volume of blood pumped per minute by each ventricle, which is also the total blood flow through the pulmonary circuit. At rest, normal human cardiac output is approximately 5 litres per minute, rising to 22 litres per minute during maximum physical exertion. The cardiac output can be calculated from heart rate (number of beats per minute) and stroke volume (volume of blood expelled from the heart per beat).

cardiovascular centre One of the areas in the brain that are responsible for the modification of the cardiovascular system based upon the integration of sensory information from the autonomic nervous system. These centres influence the heart rate via the sympathetic and parasympathetic nerves and by the action of certain hormones.

carnassial teeth Molar and premolar teeth modified for shearing flesh by having cusps with sharp cutting edges. They are typical of animals of the order *Carnivora (e.g. tigers, wolves), in which they are the first molars in the lower jaw and the last premolars in the upper.

Carnivora An order of mainly flesh-eating mammals that includes the dogs, wolves, bears, badgers, weasels, and cats. Carnivores typically have very keen sight, smell, and hearing. The hinge joint between the lower jaw and skull is very tight, allowing no lateral movement of the lower jaw. This – together with the arrangement of jaw muscles – enables a very powerful bite. The teeth are specialized for stabbing and tearing flesh: canines are large and pointed and some of the cheek teeth are modified for shearing (*see* **carnassial teeth**).

carnivore An animal that eats meat, especially a member of the order *Carnivora (e.g. tigers, wolves). Carnivores are specialized by having strong powerful jaws and well-developed canine teeth. They may be *predators or carrion eaters. *See also* **consumer**. *Compare* **herbivore; omnivore**.

carnivorous plant (**insectivorous plant**) Any plant that supplements its supply of nitrates in conditions of nitrate deficiency by digesting small animals, especially insects. Such plants are adapted in various ways to attract and trap the insects and produce proteolytic enzymes to digest them. Venus' fly trap (*Dionaea*), for example, has spiny-margined hinged leaves that snap shut on an alighting insect. Sundews (*Drosera*) trap and digest insects by means of glandular leaves that secrete a sticky substance, and pitcher plants (families Nepenthaceae and Sarraceniaceae) have leaves modified as pitchers into which insects fall, drowning in the water and digestive enzymes at the bottom.

carotene A member of a class of *carotenoid pigments. Examples are β-carotene and lycopene, which colour carrot roots and ripe tomato fruits respectively. α- and β-carotene yield vitamin A when they are broken down during animal digestion; β-carotene has *antioxidant properties.

carotenoid Any of a group of yellow, orange, red, or brown pigments chemically related to terpenes. Carotenoids are responsible for the characteristic colour of many plant organs, such as ripe tomatoes, carrots, and autumn leaves. They also occur in certain algae and other photosynthesizing organisms, functioning as accessory pigments in the light reaction of *photosynthesis (*see* **photosynthetic pigments**). *See* **carotene; xanthophyll**.

carotid artery The major artery that supplies blood to the head. A pair of *common carotid arteries* arise from the aorta (on the left) and the innominate artery (on the right) and run up the neck; each branches into an *external* and an *internal carotid artery*, which supply the head.

carotid body One of a pair of tissue masses adjacent to the *carotid sinus. Each contains receptors that are sensitive to oxygen and pH levels (acidity) in the blood. High levels of carbon dioxide in the blood lower the pH (i.e. increase the acidity). By responding to fluctuations in pH, the carotid body coordinates reflex changes in respiration rate. *See also* **ventilation centre**.

carotid sinus An enlarged region of the *carotid artery at its major branching point in the neck. Its walls contain many receptors that are sensitive to changes in pressure and it regulates blood pressure by initiating reflex changes in heart rate and dilation of blood vessels.

carpal (carpal bone) One of the bones that form the wrist (*see* **carpus**) in terrestrial vertebrates.

carpel The female reproductive organ of a flower. Typically it consists of a *stigma, *style, and *ovary. It is thought to have evolved by the fusion of the two edges of a flattened megasporophyll (*see* **sporophyll**). Each flower may have one carpel (*monocarpellary*) or many (*polycarpellary*), either free (*apocarpous*) or fused together (*syncarpous*). *See also* **pistil**.

carpogonium (*pl.* **carpogonia**) The female gametangium of the red algae (*see* **Rhodophyta**). Carpegonia, which are found at the tips of the female gametophyte, are flask-shaped structures from which protrudes a slender elongation called a *trichogyne*. The nonmotile male spermatia, released from the antheridia, attach to the trichogyne during sexual reproduction. After fertilization the carpogonium develops into the *cystocarp*, which contains the *carpospores.

carpospore Any of the spores of the red algae (*see* **Rhodophyta**), which are produced in the fertilized *carpogonium. After release the spores develop into a sporophyte (*see* **tetraspore**) or a new gametophyte.

carpus The wrist (or corresponding part of the forelimb) in terrestrial vertebrates, consisting of a number of small bones (*carpals*). The number of carpal bones varies with the species. The rabbit, for example, has two rows of carpals, the first (proximal) row containing three bones and the second (distal) row five. In humans there are also eight carpals. This large number of bones enables flexibility at the wrist joint, between the hand and forelimb. *See also* **pentadactyl limb**.

carrageenan A naturally occurring polysaccharide isolated from red algae (Rhodophyta). The polymer is composed of D-galactose units, many of which are sulphated. *K-carrageenan* is a gelling agent with uses similar to those of *agar.

carrier 1. *(in medicine)* An individual who harbours a particular disease-causing microorganism without ill-effects and who can transmit the microorganism to others. *Compare* **vector**. 2. *(in genetics)* An individual with an *allele for some defective condition that is masked by a normal *dominant allele. Such individuals therefore do not suffer from the condition themselves but they may pass on the defective allele to their offspring. In humans, women may be carriers of such conditions as red–green colour blindness and haemophilia, the alleles for which are carried on the X chromosomes (*see* **sex linkage**). **3.** *(in biochemistry)* See **carrier molecule; hydrogen carrier**.

carrier molecule 1. A molecule that plays a role in transporting electrons through the *electron transport chain. Carrier molecules are usually proteins bound to a nonprotein group; they can undergo oxidation and reduction relatively easily, thus allowing electrons to flow through the system. There are four types of carrier: flavoproteins (e.g. *FAD), *cytochromes, iron–sulphur proteins (e.g. *ferredoxin), and *coenzyme Q. **2.** A lipid-soluble molecule that can bind to lipid-insoluble molecules and transport them across membranes. Carrier molecules have specific sites that interact with the molecules they transport. Several different molecules may compete for transport by the same carrier.

cartilage (gristle) A firm flexible connective tissue that forms the adult skeleton of cartilaginous fish (e.g. sharks). In other vertebrates cartilage forms the skeleton of the embryo, being largely replaced by *bone in mature animals (although it persists in certain areas). Cartilage comprises a matrix consisting chiefly of a glycosaminoglycan (mucopolysaccharide) called *chondroitin sulphate* secreted by cells (*chondroblasts*) that become embedded in the matrix as *chondrocytes*. It also contains collagenous and elastic fibres. *Hyaline cartilage* consists largely of glycosaminoglycan, giving it a shiny glasslike appearance; this type of cartilage gives flexibility and support at the joints. *Fibrocartilage*, in which bundles of collagen fibres predominate, is stronger and less elastic than hyaline cartilage; it is found in such areas as the intervertebral discs. *Elastic cartilage* has a yellow appearance due to the presence of numerous elastic fibres (*see* **elastin**). This cartilage maintains the shape of certain organs, such as the pinna of the ear.

cartilage bone (replacing bone) *Bone that is formed by replacing the cartilage of an embryo skeleton. The process, called *endochondral ossification*, is brought about by the cells (*osteoblasts) that secrete bone. *Compare* **membrane bone**.

cartilaginous fishes *See* **Chondrichthyes**.

caruncle A small outgrowth from the testa of a seed that develops from the placenta, funicle, or micropyle. Examples include the warty outgrowth from the castor-oil seed and the tuft of hairs on the testa of the seed of willowherb. *See also* **aril**.

caryopsis A dry single-seeded indehiscent fruit that differs from an *achene in that the fruit wall is fused to the testa of the seed. It is the grain of cereals and grasses.

casein One of a group of phosphate-containing proteins (phosphoproteins) found in milk; it is also the principal protein of cheese (*see* **curd**). Caseins are easily digested by the enzymes of young mammals and represent a major source of phosphorus. *See* **rennin**.

Casparian strip A band of *suberin, an impermeable substance, found in the endodermal cell walls of plant roots; it was named after R. Caspary. Movement of water through the *apoplast pathway is diverted from the cell wall to the cytoplasm, where it then follows the *symplast pathway. The reason for this diversion is not fully understood but it is thought that it may allow the endodermal cells to actively secrete salts into the vascular tissue. This results in a low (more negative) water potential, allowing water to move down a water potential gradient from the endodermis into the vascular tissue.

caste A division found in social insects, such as the *Hymenoptera (ants, bees, wasps) and the *Isoptera (termites), in which the individuals are structurally and physiologically specialized to perform a particular function. For example, in honeybees there are queens (fertile females), workers (sterile females), and drones (males). There are several different castes of workers (all sterile females) among ants.

catabolism The metabolic breakdown of large molecules in living organisms to smaller ones, with the release of energy. Respiration is an example of a catabolic series of reactions. *See* **metabolism**. *Compare* **anabolism**.

catalase An enzyme, found in *peroxisomes, that catalyses the decomposition of hydrogen peroxide, which results from oxidation reactions in the cell (*see also* **superoxide dismutase**), into water and oxygen. High concentrations of catalase are found in the liver. Catalase is the fastest known enzyme, with an extremely high *turnover number; it is employed in the rubber industry to form oxygen, which converts latex to foam rubber.

catalysis The process of changing the rate of a chemical reaction by use of a *catalyst.

catalyst A substance that increases the rate of a chemical reaction without itself undergoing any permanent chemical change. The catalyst provides an alternative pathway by which the reaction can proceed, in which the *activation energy is lower. It thus increases the rate at which the reaction comes to equilibrium, although it does not alter the position of the equilibrium. *Enzymes are the catalysts in biochemical reactions; they are highly specific in the type of reaction they catalyse.

catecholamine Any of a class of amines that possess a catechol ($C_6H_4(OH)_2$) ring. Including *dopamine, *adrenaline, and *noradrenaline, they function as *neurotransmitters and/or hormones.

category *See* **rank**.

cathepsins Protein-digesting enzymes that are found in *lysosomes.

catkin A type of flowering shoot (*see* **racemose inflorescence**) in which the axis, which is often long, bears many small stalkless unisexual flowers. Usually the male catkins hang down from the stem; the female catkins are shorter and often erect. Examples include birch and hazel. Most plants with catkins are adapted for wind pollination, the male flowers producing large quantities of pollen; willows are an exception, having nectar-secreting flowers and being pollinated by insects.

caudal vertebrae The bones (*see* **vertebra**) of the tail, which articulate with the *sacral vertebrae. The number of caudal vertebrae varies with the species. Rabbits, for example, have 15 caudal vertebrae, while in man these vertebrae are fused to form a single bone, the *coccyx.

C cell (parafollicular cell) Any one of a group of cells in vertebrates that are derived from the terminal pair of gill pouches. In mammals these cells are incorporated into the *thyroid and *parathyroid glands.

CCK *See* **cholecystokinin**.

CD (cluster of differentiation) Any group of antigens that is associated with a specific subpopulation of human *T cells. The differentiation antigens expressed by a T cell vary with its stage of development and thus with its role in the immune response. Hence, for example, CD4 antigens are expressed by helper T cells, whereas CD8 antigens are expressed by cytotoxic and suppressor T cells. The antigens are characterized using *monoclonal antibodies.

cDNA *See* **complementary DNA**.

Generalized eukaryotic cells

cell The structural and functional unit of most living organisms (*compare* **coenocyte**; **syncytium**). Cell size varies, but most cells are microscopic (average diameter 0.01–0.1 mm). Cells may exist as independent units of life, as in bacteria and protozoans, or they may form colonies or tissues, as in all plants and animals. Each cell consists of a mass of protein material (*see* **protoplasm**) that is differentiated into a jelly-like substance (*see* **cytoplasm**) and a *nucleus, which contains DNA. The protoplasm is bounded by a *cell membrane, which in the cells of plants, fungi, algae, and bacteria is surrounded by a *cell wall. There are two main types of cell. *Prokaryotic cells* (bacteria) are the more primitive. The nuclear material is not bounded by a membrane and chemicals involved in cell metabolism are associated with the cell membrane. Reproduction is asexual and involves simple cell cleavage. In *eukaryotic cells* the nucleus is bounded by a nuclear membrane and the cytoplasm is divided by membranes into a system of interconnected cavities and separate compartments (*organelles*), e.g. *mitochondria, *endoplasmic reticulum, *Golgi apparatus, *lysosomes, and *ribosomes (see illustration). Reproduction can be either asexual (*see* **mitosis**) or sexual (*see* **meiosis**). Plants and animals consist of eukaryotic cells but plant cells possess *chloroplasts and other *plastids and bear a rigid cellulose cell wall.

cell body (perikaryon) The part of a *neurone that contains the nucleus. The cell processes that are involved in the transmission and reception of nervous impulses (the axon and the dendrites respectively) develop as extensions from the cell body.

cell culture The culturing of cells from plants, animals, or microorganisms on a culture medium *in vitro*. *See* **culture**.

Cell cycle

cell cycle The sequence of stages that a cell passes through between one cell division and the next. The cell cycle can be divided into four main stages: (1) the *M phase*, which consists of *mitosis (nuclear division) and *cytokinesis* (cytoplasmic division); (2) the G_1 *phase*, in which there is a high rate of biosynthesis and growth; (3) the S *phase*, in which the DNA content of the cell doubles and the chromosomes replicate; (4) the G_2 *phase*, during which the final preparations for cell division are made. *Interphase consists of the G_1, S, and G_2 phases, which comprise about 90% (16–24 hours) of the total time of the cell cycle in rapidly dividing cells. The M phase lasts about 1–2 hours. A point is reached in the G_1 phase, known as the *restriction point*, after which the cell becomes committed to passing through the remainder of the cell cycle regardless of the external conditions.

cell division The formation of two or more daughter cells from a single mother cell. The nucleus divides first; this is followed by division of the cytoplasm (*cytokinesis*) with the formation of a cell membrane between the daughter nuclei. *Mitosis produces two daughter nuclei that are identical to the original nucleus; in plants a *cell plate forms between the new cells. *Meiosis results in four daughter nuclei each with half the number of chromosomes in the mother cell nucleus. *See also* **cell cycle**.

cell fusion (somatic cell hybridization) The technique of combining two cells from different tissues or species in a cell culture. The cells are fused (*see* **chemical fusogen**) and coalesce but their nuclei generally remain separate. However, during cell division a single spindle is formed so that each daughter cell has a single nucleus containing sets of chromosomes from each parental line. Subsequent division of the hybrid cells often results in the loss of chromosomes (and therefore genes), so that absence of a gene product in the culture can be related to the loss of a particular chromosome. Thus the technique is used to determine the control of characteristics exerted by specific chromosomes. Hybrid cells (*see* **hybridoma**) resulting from cell fusion have also been used to produce *monoclonal antibodies.

cell membrane (plasmalemma; plasma membrane) The partially permeable *membrane forming the outer limit of a *cell. It consists mostly of protein and lipid (*see* **lipid bilayer; fluid mosaic model**). The membrane regulates the flow of materials in and out of the cell and plays a role in the production or assembly of the *cell wall or *capsule (when this is present).

cell plate The structure that forms in a dividing plant cell at the end of mitosis; it separates the cytoplasm of the two daughter protoplasts. The cell plate is formed from vesicles deposited by the *Golgi apparatus in the equatorial region of the *spindle. These vesicles contain pectin, cellulose, and *hemicellulose, which contribute to the *middle lamella and the primary wall of the new cell wall. The cell plate eventually fuses with the cell wall of the parent cell, dividing it into two daughter cells.

cell sap The solution that fills the vacuoles of plant cells. It contains sugars, amino acids, waste substances (such as tannins), and mineral salts.

cell theory The theory that was born of the findings of M. J. Schleiden (1804–81) in 1838 and T. Schwann (1810–82) in 1839, who postulated, respectively, that plants and animals were made up of cells and that these units were basic to the structure and function of all organisms. Previously, in 1665, Robert Hooke (1635–1703), while examining cork under the microscope, had observed that its structure consisted of hollow boxlike units, which he called 'cells'. At the time, however, he did not realize the significance of these units, which were in fact dead cells.

cellular respiration *See* respiration.

cellulase A carbohydrate-digesting enzyme (a *carbohydrase*) that hydrolyses cellulose to sugars, including *cellobiose* (a disaccharide consisting of two β-(1,4) linked molecules of glucose) and glucose. Cellulase breaks the β-glycosidic links that join the constituent sugar units of cellulose. *See also* **Ruminantia**.

cellulolytic Able to digest cellulose. For example, cellulolytic bacteria, in the stomach of ruminants, digest cellulose by means of the enzyme *cellulase.

1-4 ß-glycosidic bonds

Cellulose

cellulose A polysaccharide that consists of a long unbranched chain of glucose units (see formula). It is the main constituent of the cell walls of all plants, many algae, and some fungi and is responsible for providing the rigidity of the cell wall. It is an important constituent of dietary *fibre. The fibrous nature of extracted cellulose has led to its use in the textile industry for the production of cotton, artificial silk, etc.

cell wall A rigid outer layer that surrounds the cell membrane of plant, fungal, algal, and bacterial (but not animal) cells. It protects and/or gives shape to a cell, and in herbaceous plants provides mechanical support for the plant body. The cell walls in most plants and algae are composed of the polysaccharide *cellulose and plant cell walls may be secondarily thickened by the addition of *lignin. The cell walls of fungi consist mainly of *chitin. Bacterial cell walls consist of complex polymers of polysaccharides and amino acids (*see* **peptidoglycan; Gram's stain**).

Celsius scale A temperature scale in which the fixed points are the temperatures at standard pressure of ice in equilibrium with water (0°C) and water in equilibrium with steam (100°C). The scale, between these two temperatures, is divided in 100 degrees. The degree Celsius (°C) is equal in magnitude to the *kelvin. This scale was formerly known as the *centigrade scale*; the name was officially changed in 1948 to avoid confusion with a hundredth part of a grade. It is named after the Swedish astronomer Anders Celsius (1701–44), who devised the inverted form of this scale (ice point 100°, steam point 0°) in 1742.

cement (cementum) A thin layer of bony material that fixes teeth to the jaw. It covers the dentine of the root of a *tooth, below the level of the gum, and is attached to the *periodontal membrane lining the tooth socket in the jawbone.

Cenozoic (Cainozoic; Kainozoic) The geological era that began about 65 million years ago and extends to the present. It followed the *Mesozoic era and is subdivided into the *Tertiary and *Quaternary periods. The Cenozoic is often known as the *Age of Mammals* as these animals evolved to become an abundant, diverse, and dominant group. Birds and flowering plants also flourished. The era saw the formation of the major mountain ranges of the Himalayas and the Alps.

centi- Symbol c. A prefix used in the metric system to denote one hundredth. For example, 0.01 metre = 1 centimetre (cm).

centigrade scale *See* **Celsius scale**.

centipedes *See* **Chilopoda**.

Central Dogma The basic belief originally held by molecular geneticists, that flow of genetic information can only occur from *DNA to *RNA to proteins. It is now known, however, that information contained within RNA molecules of viruses can also flow back to DNA. *See also* **genetic code**.

central nervous system (CNS) The part of the nervous system that coordinates all neural functions. In invertebrates it may comprise simply a few *nerve cords and their associated *ganglia. In vertebrates it consists of the *brain and the *spinal cord. The vertebrate CNS contains *reflex arcs, which produce automatic and rapid responses to particular stimuli.

centre (in neurology) A part of the nervous system, consisting of a group of nerve cells, that coordinates a particular process. An example is the respiratory centre in the vertebrate brainstem, which controls breathing movements. The stimulation of a centre will initiate the process, while destruction of the centre will prevent or impair it.

centrifuge A device in which solid or liquid particles of different densities are separated by rotating them in a tube in a horizontal circle. The denser particles tend to move along the length of the tube to a greater radius of rotation, displacing the lighter particles to the other end.

centriole A cylindrical structure in a cell that has a role in cell division. Centrioles are not normally found in plant cells but are commonly present in pairs in animal cells (*see* **centrosome**). During cell division they move to opposite sides of the nucleus to form the ends of the *spindle. The spindle fibres develop from them.

centromere (kinomere; spindle attachment) The part of a *chromosome that attaches to the *spindle during cell division (*see* **meiosis; mitosis**). The position of the centromere is a distinguishing feature of individual chromosomes: a chromosome with the centromere at its centre is described as *metacentric*; one with the centromere towards one end is *acrocentric*; and a chromosome with the centromere at either end is *telocentric*. The centromere usually appears as a constriction when chromosomes contract during cell division; centromeres are often associated with *repetitive DNA. *See also* **kinetochore**.

centrosome (cell centre; centrosphere) A specialized region of a living cell, situated next to the nucleus, where *microtubules are assembled and broken down during cell division. The centrosomes of most animal cells contain a pair of *centrioles. During *metaphase of mitosis and meiosis, the centrosome separates into two regions, each containing one of the centrioles (where present). The two regions move to opposite ends of the cell and a *spindle of microtubules forms between them.

centrum *See* **vertebra**.

cephalization The tendency among animal groups for the major sense organs, mouth, and brain to be grouped together at the front (anterior) end of the body. These are usually contained in a specialized cephalic region – the head.

Cephalochordata A subphylum of the Chordata that contains only the lancelets, a group of small invertebrate marine fishlike animals. Lancelets have gill slits and a notochord that persist into adulthood, the notochord extending the whole length of the body and providing the main skeletal support. The pharynx is modified for filter feeding (*see* **endostyle**).

Cephalopoda The most advanced class of molluscs, containing the squids, cuttlefishes, octopuses, and the extinct *ammonites. Cephalopods have a highly concentrated central nervous system within a protective cartilaginous case. The eye has a well-developed retina and is comparable to that of vertebrates. All cephalopods are predacious carnivores capable of swimming by jet propulsion; they have highly mobile tentacles for catching and holding prey.

cephalothorax The fused head and thorax of many crustaceans and of arachnids (*see* **prosoma**), which is connected to the abdomen. *See also* **tagma**.

cercaria A tadpole-like larval stage of trematode worms (flukes) that develops in the body of a mollusc (the secondary host). It subsequently infects a primary host, in which it matures into an adult worm.

cerci A pair of many-jointed appendages that are present on the last segment of the abdomen in certain insects, such as mayflies, earwigs, and cockroaches.

cerebellum The part of the vertebrate *brain concerned with the coordination and regulation of muscle activity and the maintenance of muscle tone and balance. In mammals it consists of two connected hemispheres, composed of a core of white matter and a much-folded outer cortex of grey matter containing numerous *Purkyne cells, and it is situated above the medulla oblongata and partly beneath the cerebrum.

cerebral cortex (pallium) The layer of *grey matter that forms the outer layer of the hemispheres of the *cerebrum in many vertebrates. It is most highly developed in mammals. The cortex is responsible for the control and integration of voluntary movement and the senses of vision, hearing, touch, etc.; it also contains centres concerned with memory, language, thought, and intellect.

cerebral hemisphere Either of the two halves of the vertebrate *cerebrum.

cerebroside Any one of a class of *glycolipids in which a single sugar unit is bound to a sphingolipid (*see* **phospholipid**). The most common cerebrosides are *galactocerebrosides*, containing the sugar group galactose; they are found in the plasma membranes of neural tissue and are abundant in the myelin sheaths of neurones.

cerebrospinal fluid (CSF) The fluid, similar in composition to *lymph, that bathes the central nervous system of vertebrates. It is secreted by the *choroid plexus into the *ventricles of the brain, filling these and other cavities in the brain and spinal cord, and is reabsorbed by veins on the brain surface. Its function is to protect the central nervous system from mechanical injury.

cerebrum The largest part of the vertebrate *brain. It consists of two *cerebral hemispheres*, which develop from the embryonic *forebrain. The hemispheres have an outer convoluted layer of grey matter – the *cerebral cortex – which contains an estimated ten billion nerve cells. Underneath this is *white matter. The two halves of the cerebrum are linked by the *corpus callosum. The function of the cerebrum is to integrate complex sensory and neural functions. The cerebrum is thought to play a critical role in the process of learning, which involves both short-term and long-term memory.

cerumen Ear wax, which is secreted by *ceruminous glands* in the auditory meatus of the outer ear. Cerumen protects the delicate lining of the meatus and helps to prevent microorganisms from entering the ear.

cervical vertebrae The *vertebrae of the neck. The number of cervical vertebrae varies: for example, rabbits have 7 and humans have 12. Their main functions are to support the head and to provide articulating surfaces against which it can move relative to the backbone. *See* **atlas; axis**.

cervix A narrow or necklike part of an organ. The cervix of the uterus (*cervix uteri*) leads to the vagina. Glands in its walls produce mucus, whose viscosity changes according to the oestrous cycle. During labour, the cervix enlarges greatly to allow passage of the fetus.

Cestoda A class of flatworms (*see* **Platyhelminthes**) comprising the tapeworms – ribbon-like parasites within the gut of vertebrates. Tapeworms are surrounded by partially digested food in the host gut so they are able to absorb nutrients through their whole body surface. The body consists of a head (*scolex*), bearing suckers and hooks for attachment, and a series of segments (*proglottids* or *proglottides*), containing male and female reproductive systems. The life cycle of a tapeworm requires two hosts, the primary host usually being a predator of the secondary host. *Taenia solium* has man for its primary host and the pig as its secondary host. Mature segments, containing thousands of fertilized eggs, leave the primary host with its faeces and develop into embryos and then larvae that continue the life cycle in the gut and other tissues of a secondary host (*see* **bladderworm**).

Cetacea An order of marine mammals comprising the whales, which includes what is probably the largest known animal – the blue whale (*Balaenoptera musculus*), over 30 m long and over 150 tonnes in weight. The forelimbs of whales are modified as short stabilizing flippers and the skin is very thin and almost hairless. A thick layer of blubber insulates the body against heat loss and is an important food store. Whales breathe through a dorsal blowhole, which is closed when the animal is submerged. The toothed whales (suborder Odontoceti), such as the dolphins and killer whale, are carnivorous; whalebone whales (suborder Mysticeti), such as the blue whale, feed on plankton filtered by *whalebone plates.

CFCs *See* **chlorofluorocarbons**.

c.g.s. units A system of units based on the centimetre, gram, and second. Derived from the metric system, it was badly adapted to use with thermal quantities (based on the inconsistently defined *calorie) and with electrical quantities (in which two systems, based respectively on unit permittivity and unit permeability of free space, were used). For scientific purposes c.g.s. units have now been replaced by *SI units.

chaeta (*pl.* **chaetae**) A bristle, made of *chitin, occurring in annelid worms. In the earthworm they occur in small groups projecting from the skin in each segment and function in locomotion. The chaetae of polychaete worms (e.g. ragworm) are borne in larger groups on paddle-like appendages (*parapodia*).

Chaetognatha A phylum of marine coelomate invertebrates, the arrow worms, in which the head bears hooks for catching prey and the trunk and tail support paired lateral and tail fins. They lack organs for excretion, circulation, and respiration and are hermaphrodite.

chalaza **1.** A twisted strand of fibrous albumen in a bird's egg that is attached to the membrane at either end of the yolk and thus holds the yolk in position in the albumen. **2.** The part of a plant *ovule where the nucellus and integuments merge.

chalk A very fine-grained white rock composed of the fossilized skeletal remains of marine plankton known as *coccoliths* and consisting largely of calcium carbonate ($CaCO_3$). It is the characteristic rock of the *Cretaceous period. It should not be confused with blackboard 'chalk', which is made from calcium sulphate.

chalone Any one of a group of compounds that inhibit mitosis. This action was originally demonstrated *in vitro*, using preparations of mammalian cells and tissues; chalones are now thought to be inhibitory *growth factors.

channel *(in cell biology)* A pore formed by a protein molecule in a cell membrane that aids the diffusion of certain substances into and out of the cell. These substances are usually charged ions or lipid-insoluble molecules.

chaparral A type of stunted (scrub) woodland found in temperate regions with little summer rainfall. It is dominated by drought-resistant evergreen shrubs, forming dense thickets, interspersed with dwarfed trees, such as oaks and eucalyptus. It is the typical vegetation found in the western United States and the Mediterranean region (where it is called *maquis*).

character A distinctive inherited feature of an organism. Organisms in a population may display different aspects of a particular character, e.g. the A, B, and O human blood groups (*see* **ABO system**) are different aspects of the blood group character.

chela The terminal segment of an arthropod appendage when this can be opposed to the segment that precedes it. The chela is often enlarged and modified to act as a pincer, as in the lobster. Any appendage possessing a chela is described as *chelate*.

chelicerae The first pair of appendages on the head of arachnids and other arthropods of the subphylum *Chelicerata. These appendages take the form of pincers or claws and are used for grasping or tearing food.

Chelicerata A subphylum or superclass of the *Arthropoda that includes the classes *Arachnida (spiders and scorpions), Merostomata (horseshoe crabs), and Pycnogonida (sea spiders). The body of these arthropods is divided into an anterior *prosoma and a posterior *opisthosoma and bears six pairs of appendages, typically comprising *chelicerae, *pedipalps, and four pairs of walking legs.

chemical bond A strong force of attraction holding atoms together in a molecule or crystal. In general, atoms combine to form molecules by sharing or transferring electrons in their outer shells. Typically chemical bonds have energies of about 1000 kJ mol^{-1} and are distinguished from the much weaker forces between molecules. *See also* **covalent bond; electrovalent bond; hydrogen bond.**

chemical control The use of chemicals to kill pests (*see* **pesticide**). *Compare* **biological control**.

chemical dating An absolute *dating technique that depends on measuring the chemical composition of a specimen. Chemical dating can be used when the specimen is known to undergo slow chemical change at a known rate. For instance, phosphate in buried bones is slowly replaced by fluoride ions from the ground water. Measurement of the proportion of fluorine present gives a rough estimate of the time that the bones have been in the ground. Another, more accurate, method depends on the fact that amino acids in living organisms are L-optical isomers. After death, these racemize and the age of bones can be estimated by measuring the relative amounts of D- and L-amino acids present.

chemical fossil Any of various organic compounds found in ancient geological strata that appear to be biological in origin and are assumed to indicate that life existed when the rocks were formed. The presence of chemical fossils in Precambrian strata indicates that life existed over 3000 million years ago.

chemical fusogen Any chemical that is used in the fusing of two cells or protoplasts (*see* **cell fusion**). Polyethylene glycol (PEG) is used in the formation of *hybridomas, and sodium nitrate in the fusion of plant protoplasts in solution.

chemical reaction A change in which one or more chemical elements or compounds (the *reactants*) form new compounds (the *products*). All reactions are to some extent *reversible*; i.e. the products can also react to give the original reactants. However, in many cases the extent of this back reaction is negligibly small, and the reaction is regarded as *irreversible*. *See also* **endergonic reaction; exergonic reaction**.

ATP production in mitochondria according to the chemiosmotic theory

chemiosmotic theory A theory postulated by the British biochemist Peter Mitchell (1920–   ) to explain the formation of ATP in the mitochondrial *electron transport chain. As electrons are transferred along the electron carrier system in the inner mitochondrial membrane, hydrogen ions (protons) are actively transported (via *hydrogen carriers) into the space between the inner and outer mitochondrial membranes, which thus contains a higher concentration of protons than the matrix. This creates an electrochemical gradient across the inner membrane, down which protons move back into the matrix. This movement occurs through special channels associated with ATP synthetase, the enzyme that catalyses the conversion of ADP to ATP, and is coupled with the phosphorylation of ADP (see illustration). A similar gradient is created across the thylakoid membranes of chloroplasts during the light reaction of *photosynthesis (*see* **photophosphorylation**).

chemoautotroph *See* **autotrophic nutrition; chemosynthesis**.

chemolithotroph Any bacterium that is able to obtain its energy by the oxidation of inorganic compounds, including those of iron, nitrogen, and sulphur.

chemoreceptor A *receptor that detects the presence of particular chemicals and (in multicellular organisms) transmits this information to sensory nerves. Examples include the *taste buds and the receptors in the *carotid body.

chemosynthesis A type of *autotrophic nutrition in which organisms (called *chemoautotrophs*) synthesize organic materials using energy derived from the oxidation of inorganic chemicals, rather than from sunlight. Most chemoautotrophs are bacteria, including *Nitrosomonas*, which oxidizes ammonium to nitrite; and *Thiobacillus*, which oxidizes sulphur to sulphate.

chemosystematics *See* **systematics**.

chemotaxis *See* **taxis**.

chemotaxonomy The *classification of plants and microorganisms based on similarities and differences in their natural products and the biochemical pathways involved in their manufacture. *See also* **taxonomy**.

chemotropism The growth or movement of a plant or plant part in response to a chemical stimulus. An example is the growth of a pollen tube down the style during fertilization in response to the presence of sugars in the style.

chiasma (*pl.* **chiasmata**) The point at which paired *homologous chromosomes remain in contact as they begin to separate during the first prophase of *meiosis, forming a cross shape. A number of chiasmata can usually be identified and at these points *crossing over occurs.

chill haze The precipitation that occurs when beer is stored at cold temperatures. Chill haze consists of proteins that can be broken down by using protease enzymes.

Chilopoda A class of wormlike terrestrial arthropods comprising the centipedes, characterized by a distinct head, bearing a single pair of relatively long antennae and one pair of poison jaws, and 15–177 body segments, each bearing one pair of similar legs. Centipedes are fast-moving predators found in damp environments. *See also* **Uniramia**.

chimaera An organism composed of tissues that are genetically different. Chimaeras can develop if a *mutation occurs in a cell of a developing embryo. All the cells arising from it have the mutation and therefore produce tissue that is genetically different from adjacent tissue, e.g. brown patches in otherwise blue eyes in humans. *Graft hybrids are examples of plant chimaeras.

Chiroptera An order of flying mammals comprising the bats. Their membranous wings are supported by very elongated forelimbs and digits and stretch along the sides of the body to the hindlimbs and tail. Whenever bats rest they allow their body temperature to fall, hibernating in winter when food is scarce. Most bats are nocturnal; their ears are enlarged and specialized for *echolocation, which they use to hunt prey and avoid obstacles. Bats feed variously on insects, fruit, nectar, or blood.

chitin A *polysaccharide comprising chains of N-acetyl-D-glucosamine, a derivative of glucose. Chitin is structurally very similar to cellulose and serves to strengthen the supporting structures of various invertebrates. It also occurs in fungi.

chlamydospore A thick-walled asexual spore that is produced from a fungal hypha. It is a resting spore, capable of withstanding unfavourable growing conditions.

chlorenchyma *Parenchyma tissue that contains chloroplasts and is photosynthetic. Chlorenchyma makes up the mesophyll tissue of plant leaves and is also found in the stems of certain plant species. *Compare* collenchyma; sclerenchyma.

chloride secretory cell Any of the cells in the gills of marine bony fish that remove chloride from the blood and secrete it into the sea, thereby maintaining the osmotic composition of the body. The excretion of chloride is accompanied by movement of sodium ions from the blood to the sea, which keeps the electrochemical balance of the fish's body neutral.

chloride shift The movement of chloride ions (Cl^-) into red blood cells. Carbon dioxide reacts with water to form carbonic acid in the red blood cells (*see* **carbonic anhydrase**). The carbonic acid then dissociates into hydrogencarbonate ions (HCO_3^-) and hydrogen ions (H^+). The cell membrane is relatively permeable to negative ions. Therefore the hydrogencarbonate ions diffuse out of the cell into the plasma, leaving the hydrogen ions, which create a net positive charge; this is neutralized by the diffusion of chloride ions from the plasma into the cell.

chlorocruorin A greenish iron-containing *respiratory pigment that occurs in the blood of polychaete worms. It closely resembles *haemoglobin.

chlorofluorocarbons (CFCs) Compounds obtained by replacing the hydrogen atoms of hydrocarbons by chlorine and fluorine atoms. Their high stability to temperature makes them suitable for a variety of uses, including aerosol propellants, oils, polymers, etc. They are often known as *freons*. There has been concern that their use in aerosols may cause depletion of the *ozone layer. *See also* **pollution**.

Chlorophyll a

chlorophyll Any one of a class of pigments found in all photosynthetic organisms; the most important members are chlorophyll *a* (see formula) and chlorophyll *b*, which occur in all land plants and are responsible for their green colour. Chlorophyll molecules are the principal sites of light absorption in the light reaction of *photosynthesis (*see* **photosystems I and II**). They are magnesium-containing *porphyrins, chemically related to *cytochrome and *haemoglobin. *See also* **bacteriochlorophyll**.

Chlorophyta (green algae) A large phylum of *algae, the members of which possess chlorophylls *a* and *b*, store food reserves as starch, and have cellulose cell walls. In these respects they resemble plants more closely than do any of the other algal phyla. The Chlorophyta are widely distributed and diverse in form. Unicellular forms may occur singly (sometimes with *flagella for motility) or in colonies, while multicellular forms may be filamentous (e.g. *Spirogyra*) or platelike (e.g. *Ulva*).

chloroplast Any of the chlorophyll-containing organelles (*see* **plastid**) that are found in large numbers in those plant and algal cells undergoing *photosynthesis. Plant chloroplasts are typically lens-shaped and bounded by a double membrane. They contain membranous structures called *thylakoids, which are piled up into stacks (*see* **granum**), surrounded by a gel-like matrix (*stroma*). The light reaction of photosynthesis occurs on the thylakoid membranes while the dark reaction takes place in the stroma.

chlorosis The abnormal condition in plant stems and leaves in which synthesis of the green pigment chlorophyll is inhibited, resulting in a pale yellow coloration. This may be caused by lack of light, mineral deficiency, infection (particularly by viruses), or genetic factors.

choanae (internal nares) *See* nares.

cholecalciferol *See* vitamin D.

cholecystokinin (CCK; pancreozymin) A hormone, produced by the duodenal region of the small intestine, that induces the gall bladder to contract and eject bile into the intestine and stimulates the pancreas to secrete its digestive enzymes. Cholecystokinin output is stimulated by contact with the contents of the stomach.

cholesterol A *sterol (*see also* **steroid**) occurring widely in animal tissues and also in some plants and algae. It can exist as a free sterol or esterified with a long-chain fatty acid. Cholesterol is absorbed through the intestine or manufactured in the liver. It serves principally as a constituent of blood plasma *lipoproteins and of the lipid–protein complexes that form cell membranes. It is also important as a precursor of various steroids, especially the bile acids, sex hormones, and adrenocorticoid hormones. The derivative 7-dehydrocholesterol is converted to vitamin D_3 by the action of sunlight on skin. Increased levels of dietary and blood cholesterol have been associated with *atherosclerosis.

choline An amino alcohol, $CH_2OHCH_2N(CH_3)_3OH$. It occurs widely in living organisms as a constituent of certain types of phospholipids – the *lecithins and sphingomyelins. It is sometimes classified as a member of the *vitamin B complex.

cholinergic Describing a nerve fibre that either releases *acetylcholine when stimulated or is itself stimulated by acetylcholine. *Compare* **adrenergic**.

cholinesterase (acetylcholinesterase) An enzyme that hydrolyses the neurotransmitter *acetylcholine to choline and acetate. Cholinesterase is secreted by nerve cells at *synapses and by muscle cells at *neuromuscular junctions. Organophosphorus insecticides (*see* **pesticide**) act by inhibiting the action of cholinesterase.

Chondrichthyes A class of vertebrate chordates (*see* **Chordata**) comprising the fishes with cartilaginous skeletons. The majority belong to the subclass Elasmobranchii (skates, rays, and sharks – *see* **Selachii**). Most cartilaginous fishes are marine carnivores with powerful jaws. Unlike bony fishes, they have no swim bladder, and therefore avoid sinking only by constant swimming with the aid of an asymmetrical (*heterocercal*) tail. There is no operculum covering the gill slits, the first of which is modified as a *spiracle. Fertilization is internal so the few eggs produced are consequently yolky, large, and well-protected. Some cartilaginous fishes show viviparous development of the young (*see* **viviparity**).

chondrin The matrix of *cartilage, which is made up of chondrocytes embedded in chondroitin sulphate.

chondrocyte The cells that make up the matrix of *cartilage.

Chordata A phylum of animals characterized by a hollow dorsal nerve cord and, at some stage in their development, a flexible skeletal rod (the *notochord) and *gill slits opening from the pharynx. There are four subphyla: the *Tunicata (sea squirts), *Cephalochordata (lancelets), *Agnatha (jawless chordates), and *Gnathostomata (jawed chordates). In the Agnatha and Gnathostomata, commonly known as vertebrates or craniates, the notochord is present only in the embryo or larva and becomes replaced by the *vertebral column (backbone) before birth or metamorphosis. This has permitted the vertebrates a greater degree of movement and subsequent improvement in the sense organs and enlargement of the brain, which is enclosed in a skeletal case, the cranium.

chordotonal organs Sensory receptors that detect alterations in the tension of insect muscles.

chorion 1. A membrane enclosing the embryo, yolk sac, and allantois of reptiles, birds, and mammals. In mammals a section of the chorion becomes the embryonic part of the *placenta. *See* **extraembryonic membranes.**
2. The protective shell of an insect egg, produced by the ovary. It is pierced by a small pore (*micropyle*) that allows the entry of spermatozoa for fertilization. *See also* **egg membrane.**

chorionic gonadotrophin *See* **gonadotrophin.**

choroid A pigmented layer, rich in blood vessels, that lies between the retina and the sclerotic of the vertebrate eye. At the front of the eye the choroid is modified to form the *ciliary body and the *iris.

choroid plexus A membrane rich in blood vessels that lines the *ventricles of the brain. It is an extension of the *pia mater and secretes *cerebrospinal fluid into the ventricles; it also controls exchange of materials between the blood and cerebrospinal fluid.

chromaffin tissue A group of cells in the adrenal medulla (*see* **adrenal glands**) that synthesizes *noradrenaline. Chromaffin tissue also contains the enzyme that converts noradrenaline into *adrenaline.

chromatid A threadlike strand formed from a *chromosome during the early stages of cell division. Each chromosome divides along its length into two chromatids, which are at first held together at the centromere. They separate completely at a later stage. The DNA of the chromosome reproduces itself exactly so that each chromatid has the complete amount of DNA and becomes a daughter chromosome with exactly the same genes as the original chromosome from which it was formed.

chromatin The substance of which eukaryotic *chromosomes are composed. It consists of proteins (principally histones), DNA, and small amounts of RNA (*see* **nucleosome**). In a metabolically inactive nucleus, chromatin is mainly in a condensed form, *heterochromatin*, but in an active nucleus most of the chromatin is in an expanded form, *euchromatin*. Messenger RNA molecules are formed in euchromatic regions.

chromatogram A record obtained by chromatography. The term is applied to the developed records of *paper chromatography and *thin-layer chromatography and also to the graphical record produced in *gas–liquid chromatography.

chromatography A technique for analysing or separating mixtures of gases, liquids, or dissolved substances, such as mixtures of amino acids or chlorophyll pigments. The original technique (invented by the Russian botanist Mikhail Tsvet in 1906) is a good example of *column chromatography*. A vertical glass tube is packed with an adsorbing material, such as alumina. The sample is poured into the column and continuously washed through with a solvent (a process known as *elution*). Different components of the sample are adsorbed to different extents and move down the column at different rates. In Tsvet's original application, plant pigments were used and these separated into coloured bands in passing down the column (hence the name chromatography). The usual method is to collect the liquid (the *eluate*) as it passes out from the column in fractions.

In general, all types of chromatography involve two distinct phases – the *stationary phase* (the adsorbent material in the column in the example above) and the *moving phase* (the solution in the example). The separation depends on competition for molecules of sample between the moving phase and the stationary phase. The form of column chromatography above is an example of *adsorption chromatography*, in which the sample molecules are adsorbed on the alumina. In *partition chromatography*, a liquid (e.g. water) is first absorbed by the stationary phase and the moving phase is an immiscible liquid. The separation is then by partition between the two liquids. In *ion-exchange chromatography* the process involves competition between different ions for ionic sites on the stationary phase (*see* **ion exchange**). *Gel filtration is another chromatographic technique in which the size of the sample molecules is important.

See also **affinity chromatography; gas–liquid chromatography; paper chromatography; thin-layer chromatography**.

chromatophore 1. A pigment-containing cell found in the skin of many lower vertebrates (e.g. chameleon) and in the integument of crustaceans. Concentration or dispersion of the pigment granules in the cytoplasm of the cell causes the colour of the animal to alter to match its surroundings. A common type of chromatophore is the *melanophore*, which contains the pigment *melanin. **2.** *See* **chromoplast**.

chromophore Any group of atoms in a molecule that includes an unsaturated chemical group (such as C=C) capable of absorbing *ultraviolet light; the chromophore is responsible for the colour of the compound. For example, retinal is the chromophore of *rhodopsin, the light-sensitive pigment in the rod cells of the retina.

chromoplast (chromatophore) Any of various organelles in plant cells that contain pigments. Red, orange, and yellow chromoplasts contain carotenoid pigments, while green chromoplasts (called *chloroplasts) contain the pigment chlorophyll and provide the green coloration of most plants. *Compare* **leucoplast**. *See* **plastid**.

chromosome A threadlike structure several to many of which are found in the nucleus of plant and animal (eukaryotic) cells. Chromosomes are composed of *chromatin and carry the *genes in a linear sequence; these determine the individual characteristics of an organism. When the nucleus is not dividing, individual chromosomes cannot be identified with a light microscope. During the first stage of nuclear division, however, the chromosomes contract and, when stained, can be clearly seen under a microscope. Each consists of two *chromatids held together at the *centromere (*see also* **meiosis; mitosis**). The number of chromosomes in each cell is constant for and characteristic of the species concerned. In the normal body cells of *diploid organisms the chromosomes occur in pairs (*see* **homologous chromosomes**); in the gamete-forming germ cells, however, the diploid number is halved and each cell contains only one member of each chromosome pair. Thus in man each body cell contains 46 chromosomes (22 matched pairs and one pair of *sex chromosomes) and each germ cell 23. Abnormalities in the number or structure of chromosomes may give rise to abnormalities in the individual; *Down's syndrome is the result of one such abnormality. *See* **chromosome mutation**.

Bacterial and viral cells contain only a single chromosome; it differs from the eukaryotic chromosome in being much simpler, lacking histones and consisting simply of a single or double strand of DNA or (in some viruses) RNA.

chromosome diminution A phenomenon seen in embryonic nematode worms in which the germ cells become separated from the somatic cells at an early stage of development. The chromosomes of the somatic cells tend to break down, with only a few remaining intact, whereas the integrity of the germ-cell chromosomes is maintained.

chromosome map (linkage map) A plan showing the relative positions of *genes along the length of the chromosomes of an organism. It is constructed by making crosses and observing whether certain characteristics tend to be inherited together. The closer together two allele pairs are situated on *homologous chromosomes, the less often will they be separated and rearranged as the reproductive cells are formed (*see* **chiasma; crossing over**). The proportion of offspring that show *recombination of the alleles concerned thus reflects their spacing and is used as a unit of length in mapping chromosomes (*see* **crossover value**). *See also* **restriction mapping**.

chromosome mutation A change in the gross structure of a chromosome, which usually causes severely deleterious effects in the organism. Chromosome mutations often occur due to an error in pairing during the *crossing over stage of meiosis. The main types of chromosome mutation include *translocation, *duplication, *deletion, and *inversion. *Compare* **point mutation**. *See also* **mutation**.

chrysalis *See* **pupa**.

Chrysophyta (golden-brown algae) A large phylum of mostly freshwater *algae that possess *carotenoid pigments (responsible for their colour), in addition to chlorophylls. Their main storage products are oils and the polysaccharide *chrysolaminarin*. Most chrysophytes are unicellular, with two unequal-sized flagella, but some form colonies or are filamentous; one group of marine chrysophytes construct tests of silica.

chyle A milky fluid consisting of *lymph that contains absorbed food materials (especially emulsified fats). Most chyle occurs in the lymphatic ducts (*lacteals) in the *villi of the small intestine during the absorption of fat.

chylomicron Any of the *lipoprotein particles synthesized by intestinal epithelial cells and consisting mainly of triglycerides. Chylomicrons are the form in which dietary fat is transported in the circulatory system.

chymase *See* **rennin**.

chyme The semisolid and partly digested food that is discharged from the stomach into the duodenum.

chymotrypsin An *endopeptidase enzyme in pancreatic juice that is secreted into the duodenum. The enzyme is secreted as an inactive precursor, *chymotrypsinogen*, which is activated by another pancreatic protease, *trypsin.

chymotrypsinogen *See* **chymotrypsin**.

ciliary body The circular band of tissue surrounding and supporting the *lens of the vertebrate eye. It contains the *ciliary muscles*, which bring about changes in the shape of the lens (*see also* **accommodation**). The ciliary body produces the *aqueous humour.

ciliary feeding A method of feeding used by lancelets and many other aquatic invertebrates. The movement of cilia causes a current of water to be drawn towards and through the animal, and microorganisms in the water are filtered out by the cilia.

ciliary muscle *See* **ciliary body**.

ciliated epithelium A region of *epithelium consisting of columnar or cuboidal cells bearing hairlike appendages (*see* **cilium**) that are capable of beating rapidly. Ciliated epithelium performs the function of moving particles or fluid over the epithelial surface in such structures as the trachea, bronchial tubes, and nasal cavities. It often occurs in the vicinity of mucus-secreting *goblet cells.

Ciliophora A phylum of the Protoctista containing ciliated *protozoa (including *Paramecium*) that possess two types of nuclei, a micronucleus and macronucleus (*see* **nucleus**). The cilia are used for feeding and locomotion. Ciliates reproduce sexually by *conjugation.

cilium (*pl.* **cilia**) A short minute hairlike structure (up to 10 μm long) present on the surface of many cells, notably in certain protozoans and some types of vertebrate *epithelium. Cilia usually occur in large groups. Beating of cilia can produce cell movement or create a current in fluid surrounding a cell. *See also* **axoneme**. *Compare* **flagellum**.

circadian rhythm (diurnal rhythm) Any 24-hour periodicity in the behaviour or physiology of animals or plants. Examples are the sleep/activity cycle in many animals and the growth movements of plants. Circadian rhythms are generally controlled by *biological clocks.

circulation The mass flow of fluid (e.g. blood or lymph) through the tissues and organs of an animal, allowing the transport and exchange of such materials as oxygen, nutrients, and waste products (*see also* **vascular system; lymphatic system**). Smaller animals (e.g. arthropods and most molluscs) have an *open circulation*, i.e. the blood is pumped into the body cavity, in which the internal organs are suspended. In open circulatory systems the tissues are in direct contact with the blood and materials are exchanged directly by diffusion. In a *closed circulation*, found in larger animals, the blood flows in vessels, which usually contain a series of one-way valves to maintain the flow in one direction. *See also* **double circulation; single circulation**.

circulatory system The heart, blood vessels, blood, lymphatic vessels, and lymph, which together serve to transport materials throughout the body. *See also* **double circulation; single circulation; vascular system**.

cisterna (*pl.* **cisternae**) Any of the membrane-bound sacs that form the branches of the *endoplasmic reticulum (ER) and *Golgi apparatus. The cisternae of the ER are continuous with each other and with the nuclear envelope; in rough ER the cisternae are flattened sacs, whereas in smooth ER they are more tubular in shape.

cis-trans test A test used to determine whether two independent mutations, affecting the same phenotypic function, occur within the same *cistron or on different cistrons. A pair of homologous chromosomes are brought together in the same cell (normally a bacterial cell). Mutations located on the same chromosome are described as being in the *cis* position; if they occur one on each homologue they are in the *trans* position. If the cell is phenotypically normal for both *cis* and *trans* positions of the mutations, these mutations are considered to be present in different cistrons. If the resultant cell is normal when the mutations are in the *cis* position but mutant when the mutations are in the *trans* position, the two mutations are considered to lie in the same cistron. See illustration.

Cis-trans test

cistron A length of DNA that contains the information for coding a specific polypeptide chain. A cistron codes for a messenger *RNA (mRNA) molecule; it does not include sections of DNA that may have noncoding functions or code for ribosomal or transfer RNA. The cistron as a unit of function was defined by the *cis-trans test: mutations occurring within a cistron affect the same function.

citric acid A white crystalline hydroxy carboxylic acid, $HOOCCH_2C(OH)(COOH)CH_2COOH$. It is present in citrus fruits and is an intermediate in the *Krebs cycle in plant and animal cells.

citric-acid cycle *See* **Krebs cycle.**

clade A group of organisms that share a common ancestor. *See* **cladistics.**

cladistics A controversial method of classification in which animals and plants are placed into taxonomic groups called *clades* when they share characteristics (known as *homologies*) that are thought to indicate common ancestry. It is based on the assumption that two new species are formed suddenly, by splitting from a common ancestor, and not by gradual evolutionary change. A diagram indicating these relationships (called a *cladogram*) therefore consists of a system of dichotomous branches: each point of branching represents divergence from a common ancestor, as shown in the diagram. Thus the species A to F form a clade as they share the common ancestor X, and species A to D form a clade of a different taxonomic rank, sharing the ancestor X_2. Species C to F do not form a clade, since the latter must include *all* the descendants of a common ancestor.

A cladogram showing the relationships of six species (A–F)

cladode (cladophyll) A flattened stem or internode that resembles and functions as a leaf. It is an adaptation to reduce water loss, since it contains fewer *stomata than a leaf. An example of a plant with cladodes is asparagus.

cladogram *See* **cladistics**.

cladophyll *See* **cladode**.

class A category used in the *classification of organisms that consists of similar or closely related orders. Similar classes are grouped into a phylum. Examples include Mammalia (mammals), Aves (birds), and Musci (mosses).

classification The arrangement of organisms into a series of groups based on physiological, biochemical, anatomical, or other relationships. An *artificial classification* is based on one or a few characters simply for ease of identification or for a specific purpose; for example, birds are often arranged according to habit and habitat (seabirds, songbirds, birds of prey, etc.) while fungi may be classified as edible or poisonous. Such systems do not reflect evolutionary relationships. A *natural classification* is based on resemblances and is a hierarchical arrangement. The smallest group commonly used is the *species. Species are grouped into genera (*see* **genus**), the hierarchy continuing up through *tribes, *families, *orders, *classes, and phyla (*see* **phylum**) to *kingdom. (In traditional systems of plant classification the phylum was replaced by the *division.) Higher up in the hierarchy the similarities between members of a group become fewer. Present-day natural classifications try to take into account as many features as possible and in so doing aim to reflect evolutionary relationships (*see* **cladistics**). Natural classifications are also predictive. Thus if an organism is placed in a particular genus because it shows certain features characteristic of the genus, then it can be assumed it is very likely to possess most (if not all) of the other features of that genus. *See also* **binomial nomenclature; taxonomy**.

clavicle A bone that forms part of the *pectoral (shoulder) girdle, linking the *scapula (shoulder blade) to the sternum (breastbone). In man it forms the collar bone and serves as a brace for the shoulders.

clay An inorganic constituent of *soils consisting chiefly of clay minerals (mainly hydrous silicates of aluminium) in the form of particles less than 0.002 mm in diameter. *See also* **flocculation**.

cleavage *(in embryology)* The series of cell divisions by which a single fertilized egg cell is transformed into a multicellular body, the *blastula. Characteristically no growth occurs during cleavage, the shape of the embryo is unchanged except for the formation of a central cavity (the blastocoel), and the ratio of nuclear material (DNA) to cytoplasm increases.

cleistothecium *See* ascocarp.

climate The characteristic pattern of weather elements in an area over a period. The weather elements include temperature, rainfall, humidity, solar insolation, wind, etc. The climate of a large area is determined by several climatic controls: (1) the latitude of the area, which accounts for the amount of solar radiation it receives; (2) the distribution of land and sea masses; (3) the altitude and topography of the area; and (4) the location of the area in relation to the ocean currents. Weather elements are important *abiotic factors.

climax community A relatively stable ecological *community that is achieved at the end of a *succession.

cline A gradual variation in the characteristics of a species or population over its geographical range. It occurs in response to varying environmental factors, such as soil type or climate.

clinostat A mechanical device that rotates whole plants (usually seedlings), so removing the effect of any stimulus that normally acts in one particular direction. It is most often used to study the growth of plant organs when the influence of gravity has been removed.

clitellum A thickened glandular region of the body in certain annelid worms that is prominent when the worm becomes sexually mature. The clitellum possesses long slender chaetae. Goblet cells in the clitellum secrete mucus, which forms the *cocoon in which fertilized eggs develop.

clitoris An erectile rod of tissue in female mammals (and also some reptiles and birds) that is the equivalent of the male penis. It lies in front of the *urethra and *vagina.

cloaca The cavity in the pelvic region into which the terminal parts of the alimentary canal and the urinogenital ducts open in most vertebrates. Placental mammals, however, have a separate anus and urinogenital opening.

clone 1. A group of organisms or cells that have arisen from a single individual by asexual reproduction and are therefore all genetically identical, e.g. a group of plants propagated by bulbs or cuttings. **2. (gene clone)** An exact replica of a gene. *See* **gene cloning**.

cloning vector *See* **vector**.

closed circulation *See* **circulation**.

clotting factors (coagulation factors) A group of substances present in blood plasma that, under certain circumstances, undergo a series of chemical reactions leading to the conversion of blood from a liquid to a solid state (*see* **blood clotting**). Although they have specific names, most coagulation factors are referred to by an agreed set of Roman numerals (e.g. *Factor VIII, Factor IX). Lack of any of these factors in the blood results in the inability of the blood to clot. *See also* **haemophilia**.

clubmoss *See* **Lycopodophyta**.

cluster of differentiation *See* **CD**.

Cnidaria A phylum of aquatic invertebrates (commonly known as coelenterates) that includes *Hydra*, jellyfish, sea anemones, and *corals. A coelenterate's body is *diploblastic, with two cell layers of the body wall separated by *mesoglea, and shows *radial symmetry. The body cavity (*enteron* or *coelenteron*) is sac-shaped, with one opening acting as both mouth and anus. This opening is surrounded by tentacles bearing *thread cells. Coelenterates exist both as free-swimming *medusae (e.g. jellyfish) and as sedentary *polyps. The latter may be colonial (e.g. corals) or solitary (e.g. sea anemones and *Hydra*). In many coelenterates the life cycle alternates between these two forms (*see* **alternation of generations**). The phylum contains the classes Hydrozoa (e.g. *Hydra*, *Obelia*), most members of which show alternation of generations; Scyphozoa (jellyfish), in which the medusa phase is dominant; and Anthozoa (corals and sea anemones), in which medusae are absent. *See also* **Ctenophora**.

cnidoblast *See* **thread cell**.

CNS *See* **central nervous system**.

CoA *See* **coenzyme A**.

coacervate An aggregate of macromolecules, such as proteins, lipids, and nucleic acids, that form a stable *colloid unit with properties that resemble living matter. Many are coated with a lipid membrane and contain enzymes that are capable of converting such substances as glucose into more complex molecules, such as starch. Coacervate droplets arise spontaneously under appropriate conditions and may have been the prebiological systems from which living organisms originated.

coagulation The process in which colloidal particles come together irreversibly to form larger masses. Coagulation can be brought about by adding ions to change the ionic strength of the solution and thus destabilize the colloid (*see* **flocculation**). Ions with a high charge are particularly effective (e.g. alum, containing Al^{3+}, is used in styptics to coagulate blood). Alum and iron(III) sulphate are also used for coagulation in *sewage treatment. Heating is another way of coagulating certain colloids (e.g. boiling an egg coagulates the albumin). *See also* **blood clotting**.

coal A brown or black carbonaceous deposit derived from the accumulation and alteration of ancient vegetation, which originated largely in swamps or other moist environments. As the vegetation decomposed it formed layers of peat, which were subsequently buried (for example, by marine sediments following a rise in sea level or subsidence of the land). Under the increased pressure and resulting higher temperatures the peat was transformed into coal. Two types of coal are recognized: *humic* (or *woody*) *coals*, derived from plant remains; and *sapropelic coals*, which are derived from algae, spores, and finely divided plant material.

As the processes of coalification (i.e. the transformation resulting from the high temperatures and pressures) continue, there is a progressive transformation of the deposit: the proportion of carbon relative to oxygen rises and volatile substances and water are driven out. The various stages in this process are referred to as the *ranks* of the coal. In ascending order, the main ranks of coal are: *lignite* (or *brown coal*), which is soft, brown, and has a high moisture content; *subbituminous coal*, which is used chiefly by generating stations; *bituminous coal*, which is the most abundant rank of coal; *semibituminous coal*; *semianthracite coal*, which has a fixed carbon content of between 86% and 92%; and *anthracite coal*, which is hard and black with a fixed carbon content of between 92% and 98%.

Most deposits of coal were formed during the Carboniferous and Permian periods. More recent periods of coal formation occurred during the early Jurassic and Tertiary periods. Coal deposits occur in all the major continents; the leading producers include the USA, China, Ukraine, Poland, UK, South Africa, India, Australia, and Germany. Coal is used as a fuel and in the chemical industry; by-products include coke and coal tar.

cobalamin (vitamin B$_{12}$) *See* **vitamin B complex**.

cobalt Symbol Co. A light-grey metallic element that is a trace element (*see* **essential element**) required by animals. It is present in foods of animal origin and is a constituent of vitamin B$_{12}$. Cobalt is also a *micronutrient for plants.

coccus (*pl.* **cocci**) Any spherical bacterium. Cocci may occur singly, in pairs, in groups of four or more, in cubical packets, in grapelike clusters (*Staphylococcus*), or in chains (*Streptococcus*). Staphylococci and streptococci include pathogenic species. They are generally nonmotile and do not form spores.

coccyx The last bone in the *vertebral column in apes and man (i.e. tailless primates). It is formed by the fusion of 3–5 *caudal vertebrae.

cochlea Part of the *inner ear of mammals, birds, and some reptiles that transforms sound waves into nerve impulses. In mammals it is coiled, resembling a snail shell, and is divided by membranes into three parallel canals (see illustration): the vestibular canal (scala vestibule), cochlear duct (scala media), and tympanic canal (scala tympani). The cochlear duct is filled with a fluid (*see* **endolymph**) and contains the *organ of Corti, which houses the sound receptors. The other two canals also contain a fluid (*see* **perilymph**). Sound-induced vibrations of the *oval window are transmitted through the perilymph and endolymph and stimulate hair cells in the organ of Corti. These in turn stimulate nerve cells that transmit information, via the *auditory nerve, to the brain for interpretation of the sounds.

Diagram of the cochlea (coiling reduced for simplicity)

cockroaches *See* **Dictyoptera**.

cocoon A protective covering for eggs and/or larvae produced by many invertebrates. For example, the larvae of many insects spin a cocoon in which the pupae develop (that of the silkworm moth produces silk), and earthworms secrete a cocoon for the developing eggs (*see* **clitellum**).

codeine A pain-relieving drug that is derived from the plant *Papaver somniferum*. *See* **opiate**; **analgesic**.

co-dominance The condition that arises when both alleles in a *heterozygous organism are dominant and are fully expressed in the *phenotype. For example, the human blood group AB is the result of two alleles, A and B, both being expressed. A is not dominant to B, nor vice versa. *Compare* **incomplete dominance**.

codon A triplet of nucleotides within a molecule of messenger *RNA that functions as a unit of genetic coding (the *triplet code*), usually by specifying a particular amino acid during the synthesis of proteins in a cell (*see* **genetic code**). A few codons specify instructions during this process (*see* **start codon**; **stop codon**). The term codon may also refer to any of the corresponding nucleotide triplets of DNA that are transcribed into codons. *Compare* **anticodon**.

coelacanth A bony fish, *Latimeria chalumnae*, that was believed to have been extinct until 1938, when the first specimen of modern times was discovered. The coelacanth belongs to the same order (Crossopterygii – lobe-finned fishes) as the ancestors of the amphibians. It lives at a depth of 200–300 m in the Indian Ocean. The coelacanth is a large fish, 1–2 m long and weighing 80 kg or more, with a three-lobed tail fin. The body is covered in rough heavy scales and the pectoral fins can be used like crutches to help movement across the sea bed. The young are born alive. Fossil coelacanths are most abundant in deposits about 400 million years old and no fossils less than 70 million years old have been found.

coelenterates *See* Cnidaria.

coelom A fluid-filled cavity that forms the main *body cavity of vertebrate and most invertebrate animals. It is formed by the splitting of the *mesoderm. Ciliated ducts (*coelomoducts*) connect the coelom to the exterior allowing the exit of waste products and gametes; in higher animals these are specialized as oviducts, etc. The coelom is large and often subdivided in annelid worms (in which it functions as a hydrostatic skeleton) and vertebrates. In arthropods it is restricted to the cavities of the gonads and excretory organs, the body cavity being a blood-filled *haemocoel.

coelomoduct *See* coelom.

coenobium (*pl.* **coenobia**) A loose association of unicellular organisms that live in a colony and may be surrounded by a common membrane. Often the cells are held together in a jelly excreted by all individuals in the colony. Both algae and bacteria form coenobia. For example, *Volvox* (a green alga) forms a hollow sphere in which 20 000 cells may live; some of these are reproductive and others are concerned with photosynthesis. The cells in this coenobium are interconnected by protoplasmic strands.

coenocyte A mass of protoplasm containing many nuclei and enclosed by a cell wall. It is found in certain algae and fungi. *Compare* cell; **plasmodium; syncytium.**

Coenzyme A

coenzyme An organic nonprotein molecule that associates with an enzyme molecule in catalysing biochemical reactions. Coenzymes usually participate in the substrate–enzyme interaction by donating or accepting certain chemical groups. Many vitamins are precursors of coenzymes. *See also* **cofactor**.

coenzyme A (CoA) A complex organic compound that acts in conjunction with enzymes involved in various biochemical reactions, notably the oxidation of pyruvate via the *Krebs cycle and fatty-acid oxidation and synthesis (*see* **acetyl coenzyme A**). It comprises principally the B vitamin *pantothenic acid, the nucleotide *adenine, and a ribose–phosphate group (see formula).

coenzyme Q (ubiquinone) Any of a group of related quinone-derived compounds that serve as electron carriers in the *electron transport chain reactions of cellular respiration. Coenzyme Q molecules have side chains of different lengths in different types of organisms but function in similar ways.

coevolution The evolution of complementary adaptations in two species caused by the *selection pressures that each exerts on the other. It is common in symbiotic associations (*see* **symbiosis**). For example, many insect-pollinated plants have evolved flowers whose shapes, colours, etc., make them attractive to particular insects; at the same time the pollinating insects have evolved sense organs and mouthparts specialized for quickly locating, and extracting nectar from, particular species of plants.

cofactor A nonprotein component essential for the normal catalytic activity of an enzyme. Cofactors may be organic molecules (*coenzymes) or inorganic ions. They may activate the enzyme by altering its shape or they may actually participate in the chemical reaction.

cohesion 1. The force of attraction between like molecules. Cohesion provides the force that holds up a column of water in the xylem tissue of plants without it breaking. The *cohesion–tension theory* offers an explanation for the continual flow of water upwards through the xylem of a plant. Water is removed from the plant by the process of *transpiration, which creates a tension that pulls the water in the xylem upwards as a single column held together by cohesive forces. **2. (*in botany*)** The union of like parts, such as the fusion of petals that occurs in some flowers.

coitus *See* **sexual intercourse**.

colchicine An *alkaloid derived from the autumn crocus, *Colchicum autumnale*. It inhibits *spindle formation in cells during mitosis so that chromosomes cannot separate during anaphase, thus inducing multiple sets of chromosomes (*see* **polyploid**). Colchicine is used in genetics, cytology, and plant breeding research and also in cancer therapy to inhibit cell division.

cold-blooded animal *See* **ectotherm**.

Coleoptera An order of insects comprising the beetles and weevils and containing about 330 000 known species – the largest order in the animal kingdom. The forewings are hardened and thickened to form *elytra*, which meet at a precise mid-dorsal line and protect the underlying pair of hindwings and abdomen. The mouthparts are generally modified for biting and in some species assume antler-like proportions. Beetles occur in a wide variety of terrestrial and aquatic habitats; many feed on decaying organic matter, some eat living vegetation, while others prey on other arthropods. A number of beetles and weevils are economically important pests of stored grain, timber, and crops. The young emerge as larvae and generally undergo metamorphosis via a pupal stage to form the adult beetle.

coleoptile A protective sheath that covers the young shoot of the embryo in plants of the grass family. It bursts open when the first leaves develop. Experiments investigating growth movements of the oat coleoptile led to the discovery of the plant growth substance indoleacetic acid (IAA).

coleorhiza A protective sheath that covers the young root of the embryo in plants of the grass family.

coliform bacteria A group of Gram-negative rod-shaped bacteria that are found in the vertebrate gastrointestinal tract; their presence in water is an indicator of faecal pollution. They obtain their energy by aerobic respiration or fermentation; some of them can ferment lactose. Well-known coliform bacteria include *Escherichia coli and *Salmonella*.

collagen An insoluble fibrous protein found extensively in the connective tissue of skin, tendons, and bone. The polypeptide chains of collagen (containing the amino acids glycine and proline predominantly) form triple-stranded helical coils that are bound together to form fibrils, which have great strength and limited elasticity. Collagen accounts for over 30% of the total body protein of mammals.

collecting duct Any of the ducts in the mammalian *kidney that drains into the renal pelvis, which leads to the ureter. They are the main sites of water reabsorption from the glomerular filtrate, which drains into the ducts from the *distal convoluted tubules of the *nephrons. The cells of the collecting ducts are relatively impermeable to water. However, the influence of *antidiuretic hormone increases the permeability of the collecting ducts, allowing the reabsorption of water and controlling the final urine concentration according to the body's state of hydration.

Collembola An order of primitive wingless insects, the springtails, which leap by means of a specialized forked organ (*furcula*) that is attached on the underside of the abdomen by a special catch (*retinaculum*) and acts as a spring. Most springtails are scavengers but some are pests of leguminous plants.

collenchyma A plant tissue (*see* **ground tissues**) that consists of living cells with additional cellulose thickening in their walls, providing support and protection for young stems and leaves. It is most commonly found in the stem cortex. *Compare* **parenchyma**; **sclerenchyma**.

colloblast *See* lasso cell.

colloids Colloids were originally defined by Thomas Graham in 1861 as substances, such as starch or gelatin, which will not diffuse through a membrane. He distinguished them from *crystalloids* (e.g. inorganic salts), which would pass through membranes. Later it was recognized that colloids were distinguished from true solutions by the presence of particles that were too small to be observed with a normal microscope yet were much larger than normal molecules. Colloids are now regarded as systems in which there are two or more phases, with one (the *dispersed phase*) distributed in the other (the *continuous phase*). Moreover, at least one of the phases has small dimensions (in the range 10^{-9}–10^{-6} m). Colloids are classified in various ways.

Sols are dispersions of small solid particles in a liquid. The particles may be macromolecules or clusters of small molecules. *Lyophobic sols* are those in which there is no affinity between the dispersed phase and the liquid. An example is silver chloride dispersed in water. In such colloids the solid particles have a surface charge, which tends to stop them coming together. Lyophobic sols are inherently unstable and in time the particles aggregate and form a precipitate (*see* **flocculation**). *Lyophilic sols*, on the other hand, are more like true solutions in which the solute molecules are large and have an affinity for the solvent. Starch in water is an example of such a system. *Association colloids* are systems in which the dispersed phase consists of clusters of molecules that have lyophobic and lyophilic parts. Soap in water is an association colloid (*see* **micelle**).

Emulsions are colloidal systems in which the dispersed and continuous phases are both liquids, e.g. oil-in-water or water-in-oil. Such systems require an emulsifying agent to stabilize the dispersed particles.

Gels are colloids in which both dispersed and continuous phases have a three-dimensional network throughout the material, so that it forms a jelly-like mass. Gelatin is a common example. One component may sometimes be removed (e.g. by heating) to leave a rigid gel (e.g. silica gel).

Other types of colloid include *aerosols* (dispersions of liquid or solid particles in a gas, as in a mist or smoke) and *foams* (dispersions of gases in liquids or solids).

colon The section of the vertebrate *large intestine that lies between the *caecum and the *rectum. Its prime function is to absorb water and minerals from indigestible food residues passing from the small intestine, which results in the formation of *faeces.

colony 1. *(in zoology)* A group of animals of the same species living together and dependent upon each other. Some, such as the corals and sponges, are physically connected and function as a single unit. Others, such as insect colonies, are not physically joined but show a high level of social organization with members specialized for different functions (*see* **caste**). **2.** *(in microbiology)* A group of microorganisms, usually bacteria or yeasts, that are considered to have developed from a single parent cell.

Colonies that grow on *agar plates differ in shape, colour, surface texture, and translucency and can therefore be used as a means of identification.

colony-stimulating factor (CSF) One of several molecules that regulate the development of blood cells. *GM-CSF* is a glycoprotein manufactured in all tissues, large amounts being found in the lungs. It induces haemopoietic stem cells (the precursors of mature blood cells) to develop into granulocytes or monocytes (hence GM). *G-CSF* is a protein that induces stem cells to form granulocytes only. *M-CSF*, a glycoprotein, causes stem cells to form monocytes only.

colostrum A liquid with a high content of nitrogen, antibodies, and vitamins that is secreted from the mammary glands before and just after giving birth. The change of secretion from colostrum to proper milk takes place gradually during the days after birth.

colour blindness Any disorder of vision in which colours are confused. The most common type is red–green colour blindness. This is due to a recessive gene carried on the X chromosome (*see* **sex linkage**), and therefore men are more likely to show the defect although women may be *carriers. It results in absence or malfunctioning of one or more of the three types of cone cell responsible for *colour vision. In *protanopia* the individual lacks cones sensitive to red light; in *deuteranopia* cones sensitive to green light are absent. *Tritanopia* is a rare form of colour blindness in which the individual cannot distinguish between blue and green due to a lack of cones sensitive to blue light.

colour vision The ability of the eye to detect different wavelengths of light and to distinguish between these different wavelengths and their corresponding colours. In the mammalian eye this is achieved by the *cone cells, which are located in and around the *fovea near to the centre of the retina. The cone cells contain the light-sensitive pigment *iodopsin*, which – according to the *trichromatic theory* – exists in three forms, each form occurring in a different cone cell. Each form of iodopsin is sensitive to either red, blue, or green light. The relative stimulation of each type of cone will determine the colour that is interpreted by the brain. For example, if red cones and green cones are stimulated to an equal extent, the colour interpreted by the brain is yellow. *See also* **colour blindness**.
  The *compound eye of certain insects is also capable of colour vision.

columnar epithelium *See* epithelium.

commensalism An interaction between two animal or plant species that habitually live together in which one species (the *commensal*) benefits from the association while the other is not significantly affected. For example, the burrows of many marine worms contain commensals that take advantage of the shelter provided but do not affect the worm.

communicating junction *See* gap junction.

communication An interaction between two organisms in which information is conveyed from one to the other. Communication can occur between individuals of the same species (*intraspecific communication*) or between members of different species (*interspecific communication*). It generally involves the transmission of a signal from one organism to another; signals can be visual, chemical, or tactile or they can take the form of sounds. Visual signals between members of the same species are widely used by animals in such activities as defining and protecting *territories and finding suitable mates (*see* **courtship; display behaviour; bioluminescence**). Chemical and tactile signals also play an important role in these activities (*see* **pheromone**). Social species rely heavily on all three types of signalling, the classic example being provided by the *dance of the bees, in which information about the distance and direction of a food source is conveyed to other members of the colony. Visual signals, in the form of body coloration, are the principal means of communication between animals of different species (*see* **mimicry; warning coloration**). Sounds are more effective than visual signals for intraspecific communication over long distances and at night. Certain insects produce sounds by *stridulation, while birdsong and language are sophisticated examples of sound signals in birds and humans, respectively.

Among plants, visual and chemical signals are important in communication. Flowering plants whose flowers are pollinated by insects or other animals depend on the colour, shape, and scent of their flowers to attract suitable pollinating agents. Some plants produce chemical signals to deter competitors and predators (*see* **allelopathy**).

community A naturally occurring assemblage of plant and animal species living within a defined area or habitat. Communities are named after one of their *dominant species (e.g. a pine community) or the major physical characteristics of the area (e.g. a freshwater pond community). Members of a community interact in various ways (e.g. through *food chains and *competition). Large communities may be divided into smaller component communities. *See* **association**.

companion cell A type of cell found within the *phloem of flowering plants. Each companion cell is usually closely associated with a *sieve element. Its function is uncertain, though it appears to regulate the activity of the adjacent sieve element. In gymnosperms a similar function is attributed to *albuminous cells*, which are found closely associated with gymnosperm sieve elements.

compass plant A plant that has its leaves permanently orientated in a north–south direction. Such an arrangement enables the plant to take full advantage of morning and evening sun, while avoiding the stronger midday sunlight. An example is the compass plant of the prairies (*Silphium laciniatum*).

compensation point The point reached in a plant when the rate of photosynthesis is equal to the rate of respiration. This means that the carbon dioxide released from respiration is equivalent to that which is taken up during photosynthesis. The compensation point is reached as light intensity increases. If the light intensity is increased beyond the compensation point, the rate of photosynthesis increases proportionally until the point of *light saturation* is reached, beyond which the rate of photosynthesis is no longer affected by light intensity.

competent Describing embryonic tissue that is capable of developing into a specialized tissue when suitably stimulated. *See* **induction**; **evocation**.

competition The interaction that occurs between two or more organisms, populations, or species that share some environmental resource when this is in short supply. Competition is an important force in evolution: plants, for example, become tall to compete for light, and animals evolve various foraging methods to compete for food. There may be a direct confrontation between competitors, as occurs between barnacles competing for space on a rock, or the numbers or fecundity of the competitors are indirectly reduced through joint dependence on limited resources. Competition occurs both between members of a species (*intraspecific competition*) and between different species (*interspecific competition*). Interspecific competition often results in the dominance of one species over another (*see* **dominant**). Since competition ultimately results in the displacement by one competitor of the others, it is to the advantage of the competitors to avoid one another wherever possible. Thus in time the competitors become separated from each other geographically or ecologically, which promotes evolutionary change. Competition for mates may lead to *sexual selection.

competitive exclusion principle A rule, derived by G. F. Gause in 1934, stating that two species that occupy the same habitat cannot also occupy the same *ecological niche. Any two species that occupy the same niche will compete with each other to the detriment of one of the species, which will thus be excluded.

competitive inhibition *See* **inhibition**.

complement A group of proteins present in blood plasma and tissue fluid that aids the body's defences following an *immune response; the genes encoding it form part of the *major histocompatibility complex. Following an antibody–antigen reaction, complement is activated chemically and becomes bound to the antibody–antigen complex (*complement fixation*); it becomes involved in destroying foreign cells and attracting scavenging white blood cells (*phagocytes) to the area of conflict in the body.

complemental males The small males of certain animals that live in or on the females and are usually more or less degenerate apart from the reproductive organs. They occur in certain crustaceans (e.g. some barnacles), in which the normal individuals are hermaphrodite but the complemental males have suppressed ovaries, lose their alimentary canal, and lead a semiparasitic existence in the mantle cavity of the larger partner. This may ensure that cross fertilization occurs.

complementary DNA (cDNA) A form of DNA prepared in the laboratory using messenger *RNA (mRNA) as template, i.e. the reverse of the usual process of *transcription in cells; the synthesis is catalysed by *reverse transcriptase. cDNA thus has a base sequence that is complementary to that of the mRNA template; unlike genomic DNA, it contains no noncoding sequences (*introns). cDNA is used in *gene cloning for the expression of eukaryote genes in prokaryote host cells.

complementary genes Two (or more) genes that are interdependent, such that the dominant *allele from either gene can only produce an effect on the *phenotype of an organism if the dominant allele from the other gene is also present.

composite fruit A type of fruit that develops from an inflorescence rather than from a single flower. *See* **pseudocarp; sorosis; strobilus; syconus**.

compost A mixture of decaying organic matter, such as vegetation and manure, that is used as a *fertilizer. The organic material is decomposed by aerobic saprotrophic organisms, mostly fungi and bacteria. Some decomposition is also carried out by *detritivores. Compost is used mainly on a domestic scale.

compound eye The eye of insects and crustaceans, which consists of numerous visual units, the *ommatidia*. Each ommatidium consists of an outer cuticle covering a lens, beneath which are 6–8 retinal cells surrounding a light-sensitive *rhabdom*. Adjacent ommatidia are separated by pigment cells. The eye is convex, with nerve fibres from the retinal cells converging onto the optic nerve. There are two types of compound eye. In *apposition eyes*, typical of diurnal insects, each ommatidium focuses rays parallel to its long axis so that each gives an image of a minute part of the visual field, producing a detailed mosaic image. In *superposition eyes*, typical of nocturnal insects, the pigment separating ommatidia migrates to the ends of the cells, so that each ommatidium receives light from a larger part of the visual field and the image may overlap with those received by many neighbouring ommatidia. This produces an image that is bright but lacks sharpness of detail.

compound microscope *See* **microscope**.

concentration The quantity of dissolved substance per unit quantity of solvent in a solution. Concentration is measured in various ways. The amount of substance dissolved per unit volume (symbol c) has units of $mol\,dm^{-3}$ or $mol\,l^{-1}$. It is now called 'concentration' (formerly *molarity*). The *mass concentration* (symbol ρ) is the mass of solute per unit volume of solvent. It has units of $kg\,dm^{-3}$, $g\,cm^{-3}$, etc. The *molal concentration* (or *molality*; symbol m) is the amount of substance per unit mass of solvent, commonly given in units of $mol\,kg^{-1}$.

concentration gradient (diffusion gradient) The difference in concentration between a region of a solution or gas that has a high density of particles and a region that has a relatively lower density of particles. By random motion, particles will move from the area of high concentration towards the area of low concentration, by the process of *diffusion, until the particles are evenly distributed in the solution or gas.

conceptacle A flask-shaped cavity with a small opening (the *ostiole*) that is found in the swollen tip of certain brown algae, such as *Fucus*. It contains the sex organs.

conception The fertilization of a mammalian egg cell by a sperm cell, which occurs in the fallopian tube. Conception is followed by *implantation.

condensation reaction A chemical reaction in which two molecules combine to form a larger molecule with the elimination of a small molecule (H_2O in biological systems). *See* **esterification; glycosidic bond; peptide**.

conditional response (conditioned reflex) A learned response that develops to an initially ineffective stimulus in classical *conditioning.

conditioning A process by which animals learn about a relation between two events. In *classical* (or *Pavlovian*) *conditioning*, repeated presentations of a neutral stimulus (e.g. the sound of a bell or buzzer) are followed each time by a biologically important stimulus (such as food or electric shock), which elicits a response (e.g. salivation). Eventually the neutral stimulus presented by itself produces a response (the *conditional response*, or *conditioned reflex*) similar to that originally evoked by the biologically important stimulus. For example, Pavlov's dogs learned to salivate in response to the sound of a metronome that preceded the presentation of food. In *instrumental* (or *operant*) *conditioning* the animal is rewarded (or punished) each time it makes a particular response; this eventually causes the frequency of the response to increase (or decrease). For example, a rat will learn to press a lever in order to obtain food. *See* **reinforcement**.

condyle A smooth round knob of bone that fits into a socket on an adjoining bone, forming a *joint. Such a joint permits up-and-down or side-to-side movement but does not allow rotation. There are condyles where the lower jawbone (mandible) is attached to the skull, which permits chewing movements. *See also* **occipital condyle**.

cone 1. *(in botany)* A reproductive structure occurring in gymnosperms, known technically as a *strobilus*. It consists of *sporophylls bearing the spore-producing sporangia. Gymnosperms produce different male and female cones. The large woody female cones of pines, firs, and other conifers are made up of structures called *ovuliferous scales*, which bear the ovules. Cones are also produced by clubmosses and horsetails. 2. *(in animal anatomy)* A type of light-sensitive receptor cell, found in the *retinas of all diurnal vertebrates. Cones are specialized to transmit information about colour (*see* **colour vision**) and are responsible for the *visual acuity of the eye. They function best in bright light. They are not evenly distributed on the retina, being concentrated in the *fovea and absent on the margin of the retina. *Compare* **rod**.

conformer Any organism whose *internal environment is highly influenced by external factors. Many marine invertebrates are conformers: they have no need to control their internal environment since the external environment is fairly constant in terms of temperature, oxygen tension, and nutrients. *Compare* **regulator**.

congenital Present at birth. Congenital disorders of the body may be due to genetic factors, e.g. *Down's syndrome, or caused by injury or environmental factors, e.g. drugs (such as thalidomide), chemicals (such as dioxin), and infections (such as those caused by *Listeria and *cytomegalovirus).

conidiospore *See* conidium.

conidium (conidiospore) A spore of certain fungi, such as mildews and moulds, that is produced by the constriction of the tip of a specialized hypha, the *conidiophore*. Chains of conidia may be formed in this way; they are cut off, one at a time, from the tip of the hypha.

Coniferophyta A phylum of seed-bearing plants comprising the conifers, including the pines, firs, and spruces. Conifers have an extensive fossil record going back to the late Devonian. The gametes are carried in male and female *cones, fertilization usually being achieved by wind-borne pollen. The ovules and the seeds into which they develop are borne unprotected (rather than enclosed in a carpel, as are those of the *Angiospermophyta). Internal tissue and cell structure of these species is not as advanced as in the angiosperms. Conifers are typically evergreen trees inhabiting cool temperate regions and have leaves reduced to needles or scales. The wood of conifers, which is called *softwood* in contrast to the *hardwood* of angiosperm trees, is widely used for timber and pulp. *See also* **gymnosperm**.

conjugation 1. The fusion of two reproductive cells, particularly when these are both the same size (*see* **isogamy**). **2.** A form of sexual reproduction seen in some algae (e.g. *Spirogyra*), some bacteria (e.g. *Escherichia coli*), and ciliate protozoans. Two individuals are united by a tube formed by outgrowths from one or both of the cells. Genetic material from one cell (designated the male) then passes through the tube to unite with that in the other (female) cell. In bacteria conjugation is initiated and directed by *sex factors.

conjunctiva The delicate membrane that covers the cornea and lines the inside of the eyelid of a vertebrate eye. It is kept clean by secretions of the *lacrimal (tear) gland and the reflex blink mechanism.

connective tissue An animal tissue consisting of a small number of cells (e.g. *fibroblasts and *mast cells) and fibres (*see* **collagen; elastin**) and a large amount of intercellular material (*matrix*, or *ground substance). It is widely distributed and has many functions, including support, packing, defence, and repair. The individual constituents vary, depending on the function of the tissue. Different types of connective tissue include *mesenchyme* in the embryo, *adipose tissue, *areolar connective tissue, blood, lymph, cartilage, and bone.

consensual Of or relating to movements that take place through the action of the nervous system independently of the will. These movements generally involve the involuntary reaction being correlated with a voluntary action. For example, the reflex action of both pupils contracting will occur even if only one retina has been stimulated by bright light (*see* **pupillary reflex**).

conservation The sensible use of the earth's natural resources in order to avoid excessive degradation and impoverishment of the environment (*see* **desertification**). It should include the search for alternative food and fuel supplies when these are endangered (as by *deforestation and overfishing); an awareness of the dangers of *pollution; and the maintenance and preservation of natural habitats and the creation of new ones (e.g. nature reserves, national parks, and *SSSIs).

conservative replication A hypothesis suggesting that *DNA replication occurs by one DNA molecule initiating the synthesis of a new molecule while remaining intact. The preferred theory (*see* **semiconservative replication**) proposes that the DNA molecule divides to provide two templates for synthesizing the other half of the molecule. *Compare* **dispersive replication**.

consociation A climax plant *community that is dominated by one particular species, e.g. a pine forest. *See* **dominant**. *Compare* **association**.

consumer An organism that feeds upon those below it in a *food chain (i.e. at the preceding *trophic level). Herbivores, which feed upon green plants, are *primary consumers*; a carnivore that feeds only upon herbivores is a *secondary consumer*; a *tertiary consumer* is a carnivore that feeds on other carnivores. The consumer at the end of a food chain is known as the *top carnivore. Compare* **producer**.

contact insecticide Any insecticide (*see* **pesticide**) that kills its target insect by being absorbed through the cuticle or by blocking the spiracles, rather than by being ingested.

continental drift The theory that the earth's continents once formed a single mass and have since moved relative to each other. It was first postulated by A. Snider in 1858 and greatly developed by Alfred Wegener in 1912. He used evidence, such as the fit of South America into Africa and the distribution of rock types, flora, fauna, and geological structures, to suggest that the present distribution of the continents results from the breaking up of one or two greater land masses. The original land mass was named

Pangaea and it was suggested that this broke up into the northerly *Laurasia* and the southerly *Gondwanaland* (see illustration). The theory was not accepted for about 50 years by the majority of geologists but during the early 1960s, H. H. Hess's seafloor-spreading hypothesis and the subsequent development of *plate tectonics produced a mechanism to explain the drift of the continents.

(a) 200 million years ago

(b) 135 million years ago

(c) 65 million years ago

Continental drift

continuous culture A technique used to grow microorganisms or cells continually in a particular phase of growth. For example, if a constant supply of cells is required, a cell culture maintained in the log phase is best; the conditions must therefore be continually monitored and adjusted accordingly so that the cells do not enter the stationary phase (*see* **bacterial growth curve**). Growth may also have to be maintained in a particular growth phase if an enzyme or chemical product is produced only during that phase.

continuous variation (quantitative variation) The range of differences that can be observed in many characteristics in a population. Characteristics resulting from *multifactorial inheritance show continuous variation, e.g. the wide range of foot sizes in an adult human population. *Compare* **discontinuous variation**.

contour feathers *Feathers that are arranged in regular rows on a bird's body, giving the body its streamlined shape. Each has a central horny shaft (the *rachis*) with a flattened *vane* on each side. Each vane is composed of two rows of filament-like *barbs, which are connected to each other by means of hooked *barbules to form a smooth surface. There is often a small second vane, the *aftershaft*, near the base of the feather.

contraception *See* **birth control**.

contractile root Any of the modified adventitious roots that develop from the base of the stem of a bulb or corm. The new bulb or corm develops at a higher level in the soil than the old one. The contractile roots shorten and pull it down to a suitable level.

contractile vacuole A membrane-surrounded cavity in a cell that periodically expands, filling with water, and then suddenly contracts, expelling its contents to the cell's exterior. It is thus an organ of *osmoregulation and excretion. Contractile vacuoles are common in freshwater sponges and typical of freshwater protozoans, such as *Amoeba* (which has one spherical vacuole) and *Paramecium* (in which a number of accessory vacuoles are attached to a main vacuole).

contraction *(in animal physiology)* The shortening of muscle fibres in order to exert a force on a tissue or organ of the body. In striated muscle contraction is brought about by interaction of actin and myosin filaments (*see* **sarcomere; sliding filament theory; voluntary muscle**): it provides a force for *locomotion and plays a role in maintaining the balance and posture of the animal. *See also* **involuntary muscle**.

control 1. The part of an experiment that acts as a standard by which to compare experimental observations. **2.** The natural regulation of biological processes. *See* **control mechanism**. **3.** *See* **biological control; chemical control**.

control mechanism Any mechanism that regulates a biological process, such as a metabolic pathway or enzyme-controlled reaction, or that helps to maintain the *internal environment (*see* **homeostasis**). *See also* **feedback**.

conus arteriosus A small thick-walled chamber of the heart of vertebrate embryos that receives blood from the single ventricle and leads to the ventral aorta. It is retained as a distinct structure in adult fishes and (with modifications) amphibians, but in higher vertebrates it is incorporated into the roots of the aorta and pulmonary arteries.

convergent evolution The development of superficially similar structures in unrelated organisms, usually because the organisms live in the same kind of environment. Examples are the wings of insects and birds and the streamlined bodies of whales and fish. *Compare* **adaptive radiation**.

convoluted tubule *See* **distal convoluted tubule; proximal convoluted tubule; nephron**.

coomassie blue A biological dye used for staining proteins.

cooperation An association between two or more members of the same species (*intraspecific cooperation*), or between individuals of different species (*interspecific cooperation*), in which all members benefit. An example of interspecific cooperation is the relationship formed between ants and aphids: the aphids gain protection by living in the ant colonies, while the ants feed on secretions from the aphids. Interspecific cooperation is a looser association than *mutualism.

coordinate bond *See* covalent bond.

coordination *(in animal physiology)* The processes involved in the reception of sensory information, the integration of that information, and the subsequent response of the organism. Coordination is controlled by regions of the brain that deal with specific functions, such as locomotion and breathing, and is carried out by the nervous system.

Copepoda A class of crustaceans occurring in marine and freshwater habitats. Copepods are usually 0.5–2 mm long and lack both a carapace and compound eyes. Copepods are important members of plankton: some are free-living, feeding on microscopic organisms; others are parasitic. A familiar freshwater genus is *Cyclops*, so named because the members have a single median eye.

coprophagy The ingestion of faeces as a means of obtaining nutrients. Coprophagous animals include dung beetles, which eat cow dung, and rabbits, which ingest their own faeces.

copulation *See* sexual intercourse.

coral Any of a group of sedentary colonial marine invertebrates belonging to the class Anthozoa of the phylum *Cnidaria. A coral colony consists of individual *polyps within a protective skeleton that they secrete: this skeleton may be soft and jelly-like, horny, or stony. The horny skeleton secreted by corals of the genus *Corallium*, especially *C. rubrum*, constitutes the red, or precious, coral used as a gemstone. The skeleton of stony, or true, corals consists of almost pure calcium carbonate and forms the coral reefs common in tropical seas.

cork (phellem) A protective waterproof plant tissue produced by the *cork cambium. It develops in plants undergoing *secondary growth and replaces the epidermis. Its cells, whose walls are impregnated with *suberin, are arranged in radial rows and fit closely together except where the cork is interrupted by *lenticels. Some cork cells become air-filled while others contain deposits of lignin, tannins, and fatty acids, which give the cork a particular colour. The cork oak (*Quercus suber*) produces cork that can be used commercially.

cork cambium (phellogen) A type of *cambium arising within the outer layers of the stems of woody plants, usually as a complete ring surrounding the inner tissues. The cells of the cork cambium divide to produce an outer corky tissue (*cork or *phellem*) and an inner secondary cortex (*phelloderm*). Cork, cork cambium, and phelloderm together make up the *periderm*, an impermeable outer layer that protects the inner stem tissues if the outer tissues split as the stem girth increases with age. It thus takes over the functions of the epidermis.

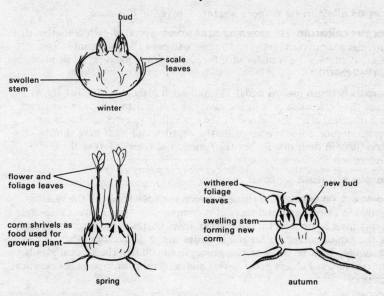

Development of a corm

corm An underground organ formed by certain plants, e.g. crocus and gladiolus, that enables them to survive from one growing season to the next (see illustration). It consists of a short swollen food-storing stem surrounded by protective scale leaves. One or more buds in the axils of scale leaves produce new foliage leaves and flowers in the subsequent season, using up the food stored in the stem. *Compare* **bulb**.

cornea A transparent layer of tissue, continuous with the *sclerotic, that forms the front part of the vertebrate eye, over the iris and lens. The cornea refracts light entering the eye onto the lens, thus assisting in the focusing of images onto the *retina. *See also* **astigmatism**.

cornification *See* **keratinization**.

corolla The *petals of a flower, collectively, forming the inner whorl of the *perianth. It encircles the stamens and carpels. The form of the corolla is very variable. The petals may either be free (*polypetalous*) or united to form a tube (*gamopetalous* or *sympetalous*).

coronary vessels Two pairs of blood vessels (the coronary arteries and coronary veins) that supply the muscles of the heart itself. The coronary arteries arise from the aorta and divide into branches that encircle the heart. A blood clot in a coronary artery (*coronary thrombosis*) is one of the causes of a 'heart attack'.

corpus allatum (*pl.* **corpora allata**) *See* **juvenile hormone**.

corpus callosum The sweeping band (commissure) of *white matter that provides a connection between the two halves of the cerebrum in the brain. It enables the transfer of information from one cerebral hemisphere to the other.

corpus luteum (yellow body) The yellowish mass of tissue that forms from the *granulosa cells in the cavity of a *Graafian follicle in the ovary of a mammal after the release of the egg cell. It secretes the hormone *progesterone. Some species of sharks, reptiles, and birds have similar structures in their ovaries but the function of these is less well understood.

corpus striatum *See* **basal ganglia**.

cortex 1. *(in botany)* The tissue between the epidermis and the vascular system in plant stems and roots. It is composed of *parenchyma cells and shows little or no structural differentiation. Cortex is produced by activity of the *apical meristem. *See also* **endodermis**. **2.** *(in zoology)* The outermost layer of tissue of various organs, including the adrenal glands (*adrenal cortex*), kidneys (*renal cortex*), and cerebral hemispheres (*see* **cerebral cortex**).

Corti cell Any of the cells found in the *organ of Corti in the inner ear. Corti cells are epithelial cells from which hairlike structures project; they play a role in the detection of sound.

corticosteroid Any of several hormones produced by the cortex of the *adrenal glands. *Glucocorticoids* regulate the use of carbohydrates, proteins, and fats in the body and include *cortisol and *cortisone. *Mineralocorticoids* regulate salt and water balance (*see* **aldosterone**).

corticotrophin *See* ACTH.

cortisol (hydrocortisone) A hormone (*see* **corticosteroid**), produced by the adrenal glands, that promotes the synthesis and storage of glucose and is therefore important in the normal response to stress, suppresses or prevents inflammation, and regulates deposition of fat in the body. It is used as treatment for various allergies and for rheumatic fever, certain skin conditions, and adrenal failure (Addison's disease).

cortisone A hormone (*see* **corticosteroid**), produced by the adrenal glands, that is closely related to, and has an action similar to that of, *cortisol. Cortisone is also the name for a synthetic version of cortisol.

corymb A type of flowering shoot (*see* **racemose inflorescence**) in which the lower flower stalks are longer than the higher ones, resulting in a flat-topped cluster of flowers. Examples are candytuft and wallflower.

cosmid A hybrid *vector, used in *gene cloning, that includes the *cos* gene (from the lambda bacteriophage). It also contains drug resistance *marker genes and other plasmid genes. Cosmids can incorporate larger DNA fragments than either phage or plasmid vectors alone and are especially suitable for cloning large mammalian genes or multigene fragments.

cosmoid scale *See* scales.

cotyledon (seed leaf) A part of the embryo in a seed plant. The number of cotyledons is an important feature in classifying plants. Among the flowering plants, the class known as *Monocotyledonae have a single cotyledon and *Dicotyledonae have two. Conifers have either two cotyledons, as in *Taxus* (yews), or five to ten, as in *Pinus* (pines). In seeds without an *endosperm, e.g. garden pea and broad bean, the cotyledons store food, which is used in germination. In seeds showing *epigeal germination, e.g. runner bean, they emerge above the soil surface and become the first photosynthetic leaves.

countercurrent heat exchange A *counterflow mechanism that enables fluids at different temperatures flowing in channels in opposite directions to exchange their heat content without mixing. An example of countercurrent heat exchange occurs in the feet of penguins, in which heat from blood in the arteries supplying the feet is transferred to blood returning to the body's core in veins that lie close to these arteries. This helps to maintain the core temperature in freezing conditions.

countercurrent multiplier system An active process occurring in the *loops of Henle in the kidney, which is responsible for the production of concentrated urine in the collecting ducts of the nephrons. Sodium and chloride ions are actively pumped from the ascending limb of the loop but

Countercurrent multiplier system in the kidney

water is retained, since the ascending limb is impermeable to water. This creates a concentration gradient in the medulla in which the concentration of sodium and chloride is greatest in the region of the bend of the loop. Fluid passing from the loop of Henle to the distal tubule is less concentrated than that entering the loop, but because of the high osmotic pressure in the medulla water diffuses out of the collecting ducts, producing a concentrated urine. See illustration.

counterflow The flow of two fluids in apposed vessels in opposite directions. In biological systems such an arrangement enables the efficient transfer of heat, ions, molecules, etc., from fluids that are rich in these resources to fluids that are deficient in them.

courtship Behaviour in animals that plays a part in the initial attraction of a mate or as a prelude to copulation. Courtship often takes the form of *displays that have evolved through *ritualization; some are derived from other contexts (e.g. food begging in some birds). Chemical stimuli (*see* **pheromone**) are also important in many mammals and insects.

As well as ensuring that the prospective mate is of the same species, the male's courtship performance allows females to choose between different males. The later stages of courtship may involve both partners in an alternating series of displays that inhibit *aggression and fear responses and ensure synchrony of sexual arousal.

COV *See* **crossover value**.

covalent bond A type of *chemical bond in which atoms are held together in a molecule by sharing one or more pairs of electrons in their outer shells. For example, in the water molecule (H_2O) each hydrogen atom forms a covalent bond by sharing its only electron with one of the six electrons in the outer shell of the oxygen atom. *Coordinate* (or *dative*) bonds are covalent bonds in which one of the atoms supplies both the electrons. *Single bonds* are those in which one pair of electrons is shared; in *double* or *triple bonds* two or three pairs, respectively, are shared.

Cowper's glands (bulbourethral glands) A pair of pea-sized glands that lie beneath the prostate gland, named after the English surgeon William Cowper (1660–1709). Cowper's glands secrete an alkaline fluid that forms part of the *semen. This fluid neutralizes the acidic environment of the urethra, thereby protecting the sperm. *See also* **seminal vesicle**.

coxa The first (basal) segment, attached to the thorax, of the leg of an insect, arachnid, or of certain other arthropods. *See also* **femur; trochanter**.

coxal glands Paired ducts (coelomoducts) in arthropods that lead from the *coelom to the exterior and are normally involved in excretion. In arachnids one or two pairs of coxal glands in the cephalothorax open at the bases (coxae) of one or two pairs of legs.

C₃ pathway The metabolic pathway followed in the dark stage of *photosynthesis by most plants of temperate regions, in which the first product is the three-carbon compound glycerate 3-phosphate. This is formed when carbon dioxide combines with *ribulose bisphosphate in the first reaction of the *Calvin cycle. Plants that follow this pathway are referred to as C_3 *plants. Compare* **C_4 pathway**.

C_4 pathway

C_4 pathway (Hatch–Slack pathway) The metabolic pathway followed in the dark stage of *photosynthesis by tropical plants, such as sugar cane and maize, and by plants that live in arid environments; these plants are known as C_4 *plants*. The initial fixation of atmospheric carbon dioxide occurs with phosphoenolpyruvate (PEP), through the mediation of the enzyme PEP carboxylase, to form the four-carbon compound oxaloacetic acid (see illustration; *compare* **C_3 pathway**). Oxaloacetic acid is then converted into malic acid; this breaks down to give carbon dioxide, which passes to the *Calvin cycle, and pyruvic acid, which generates more PEP, thereby ensuring a constant supply of this compound for further carbon dioxide fixation. The C_4 pathway allows photosynthesis to occur at very low concentrations of carbon dioxide as PEP carboxylase has an extremely high affinity for carbon dioxide. This pathway also works well at high temperatures and light intensity, enabling efficient photosynthesis in tropical plants. In addition, the malic acid formed can be stored before being used, to be later broken down to carbon dioxide when required in the Calvin cycle. This is important in desert plants, which need to close their stomata during the day to reduce water loss (*see* **crassulacean acid metabolism**). The anatomy of the C_4 plant leaf is adapted to suit this metabolic pathway (*see* **bundle sheath cells**).

cranial nerves Ten to twelve pairs of nerves in vertebrates that emerge directly from the brain. They supply the sense organs and muscles of the head, neck, and viscera. Examples of cranial nerves include the *optic nerve (II) and the *vagus nerve (X). With the *spinal nerves, the cranial nerves form an important part of the *peripheral nervous system.

cranial reflex *See* **reflex**.

craniates *See* **Chordata**.

cranium (brain case) The part of the vertebrate *skull that encloses and protects the brain. It is formed by the fusion of several flattened bones, which have immovable joints (sutures) between them.

crassulacean acid metabolism Photosynthesis by the $*C_4$ pathway in which carbon dioxide is taken up during the night, when the plant's stomata are open, and fixed into malic acid. During the day, when the stomata are closed, carbon dioxide is released from malic acid for use in the *Calvin cycle. This is important for plants that live in arid conditions as it enables them to keep their stomata closed during the day to reduce water loss from evaporation. Crassulacean acid metabolism is common in succulent plants of desert regions, many of which belong to the family Crassulaceae.

creatine A compound, synthesized from the amino acids arginine, glycine, and methionine, that occurs in muscle. In the form of *creatine phosphate* (or *phosphocreatine*), it is an important reserve of energy for muscle contraction, which is released when creatine phosphate loses its phosphate and is converted to *creatinine*, which is excreted in the urine (at a rate of 1.2–1.5 g/day in humans). *See also* **phosphagen**.

creatinine *See* creatine.

creationist A proponent of the theory of *special creation.

cremocarp A dry fruit that is a type of *schizocarp formed from two one-seeded carpels. The carpels remain separate and form indehiscent *mericarps* that are attached to a central supporting strand (*carpophore*) for some time before dispersal. It is characteristic of the Umbelliferae (carrot family).

crenation The shrinkage of cells that occurs when the surrounding solution is *hypertonic to the cellular cytoplasm. Water leaves the cells by *osmosis, which causes the plasma membrane to wrinkle and the cellular contents to condense.

Cretaceous The final geological period of the Mesozoic era. It extended from about 136 million years ago, following the Jurassic, to about 65 million years ago, when it was succeeded by the Palaeocene epoch. The name of the period is derived from *creta* (Latin: chalk) and the Cretaceous was characterized by the deposition of large amounts of *chalk in western Europe. The Cretaceous was the time of greatest flooding in the Mesozoic. Angiosperm plants made their first appearance on land and in the early Cretaceous Mesozoic reptiles reached their peak. At the end of the period there was a widespread extinction of the dinosaurs, flying reptiles, and ammonites, the cause of which may be related to environmental changes resulting from collisions of the earth with large meteorites (*see* **iridium anomaly**).

cretinism The condition that results from inadequate secretion of thyroid hormones during fetal life or early infancy. The brain and skeleton fail to develop properly, resulting in mental retardation and dwarfism.

crista 1. *See* **semicircular canals. 2.** *See* **mitochondrion**.

critical group A large group of related organisms that, although variations exist between them, cannot be divided into smaller groups of equivalent taxonomic rank to the parent group. Critical groups are found among plants that reproduce by *apomixis; for example, the 400 or so species of *Rubus* (brambles, etc.) are regarded as a critical group.

Cromagnon man The earliest form of modern man (*Homo sapiens*), which is believed to have appeared in Europe about 35 000 years ago and possibly earlier in Africa and Asia. Fossils indicate that these hominids were taller and more delicate than *Neanderthal man, which they replaced. They used intricately worked tools of stone and bone and left the famous cave drawings at Lascaux in the Dordogne. The name is derived from the site at Cromagnon, France, where the first fossils were found in 1868.

crop 1. A plant that is cultivated for the purpose of harvesting its seeds, roots, leaves, or other parts that are useful to humans. *See* **agriculture. 2.** An enlarged portion of the anterior section of the alimentary canal in some animals, in which food may be stored and/or undergo preliminary digestion. The term is most commonly applied to the thin-walled sac in birds between the oesophagus and the *proventriculus. In female pigeons the crop contains glands that secrete *crop milk*, used to feed nestlings.

crop rotation An agricultural practice in which different crops are cultivated in succession on the same area of land over a period of time so as to maintain soil fertility and reduce the adverse effects of pests. Legumes are important in the rotation as they are a source of nitrogen for the soil (*see* **nitrogen fixation; root nodule**). In the UK, other crops that may be included in a typical four-stage rotation are wheat, barley, and root crops. However, the use of pesticides enables the monoculture of crops in modern farming systems (*see* **agriculture**).

cross 1. A mating between two selected individuals. Controlled crosses are made for many reasons, e.g. to investigate the inheritance of a particular characteristic or to improve a livestock or crop variety. *See also* **back cross; reciprocal cross; test cross. 2.** An organism resulting from such a mating.

cross-fertilization *See* **fertilization**.

crossing over An exchange of portions of chromatids between *homologous chromosomes. As the chromosomes begin to move apart at the end of the first prophase of *meiosis, they remain in contact at a number of points (*see* **chiasma**). At these points the chromatids break and rejoin in such a way that sections are exchanged (see illustration). Crossing over thus alters the pattern of genes in the chromosomes. *See* **recombination**.

Crossing over at two chiasmata in a pair of homologous chromosomes

crossover value (COV) The percentage of linked genes (*see* **linkage**) that are exchanged during the process of *crossing over during the first prophase of *meiosis. The COV can be calculated by the percentage of offspring that show *recombination and is used to map the genes on a chromosome (*see* **chromosome map**). A small COV for a given pair of genes indicates that the genes are situated close together on the chromosome.

cross-pollination *See* **pollination**.

Crustacea A subphylum or superclass of arthropods containing over 35 000 species distributed worldwide, mainly in freshwater and marine habitats, where they constitute a major component of plankton. Crustaceans include shrimps, crabs, lobsters, etc. (*see* **Decapoda**) and the terrestrial woodlice, all of which belong to the class Malacostraca; the barnacles (class Cirripedia); the water fleas (*see* **Daphnia**), fairy shrimps, and tadpole shrimps (class Branchiopoda); and the copepods (*see* **Copepoda**). The segmented body usually has a distinct head (bearing *compound eyes, two pairs of *antennae, and various mouthparts), thorax, and abdomen, and is protected by a shell-like carapace. Each body segment may bear a pair of *biramous appendages used for locomotion, as gills, and for filtering food particles from the water. Appendages in the head region are modified to form jaws and in the abdominal region are often reduced or absent. Typically, the eggs hatch to produce a free-swimming *nauplius larva. This develops either by a series of moults or undergoes metamorphosis to the adult form.

cryobiology The study of the effects of very low temperatures on organisms, tissues, and cells. The ability of some animal tissues to remain viable in a frozen state enables them to be preserved by freezing for future use as *grafts.

cryophyte An organism that can live in ice and snow. Most cryophytes are algae, including the green alga *Chlamydomonas nivalis* and some diatoms, but they also include dinoflagellates, certain mosses, bacteria, and fungi.

cryptic coloration The type of colouring or marking of an animal that helps to camouflage it in its natural environment. It may enable the animal to blend with its background or, like the stripes of zebras and tigers, help to break up the outline of its body.

Cryptophyta A phylum of the Protoctista comprising unicellular organisms with flattened ovoid cells and two unequal flagella arising from an obliquely situated gullet (*crypt*). Cryptophytes (or cryptomonads) include freshwater and marine algae, which contain chlorophyll and *phycobiliprotein photosynthetic pigments; and heterotrophic protozoa, which may be carnivorous or parasitic (in animal intestines).

crypts of Lieberkühn (intestinal glands) Tubular glands that lie between the finger-like projections (*see* **villus**) of the inner surface of the small intestine. The cells of these glands (called *Paneth cells*) secrete *intestinal juice as they gradually migrate along the side of the crypt and the villus; they are eventually shed into the lumen of the intestine. The glands are named after German anatomist J. N. Lieberkühn (1711–56).

CSF 1. *See* **cerebrospinal fluid. 2.** *See* **colony-stimulating factor**.

ctenidia The gills of aquatic molluscs, which are present on both sides of the mantle cavity and are held in place by specialized membranes. The gills are involved in both filter feeding and the exchange of respiratory gases.

Ctenophora A phylum of marine invertebrates, closely related to the *Cnidaria, that contains the comb jellies (e.g. *Pleurobrachia*). They possess tentacles armed with *lasso cells, for catching prey, and many hundreds of thousands of cilia, which are fused at their bases and grouped together into longitudinal rows (*comb plates* or *ctenes*). The beating of the cilia enables these animals to swim among the plankton.

CT scanner (computerized tomography scanner) *See* **tomography**.

cuboidal epithelium *See* **epithelium**.

cultivar A plant that has been developed and maintained by cultivation as a result of agricultural or horticultural practices. The term is derived from *culti*vated *var*iety.

cultivation The planting and breeding of crop plants in *agriculture and horticulture. It involves the investigation of new means of increasing crop yield and quality.

culture A batch of cells, which can be microorganisms or of animal or plant origin, that are grown under specific conditions of nutrient levels, temperature, pH, oxygen levels, osmotic factors, light, pressure, and water content. Cultures of cells are prepared in the laboratory for a wide spectrum of scientific research. A *culture medium provides the appropriate conditions for growth. *See also* **batch culture; continuous culture; monolayer culture; organ culture; suspension culture; tissue culture**.

culture medium A nutrient material, either solid or liquid, used to support the growth and reproduction of microorganisms or to maintain tissue or organ cultures. *See also* **agar**.

cupula *See* **ampulla**.

cupule **1.** A hard or membranous cup-shaped structure formed from bracts and enclosing various fruits, such as the hazelnut and acorn. **2.** A structure in clubmosses (*Lycopodium* species) that protects the gemma (resting bud) during its development. It is composed of six leaflike structures. **3.** The bright red tissue around the seed of yew (*Taxus*), forming the yew 'berry'.

curare A resin obtained from the bark of South American trees of the genera *Strychnos* and *Chondrodendron* that causes paralysis of voluntary muscle. It acts by blocking the action of the neurotransmitter *acetylcholine at *neuromuscular junctions. Curare is used as an arrow poison by South American Indians and was formerly used as a muscle relaxant in surgery.

curd The solid component produced by the coagulation of milk during the manufacture of cheese. After being pasteurized, milk is cooled down and a culture of lactic acid bacteria is added to ferment the milk sugar, lactose, to lactic acid. The resulting decrease in pH causes casein, a milk protein, to coagulate, a process known as *curdling*. The solid curds are then separated from the liquid component, known as *whey*, and inoculated with different types of microbes to produce different cheeses.

cusp **1.** A sharp raised protuberance on the surface of a *molar tooth. The cusps of opposing molars (i.e. on opposite jaws) are complementary to each other, which increases the efficiency of grinding food during chewing. **2.** A flap forming part of a *valve.

cuticle **1.** *(in botany)* The continuous waxy layer that covers the aerial parts of a plant. Composed of *cutin, it is secreted by the *epidermis and its primary function is to prevent water loss. **2.** *(in zoology)* A layer of horny noncellular material covering, and secreted by, the epidermis of many invertebrates. It is usually made of a collagen-like protein or of *chitin and its main function is protection. In arthropods it is also strong enough to act as a skeleton (*see* **exoskeleton**) and in insects it reduces water loss. Growth is allowed by moulting of the cuticle (*see* **ecdysis**).

cuticularization The secretion by the outer (epidermal) layer of cells of plants and many invertebrates of substances that then harden to form a *cuticle.

cutin A mixture of waxy substances that impregnates the cell wall and covers the outer surface of mature epidermal plant cells, forming a *cuticle. The deposition of cutin (*cutinization*) reduces water loss by the plant and helps prevent the entry of pathogens. *See also* **suberin**.

cutinization The deposition of *cutin in plant cell walls, principally in the outermost layers of leaves and young stems.

cutis *See* **dermis**.

cutting A part of a plant, such as a bud, leaf, or a portion of a root or shoot, that, when detached from the plant and inserted in soil, can take root and give rise to a new daughter plant. Taking or striking cuttings is a horticultural method for propagating plants. *See also* **vegetative propagation**.

Cyanobacteria (blue-green bacteria) A phylum of photosynthetic bacteria formerly classified as algae (blue-green algae: division Cyanophyta). The colour is due to the presence of the photosynthetic pigment phycocyanin; the red pigment phycoerythrin is often also present. Cyanobacteria are unicellular but sometimes become joined in colonies or filaments by a sheath of mucilage. They occur in all aquatic habitats. A few species fix atmospheric nitrogen and thus contribute to soil fertility (*see* **heterocyst**; **nitrogen fixation**). Others exhibit symbiosis (*see* **lichens**).

Cycadophyta A phylum of seed plants (*see* **gymnosperm**) that contains many extinct species; the few modern representatives of the group include *Cycas* and *Zamia*. Cycads inhabit tropical and subtropical regions, sometimes growing to a height of 20 m. The stem bears a crown of fernlike leaves. These species are among the most primitive of living seed plants.

cyclamates Salts of the acid $C_6H_{11}.NH.SO_3H$, where $C_6H_{11}-$ is a cyclohexyl group. Sodium and calcium cyclamates were formerly used as sweetening agents in soft drinks, etc, until their use was banned when they were suspected of causing cancer.

cyclic AMP A derivative of *ATP that is widespread in animal cells as a *second messenger in many biochemical reactions induced by hormones. Upon reaching their target cells, the hormones activate *adenylate cyclase, the enzyme that catalyses cyclic AMP production. Cyclic AMP ultimately activates the enzymes of the reaction induced by the hormone concerned. Cyclic AMP is also involved in controlling gene expression and cell division, in immune responses, and in nervous transmission.

cyclic phosphorylation (cyclic photophosphorylation) *See* photophosphorylation; photosystems I and II.

Cyclostomata *See* **Agnatha**.

cyme *See* **cymose inflorescence**.

cymose inflorescence (cyme; definite inflorescence) A type of flowering shoot (*see* **inflorescence**) in which the first-formed flower develops from the growing region at the top of the flower stalk (see illustration). Thus no new flower buds can be produced at the tip and other flowers are produced from lateral buds beneath. In a *monochasial cyme* (or *monochasium*), the development of the flower at the tip is followed by a new flower axis growing from a single lateral bud. Subsequent new flowers may develop from the same side of the lateral shoots, as in the buttercup, or alternately on opposite sides, as in forget-me-not. In a *dichasial cyme* (or *dichasium*), the development of the flower at the apex is followed by two new flower axes developing from buds opposite one another, as in plants of the family Caryophyllaceae (such as stitchwort). *Compare* **racemose inflorescence**.

buttercup   forget-me-not

monochasial cymes

1 = oldest flower

dichasial cyme

Types of cymose inflorescence

cypsela A dry single-seeded fruit that does not split open during seed dispersal and is formed from a double ovary in which only one ovule develops into a seed. It is similar to an *achene and characteristic of members of the family Compositae, such as the dandelion. *See also* **pappus**.

cysteine *See* **amino acid**.

cysticercus *See* **bladderworm**.

cystine A molecule resulting from the oxidation reaction between the sulphydryl (–SH) groups of two cysteine molecules (*see* **amino acid**). This often occurs between adjacent cysteine residues in polypeptides. The resultant *disulphide bridges (–S–S–) are important in stabilizing the structure of protein molecules.

cystocarp *See* **carpogonium**.

cytidine A nucleoside comprising one cytosine molecule linked to a D-ribose sugar molecule. The derived nucleotides, cytidine mono-, di-, and triphosphate (CMP, CDP, and CTP respectively), participate in various biochemical reactions, notably in phospholipid synthesis.

cytochrome Any of a group of proteins, each with an iron-containing *haem group, that form part of the *electron transport chain in mitochondria and chloroplasts. Electrons are transferred by reversible changes in the iron atom between the reduced Fe(II) and oxidized Fe(III) states. *See also* **cytochrome oxidase**.

cytochrome oxidase An enzyme complex comprising the terminal two cytochromes (cytochromes a and a_3) of the respiratory chain in the mitochondria (*see* **electron transport chain**). It is responsible for the reduction of oxygen, receiving two electrons from cytochrome c (which precedes it in the chain) and combining them with two hydrogen ions and an oxygen atom to form water.

cytogenetics The study of inheritance in relation to the structure and function of cells. For example, the results of breeding experiments can be explained in terms of the behaviour of chromosomes during the formation of the reproductive cells.

cytokine Any soluble factor secreted by cells of the lymphoid system that acts as a signal to other lymphoid cells. There are two categories: *lymphokines, secreted by lymphocytes; and *monokines*, secreted by macrophages. However, certain cytokines, notably *interferons and *interleukins, are secreted by both lymphocytes and macrophages.

cytokinesis *See* **cell cycle; cell division**.

cytokinin (kinin) Any of a group of plant *growth substances chemically related to the purine adenine. Cytokinins stimulate cell division in the presence of *auxin and have also been found to delay senescence, overcome *apical dominance, and promote cell expansion. Zeatin is a naturally occurring cytokinin.

cytology The study of the structure and function of cells. The development of the light and electron microscopes has enabled the detailed structure of the nucleus (including the chromosomes) and other organelles to be elucidated. Microscopic examination of cells, either live or as stained sections on a slide, is also used in the detection and diagnosis of various diseases, especially *cancer.

cytolysis The breakdown of cells, usually as a result of destruction or dissolution of their outer membranes.

cytomegalovirus A virus belonging to the herpes group (*see* **herpesvirus**). In man it normally causes symptoms that are milder than the common cold, but it can produce more serious symptoms in those whose *immune response is disturbed (e.g. cancer patients). Infection in pregnant women may cause congenital handicap in their children.

cytoplasm The jelly-like material surrounding the nucleus of a *cell. It consists of a matrix (*see* **cytosol**) in which the cell's organelles are suspended. The cytoplasm may be differentiated into dense outer *ectoplasm*, which is concerned primarily with cell movement, and less dense *endoplasm*, which contains most of the cell's structures.

cytoplasmic inheritance The inheritance of genes contained in the cytoplasm of a cell, rather than the nucleus. Only a very small number of genes are inherited in this way. The phenomenon occurs because certain organelles, the *mitochondria and (in plants) the *chloroplasts, contain their own genes and can reproduce independently. The female reproductive cell (the egg) has a large amount of cytoplasm containing many such organelles, which are consequently incorporated into the cytoplasm of all the cells of the embryo. The male reproductive cells (sperm or pollen), however, consist almost solely of a nucleus. Cytoplasmic organelles are thus not inherited from the male parent. In plants, male sterility can be inherited via the cytoplasm. The inheritance of any such factors does not follow Mendelian laws.

cytoplasmic streaming The directional movement of cytoplasm in certain cells, which allows movement of substances through the cell. It has been observed most clearly in large cells, such as plant sieve elements and unicellular algae, in which simple diffusion is ineffective as a means of local transport in the cell. The exact mechanism of streaming is unknown but it is thought to involve protein filaments; these have been observed to extend throughout the length of sieve elements and are thought to be instrumental in the *translocation of the products of photosynthesis. In the cells of the green alga *Chara*, streams of cytoplasm follow helical pathways along both sides of the cell and across at either end. It is thought that actin and myosin filaments are involved in this process. A similar streaming of cytoplasm is responsible for *amoeboid movement.

cytosine A *pyrimidine derivative. It is one of the principal component bases of *nucleotides and the nucleic acids *DNA and *RNA.

cytoskeleton A network of fibres permeating the matrix of living cells that provides a supporting framework for organelles, anchors the cell membrane, facilitates cellular movement, and provides a suitable surface for chemical reactions to take place. The fibres are composed of *microtubules and *actin *microfilaments*.

cytosol The semifluid soluble part of the cytoplasm of cells, which contains the microfilaments of the *cytoskeleton. The cell's organelles are suspended in the cytosol.

cytostome A mouthlike structure of certain protozoa, through which particulate food is ingested. The cytostome is typically located at the base of an indentation in the cell.

cytotaxonomy *See* taxonomy.

cytotoxic Destructive to living cells. The term is applied particularly to a class of drugs that inhibit cell division and are therefore used to destroy cancer cells and to a group of *T cells that destroy virus-infected cells.

2,4-D 2,4-dichlorophenoxyacetic acid (2,4-dichlorophenoxyethanoic acid): a synthetic *auxin frequently used as a weedkiller of broad-leaved weeds. *See also* **pesticide.**

dance of the bees A celebrated example of *communication in animals, first investigated by Karl von Frisch (1886–1982). Honeybee workers on returning to the hive after a successful foraging expedition perform a 'dance' on the comb that contains coded information about the distance and direction of the food source. For example the *waggle dance*, characterized by tail-wagging movements, indicates the direction of a food source at a distance of more than 100 metres. Other workers, sensing vibrations from the dance, follow the instructions to find the food source.

Daphnia A genus of crustaceans belonging to the class Branchiopoda and order Cladocera (water fleas). *Daphnia* species have a transparent carapace and a protruding head with a pair of highly branched antennae for swimming and a single median compound eye. The five pairs of thoracic appendages form an efficient filter-feeding mechanism. Reproduction can take place without mating, i.e. by *parthenogenesis.

dark adaptation The changes that need to take place in the eye when an animal moves from a brightly lit environment to a relatively dark one to enable objects to be seen clearly. On moving to a darker environment, the pupils dilate and *rhodopsin – the pigment in the *rod cells that is broken down in bright light – is regenerated from its constituents.

dark period *(in botany)* The period considered to be critical in the responses of plants to changes in day length (*see* **photoperiodism**). It is believed that such responses, which include the onset of flowering, are determined by the length of the period of darkness that occurs between two periods of light.

dark reaction (light-independent reaction) *See* **photosynthesis.**

Darwinism The theory of *evolution proposed by Charles Darwin (1809–82) in *On the Origin of Species* (1859), which postulated that present-day species have evolved from simpler ancestral types by the process of *natural selection acting on the variability found within populations. *On the Origin of Species* caused a furore when it was first published because it suggested that species are not immutable nor were they specially created – a view directly opposed to the doctrine of *special creation. However the wealth of evidence presented by Darwin gradually convinced most people and the only major unresolved problem was to explain how the variations in populations arose and were maintained from one generation to the next.

This became clear with the rediscovery of Mendel's work on classical genetics in the 1900s and led to the present theory of *neo-Darwinism.

Darwin's finches (Galapagos finches) The 14 species of finch, unique to the Galapagos Islands, that Charles Darwin studied during his journey on HMS *Beagle*. Each is adapted to exploit a different food source. They are not found on the mainland because competition there for these food sources from other birds is fiercer. Darwin believed all the Galapagos finches to be descendants of a few that strayed from the mainland, and this provided important evidence for his theory of evolution. *See also* **adaptive radiation**.

dating techniques Methods of estimating the age of rocks, palaeontological specimens, archaeological sites, etc. *Relative dating techniques* date specimens in relation to one another; for example, stratigraphy is used to establish the succession of fossils. *Absolute* (or *chronometric*) *techniques* give an absolute estimate of the age and fall into two main groups. The first depends on the existence of something that develops at a seasonally varying rate, as in *dendrochronology and *varve dating. The other uses some measurable change that occurs at a known rate, as in *chemical dating, *radioactive* (or *radiometric*) *dating* (*see* **carbon dating; fission-track dating; potassium–argon dating; rubidium–strontium dating; uranium–lead dating**), and *thermoluminescence.

day-neutral plant A plant in which flowering can occur irrespective of the day length. Examples are cucumber and maize. *See* **photoperiodism**. *Compare* **long-day plant; short-day plant**.

DDT Dichlorodiphenyltrichloroethane; a colourless organic crystalline compound, $(ClC_6H_4)_2CH(CCl_3)$, made by the reaction of trichloromethanal with chlorobenzene. DDT is the best known of a number of chlorine-containing *pesticides used extensively in agriculture in the 1940s and 1950s. The compound is stable, accumulates in the soil, and concentrates in fatty tissue, reaching dangerous levels in carnivores high in the food chain. Restrictions are now placed on the use of DDT and similar pesticides.

deacetylation The removal of an acetyl group ($-COCH_3$) from a molecule. Deacetylation is an important reaction in several chemical pathways, including the *Krebs cycle.

deamination The removal of an amino group ($-NH_2$) from a compound. Enzymatic deamination occurs in the liver and is important in amino-acid metabolism, especially in their degradation and subsequent oxidation (*see also* **oxidative deamination**). The amino group is removed as ammonia and excreted, either unchanged or as urea or uric acid.

death The point at which the processes that maintain an organism alive no longer function. In humans it is diagnosed by permanent cessation of the heartbeat; however, the heart can continue beating after a large part of the brain ceases to function (*see* **brain death**). The death of a cell due to external damage or the action of toxic substances is known as *necrosis.

This must be distinguished from programmed cell death (*see* **apoptosis**), which is a normal part of the developmental process.

death phase *See* **bacterial growth curve**.

death rate (**mortality**) The rate at which a particular species or population dies, whatever the cause. The death rate is an important factor in controlling the size of a population. *Compare* **birth rate**.

deca- Symbol da. A prefix used in the metric system to denote ten times. For example, 10 hertz = 1 decahertz (daHz).

Decapoda An order of crustaceans of the class Malacostraca that are distributed worldwide, mainly in marine habitats. Decapods comprise swimming forms (shrimps and prawns) and crawling forms (crabs, lobsters, and crayfish). All are characterized by five pairs of walking legs, the first pair of which are highly modified in crawling forms to form powerful grasping pincers. The carapace is fused with the thorax and head forming a *cephalothorax. The antennae are especially long in shrimps and prawns, which also possess several pairs of well-developed swimming appendages (*pleopods*) posterior to the walking legs. Following fertilization by the male, females usually carry the eggs until they hatch. The larvae undergo several transformations before attaining adult form.

decarboxylation The removal of carbon dioxide from a molecule. Decarboxylation is an important reaction in many biochemical processes, such as the *Krebs cycle and the synthesis of *fatty acids. *See also* **oxidative decarboxylation**.

decay *See* **decomposition**.

deci- Symbol d. A prefix used in the metric system to denote one tenth. For example, 0.1 metre = 1 decimetre (dm).

decibel A unit used to compare two power levels, usually applied to sound or electrical signals. Although the decibel is one tenth of a *bel*, it is the decibel, not the bel, that is invariably used. Two power levels P and P_0 differ by n decibels when $n = 10\log_{10}P/P_0$. If P is the level of sound intensity to be measured, P_0 is a reference level, usually the intensity of a note of the same frequency at the threshold of audibility.

The logarithmic scale is convenient as human audibility has a range of 1 (just audible) to 10^{12} (just causing pain) and one decibel, representing an increase of some 26%, is about the smallest change the ear can detect.

deciduous Describing plants in which all the leaves are shed at the end of each growing season, usually the autumn in temperate regions or at the beginning of a dry season in the tropics. This seasonal leaf fall helps the plant retain water that would otherwise be lost by transpiration from the leaves. Examples of deciduous plants are rose and horse chestnut. *Compare* **evergreen**.

deciduous teeth (**milk teeth**) The first of two sets of teeth of a mammal.

These teeth are smaller than those that replace them (the *permanent teeth) and fewer in number, since there are no deciduous *molars. *See also* **diphyodont**.

decomposer An organism that obtains energy from the chemical breakdown of dead organisms or animal or plant wastes. Decomposers, most of which are bacteria and fungi, secrete enzymes onto dead matter and then absorb the breakdown products (*see* **saprotroph**). Many decomposers (e.g. nitrifying bacteria) are specialized to break down organic materials that are difficult for other organisms to digest. Decomposers fulfil a vital role in the *ecosystem, returning the constituents of organic matter to the environment in inorganic form so that they can again be assimilated by plants. *Compare* **detritivore**. *See also* **carbon cycle; nitrogen cycle**.

decomposition (decay) The chemical breakdown of organic matter into its constituents by the action of *decomposers.

defecation The expulsion of faeces from the rectum due to contractions of muscles in the rectal wall. A sphincter muscle, which is under voluntary control, is situated at the end of the rectum (the anus); relaxation of this muscle allows defecation to occur. In babies control of the anal sphincter muscle has not been developed and defecation occurs automatically as a reflex response to the presence of faeces in the rectum.

deficiency disease Any disease caused by an inadequate intake of an essential nutrient in the diet, primarily vitamins, minerals, and amino acids. Examples are *scurvy (lack of vitamin C), *rickets (lack of vitamin D), and iron-deficiency *anaemia. *See also* **mineral deficiency**.

definite inflorescence *See* **cymose inflorescence**.

deforestation The extensive cutting down of forests for the purpose of extracting timber or fuel wood or to clear the land for mining or agriculture. Forests are often situated in upland areas and are important in trapping rainwater. Deforestation in these areas, particularly in India and Bangladesh, has resulted in the flooding of low-lying plains; it has also led to an increase in soil erosion and hence desert formation (*see* **desertification**), resulting in crop loss and economic problems for local communities. The felling and burning of trees releases large amounts of carbon dioxide, thereby increasing global carbon dioxide levels and possibly contributing to the *greenhouse effect. Rainforests, particularly those of South America, are rich in both fauna and flora; their removal leads to an overall decrease in *biodiversity and the loss of plant species that have potentially beneficial pharmaceutical effects. Despite movements to reduce deforestation, economic pressures ensure that the process still continues.

degeneration 1. Changes in cells, tissues, or organs due to disease, etc., that result in an impairment or loss of function and possibly death and breakdown of the affected part. **2.** The reduction in size or complete loss of organs during evolution. The human appendix has undergone this process and performs no function in man. Degeneration of external organs may cause animals to appear to be more primitive than they really are; for example, early zoologists believed whales were fish rather than mammals because of the degeneration of their limbs. *See also* **vestigial organ**.

deglutition (swallowing) A reflex action initiated by the presence of food in the pharynx. During deglutition, the soft *palate is raised, which prevents food from entering the nasal cavity; the *epiglottis closes, which blocks the entrance to the windpipe; and the oesophagus starts to contract (*see* **peristalsis**), which ensures that food is conveyed to the stomach.

dehiscence The spontaneous and often violent opening of a fruit, seed pod, or anther to release and disperse the seeds or pollen. Examples are the splitting of laburnum pods and primrose capsules; such structures are described as *dehiscent* (*compare* **indehiscent**).

dehydrogenase Any enzyme that catalyses the removal of hydrogen atoms (*dehydrogenation) in biological reactions. Dehydrogenases occur in many biochemical pathways but are particularly important in driving the *electron-transport-chain reactions of cell respiration. They work in conjunction with the hydrogen-accepting coenzymes *NAD and *FAD.

dehydrogenation A chemical reaction in which hydrogen is removed from a compound. Dehydrogenation of organic compounds converts single carbon–carbon bonds into double bonds. In biological systems it is usually effected by *dehydrogenases.

deletion *(in genetics)* **1.** A *point mutation involving the removal of one or more base pairs in the DNA sequence. **2.** A frequently lethal *chromosome mutation that arises from an inequality in *crossing over during meiosis such that one of the chromatids loses more genetic information than it receives.

deme A group of organisms in the same *taxon. The term is used with various prefixes that denote how the group differs from other groups. For example, an *ecodeme* occurs in a particular ecological habitat, *cytodemes* differ from each other cytologically, and *genodemes* differ genetically.

denature To produce a structural change in a protein or nucleic acid that results in the reduction or loss of its biological properties. Denaturation involves unfolding of the polypeptide chains of proteins and of the double helix of nucleic acids, with loss of secondary and tertiary structure; it is caused by heat (*thermal denaturation*), chemicals, and extremes of pH. The differences between raw and boiled eggs are largely a result of denaturation. *Compare* **renaturation**.

dendrite Any of the slender branching processes that arise from the *dendrons of the cell body of a motor *neurone. A dendrite forms connections (*see* **synapse**) with the axons of other neurones and transmits nerve impulses from these to the cell body.

dendrochronology An absolute *dating technique using the *growth

rings of trees. It depends on the fact that trees in the same locality show a characteristic pattern of growth rings resulting from climatic conditions. Thus it is possible to assign a definite date for each growth ring in living trees, and to use the ring patterns to date fossil trees or specimens of wood (e.g. used for buildings or objects on archaeological sites) with lifespans that overlap those of living trees. The bristlecone pine (*Pinus aristata*), which lives for up to 5000 years, has been used to date specimens over 8000 years old. Fossil specimens accurately dated by dendrochronology have been used to make corrections to the *carbon-dating technique.
Dendrochronology is also helpful in studying past climatic conditions. Analysis of trace elements in sections of rings can also provide information on past atmospheric pollution.

dendrogram A diagram, similar to a family tree, that indicates some type of similarity between different organisms. Dendrograms can be based on *phenetic or *phylogenetic similarities; a *cladogram* shows similarities according to the system of *cladistics.

dendron Any of the major cytoplasmic processes that arise from the cell body of a motor neurone. A dendron usually branches into *dendrites.

denitrification A chemical process in which nitrates in the soil are reduced to molecular nitrogen, which is released into the atmosphere. This process is effected by denitrifying bacteria (e.g. *Pseudomonas denitrificans*), which use nitrates as a source of energy for other chemical reactions in a manner similar to respiration in other organisms. *Compare* **nitrification**. *See* **nitrogen cycle**.

de novo **pathway** Any metabolic pathway in which a *biomolecule is synthesized from simple precursor molecules. Nucleotide synthesis is an example.

density-dependent factor Any factor limiting the size of a population whose effect is dependent on the number of individuals in the population. For example, disease will have a greater effect in limiting the growth of a large population, since overcrowding facilitates its spread. *See also* **environmental resistance**.

density-independent factor Any factor limiting the size of a population whose effect is not dependent on the number of individuals in the population. An example of such a factor is an earthquake, which will kill all members of the population regardless of whether the population is small or large.

dental caries Tooth decay, which involves the destruction of the enamel layer of the tooth by acids produced by the action of bacteria on sugar. Bacteria can bind to teeth on *dextran*, a sticky substance derived from sucrose. The bacterial cells and other waste attached to dextran gives rise to *plaque. If dental caries is not treated it can spread to the dentine and pulp of the tooth, which leads to infection and death of the tooth.

2	1	2	3	man	2	0	3	3	rabbit	3	1	4	2	bear
2	1	2	3	(32 teeth)	1	0	2	3	(28 teeth)	3	1	4	3	(42 teeth)

Representative dental formulas

dental formula A representation of the dentition of an animal. A dental formula consists of eight numbers, four above and four below a horizontal line. The numbers represent (from left to right) the numbers of incisors, canines, premolars, and molars in either half of the upper and lower jaws. The total number of teeth in both jaws is therefore obtained by adding up all the numbers in the dental formula and multiplying by 2. Representative dental formulas are shown in the illustration. *See also* **permanent teeth**.

dentary A *membrane bone, present in the lower jaw of the vertebrates, that supports the teeth. In mammals the dentary is the sole bone of the lower jaw.

denticle (placoid scale) *See* scales.

dentine The bony material that forms the bulk of a *tooth. Dentine is similar in composition to bone but is perforated with many tiny canals for nerve fibres, blood capillaries, and processes of the dentine-forming cells (*odontoblasts). Ivory, the material that forms elephant tusks, is made of dentine.

dentition The type, number, and arrangement of teeth in a species. This can be represented concisely by a *dental formula. *See also* **permanent teeth; diphyodont; monophyodont; polyphyodont; heterodont; homodont.**

deoxyribonuclease *See* DNAase.

deoxyribonucleic acid *See* DNA.

deoxyribose (2-deoxyribose) A pentose (five-carbon) sugar, a derivative of *ribose, that is a component of the nucleotides (deoxyribonucleotides) that form the building blocks of *DNA.

depolarization A reduction in the difference of electrical potential that exists across the cell membrane of a nerve or muscle cell. Depolarization of a nerve-cell membrane occurs during the passage of an *action potential along the axon when the nerve is transmitting an impulse.

dermal bone *See* membrane bone.

Dermaptera An order of insects comprising the earwigs. Earwigs typically have long thin cylindrical bodies with biting mouthparts and a stout pair of curved forceps (*cerci*) at the tip of the abdomen, used for catching prey and in courtship. Some species have a single pair of wings, which at rest are folded back over the abdomen like a fan; others are wingless. Most earwigs are nocturnal and omnivorous.

dermis (corium; cutis) The thicker and innermost layer of the *skin of vertebrates, the other layer being the *epidermis. The dermis consists of fibrous connective tissue in which are embedded blood vessels, sensory nerve endings, and (in mammals) hair follicles, sebaceous glands, and sweat ducts. Beneath the dermis lies the *subcutaneous tissue.

descending tracts *See* **spinal cord.**

desert A major terrestrial *biome characterized by low rainfall. Hot deserts, such as the Sahara and Kalahari deserts of Africa, have a rainfall of less than 25 cm a year and extremely high daytime temperatures (up to 36°C). Vegetation is sparse, and desert plants are adapted to conserve water and take advantage of the rain when it falls. The perennials include xerophytic trees and shrubs (*see* **xerophyte**) and *succulents, such as cacti; many succulents show *crassulacean acid metabolism. Annual plants are *ephemerals, lying dormant as seeds for most of the year and completing their life cycle in the brief rainy periods. Desert animals are typically nocturnal or active at dawn and dusk, thus avoiding the extreme daytime temperatures.

desertification The gradual conversion of fertile land into desert, usually as a result of human activities. Loss of topsoil leads to further soil erosion until the land can no longer be used to grow crops or support livestock. A major factor contributing to desertification is bad management of farmland. Overgrazing of livestock removes the plant cover and exposes the soil, making it vulnerable to erosion. Overintensive cultivation of crop plants, especially monoculture (*see* **agriculture**), depletes the soil of nutrients and organic matter, resulting in loss of fertility and increasing its susceptibility to erosion. In many Third World countries it is difficult to control the process of desertification as the livelihood of the people often depends on practices that contribute to soil erosion. Another major cause of desertification is *deforestation.

desiccation A method of preserving organic material by the removal of its water content. Cells and tissues can be preserved by desiccation after lowering the samples to freezing temperatures; thereafter they can be stored at room temperature.

desiccator A container for drying substances or for keeping them free from moisture. Simple laboratory desiccators are glass vessels containing a drying agent, such as silica gel. They can be evacuated through a tap in the lid.

desmids Unicellular green algae of the class Desmidioideae. Like *Spirogyra*, they have an elaborate chloroplast. The cells of desmids are characteristically split into two halves joined by a narrow neck, each half being a mirror image of the other. The outer wall of the cell is patterned with various protuberances. Desmids are an important component of the plankton.

desmosome The band of material that holds adjacent eukaryote cells together, enabling groups of cells to function as a unit. Tissues that are subject to severe mechanical stress, such as skin epithelium and cardiac muscle, are rich in desmosomes. Desmosomes comprise one of several types of intercellular junction (*compare* **gap junction; tight junction**).

desorption The removal of adsorbed atoms, molecules, or ions from a surface.

desulphuration The removal of sulphur from a compound, which can occur by a variety of metabolic pathways. Desulphuration has been implicated in the toxicity of a number of sulphur-containing organic compounds: it is postulated that the atomic sulphur released into a cell by desulphuration, which is highly electrophilic, can bind to proteins and thereby alter their function.

determined Describing embryonic tissue at a stage when it can develop only as a certain kind of tissue (rather than as any kind).

detoxification (detoxication) The process by which harmful compounds, such as drugs and poisons, are converted to less toxic compounds in the body. Detoxification is an important function of the *liver. *See also* **phase I metabolism; phase II metabolism**.

detritivore An animal that feeds on *detritus. Examples of detritivores are earthworms, blowflies, maggots, and woodlice. Detritivores play an important role in the breakdown of organic matter from decomposing animals and plants (*see* **decomposer**).

detritus Particles of organic material derived from dead and decomposing organisms, resulting from the activities of the *decomposers. Detritus is the source of food for *detritivores, which can themselves be eaten by carnivores in a *detritus food chain*:

detritus → detritivore → carnivore.

detrusor muscle The smooth muscle of the *bladder wall, which is innervated with sympathetic fibres and contracts as a result of a reflex action when the bladder wall tension has reached a certain level.

deuteranopia *See* **colour blindness**.

Deuteromycota A phylum of fungi in which sexual reproduction is absent. Otherwise they are generally similar to the *Ascomycota and are usually regarded as ascomycetes that have lost the ability to produce ascospores. They are also known as Fungi Imperfecti ('imperfect fungi'). Examples of these fungi are the *Penicillium* moulds.

deutoplasm The nutritional material found in the *yolk of eggs.

development The complex process of growth and maturation that occurs in living organisms. Cell division and differentiation are important processes in development. In vertebrate animals there are three developmental stages: (1) *cleavage, in which the zygote divides to form a ball of cells, the *blastula; (2) gastrulation, in which the cells become arranged in three primary *germ layers (*see* **gastrula**); (3) *organogenesis* (or *organogeny*), in which further cell division and differentiation results in the formation of organs. The development of many invertebrates (e.g. insects) and amphibians involves the process of *metamorphosis. *See also* **morphogenesis; primary growth**.

Devonian A geological period in the Palaeozoic era that extended from the end of the Silurian (about 395 million years ago) to the beginning of the Carboniferous (about 345 million years ago). It was named by Adam Sedgwick (1785–1873) and Roderick Murchison (1792–1871) in 1839. The Devonian is divided into seven stages based on invertebrate fossil remains, such as corals, brachiopods, ammonoids, and crinoids, found in marine deposits. There were also extensive continental deposits consisting of conglomerates, red silts, and sandstones, forming the Old Red Sandstone facies. Fossils in the Old Red Sandstone include fishes and the earliest land plants. Graptolites became extinct early in the Devonian and the trilobites declined.

dextrin An intermediate polysaccharide compound resulting from the hydrolysis of starch to maltose by amylase enzymes.

dextrorotatory Denoting a chemical compound that rotates the plane of polarization of plane-polarized light to the right (clockwise as observed by someone facing the oncoming radiation). *See* **optical activity**.

dextrose *See* **glucose**.

***d*-form** *See* **optical activity**.

diabetes *See* **antidiuretic hormone; insulin**.

diakinesis The period at the end of the first prophase of *meiosis when the separation of *homologous chromosomes is almost complete and *crossing over has occurred.

dialysis A method by which large molecules (such as starch or protein) and small molecules (such as glucose or amino acids) in solution may be separated by selective diffusion through a semipermeable membrane. For example, if a mixed solution of starch and glucose is placed in a closed container made of a semipermeable substance (such as Cellophane), which is then immersed in a beaker of water, the smaller glucose molecules will pass through the membrane into the water while the starch molecules remain behind. The cell membranes of living organisms are *partially permeable, and dialysis takes place naturally in the kidneys for the excretion of nitrogenous waste. An artificial kidney (*dialyser*) utilizes the principle of dialysis by taking over the functions of diseased kidneys.

diapause A period of suspended development or growth occurring in some insects during which metabolism is greatly decreased. It is often triggered by seasonal changes and regulated by an inborn rhythm and enables the insect to survive unfavourable environmental conditions so that its offspring may be produced in more favourable ones.

diaphragm The muscular membrane that divides the thorax (chest) from the abdomen in mammals. It plays an essential role in breathing (*see also* **respiratory movement**), being depressed during *inspiration and raised during *expiration.

diaphysis The shaft of a mammalian limb bone, which in immature animals is separated from the ends of the bone (*see* **epiphysis**) by cartilage.

diastase *See* **amylase**.

diastema The gap that separates the biting teeth from the grinding teeth in herbivores. It creates a space in which food can be held in readiness for the grinding action of the teeth. This space is filled by large canine teeth in carnivores.

diastole The phase of a heart beat that occurs between two contractions of the heart, during which the heart muscles relax and the ventricles fill with blood. *Compare* **systole**. *See* **blood pressure**.

diatoms *See* **Bacillariophyta**.

dicarboxylic acid A *carboxylic acid having two carboxyl groups in its molecules. In systematic chemical nomenclature, dicarboxylic acids are denoted by the suffix *-dioic*; e.g. hexanedioic acid, $HOOC(CH_2)_4COOH$.

dichasium *See* **cymose inflorescence**.

2,4-dichlorophenoxyacetic acid *See* **2,4-D**.

dichogamy The condition in which the male and female reproductive organs of a flower mature at different times, thereby ensuring that self-fertilization does not occur. *Compare* **homogamy**. *See also* **protandry**; **protogyny**.

dichotomous 1. Describing the type of branching in plants that results when the growing point (apical bud) divides into two equal growing points, which in turn divide in a similar manner after a period of growth, and so on. Dichotomous branching is common is ferns and mosses. **2.** *See* **key**.

Dicotyledonae One of the two classes of plants within the *Angiospermophyta, distinguished by having two seed leaves (*cotyledons) within the seed. The dicotyledons usually have leaf veins in the form of a net, a ring of vascular bundles in the stem, and flower parts in fours or fives or multiples of these. Dicotyledons include many food plants (e.g. potatoes, peas, beans), ornamentals (e.g. roses, ivies, honeysuckles), and hardwood trees (e.g. oaks, limes, beeches). *Compare* **Monocotyledonae**.

Dictyoptera An order of insects (sometimes classified as *Orthoptera) comprising the cockroaches (suborder Blattaria) and the mantids (suborder Mantodea), occurring mainly in tropical regions. Cockroaches are oval and flattened in shape; some have a single well-developed pair of wings, folded back over the abdomen at rest, while in others the wings may be reduced or absent. They are usually found in forest litter, feeding on dead organic matter, but some species, e.g. the American cockroach (*Periplaneta americana*), are major household pests, scavenging on starchy foods, fruits, etc. In most species the females produce capsules (*oothecae*) containing 16–40 eggs. These are either deposited or carried by the female during incubation.

diet The food requirements of an organism. The foods that constitute the human diet should contain vitamins, mineral salts (*see* **essential element**), and dietary *fibre as well as water, carbohydrates and fats (which provide energy), and proteins (required for growth and maintenance). A balanced diet contains of the correct proportions of these *nutrients, which will vary depending on the age, sex, body size, and the level of activity of the individual. An inadequate supply of different food types in the diet can lead to *malnutrition.

dietary fibre *See* fibre.

differentiation The changes from simple to more complex forms undergone by developing tissues and organs so that they become specialized for particular functions. Differentiation occurs during embryonic *development, *regeneration, and (in plants) meristematic activity (*see* **meristem**).

diffusion (passive transport) The random movement of particles (e.g. molecules or ions) from an area of high concentration to an area of low concentration until an even distribution of particles (i.e. uniform concentration) is obtained. Small molecules and ions (such as oxygen and Na^+) can diffuse across a plasma membrane.

diffusion gradient *See* **concentration gradient**.

digestion The breakdown by a living organism of ingested food material into chemically simpler forms that can be readily absorbed and assimilated by the body. This process requires the action of digestive enzymes and may take place extracellularly (i.e. in the *alimentary canal), as is the case in most animals; or intracellularly (e.g. by engulfing phagocytic cells), as occurs in protozoans and coelenterates.

digestive system The system of organs that are involved in the process of *digestion. The digestive system of mammals is divided into the *gastrointestinal tract* (*see* **alimentary canal**) and accessory structures, such as teeth, tongue, liver, pancreas, and gall bladder.

digit A finger or toe. In the basic limb structure of terrestrial vertebrates there are five digits (*see* **pentadactyl limb**). This number is retained in man and other primates, but in some other species the number of digits is reduced. Frogs, for example, have four fingers and five toes, and in ungulate (hooved) mammals, the digits are reduced and their tips are enclosed in horn, forming hooves.

digitalis A preparation of the dried leaves or seeds of the foxglove (*Digitalis*), used historically as a heart stimulant. Modern clinically prescribed drugs derived from digitalis include *digoxin* and *digitoxin*, both of which belong to a class of drugs known as the cardiac *glycosides. They are used to treat heart failure and some forms of *arrhythmia because of their ability to increase the force of contraction of the heart muscle. Their toxic effects arise from their capacity to disturb the normal rhythm of the heart.

digitigrade Describing the gait of most fast-running animals, such as dogs and cats, in which only the toes are on the ground and the rest of the foot is raised off the ground. *Compare* **plantigrade; unguligrade**.

dihybrid cross A genetic cross between parents that differ in two characteristics, controlled by genes at different loci. Mendel performed a dihybrid cross using pea plants and the characteristics of seed colour and texture: the parental plants had either smooth yellow seeds (*SSYY*) – the dominant characteristics – or wrinkled green seeds (*ssyy*) – the recessive characteristics. All the offspring had smooth yellow seeds, being

Dihybrid cross

heterozygous (*SsYy*) for the two alleles. Crossing between these offspring produced an F₂ generation of plants with smooth yellow, smooth green, wrinkled yellow, and wrinkled green seeds in the ratio 9:3:3:1 (see illustration). Mendel used these results as the basis for his Law of Independent Assortment (*see* **Mendel's laws**). *Compare* **monohybrid**.

dikaryon A cell of a fungal hypha or mycelium containing two haploid nuclei of different strains. The nuclei associate in pairs but do not fuse, therefore the cell is not truly diploid. Dikaryosis occurs in the Basidiomycota and Ascomycota. *See* **heterokaryosis**.

dikaryosis The most common form of *heterokaryosis. *See also* **dikaryon**.

dilation *See* **vasodilation**.

dimethylbenzenes (xylenes) Three compounds with the formula $(CH_3)_2C_6H_4$, each having two methyl groups substituted on the benzene ring. 1,2-dimethylbenzene is ortho-xylene, etc. A mixture of the isomers is obtained from petroleum and is used as a clearing agent in preparing specimens for optical microscopy.

dimictic lake A lake that is stratified by a *thermocline that is not permanent but is turned over twice during one year. The thermocline is disrupted due to seasonal changes in the climate. A *meromictic lake* is one in which there is a permanent stratification.

dimorphism The existence of two distinctly different types of individual within a species. An obvious example is *sexual dimorphism* in certain animals, in which the two sexes differ in colouring, size, etc. Dimorphism also occurs in some lower plants, such as mosses and ferns, that show an *alternation of generations.

Dinoflagellata A phylum of mostly single-celled organisms, the dinoflagellates, now classified in the kingdom Protoctista. They are abundant in the marine plankton; many are *photoautotrophs, containing brown xanthophyll pigments in addition to chlorophyll. Dinoflagellates characteristically have two flagella for locomotion and most have a rigid cell wall of cellulose encrusted with silica. Some species (e.g. *Noctiluca miliaris*) are bioluminescent.

dinosaur An extinct terrestrial reptile belonging to a group that constituted the dominant land animals of the Jurassic and Cretaceous periods, 190–65 million years ago. There were two orders. The Ornithischia were typically quadrupedal herbivores, many with heavily armoured bodies, and included *Stegosaurus*, *Triceratops*, and *Iguanodon*. They were all characterized by birdlike pelvic girdles. The Saurischia included many bipedal carnivorous forms, such as *Tyrannosaurus* (the largest known carnivore), and some quadrupedal herbivorous forms, such as *Apatosaurus* (*Brontosaurus*) and *Diplodocus*. They all had lizard-like pelvic girdles. Many of the herbivorous dinosaurs were amphibious or semiaquatic.

dinucleotide A compound consisting of two *nucleotides.

dioecious Describing plant species that have male and female flowers on separate plants. Examples of dioecious plants are willows. *Compare* **monoecious**.

dipeptide A compound consisting of two amino acid units joined at the amino ($-NH_2$) end of one and the carboxyl ($-COOH$) end of the other. This peptide bond (*see* **peptide**) is formed by a condensation reaction that involves the removal of one molecule of water.

diphyodont Describing a type of dentition that is characterized by two successive sets of teeth: the *deciduous (milk) teeth, which are followed by the *permanent (adult) teeth. Mammals have a diphyodont dentition. *Compare* **monophyodont; polyphyodont**.

diploblastic Describing an animal with a body wall composed only of two layers, *ectoderm and *endoderm, sometimes with a noncellular *mesoglea between them. Coelenterates are diploblastic; all other Metazoa are *triploblastic.

diploid Describing a nucleus, cell, or organism with twice the *haploid number of chromosomes characteristic of the species. The diploid number is designated as 2*n*. Two sets of chromosomes are present, one set being derived from the female parent and the other from the male. In animals, all the cells except the reproductive cells are diploid.

diplont An organism that is at the *diploid stage of its life cycle. *Compare* **haplont**.

Diplopoda A class of wormlike terrestrial arthropods comprising the millipedes, characterized by a distinct head, bearing a single pair of short antennae, and 20 to over 60 body segments each bearing two pairs of legs. Restricted to damp habitats, millipedes are slow moving and feed on decaying leaves. *See also* **Uniramia**.

diplotene The period in the first prophase of *meiosis when paired *homologous chromosomes begin to move apart. They remain attached at a number of points (*see* **chiasma**).

Dipnoi A subclass or order of bony fishes that contains the lungfishes, which have lungs and breathe air. They are found in Africa, Australia, and South America, where they live in freshwater lakes and marshes that tend to become stagnant or even dry up in summer. They survive in these conditions by burrowing into the mud, leaving a small hole for breathing air, and entering a state of *aestivation, in which they can remain for six months or more. The Dipnoi date from the Devonian era (395–345 million years ago) and share many features with the modern *Amphibia.

Diptera An order of insects comprising the true, or two-winged, flies. Flies possess only one pair of wings – the forewings; the hindwings are modified to form small clublike *halteres* that function as balancing organs. Typically fluid feeders, flies have mouthparts adapted for piercing and sucking or for lapping; the diet includes nectar, sap, decaying organic matter, and blood. Some species prey on insects; others are parasitic. Dipteran larvae (*maggots*) are typically wormlike with an inconspicuous head. They undergo metamorphosis via a pupal stage to the adult form. Many flies or their larvae are serious pests, either by feeding on crops (e.g. fruit flies) or as vectors of disease organisms (e.g. the house fly (*Musca domestica*) and certain mosquitoes).

directional selection *Natural selection that favours the establishment of one particular advantageous mutation within a population, resulting in a change in *phenotype in that direction. An example of directional selection is the increase in darker forms of the peppered moth (*Biston betularia*) that occurred in industrial areas, where the moths with darker wing coloration are better camouflaged than those with lighter wings against polluted tree trunks (*see* **industrial melanism**). *Compare* **disruptive selection; stabilizing selection**.

disaccharide A sugar consisting of two linked *monosaccharide molecules. For example, sucrose comprises one glucose molecule and one fructose molecule bonded together.

discontinuous variation (qualitative variation) Clearly defined differences in a characteristic that can be observed in a population. Characteristics that are determined by different *alleles at a single locus show discontinuous variation, e.g. garden peas are either wrinkled or smooth. *Compare* **continuous variation**.

disease A condition in which the normal function of some part of the body (cells, tissues, or organs) is disturbed. A variety of microorganisms and environmental agents are capable of causing disease. The functional disturbances are often accompanied by structural changes in tissue.

disinfectant Any substance that kills or inhibits the growth of disease-producing microorganisms and is in general toxic to human tissues. Disinfectants include cresol, bleaching powder, and phenol. They are used to cleanse surgical apparatus, sick-rooms, and household drains and if sufficiently diluted can be used as *antiseptics.

disinhibition *(in animal behaviour)* The tendency to exhibit displacement behaviour at a point of equilibrium in situations where there are two or more conflicting drives. *See* **displacement activity**.

dispersal The dissemination of offspring of plants or sessile animals. Dispersal provides organisms that are not mobile with a better chance of survival by reducing *competition among offspring and parents. It also promotes the colonization of new habitats. Flowering plants produce fruits or seeds that are dispersed by such agents as wind, water, or animals. Specialized structures have evolved in many species to aid dispersal (*see* **fruit**).

dispersive replication A form of *DNA replication in which the original DNA chain breaks and recombines in a random fashion before the double helix structure unwinds and separates to act as a template for messenger *RNA synthesis. *Compare* **conservative replication; semiconservative replication**.

displacement activity An activity shown by an animal that appears to be irrelevant to its situation. Displacement activities are frequently observed when there is conflict between opposing drives. For example, birds in aggressive situations, in which there are simultaneous drives to attack and to flee, may preen their feathers as a displacement activity.

display behaviour Stereotyped movement or posture that serves to influence the behaviour of another animal. Many displays in *courtship and *aggression are conspicuous and characteristic of the species; special markings or parts of the body may be prominently exhibited (for example, the male peacock spreads its tail in courtship). Other displays are cryptic and make it harder for a predator to recognize the displaying animal as potential prey. For example, geometer moth caterpillars, which look like twigs, hold themselves on plant stems with one end sticking into the air.

disruptive selection *Natural selection that favours the extremes of a phenotype in a population. It often operates when an environmental factor shows distinct variations, for example high temperatures in summer and low temperatures in winter, with no intermediate forms. In this case the population will be variously adapted to withstand both high and low temperatures.

distal Describing the part of an organ that is farthest from the organ's point of attachment to the rest of the body. For example, hands and feet are at the distal ends of arms and legs, respectively. *Compare* **proximal**.

distal convoluted tubule (second convoluted tubule) The part of a *nephron that leads from the thick ascending limb of the *loop of Henle and drains into a *collecting duct. The main function of the distal tubule is to absorb sodium chloride and other inorganic salts while retaining water.

disulphide bridge (sulphur bridge) A covalent bond (S–S) formed between the thiol groups (–SH) of two cysteine residues, usually in the polypeptide chains of proteins. Easily hydrolysed and prone to rearrangement, these bonds contribute to the tertiary structure of *proteins.

diuresis The production of watery urine in large quantities. Diuresis causes the osmotic pressure of the blood to rise and is counteracted by the *antidiuretic hormone.

diurnal Daily; denoting an event that happens once every 24 hours.

diurnal rhythm *See* **circadian rhythm**.

divergent evolution *See* **adaptive radiation**.

diverticulum A saclike or tubular outgrowth from a tubular or hollow internal organ. Diverticula may occur as normal structures (e.g. the *caecum and *appendix in the alimentary canal) or abnormally, from a weakened area of the organ.

division A category used traditionally in the *classification of plants that consists of one or several similar classes. An example is Spermatophyta (seed-bearing plants). In modern classification systems the *phylum has replaced the division.

dizygotic twins *See* fraternal twins.

***dl*-form** *See* optical activity.

DNA (deoxyribonucleic acid) The genetic material of most living organisms, which is a major constituent of the *chromosomes within the cell nucleus and plays a central role in the determination of hereditary characteristics

Detail of molecular structure of sugar–phosphate backbone. Each deoxyribose unit is attached to a phosphate group and a base, forming a nucleotide

Double helical structure of DNA

The four bases of DNA, showing the hydrogen bonding between base pairs

Molecular structure of DNA

by controlling *protein synthesis in cells (*see also* **genetic code**). It is also found in organelles other than the nucleus, such as the mitochondria (*see* **extranuclear genes; mitochondrial DNA**). DNA is a nucleic acid composed of two chains of *nucleotides in which the sugar is deoxyribose and the bases are *adenine, *cytosine, *guanine, and *thymine (*compare* **RNA**). The two chains are wound round each other and linked together by hydrogen bonds between specific complementary bases (*see* **base pairing**) to form a spiral ladder-shaped molecule (*double helix*). See illustration.

When the cell divides, its DNA also replicates in such a way that each of the two daughter molecules is identical to the parent molecule (*see* **DNA replication**). *See also* **complementary DNA**.

DNAase (DNase; deoxyribonuclease) An enzyme that catalyses the cleavage of DNA. DNAase I is a digestive enzyme, secreted by the pancreas, that degrades DNA into shorter nucleotide fragments. Many other *endonucleases and *exonucleases cleave DNA, including the *restriction enzymes and enzymes involved in DNA repair and replication.

DNA-binding proteins Proteins that are able to bind to DNA in both eukaryotes and prokaryotes and act as either activators or repressors of *gene expression by controlling the binding of RNA polymerase to DNA during the process of transcription (*see* **transcription factor**). They are also involved in *DNA replication: by binding to the nucleotides in the single DNA strands of the template that have unwound, they stabilize these strands so that they do not rewind. DNA-binding proteins do not include the *histones.

DNA cloning *See* **gene cloning**.

DNA-dependent RNA polymerase *See* **polymerase**.

DNA filter assay A method used to determine the presence of *recombinant DNA in cloned cells. This is necessary as a culture will contain many cells that have not been transformed. After the culture has been plated on Petri dishes a copy of the cell pattern is made using a special filter and the cells are then lysed. DNA from the cells binds to the filter as it is heated in a pattern equivalent to that of the cells on the Petri dish. The presence of recombinant DNA is revealed by the use of a *gene probe and *autoradiography.

DNA fingerprinting *See* **genetic fingerprinting**.

DNA hybridization A method of determining the similarity of DNA from different sources. Single strands of DNA from two sources, e.g. different bacterial species, are put together and the extent to which double hybrid strands are formed is estimated. The greater the tendency to form these hybrid molecules, the greater the extent of complementary base sequences, i.e. gene similarity. The method is one way of determining the genetic relationships of species.

DNA ligase An enzyme that is able to join together two portions of DNA and therefore plays an important role in *DNA repair. DNA ligase is also used in recombinant DNA technology (*see* **genetic engineering**) as it ensures that the foreign DNA (e.g. the complementary DNA used in *gene cloning) is bound to the plasmid into which it is incorporated.

DNA polymerase *See* polymerase.

DNA probe *See* gene probe.

DNA repair A variety of mechanisms that help to ensure that the genetic sequence, as expressed in the DNA, is maintained and that errors that occur during *DNA replication, by mutation, are not allowed to accumulate. An error in the genetic sequence could cause cell death by interfering with the replication process. The mechanisms work because DNA is made up of two complementary strands. A damaged section of a strand, or a damaged nucleotide, can be removed by enzymes and replaced by the correct form by DNA *polymerases. The phosphodiester backbone is then sealed by *DNA ligase. Another type of repair mechanism involves *DNA glycosylases*, which are able to detect and remove altered bases on the nucleotides.

DNA replication The process whereby DNA makes exact copies of itself, which is controlled by the enzyme DNA *polymerase. Replication occurs at rates of between 50 nucleotides per second (in mammals) and 500 nucleotides per second (in bacteria). The hydrogen bonds between the complementary bases on the two strands of the parent DNA molecule break and the strands unwind, each strand acting as a template for the synthesis of a new one complementary to itself (*see* **DNA-binding proteins**). DNA polymerases move down the two single strands linking free nucleotides to their complementary bases (*see* **base pairing**) on the templates. The process continues until all the nucleotides on the templates have joined with appropriate free nucleotides and two identical molecules of DNA have been formed. This process is known as *semiconservative replication* as each new molecule contains half of the original parent DNA molecule (*compare* **conservative replication; dispersive replication**). Sometimes mutations occur that may cause the exact sequence of the parent DNA not to be replicated. However, *DNA repair mechanisms reduce this possibility.

DNA sequencing *See* gene sequencing.

dodecanoic acid *See* lauric acid.

domain A functional unit of the tertiary structure of a *protein. It consists of chains of amino acids folded into *alpha helices and *beta sheets to form a globular structure. Different domains are linked together by relatively straight sections of polypeptide chain to form the protein molecule. Domains allow a degree of movement in the protein structure. *See also* finger domain.

dominance hierarchy *See* dominant.

dominant 1. *(in genetics)* Describing the *allele that is expressed in the *phenotype when two different alleles of a gene are present in the cells of an organism. For example, the height of garden peas is controlled by two alleles, 'tall' (T) and 'dwarf' (t). When both are present (Tt), i.e. when the cells are *heterozygous, the plant is tall since T is dominant and t is *recessive. *See also* **incomplete dominance. 2. (in ecology)** Describing the most conspicuously abundant and characteristic species in a *community. The term is usually used of a plant species in plant ecology; for example, pine trees in a pine forest. **3. (in animal behaviour)** Describing an animal that is allowed priority in access to food, mates, etc., by others of its species because of its success in previous aggressive encounters. Less dominant animals frequently show *appeasement behaviour towards a more dominant individual, so overt *aggression is minimized. In a stable group there may be a linear *dominance hierarchy* or *peck order* (so called because it was first observed in domestic fowl), with each animal being subservient to those above it in the hierarchy and taking precedence over those below it.

donor An individual whose tissues or organs are transferred to another (the *recipient*). Donors may provide blood for transfusion or a kidney or heart for transplantation.

dopa (dihydroxyphenylalanine) A derivative of the amino acid tyrosine. It is found in particularly high levels in the adrenal glands and is a precursor in the synthesis of *dopamine, *noradrenaline, and *adrenaline. The laevorotatory form, *L-dopa*, is administered in the treatment of Parkinson's disease, in which brain levels of dopamine are reduced.

dopamine A *catecholamine that is a precursor in the synthesis of *noradrenaline and *adrenaline. It also functions as a neurotransmitter, especially in the brain.

dormancy An inactive period in the life of an animal or plant during which growth slows or completely ceases. Physiological changes associated with dormancy help the organism survive adverse environmental conditions. Annual plants survive the winter as dormant seeds while many perennial plants survive as dormant tubers, rhizomes, or bulbs. *Hibernation and *aestivation in animals help them survive extremes of cold and heat, respectively.

dorsal Describing the surface of a plant or animal that is farthest from the ground or other support, i.e. the upper surface. In vertebrates, the dorsal surface is that down which the backbone runs. Thus in upright (bipedal) mammals, such as man and kangaroos, it is the backward-directed (*posterior) surface. *Compare* **ventral.**

dorsal aorta The artery in vertebrate embryos that transports blood from the *aortic arches to the trunk and limbs. In adult fish it is a major artery that carries oxygenated blood from the efferent branchial arteries to branches that supply the body organs; in adult tetrapods it arises from the *systemic arch (*see* **aorta**). *Compare* **ventral aorta.**

dorsal root The part of a spinal nerve that enters the *spinal cord on the dorsal side and contains only sensory fibres. The cell bodies of these fibres form the *dorsal root ganglion* (*see* **ganglion**), a swelling in the root that lies just outside the cord. *Compare* **ventral root**.

dose A measure of the extent to which matter has been exposed to *ionizing radiation. The *absorbed dose* is the energy per unit mass absorbed by matter as a result of such exposure. The SI unit is the gray, although it is often measured in rads (1 rad = 0.01 gray; *see* **radiation units**). The *maximum permissible dose* is the recommended upper limit of absorbed dose that a person or organ should receive in a specified period according to the International Commission on Radiological Protection.

dot-blot A method for detecting a specific sequence of nucleotides in a DNA or RNA molecule. The nucleic acid sample is adsorbed onto a nitrocellulose filter and an appropriate *gene probe, which matches the specific sequence being investigated, is added. After a period of incubation any excess probe is washed off and the nucleotide sequence under investigation can be detected by *autoradiography.

double circulation The type of circulatory system that occurs in mammals, in which the blood passes through the heart twice before completing a full circuit of the body (see illustration). Blood is pumped from the heart to the lungs and returns to the heart before being distributed to the other organs and tissues of the body. The heart is divided into two separate compartments to prevent oxygenated blood returning from the lungs from mixing with deoxygenated blood from the other parts of the body. *See also* **pulmonary circulation; systemic circulation**. *Compare* **single circulation**.

double fertilization A process, unique to flowering plants, in which two male gamete nuclei, which have travelled down the pollen tube, separately fuse with different female nuclei in the *embryo sac. The first male nucleus fuses with the egg cell to form the zygote; the second male nucleus fuses with the two *polar nuclei to form a triploid nucleus that develops into the endosperm.

double helix *See* DNA.

double recessive An organism with two *recessive alleles for a particular characteristic.

down feathers (plumules) Small soft feathers that cover and insulate the whole body of a bird. In nestlings they are the only feathers; in adults they lie between and beneath the *contour feathers. Down feathers have a fluffy appearance as their *barbs are not joined together to form a smooth vane.

Down's syndrome A congenital form of mental retardation due to a chromosome defect in which there are three copies of chromosome no. 21 instead of the usual two (*see* **trisomy**). The affected individual has a short broad face and slanted eyes (as in the Mongolian races), short fingers, and weak muscles. Down's syndrome can be detected before birth by *amniocentesis. It is named after the British physician John Down (1828–96), who first studied the incidence of the disorder.

dragonflies *See* **Odonata**.

drive The *motivation that results in an animal performing a particular activity. There are two types: *primary drive*, which arises as a direct result of tissue requirements (e.g. the need for food or water); and *secondary drive*, resulting from learned behaviour.

drone A fertile male in a colony of social bees, especially the honeybee (*Apis mellifera*). The drones die after mating with the queen bee as the male reproductive organs explode within the female.

Drosophila A genus of fruit flies often used in genetic research because the larvae possess giant chromosomes in their salivary glands (*see* **polyteny**). These chromosomes have conspicuous transverse bands, which can be studied microscopically to reveal chromosome mutations and gene activity. Fruit flies have a short life cycle and produce a large number of offspring, which also makes them a good model animal for genetic research.

□ = oxygenated blood   ▨ = deoxygenated blood   ──▶ = direction of blood flow

Double circulation in a mammal

drug Any chemical substance that alters the physiological state of a living organism. Drugs are widely used in medicine for the prevention, diagnosis, and treatment of diseases; they include *analgesics, *antibiotics, anaesthetics, *antihistamines, and *anticoagulants. Some drugs are taken solely for the pleasurable effects they induce; these include *narcotics; stimulants, such as cocaine and *amphetamine; *hallucinogens, such as LSD; and some tranquillizers. Many of these drugs are habit-forming and their use is illegal.

drupe (pyrenocarp) A fleshy fruit that develops from either one or several fused carpels and contains one or many seeds. The seeds are enclosed by the hard protective endocarp (*see* **pericarp**) of the fruit. Thus the stone of a peach is the endocarp containing the seed. Plums, cherries, coconuts, and almonds are other examples of one-seeded drupes; holly and elder fruits are examples of many-seeded drupes. *See also* **etaerio**.

dry mass The mass of a biological sample after the water content has been removed, usually by placing the sample in an oven. The dry mass is used as a measure of the *biomass of a sample.

Dryopithecus A genus of extinct apes, fossils of which have been found in India and dated as 30–40 million years old. Fossils of *Dryopithecus* and of the similar genus *Proconsul* are often referred to as *dryopithecines*. Dryopithecines are believed to be either the ancestors of the fossil hominids or ancestral to both the human and ape lines.

duct A tube or passage in an organism that is involved in the secretion or excretion of substances (*see* **gland**).

ductless gland *See* **endocrine gland**.

ductus arteriosus A channel that connects the pulmonary artery with the aorta in the mammalian fetus and therefore allows blood to bypass the inactive lungs of the fetus. Derived from the fifth *aortic arch, it normally closes soon after birth.

duodenum The first section of the *small intestine of vertebrates. It is the site where food from the stomach is subjected to the action of bile (from the bile duct) and pancreatic enzymes (from the pancreatic duct) as well as the enzymes secreted by digestive glands in the duodenum itself (*see* **succus entericus**), which are required in the breakdown of proteins, carbohydrates, and fats. By neutralizing the acidic secretions of the stomach, the duodenum provides an alkaline environment necessary for the action of the intestinal enzymes.

duplex Describing a biological molecule comprising two cross-linked polymeric chains; the term is applied particularly to the double-stranded structure of *DNA.

duplication (*in genetics*) The doubling or repetition of part of a chromosome, which generally originates during the *crossing over phase of meiosis. Occasionally this type of *chromosome mutation may have beneficial effects on a population. For example, a beneficial duplication resulted in the evolution of four types of haemoglobin in man and apes from a single form. One of these types of haemoglobin (gamma or fetal haemoglobin) has a greater affinity for oxygen and maximizes fetal uptake of oxygen from the mother's blood.

dura mater The outermost and toughest of the three membranes (*meninges) that surround the central nervous system in vertebrates. It lies adjacent to the skull and its purpose is to protect the delicate inner meninges (the *arachnoid membrane and the *pia mater).

duramen *See* heartwood.

dwarfism *See* growth hormone.

dynein A protein, found in many eukaryotic cells, that possesses *ATPase activity and binds to microtubules. Dynein is associated with the microtubules of cilia and flagella (*see* **axoneme**); it moves along the microtubules, using energy derived from the hydrolysis of ATP, causing the cilium to bend. Dynein in the cytoplasm of cells causes the movement of organelles along the microtubules.

dystrophic Describing a body of water, such as a lake, that contains large amounts of undecomposed organic matter derived from terrestrial plants. Dystrophic lakes are poor in dissolved nutrients and therefore unproductive; they are common in peat areas and may develop into peat bogs. *Compare* **eutrophic; mesotrophic; oligotrophic.**

ear The sense organ in vertebrates that is specialized for the detection of sound and the maintenance of balance. It can be divided into the *outer ear and *middle ear, which collect and transmit sound waves, and the *inner ear, which contains the organs of balance and (except in fish) hearing (see illustration). The term ear is often used for the *pinna of the mammalian outer ear.

Structure of the mammalian ear

eardrum *See* **tympanum**.

ear ossicles Three small bones – the *incus* (anvil), *malleus* (hammer), and *stapes* (stirrup) – that lie in the mammalian *middle ear, forming a bridge between the tympanum (eardrum) and the *oval window. The function of the ossicles is to transmit (and amplify) vibrations of the tympanum across the middle ear to the oval window, which transfers them to the *inner ear. Muscles of the middle ear constrict the movement of the ossicles. This serves to safeguard the ear from damage caused by excessively loud noise.

earwigs *See* **Dermaptera**.

eccrine secretion *See* **secretion**.

ecdysis (moulting) **1.** The periodic loss of the outer cuticle of arthropods. It starts with the reabsorption of some materials in the inner part of the old cuticle and the formation of a new soft cuticle. The remains of the old cuticle then split; the animal emerges and absorbs water or swallows air and increases in size while the new cuticle is still soft. This cuticle is then hardened with chitin and lime salts. In insects and crustaceans ecdysis is controlled by the hormone *ecdysone. **2.** The periodic shedding of the outer layer of the epidermis of reptiles (except crocodiles) to allow growth to occur.

ecdysone A steroid hormone, secreted by a pair of *prothoracic glands* in the thorax of insects and probably by *Y organs in crustaceans, that stimulates moulting (*see* **ecdysis**) and metamorphosis. It acts on specific gene loci, stimulating the synthesis of proteins involved in these bodily changes.

ECG *See* **electrocardiogram**.

Echinodermata A phylum of marine invertebrates that includes the sea urchins, starfish, brittlestars, and sea cucumbers. Echinoderms have an exoskeleton (*test*) of calcareous plates embedded in the skin. In many species (e.g. sea urchins) spines protrude from the test. A system of water-filled canals (the *water vascular system*) provides hydraulic power for thousands of *tube feet*: saclike protrusions of the body wall used for locomotion, feeding, and respiration. Echinoderms have a long history: fossils of primitive echinoderms are known from rocks 500 million years old. Extant classes include the Crinoidea (sea lilies), Holothuroidea (sea cucumbers), Echinoidea (sea urchins), and Stelleroidea (starfish and brittlestars).

echolocation A method used by some animals (such as bats, dolphins, and certain birds) to detect objects in the dark. The animal emits a series of high-pitched sounds that echo back from the object and are detected by the ear or some other sensory receptor. From the direction of the echo and from the time between emission and reception of the sounds the object is located, often very accurately.

E. coli *See* **Escherichia coli**.

ecological equivalents Unrelated organisms that occupy similar habitats and resemble each other. Ecological equivalents result from *convergent evolution. For example, sharks (fish) and dolphins (mammals) live in a marine habitat and superficially resemble each other.

ecological niche The status or role of an organism in its environment. An organism's niche is defined by the types of food it consumes, its predators, temperature tolerances, etc. Two species cannot coexist stably if they occupy identical niches.

ecological pyramid *See* **pyramid of biomass; pyramid of energy; pyramid of numbers**.

ecology The study of the interrelationships between organisms and their natural environment, both living and nonliving. For this purpose, ecologists study organisms in the context of the *populations and *communities in which they can be grouped and the *ecosystems of which they form a part. The study of ecological interactions provides important information on the nature and mechanisms of evolutionary change. Advances made in ecology over the last 25 years have led to increased concern about the effects of man's activities on the environment (notably the effects of *pollution), which has resulted in a greater awareness of the importance of *conservation.

ecosystem A biological *community and the physical environment associated with it. Nutrients pass between the different organisms in an ecosystem in definite pathways; for example, nutrients in the soil are taken up by plants, which are then eaten by herbivores, which in turn may be eaten by carnivores (*see* **food chain**). Organisms are classified on the basis of their position in an ecosystem into various *trophic levels. Nutrients and energy move round ecosystems in loops or cycles (in the case above, for example, nutrients are returned to the soil via animal wastes and decomposition). *See* **carbon cycle; nitrogen cycle**.

ectoderm The external layer of cells of the *gastrula, which will develop into the epidermis and the nervous system in the adult. *See also* **germ layers**.

ectoparasite A parasite that lives on the outside of its host's body. *See* **parasitism**.

ectoplasm *See* **cytoplasm**.

Ectoprocta A phylum of aquatic, mainly marine, invertebrates traditionally classified with the similar Entoprocta as the phylum Bryozoa (moss animals and sea mats). Ectoprocts live in colonies, 50 cm or more across, which are attached to rocks, seaweeds, or shells. The individuals (*zooids*) making up the colonies are about 1 mm long and superficially resemble coelenterate *polyps, with a mouth surrounded by ciliated tentacles of the *lophophore that trap minute particles of organic matter in the water. Some have a horny or calcareous outer skeleton into which the body can be withdrawn.

ectotherm (poikilotherm) An animal that maintains its body temperature by absorbing heat from the surrounding environment. All animals except mammals and birds are ectotherms; they are often described as being *cold-blooded* and are unable to regulate their body temperature metabolically. *See* **poikilothermy**. *Compare* **endotherm**.

edaphic factor An *abiotic factor relating to the physical or chemical composition of the soil found in a particular area. For example, very alkaline soil may be an edaphic factor limiting the variety of plants growing in a region.

EDTA Ethylenediaminetetraacetic acid, $(HOOCCH_2)_2N(CH_2)_2N(CH_2COOH)_2$: a compound that acts as a chelating agent, reversibly binding with iron, magnesium, and other metal ions. It is used in certain culture media bound with iron, which it slowly releases into the medium, and also in some forms of quantitative analysis.

EEG *See* **electroencephalogram**.

effector A cell or organ that produces a physiological response when stimulated by a nerve impulse. Examples include muscles and glands.

effector neurone A nerve cell, such as a motor neurone, that transmits impulses from the central nervous system to an *effector in order to bring about a physiological response to changes in the environment.

efferent Carrying (nerve impulses, blood, etc.) away from the centre of a body or organ towards peripheral regions. The term is usually applied to types of nerve fibres or blood vessels. *Compare* **afferent**.

egestion The expulsion from the body of waste food materials that have never left the gut, particularly the expulsion of undigested materials from the gut through the anus (*see* **defecation**). Egestion should not be confused with *excretion, in which the waste materials are produced by metabolic activity in the body's tissues.

egg 1. The fertilized ovum (*zygote) in egg-laying animals, e.g. birds and insects, after it emerges from the body. The egg is covered by *egg membranes that protect it from environmental damage, such as drying. **2.** (*or* **egg cell**) The mature female reproductive cell in animals and plants. *See* **oosphere; ovum**.

egg membrane The layer of material that covers an animal egg cell. *Primary membranes* develop in the ovary and cover the egg surface in addition to the normal cell membrane. The primary membrane is called the *vitelline membrane* in insects, molluscs, birds, and amphibians, the *chorion* in tunicates and fish, and the *zona pellucida in mammals. Insects have a second thicker membrane, also called the chorion. *Secondary membranes* are secreted by the oviducts and parts of the genital system while the egg is passing to the outside. They include the jelly coat of frogs' eggs and the albumen and shell of birds' eggs.

ejaculation The propulsion of semen out of the erect penis due to powerful rhythmic contractions of the urethra. An ejaculation coincides with the peak of sexual excitement (*orgasm*) and is accompanied by various physiological effects in the body, such as increased respiration rate and heart rate.

Elasmobranchii *See* **Chondrichthyes**.

elastic cartilage *See* **cartilage**.

elastic fibres *See* **elastin**.

elastin A fibrous protein that is the major constituent of the yellow *elastic fibres* of *connective tissue. It is rich in glycine, alanine, proline, and other nonpolar amino acids that are cross-linked, making the protein relatively insoluble. Elastic fibres can stretch to several times their length and then return to their original size. Elastin is particularly abundant in elastic *cartilage, blood-vessel walls, ligaments, and the heart.

electric organ An organ occurring on the body or tail of certain fish, such as the electric ray (*Torpedo*) and electric eel (*Electrophorus electricus*). It gives an electric shock when touched and is used either to stun prey or predators or, in some species, to maintain a weak electric field in the surrounding water that is used in navigation. The organ is composed of modified muscle cells (*electroplate* or *electroplax cells*), nervous stimulation of which greatly increases the potential difference across the cell. The electroplates are in series so a high overall voltage can be achieved.

electric potential Symbol V. The energy required to bring unit electric charge from infinity to the point in an electric field at which the potential is being specified. The unit of electric potential is the volt. The *potential difference* (*p.d.*) between two points in an electric field or circuit is the difference in the values of the electric potentials at the two points, i.e. it is the work done in moving unit charge from one point to the other.

electrocardiogram (ECG) A tracing or graph of the electrical activity of the heart. Recordings are made from electrodes fastened over the heart and usually on both arms and a leg. Changes in the normal pattern of an ECG may indicate heart irregularities or disease.

electroencephalogram (EEG) A tracing or graph of the electrical activity of the brain. Electrodes taped to the scalp record electrical waves from different parts of the brain. The pattern of an EEG reflects an individual's level of consciousness and can be used to detect such disorders as epilepsy, tumours, or brain damage. *See also* **brain death**.

electrolyte A liquid that conducts electricity as a result of the presence of positive or negative ions. An example is a sodium chloride solution, which consists of free sodium (Na^+) and chloride (Cl^-) ions. In biology and medicine 'electrolyte' usually refers to the ion itself.

electromagnetic spectrum The range of wavelengths over which electromagnetic radiation extends. The longest waves (10^5–10^{-3} metres) are radio waves, the next longest (10^{-3}–10^{-6} m) are infrared waves, then comes the narrow band (4–7×10^{-7} m) of visible light, followed by *ultraviolet radiation (10^{-7}–10^{-9} m), *X-rays (10^{-9}–10^{-11} m), and gamma rays (10^{-11}–10^{-14} m).

electron An elementary particle present in all atoms in groupings called shells around the nucleus. When electrons are detached from the atom they are called *free electrons*.

electron flow The transfer of electrons along a series of carrier molecules in the *electron transport chain.

electron microscope A form of microscope that uses a beam of electrons instead of a beam of light (as in the optical microscope) to form a large image of a very small object, such as a cell organelle, a virus, or a DNA molecule. In optical microscopes the resolution is limited by the wavelength of the light. High-energy electrons, however, can be associated with a considerably shorter wavelength than light; for example, electrons accelerated to an energy of 10^5 electronvolts have a wavelength of 0.04 nanometre, enabling a resolution of 0.2–0.5 nm to be achieved. The *transmission electron microscope* (see illustration) has an electron beam, sharply focused by electron lenses (coils producing a magnetic field or electrodes between which an electric field is created), passing through a very thin metallized specimen (less than 50 nanometres thick) onto a fluorescent screen, where a visual image is formed. This image can be photographed. The *scanning electron microscope* can be used with thicker specimens and forms a perspective image, although the resolution and magnification are lower. It is used particularly for examining surface features of small objects, such as pollen grains. In this type of instrument a beam of primary electrons scans the specimen and those that are reflected, together with any secondary electrons emitted, are collected. This current is used to modulate a separate electron beam in a TV monitor, which scans the screen at the same frequency, consequently building up a picture of the specimen. The resolution is limited to about 10–20 nm. *See also* **field-emission microscope; field-ionization microscope**.

Principle of transmission electron microscope

electron transport chain (**electron transport system**) A sequence of biochemical reduction–oxidation reactions that effects the transfer of electrons through a series of carriers. An electron transport chain, also known as the *respiratory chain*, forms the final stage of *aerobic respiration. In the mitochondria, NADH or FADH$_2$, generated by the *Krebs cycle, transfer their electrons through a chain of *carrier molecules, including *coenzyme Q (ubiquinone) and a series of *cytochromes, that undergo reversible reduction–oxidation reactions, accepting electrons and then donating them to the next carrier in the chain – a process known as *electron flow*. *Cytochrome oxidase combines electrons and hydrogen ions with oxygen to form water (see illustration). This process is coupled to the formation of ATP (*see* **chemiosmotic theory; oxidative phosphorylation**). An electron transport chain also occurs in the thylakoid membranes of chloroplasts in *photosynthesis; the carrier molecules include *plastoquinone, *plastocyanin, and *ferredoxin. *See* **photophosphorylation**.

The mitochondrial electron transport chain

electrophoresis (cataphoresis) A technique for the analysis and separation of colloids, based on the movement of charged colloidal particles in an electric field. There are various experimental methods. In one the sample is placed in a U-tube and a buffer solution added to each arm, so that there are sharp boundaries between buffer and sample. An electrode is placed in each arm, a voltage applied, and the motion of the boundaries under the influence of the field is observed. The rate of migration of the particles depends on the field, the charge on the particles, and on other factors, such as the size and shape of the particles. More simply, electrophoresis can be carried out using an adsorbent, such as a strip of filter paper, soaked in a buffer with two electrodes making contact. The sample is placed between the electrodes and a voltage applied. Different components of the mixture migrate at different rates, so the sample separates into zones. The components can be identified by the rate at which they move. This technique has also been called *electrochromatography*.

  Electrophoresis is used extensively in studying mixtures of proteins (*see* **PAGE**), nucleic acids, carbohydrates, enzymes, etc. In clinical medicine it is used for determining the protein content of body fluids.

electroplax One of the cells found in the *electric organs of certain fish.

electrovalent bond (ionic bond) A type of *chemical bond formed by the transfer of one or more electrons from one atom to another, so that oppositely charged ions are produced. For example, the bond between the sodium and chlorine atoms in sodium chloride (NaCl) is formed by the transfer of an electron from sodium to chlorine, creating Na$^+$ and Cl$^-$ ions.

The electrostatic attraction between these ions provides the bonding in NaCl.

ELISA (enzyme-linked immunosorbent assay) A sensitive technique (*see* **immunoassay**) for accurately determining the amount of protein or other antigen in a given sample by means of an enzyme-catalysed colour change. Antibody specific to the test protein is adsorbed onto a solid substrate, such as a PVC sheet, and a measured amount of the sample is added; all molecules of the test protein in the sample are bound by the antibody. A second antibody specific for a second site on the test protein is added; this is conjugated with an enzyme, which catalyses a colour change in the fourth reagent, added finally to the sheet. The colour change can be measured photometrically and compared against a standard curve to give the concentration of protein in the sample. ELISA is widely used for diagnostic and other purposes.

elongation *(in protein synthesis)* The phase in which amino acids are linked together by sequentially formed peptide bonds to form a polypeptide chain (*see* **translation**). *Elongation factors* are proteins that – by binding to a tRNA–amino-acid complex – enable the correct positioning of this complex on the ribosome, so that translation can proceed.

elytra The thickened horny forewings of the *Coleoptera (beetles), which cover and protect the membranous hindwings when the insect is at rest.

emasculation The removal of the anthers of a flower in order to prevent self-pollination or the undesirable pollination of neighbouring plants.

Embden–Meyerhof pathway *See* glycolysis.

embedding A stage in the preparation of a sample for examination by microscopy that involves impregnation of the sample with wax or plastic following dehydration. The embedded sample can then be cut into extremely thin sections to reveal cellular and subcellular structure.

embryo 1. An animal in the earliest stages of its development, from the time when the fertilized ovum starts to divide (*see* **cleavage**), while it is contained within the egg or reproductive organs of the mother, until hatching or birth. A human embryo (see illustration) is called a *fetus after the first eight weeks of pregnancy. **2.** The structure in plants that develops from the zygote prior to germination. In seed plants the zygote is situated in the *embryo sac of the ovule. It divides by mitosis to form the embryonic cell and a structure called the *suspensor*, which embeds the embryo in the surrounding nutritive tissue. The embryonic cell divides continuously and eventually gives rise to the *radicle (young root), *plumule (young shoot), and one or two *cotyledons (seed leaves). Changes also take place in the surrounding tissues of the ovule, which becomes the *seed enclosing the embryo plant.

embryology The study of the development of animals from the fertilized egg to the new adult organism. It is sometimes limited to the

A developing human embryo

fertilized egg to the new adult organism. It is sometimes limited to the period between fertilization of the egg and hatching or birth (*see* **embryo**).

embryo mother cell *See* **megaspore mother cell.**

embryo sac A large cell that develops in the *ovule of flowering plants. It is equivalent to the female *gametophyte of lower plants, although it is very much reduced. Typically, it contains eight nuclei formed by division of the *megaspore mother cell (see illustration). The *oosphere (egg cell), which is associated with two *synergid cells to form the *egg apparatus*, is fertilized by a male nucleus and becomes the *embryo. The two *polar nuclei fuse with a second male nucleus to form a triploid nucleus that gives rise to the *endosperm. The three remaining nuclei form the *antipodal cells*.

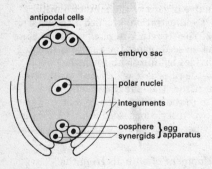

Embryo sac

emulsification *(in digestion)* The breakdown of fat globules in the duodenum into tiny droplets, which provides a larger surface area on which the enzyme pancreatic *lipase can act to digest the fats into fatty acids and glycerol. Emulsification is assisted by the action of the bile salts (*see* **bile**).

emulsion A *colloid in which small particles of one liquid are dispersed in another liquid. Usually emulsions involve a dispersion of water in an oil or a dispersion of oil in water. Dietary fats are reduced to an emulsion in the duodenum to facilitate their subsequent digestion (*see* **emulsification**).

enamel The material that forms a covering over the crown of a *tooth (i.e. the part that projects above the gum). Enamel is smooth, white, and extremely hard, being rich in minerals containing calcium, especially *apatite. It is produced by certain cells (*ameloblasts*) of the oral epithelium and protects the underlying dentine of the tooth. Enamel may also be found in the placoid *scales of certain fish, which demonstrates the common developmental origin of scales and teeth.

enamel organ A mass of soft pulpy tissue that develops on the surface of a dental *papilla. It consists of a mesh of fibres, fluid *albumin, and *ameloblasts, which deposit a layer of protective *enamel on the outer surface of the dentine.

encephalin (enkephalin) A peptide *neurotransmitter found principally in the brain. Two encephalins have so far been isolated. They bind to opiate receptors in the brain and their release is thought to control levels of pain and other sensations. Encephalins have also been found in spinal cord nerve cells and in gut epithelial cells. *Compare* **endorphin**.

endangered species A plant or animal species defined by the IUCN (International Union for the Conservation of Nature and Natural Resources) as being in immediate danger of *extinction because its population numbers have reached a critical level or its habitats have been drastically reduced. If these causal factors continue the species is unlikely to survive. A list of endangered species is published by the IUCN, which also defines other categories of threatened species.

endemic 1. Describing a plant or animal species that is restricted to one or a few localities in its distribution. Endemic species are usually confined to islands and are vulnerable to extinction. **2.** Describing a disease or a pest that is always present in an area. For example, malaria is endemic in parts of Africa.

endergonic reaction A chemical reaction in which energy is absorbed. In an *endothermic reaction the energy is in the form of heat. *Compare* **exergonic reaction**.

endocardium *See* **myocardium**.

endocarp *See* **pericarp**.

endochondral ossification *See* ossification.

endocrine gland (ductless gland) Any gland in an animal that manufactures *hormones and secretes them directly into the bloodstream to act at distant sites in the body (known as *target organs* or *cells*). Endocrine glands tend to control slow long-term activities in the body, such as growth and sexual development. In mammals they include the *pituitary, *adrenal, *thyroid, and *parathyroid glands, the *ovary and *testis, the *placenta, and part of the pancreas (*see* **islets of Langerhans**). The activity of endocrine glands is controlled by negative feedback, i.e. a rise in output of hormone inhibits a further increase in its production, either directly or indirectly via the target organ or cell. *See also* **neuroendocrine system**. *Compare* **exocrine gland**.

endocrinology The study of the structure and functions of the *endocrine glands and of the *hormones they produce.

endocytosis The process by which materials enter a cell without passing through the cell membrane. The membrane folds around material outside the cell, resulting in the formation of a saclike vesicle into which the material is incorporated. This vesicle is then pinched off from the cell surface so that it lies within the cell (*see also* **endosome**). Both *phagocytosis and *pinocytosis are forms of endocytosis. *Compare* **exocytosis**.

endoderm (entoderm) The internal layer of cells of the *gastrula, which will develop into the alimentary canal (gut) and digestive glands of the adult. *See also* **germ layers**.

endodermis The innermost layer of the root *cortex of a plant, lying immediately outside the vascular tissue. Various modifications of the endodermal cell walls indicate that they regulate the passage of materials both into and out of the vascular system (*see* **Casparian strip**). An endodermis may also be seen in the stems of some plants.

endogamy The fusion of reproductive cells from closely related parents, i.e. *inbreeding. *Compare* **exogamy**.

endogenous Describing a substance, stimulus, organ, etc., that originates from within an organism. For example, growth rhythms not directed by environmental stimuli are termed endogenous rhythms. Lateral roots, which always grow from inside the main root rather than from its surface, are said to arise endogenously. *Compare* **exogenous**.

endolymph The fluid that fills the membranous labyrinth of the vertebrate *inner ear. *See* **cochlea; semicircular canals**. *Compare* **perilymph**.

endometrium The mucous membrane that lines the *uterus of mammals. It comprises an upper mucus-secreting layer, which is shed during menstruation, and a basal layer, which proliferates to form the upper layer. *See also* **menstrual cycle**.

endonuclease An enzyme that catalyses the internal cleavage of nucleic acids. *See also* **restriction enzyme**. *Compare* **exonuclease**.

endoparasite A parasite that lives inside its host's body. *See* **parasitism**.

endopeptidase A protein-digesting enzyme that cleaves a polypeptide chain at specific sites between amino acids. For example, *chymotrypsin cleaves the chain next to aromatic amino acids, such as phenylalanine; *trypsin cleaves the chain next to basic amino acids, such as lysine or arginine; and *pepsin cleaves the chain next to tyrosine and phenylalanine. *Compare* **exopeptidase**.

endoplasm *See* **cytoplasm**.

endoplasmic reticulum (ER) A system of membranes within the cytoplasm of plant and animal *cells. It forms a link between the cell and nuclear membranes (*see* **cisterna**) and is the site of protein synthesis. It is also concerned with the transport of proteins and lipids within the cell. *Rough ER* has *ribosomes attached to its surface; proteins synthesized on the ribosomes are enclosed in vesicles and transported to the *Golgi apparatus. *Smooth ER* lacks ribosomes; it is the site of important metabolic reactions, including phospholipid and fatty-acid synthesis.

endopterygote Any winged insect in which the eggs hatch into *larvae that are dissimilar to the adults and lack wings. These develop internally before the final moult into a *pupa, from which the adult emerges; metamorphosis is described as *complete*. Endopterygote insects include the beetles (*Coleoptera), ants, bees, and wasps (*Hymenoptera), flies (*Diptera), and butterflies and moths (*Lepidoptera). *Compare* **exopterygote**.

end organ The structure at the end of a peripheral nerve. Examples of end organs are the muscle *end plate at the end of a motor neurone and the *receptor at the end of a sensory neurone.

endorphin A peptide neurotransmitter found in the pituitary gland and hypothalamus that has pain-relieving properties similar to the *encephalins.

endoscopy The study of the interior of hollow organs, such as the intestines and the reproductive organs, using a flexible probe (*endoscope*), which contains *optical fibres for transmitting images onto a screen.

endoskeleton A supporting framework that lies entirely within the body of an animal, such as the bony *skeleton of vertebrates or the spicules of a sponge. The function of an endoskeleton is to support the body and in vertebrates it also protects the organs and provides a system of levers on which the muscles can act to produce movement. *Compare* **exoskeleton**.

endosome A vesicle formed within a cell during forms of *endocytosis in which the material to be ingested first binds to *receptor sites on the cell surface. These receptors occur in areas of the plasma membrane called *coated pits*, which are pinched off to become *coated vesicles*, surrounded by a protein coat and containing the ingested material, within the cell. The vesicles then shed their coats and merge with other vesicles to form endosomes. The endosomes fuse with vesicles from the Golgi apparatus to form *lysosomes.

endosperm A nutritive tissue, characteristic of flowering plants, that surrounds the developing embryo in a seed. It develops from nuclei in the *embryo sac and its cells are triploid. In *endospermic* seeds it remains and increases in size; in *nonendospermic* seeds it disappears as the food is absorbed by the embryo, particularly the *cotyledons. Many plants with endospermic seeds, such as cereals and oil crops, are cultivated for the rich food reserves in the endosperm.

endospore The resting stage of certain bacteria, formed in response to adverse conditions. The bacterial cell becomes enclosed in a tough resistant spore coat. On return to favourable conditions the spore changes back to the normal vegetative form of the organism.

endostyle A ciliated groove in the wall of the pharynx of lancelets (Cephalochordata), some tunicates, and larval lampreys that contains mucus-secreting glands. Mucus secreted by the endostyle traps food particles in water drawn into the pharynx by the beating action of the cilia.

endothelium A single layer of thin platelike cells that line the inner surfaces of blood and lymph vessels and the heart. Endothelium is derived from the *mesoderm. *Compare* **epithelium; mesothelium**.

endotherm (homoiotherm) An animal that can generate and maintain heat within its body independently of the environmental temperature. Mammals and birds are endotherms; they are often described as being *warm-blooded*. *See* **homoiothermy**. *Compare* **ectotherm**.

endothermic 1. Of or relating to *endotherms. **2.** Denoting a chemical reaction that takes heat from its surroundings. *See* **endergonic reaction**. *Compare* **exothermic**.

endotoxin *See* **toxin**.

end plate The area of muscle cell membrane that lies immediately beneath a motor nerve ending at a *neuromuscular junction. Release of a *neurotransmitter at the end plate induces contraction of the muscle fibre.

energy A measure of a system's ability to do work. Like work itself, it is measured in joules. Energy is conveniently classified into two forms: *potential energy* is the energy stored in a body or system as a consequence of its position, shape, or state (this includes chemical energy in food substances, etc.); *kinetic energy* is energy of motion and is usually defined as the work that will be done by the body possessing the energy when it is brought to rest.

energy flow *(in ecology)* The flow of energy that occurs along a *food chain. Energy enters the food chain at the level of the *producers (usually plants) in the form of solar energy. The plants convert solar energy into chemical energy in the process of *photosynthesis. Chemical energy is passed from one trophic level to the next through feeding. Since a large proportion of energy is lost at each trophic level, mostly in the form of heat energy due to respiration, a food chain does not normally consist of more than five trophic levels: the fifth trophic level does not contain enough energy to support further levels. Energy is also lost from the food chain in excretory products and the remains of dead organisms; this is converted into heat energy by the action of *decomposers. *See also* **productivity; pyramid of energy.**

enhancer A region on a DNA molecule that can initiate the *transcription of a gene that may be some distance away from the enhancer on the same chromosome. It is activated by *DNA-binding proteins, which bind to the enhancer regions and in turn regulate the binding of RNA *polymerase to the DNA molecule.

enterokinase An enzyme in the small intestine that activates trypsinogen to *trypsin.

enteron (coelenteron) The body cavity of the coelenterates, which has one opening functioning both as mouth and anus. *See* **Cnidaria.**

enthalpy Symbol H. A thermodynamic property of a system defined by $H = U + pV$, where H is the enthalpy, U is the internal energy of the system, p its pressure, and V its volume. In a chemical reaction carried out in the atmosphere the pressure remains constant and the enthalpy of reaction, ΔH, is equal to $\Delta U + p\Delta V$. For an exothermic reaction ΔH is taken to be negative.

entoderm *See* **endoderm.**

entomology The study of insects.

entomophily Pollination of a flower in which the pollen is carried on an insect. Entomophilous flowers are usually brightly coloured and scented and often secrete nectar. In some species (e.g. primulas) there are structural differences between the flowers to ensure that cross-pollination occurs. Other examples of entomophilous flowers are orchids and antirrhinums. *Compare* **anemophily; hydrophily.**

entropy Symbol S. A measure of the unavailability of a system's energy to do work; an increase in entropy is accompanied by a decrease in energy availability. When a system undergoes a reversible change the entropy (S) changes by an amount equal to the energy (Q) absorbed by the system divided by the thermodynamic temperature (T) at which the energy is absorbed, i.e. $\Delta S = \Delta Q/T$. However, all real processes are to a certain extent irreversible changes and in any closed system an irreversible change is always accompanied by an increase in entropy.

In a wider sense entropy can be interpreted as a measure of a system's disorder; the higher the entropy the greater the disorder. As any real change to a closed system tends towards higher entropy, and therefore higher disorder, it follows that the entropy of the universe (if it can be considered a closed system) is increasing and its available energy is decreasing. This increase in the entropy of the universe is one way of stating the second law of thermodynamics.

environment *(in ecology)* The physical, chemical, and biological conditions of the region in which an organism lives. *See also* **ecology**; **ecosystem**.

environmental resistance The sum total of the factors that prevent populations from continually growing and therefore tend to keep populations at constant levels. These factors include predators, disease, and a shortage of any of the various requirements for survival, such as food, water, shelter, and light (which is particularly important for plants). *See also* **population growth**.

environmental selection *Natural selection within a population resulting from the influence exerted by the environment. Environmental selection inevitably leads to a change in the composition of genes within a population. For example, a change in the environment resulting in a drop in temperature will favour animals with an increased ability to maintain their body heat and may eventually result in a population of animals with denser fur.

enzyme A protein that acts as a *catalyst in biochemical reactions. Each enzyme is specific to a particular reaction or group of similar reactions. Many require the association of certain nonprotein *cofactors in order to function. The molecule undergoing reaction (the *substrate*) binds to a specific *active site on the enzyme molecule to form a short-lived intermediate (*see* **enzyme–substrate complex**): this greatly increases (by a factor of up to 10^{20}) the rate at which the reaction proceeds to form the product. Enzyme activity is influenced by substrate concentration and by temperature and pH, which must lie within a certain range. Other molecules may compete for the active site, causing *inhibition of the enzyme or even irreversible destruction of its catalytic properties.

Enzyme production is governed by a cell's genes. Enzyme activity is further controlled by pH changes, alterations in the concentrations of essential cofactors, feedback inhibition by the products of the reaction, and activation by another enzyme, either from a less active form or an inactive precursor (*zymogen). Such changes may themselves be under the control of hormones or the nervous system. *See also* **enzyme kinetics**.

Enzymes are classified into six major groups, according to the type of reaction they catalyse: (1) *oxidoreductases; (2) *transferases; (3) *hydrolases; (4) *lyases; (5) *isomerases; (6) *ligases. The names of most individual enzymes also end in *-ase*, which is added to the names of the substrates on which they act. Thus *lactase is the enzyme that acts to break down lactose; it is classified as a hydrolase.

enzyme inhibition *See* inhibition.

enzyme kinetics The study of the rates of enzyme-catalysed reactions. Rates of reaction are usually measured by using the purified enzyme *in vitro* with the substrate and then observing the formation of the product or disappearance of the substrate. As the concentration of the substrate is increased the rate of reaction increases proportionally up to a certain point, after which any further increase in substrate concentration no longer increases the reaction rate (*see* **Michaelis–Menton curve**). At this point, all active sites of the enzyme are saturated with substrate; any further increase in the rate of reaction will occur only if more enzyme is added. Reaction rates are also affected by the presence of inhibitors (*see* **inhibition**), temperature, and pH (*see* **enzyme**).

enzyme-linked immunosorbent assay *See* ELISA.

enzyme–substrate complex The intermediate formed when a substrate molecule interacts with the *active site of an enzyme. Following the formation of an enzyme–substrate complex, the substrate molecule undergoes a chemical reaction and is converted into a new product. Various mechanisms for the formation of enzyme–substrate complexes have been suggested, including the *induced-fit model and the *lock-and-key mechanism.

Eocene The second geological epoch of the *Tertiary period. It extended from the end of the Palaeocene epoch, about 54 million years ago, to the beginning of the Oligocene epoch, about 38 million years ago. The term was first proposed by Sir Charles Lyell (1797–1875) in 1833. In some classifications of geological time the *Palaeocene is included as part of the Eocene. Mammals were dominant in the Eocene: rodents, artiodactyls, carnivores, perissodactyls (including early horses), and whales were among the groups to make their first appearance.

eosin One of a series of acidic dyes, used in optical microscopy, that colours cytoplasm pink and cellulose red. It is frequently used as a counterstain with *haematoxylin for colouring tissue smears and sections of animal tissue.

eosinophil A type of white blood cell (*leucocyte) that has a granular cytoplasm (*see* **granulocyte**). Its function involves the regulation of allergic responses and it also produces an enzyme capable of destroying parasites.

ephemeral **1.** *(in botany)* An *annual plant that completes its life cycle in considerably less than one growing season. A number of generations can therefore occur in one year. Many troublesome weeds, such as groundsel and willowherb, are ephemerals. Certain desert plants are also ephemerals, completing their life cycles in a short period following rain. **2.** *(in zoology)* A short-lived animal, such as a mayfly.

Ephemeroptera An order of *exopterygote insects that comprises the mayflies, in which the adult stage lasts for only a few hours. The adults have two pairs of wings held vertically at rest, a pair of tail bristles (cerci), and vestigial mouthparts (they do not feed). The nymphs live for up to a number of years; they are mainly herbivorous but some possess mandibles for feeding on animal prey.

epicalyx A ring of bracts below a flower that resembles a calyx. It is seen, for example, in the strawberry flower.

epicarp *See* **pericarp**.

epicotyl The region of a seedling stem above the stalks of the seed leaves (*cotyledons) of an embryo plant. It grows rapidly in seeds showing *hypogeal germination and lifts the stem above the soil surface. *Compare* **hypocotyl**.

epidemic An outbreak of a disease (especially an infectious disease) that affects a large number of individuals within a population at the same time. *Compare* **endemic**.

epidermis 1. *(in zoology)* The outermost layer of cells of the body of an animal. In invertebrates the epidermis is normally only one cell thick and is covered by an impermeable *cuticle. In vertebrates the epidermis is the thinner of the two layers of *skin (*compare* **dermis**). It consists of a basal layer of actively dividing cells (*see* **Malpighian layer**), covered by layers of cells that become impregnated with keratin (*see* **keratinization**). The outermost layers of epidermal cells (the *stratum corneum) form a water-resistant protective layer. The epidermis may bear a variety of specialized structures (e.g. *feathers, *hairs). **2. *(in botany)*** The outermost layer of cells covering a plant. It is overlaid by a *cuticle and its functions are principally to protect the plant from injury and to reduce water loss. Some epidermal cells are modified to form guard cells (*see* **stoma**) or hairs of various types (*see* **piliferous layer**). In woody plants the functions of the shoot epidermis are taken over by the periderm tissues (*see* **cork cambium**) and in mature roots the epidermis is sloughed off and replaced by the *hypodermis.

epididymis A long coiled tube in which spermatozoa are stored in vertebrates. In reptiles, birds, and mammals it is attached at one end to the *testis and opens into the sperm duct (*vas deferens) at the other.

epigamic Serving to attract a mate. Epigamic characters include the bright plumage of some male birds.

epigeal Describing seed germination in which the seed leaves (cotyledons) emerge from the ground and function as true leaves. Examples of epigeal germination are seen in sycamore and sunflower. *Compare* **hypogeal**.

epiglottis A flexible flap of cartilage in mammals that is attached to the wall of the pharynx near the base of the tongue. During swallowing (*see* **deglutition**) it covers the *glottis (the opening to the respiratory tract) and helps to prevent food from entering the trachea (windpipe), although it is not essential for this purpose.

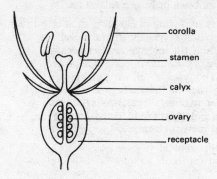

Epigyny

epigyny A floral arrangement in which the ovary is completely enclosed by the receptacle so that the stamens and perianth arise above it, from the top of the receptacle; i.e. the ovary is *inferior (see illustration). The perianth and stamens are said to be *epigynous* with respect to the ovary, as seen in the daffodil. *Compare* **hypogyny; perigyny.**

epilimnion The upper layer of water in a lake. *Compare* **hypolimnion.** *See* **thermocline.**

epinephrine *See* **adrenaline.**

epiphysis The terminal section of a growing bone (especially a long limb bone) in mammals. It is separated from the bone shaft (*diaphysis*) by cartilage. New bone is produced on the side of the cartilage facing the diaphysis, while new cartilage is produced on the other side of the cartilage disc. When the bone reaches adult length the epiphysis merges with the diaphysis.

epiphyte A plant that grows upon another plant but is neither parasitic on it nor rooted in the ground. Epiphytes include many mosses and lichens and some tropical orchids.

episome A genetic element that can exist and replicate either independently of its host cell's chromosomes or as an integrated part of the chromosomes. Examples include certain bacterial *plasmids.

epistasis A gene interaction in which one gene suppresses the effect of another gene that is situated at a different *locus on the chromosome. For example, in guinea pigs the gene that controls the production of melanin is epistatic to the gene that regulates the deposition of melanin. A dominant allele (*C*) is responsible for the production of melanin, while the amount of melanin deposited is controlled by a second gene, which determines whether the coat colour is black or brown. If an animal is homozygous recessive (*cc*) for melanin production, the coat colour will be white regardless of the alleles that produce black or brown coloration.

epithelium A tissue in vertebrates consisting of closely packed cells in a sheet with little intercellular material. It forms a membrane over the outer surfaces of the body and walls of the internal cavities (coeloms). It also forms glands and parts of sense organs. Its functions are protective, absorptive, secretory, and sensory. The types of cell vary, giving rise to *squamous epithelium; *ciliated epithelium; *cuboidal epithelium*, with cube-shaped cells; and *columnar epithelium*, with rectangular cells (*see also* **stratified epithelium**). Epithelium is derived from *ectoderm and *endoderm. *Compare* **endothelium; mesothelium.**

EPSP *See* **excitatory postsynaptic potential.**

Epstein–Barr virus *See* **herpesvirus.**

equifinality The ability of certain structures or behaviour patterns to form by more than one route in a developing embryo or young animal.

erepsin An obsolete name for a mixture of protein-degrading enzymes secreted by the small intestine.

ergocalciferol *See* **vitamin D.**

ergonomics The study of the engineering aspects of the relationship between workers and their working environment.

ergosterol A *sterol occurring in fungi, bacteria, algae, and plants. It is converted into vitamin D_2 by the action of ultraviolet light.

ergot The dark hard-walled mass of hyphae produced by the fungus *Claviceps purpurea* (*see* **Ascomycota**) in the grain of the cereals and other grasses that it parasitizes. Ergots are the *sclerotia* of this species – resting bodies that germinate in favourable conditions to produce mycelia or ascocarps. They contain alkaloids related to LSD, which cause blood vessels to constrict and are used therapeutically in the treatment of migraine and haemorrhage. Ingestion of infected grain can result in ergot poisoning (ergotism), with symptoms of gangrene and hallucinations – known as 'St Anthony's fire' in the Middle Ages.

erythroblast Any of the cells in the *myeloid tissue of red bone marrow that develop into erythrocytes (red blood cells). Erythroblasts have a nucleus and are at first colourless, but fill with *haemoglobin as they develop. In mammals the nucleus disappears. *See* **erythropoiesis.**

erythrocyte (red blood cell) The most numerous type of blood cell, which contains the red pigment *haemoglobin and is responsible for oxygen transport. Mammalian erythrocytes are disc-shaped and lack a nucleus; those of other vertebrates are oval and nucleated. In man the number of erythrocytes in the blood varies between 4.5 and 5.5 million per cubic millimetre. They survive for about four months and are then destroyed in the spleen and liver. *See also* **erythropoiesis**. *Compare* **leucocyte**.

erythropoiesis The formation of red blood cells (erythrocytes), which occurs in the red bone marrow (*see* **haemopoietic tissue**). The earliest precursor that can be distinguished microscopically is the *proerythroblast*, which develops from a haemopoietic stem cell and gives rise successively to the *early erythroblast*, *intermediate erythroblast*, and the *late erythroblast*, in which most haemoglobin is synthesized. In mammals the nucleus is then forced from the cell, which assumes a biconcave shape and is known as a *reticulocyte*. Reticulocytes are released into the blood and develop into mature erythrocytes within two days. *See also* **erythropoietin**.

erythropoietin A hormone that is released from the kidney (and to a lesser extent from the liver) in response to low concentrations of oxygen in the tissues. It speeds up the process of *erythropoiesis and is the means by which the rate of red cell production is controlled.

Escherichia coli (*E. coli*) A species of Gram-negative aerobic bacteria that is found in the intestine (*see* **coliform bacteria**) and is also widely used in microbiological and genetics research. The motile rod-shaped cells ferment lactose and are usually harmless commensals, although certain strains are pathogenic. Studies of *E. coli* laboratory cultures have revealed much about the genetics of prokaryotes; the species is also frequently used in genetic engineering, particularly as a host for *gene cloning and the expression of recombinant foreign genes in culture.

eserine (physostigmine) An alkaloid, derived from the calabar bean plant, that inhibits *cholinesterase by covalently binding with it (*see* **inhibition**). Eserine is used to treat the eye condition glaucoma.

essential amino acid An *amino acid that an organism is unable to synthesize in sufficient quantities. It must therefore be present in the diet. In man the essential amino acids are arginine, histidine, lysine, threonine, methionine, isoleucine, leucine, valine, phenylalanine, and tryptophan. These are required for protein synthesis and deficiency leads to retarded growth and other symptoms. Most of the amino acids required by man are also essential for all other multicellular animals and for most protozoans.

essential element Any of a number of elements required by living organisms to ensure normal growth, development, and maintenance. Apart from the elements found in organic compounds (i.e. carbon, hydrogen, oxygen, and nitrogen), plants, animals, and microorganisms all require a range of elements in inorganic forms in varying amounts, depending on the type of organism. The *major elements*, present in tissues in relatively large amounts (greater than 0.005%), are calcium, phosphorus, potassium, sodium, chlorine, sulphur, and magnesium (*see also* **macronutrient**). The *trace elements* occur at much lower concentrations and thus requirements are much less. The most important are iron, manganese, zinc, copper, iodine, cobalt, selenium, molybdenum, chromium, and silicon (*see also*

micronutrient). Each element may fulfil one or more of a variety of metabolic roles. Sodium, potassium, and chloride ions are the chief electrolytic components of cells and body fluids and thus determine their electrical and osmotic status. Calcium, phosphorus, and magnesium are all present in bone. Calcium is also essential to nerve and muscle activity, while phosphorus is a key constituent of the chemical energy carriers (e.g. *ATP) and the nucleic acids. Sulphur is needed primarily for amino acid synthesis (in plants and microorganisms). The trace elements may serve as *cofactors or as constituents of complex molecules, e.g. iron in haem and cobalt in vitamin B_{12}. *See also* **mineral deficiency**.

essential fatty acids *Fatty acids that must normally be present in the diet of certain animals, including man. Essential fatty acids, which include *arachidonic, *linoleic, and *linolenic acids, all possess double bonds at the same two positions along their hydrocarbon chain and so can act as precursors of *prostaglandins. Deficiency of essential fatty acids can cause dermatosis, weight loss, irregular oestrus, etc. An adult human requires 2–10 g linoleic acid or its equivalent per day.

essential oil A natural oil with a distinctive scent secreted by the glands of certain aromatic plants. *Terpenes are the main constituents. Essential oils are extracted from plants by steam distillation, extraction with cold neutral fats or solvents (e.g. alcohol), or pressing and used in perfumes, flavourings, and medicine. Examples are citrus oils, flower oils (e.g. rose, jasmine), and oil of cloves.

ester An organic compound formed by *esterification. Esters formed from carboxylic acids have the general formula RCOOR′. Esters containing simple hydrocarbon groups are volatile fragrant substances used as flavourings in the food industry. Triesters, molecules containing three ester groups, occur in nature as oils and fats. *See also* **glyceride**.

esterification A *condensation reaction of an alcohol with an acid to produce an *ester and water; e.g.

$$CH_3OH + C_6H_5COOH \leftrightarrows CH_3OOCC_6H_5 + H_2O$$

The reaction is an equilibrium and is slow under normal conditions, but can be speeded up by addition of a strong acid catalyst.

etaerio A cluster of fruits formed from the unfused carpels of a single flower. For example, the anemone has an etaerio of *achenes, larkspur an etaerio of *follicles, and blackberry an etaerio of *drupes.

ethanedioic acid *See* oxalic acid.

ethanoate *See* acetate.

ethanoic acid *See* acetic acid.

ethanol (ethyl alcohol) A colourless water-soluble *alcohol, C_2H_5OH. It is the active principle in intoxicating drinks, in which it is produced by *fermentation of sugar using yeast:

$C_6H_{12}O_6 \rightarrow 2C_2H_5OH + 2CO_2$

The ethanol produced kills the yeast and fermentation alone cannot produce ethanol solutions containing more than 15% ethanol by volume. *See also* **brewing**.

ethene *See* ethylene.

ethology The study of the biology of *animal behaviour. Central to the ethologist's approach is the principle that animal behaviour (like physical characteristics) is subject to evolution through natural selection. Ethologists therefore seek to explain how the behaviour of an animal in its natural environment may contribute to the survival of the maximum number of its relatives and offspring. This involves recognizing the stimuli that are important in nature (*see* **sign stimulus**) and how innate predispositions interact with *learning in the development of behaviour (*see* **instinct**). Studies of this sort were pioneered by Konrad Lorenz (1903–89) and Niko Tinbergen (1907–88).

ethylene (ethene) A colourless gaseous hydrocarbon, C_2H_4, that occurs naturally in many plants and acts as a *growth substance. Among its effects are the inhibition of longitudinal growth and the radial expansion of tissues, the promotion of fruit ripening and (in some species) flowering, and the breaking of seed dormancy in some species.

etiolation The abnormal form of growth observed when plants grow in darkness or severely reduced light. Such plants characteristically have blanched leaves and shoots, excessively long shoots, and reduced leaves and root systems.

Eubacteria A subphylum comprising all the so-called 'true' *bacteria, i.e. those bacteria that cannot be classified as *Archaebacteria; it includes the vast majority of bacteria.

eucarpic Denoting a fungus in which the thallus is differentiated into vegetative and reproductive regions. *Compare* **holocarpic**.

eucaryote *See* eukaryote.

euchromatin *See* chromatin.

eugenics The study of methods of improving the quality of human populations by the application of genetic principles. *Positive eugenics* would seek to do this by selective breeding programmes. *Negative eugenics* aims to eliminate harmful genes (e.g. those causing haemophilia and colour blindness) by counselling any prospective parents who are likely to be *carriers.

Euglenophyta A phylum of mostly unicellular organisms (including *Euglena*) that move by means of flagella. Most euglenoids are photosynthetic and inhabit fresh water, and in some classification systems are regarded as green algae (phylum Chlorophyta). However, they lack a cell wall and some forms are colourless and thus ingest food, since they cannot photosynthesize; they have in the past been classified as *protozoa. In modern classification systems euglenoids comprise a phylum of the *Protoctista.

eukaryote (eucaryote) An organism consisting of cells in which the genetic material is contained within a distinct nucleus. All organisms except bacteria are eukaryotes. *See* **cell**. *Compare* **prokaryote**.

euploid Describing a nucleus, cell, or organism that has an exact multiple of the haploid number (n) of chromosomes. For example, *diploid ($2n$), *triploid ($3n$), and *tetraploid ($4n$) nuclei or cells are all euploid. *Compare* **aneuploid**.

eury- A prefix denoting width of range. For example, *euryhaline* aquatic organisms can tolerate a wide range of saline conditions. *Compare* **steno-**.

Eustachian tube The tube that connects the *middle ear to the back of the throat (pharynx) in vertebrates. It is normally closed, but during swallowing it opens to allow air into the middle ear, which equalizes the pressure on each side of the *tympanum (eardrum). It was named after the Italian anatomist Bartolomeo Eustachio (?1520–74).

Eutheria (Placentalia) An infraclass of mammals in which the embryos are retained in a uterus in the mother's body and nourished by a *placenta. The young are thus fully protected during their embryonic development and kept at a constant temperature. Placental mammals evolved during the Cretaceous period (about 100 million years ago). Modern placentals are a highly diverse group that occupy all types of habitat in all parts of the world. They include the orders *Artiodactyla, *Carnivora, *Cetacea, *Chiroptera, *Insectivora, *Perissodactyla, *Primates, *Proboscidea, and *Rodentia. *Compare* **Metatheria; Prototheria**.

eutrophic Describing a body of water (e.g. a lake) with an abundant supply of nutrients and a high rate of formation of organic matter by photosynthesis. Pollution of a lake by *sewage or *fertilizers renders it eutrophic (a process called *eutrophication*). This stimulates excessive growth of algae (*see* **algal bloom**); the death and subsequent decomposition of these increases the *biochemical oxygen demand and thus depletes the oxygen content of the lake, resulting in the death of the lake's fish and other animals. *Compare* **dystrophic; mesotrophic; oligotrophic**.

evergreen (Describing) a plant that bears leaves throughout the year, each leaf being shed independently of the others after two or three years. The leaves of evergreens are often reduced or adapted in some way to prevent excessive water loss; examples are the needles of conifers and the leathery waxy leaves of holly. *Compare* **deciduous**.

evocation The ability of experimental stimuli (e.g. chemicals or tissue implants) to cause unspecialized embryonic tissue to develop into specialized tissue.

evolution The gradual process by which the present diversity of plant and animal life arose from the earliest and most primitive organisms, which is believed to have been continuing for at least the past 3000 million years. Until the middle of the 18th century it was generally believed that each species was divinely created and fixed in its form throughout its existence (*see* **special creation**). Lamarck was the first biologist to publish a theory to explain how one species could have evolved into another (*see* **Lamarckism**), but it was not until the publication of Darwin's *On the Origin of Species* in 1859 that special creation was seriously challenged. Unlike Lamarck, Darwin proposed a feasible mechanism for evolution and backed it up with evidence from the fossil record and studies of comparative anatomy and embryology (*see* **Darwinism; natural selection**). The modern version of Darwinism, which incorporates discoveries in genetics made since Darwin's time, probably remains the most acceptable theory of species evolution (*see also* **punctuated equilibrium**). More controversial, however, and still to be firmly clarified, are the relationships and evolution of groups above the species level.

exa- Symbol E. A prefix used in the metric system to denote 10^{18} times. For example, 10^{18} metres = 1 exametre (Em).

excitation 1. The act or process of stimulating a cell, especially a nerve cell. **2.** The response of a cell that has been stimulated.

excitatory postsynaptic potential (EPSP) The electric potential that is generated in a postsynaptic neurone during the transmission of a nerve impulse (*see* **synapse**). It is caused by *depolarization of the postsynaptic membrane when a *neurotransmitter (such as acetylcholine), released from the presynaptic membrane, binds to the postsynaptic membrane. This will induce an *action potential in the receiving neurone if the EPSP is large enough. *See* **facilitation**. *Compare* **inhibitory postsynaptic potential**.

excretion The elimination by an organism of the waste products that arise as a result of metabolic activity. These products include water, carbon dioxide, and nitrogenous compounds. Excretion plays an important role in maintaining the constancy of an organism's *internal environment (*see* **homeostasis**). In plants and simple animals waste products are excreted by simple diffusion from the body, but higher animals have specialized organs and organ systems devoted to this function (*see* **Malpighian tubule; nephridium**). Examples of excretory organs in vertebrates are the lungs (for carbon dioxide and water), and the *kidneys (for *nitrogenous waste and water). In addition, mammals excrete small amounts of urea, salts, and water from the skin in sweat.

exercise Increased muscular activity, which results in an increase in metabolic rate, heart rate, and oxygen uptake. Exercise also causes an increase in *anaerobic respiration in order to compensate for the *oxygen debt, which results in a build-up of lactic acid in the tissues.

exergonic reaction A chemical reaction in which energy is released (*compare* **endergonic reaction**). An *exothermic reaction is an exergonic reaction in which energy is released in the form of heat.

exhalation *See* **expiration**.

exine *See* **pollen**.

exocarp *See* **pericarp**.

exocrine gland A gland that discharges its secretion into a body cavity (such as the gut) or onto the body surface. Examples are the *sebaceous and *sweat glands, the *mammary glands, and part of the pancreas. Exocrine glands are formed in the embryo from the invagination of epithelial cells. Their secretions pass initially into a cavity (an *alveolus* or *acinus) and then out through a duct or duct network, along which the secretion may become modified by exchange with the blood across the duct epithelium.

exocytosis The passage of material from the inside of the cell to the cell surface within membrane-bound vesicles. The membranes of the vesicles fuse with the plasma membrane, releasing their contents to the exterior. Exocytosis is used both for the removal of waste material from the cell and for secretion; for example of mucus by *goblet cells. *Compare* **endocytosis**.

exodermis *See* **hypodermis**.

exogamy The fusion of reproductive cells from distantly related or unrelated organisms, i.e. *outbreeding. *Compare* **endogamy**.

exogenous Describing substances, stimuli, etc., that originate outside an organism. For example, vitamins that cannot be synthesized by an animal are said to be supplied exogenously in the diet. *Compare* **endogenous**.

exon A nucleotide sequence in a gene that codes for part or all of the gene product or for some control function. In eukaryotes, exons are separated by noncoding sequences called *introns.

exonuclease An enzyme that catalyses the cleavage of nucleotides from the end of a nucleic acid molecule. *Compare* **endonuclease**.

exopeptidase A protein-digesting enzyme that cleaves amino acids from the ends of a polypeptide chain. *Carboxypeptidase, which breaks down proteins in the small intestine, is an example of a exopeptidase. *Compare* **endopeptidase**.

exopterygote Any winged insect in which the eggs hatch into young (called *nymphs) that resemble the adults but lack wings; these develop gradually and externally in a series of stages (*instars) until the final moult produces the adult insect. There is no pupal stage and metamorphosis is described as *incomplete*. Exopterygote insects include the mayflies (*Ephemeroptera), dragonflies (*Odonata), grasshoppers (*Orthoptera), and bugs (*Hemiptera). *Compare* **endopterygote**.

exoskeleton A rigid external covering for the body in certain animals, such as the hard chitinous cuticle of arthropods. An exoskeleton protects and supports the body and provides points of attachment for muscles. The cuticle of arthropods must be shed at intervals to allow growth to occur (*see* **ecdysis**). Other examples of exoskeletons are the shells of molluscs and the bony plates of tortoises and armadillos. *Compare* **endoskeleton**.

exothermic Denoting a chemical reaction that releases heat into its surroundings. *See* **exergonic reaction**. *Compare* **endothermic**.

exotoxin *See* **toxin**.

experiment A process or trial designed to test a scientific theory.

experimental taxonomy *See* **systematics**.

expiration (**exhalation**) The process by which gas is expelled from the lungs (*see* **respiratory movement**). In mammals, the volume of the thoracic cavity is reduced by contraction of the internal *intercostal muscles and relaxation of the muscles of the diaphragm, assisted by upward pressure of the abdominal organs. As a result, pressure in the lungs exceeds atmospheric pressure and gas flows out of the lungs, allowing the pressures to equalize. *Compare* **inspiration**.

expiratory centre *See* **ventilation centre**.

explantation The removal of cells, tissues, or organs of animals and plants for observation of their growth and development in appropriate culture media. The removed parts are known as *explants*. *See also* **tissue culture; organ culture**.

exponential growth A form of *population growth in which the rate of growth is related to the number of individuals present. Increase is slow when numbers are low but rises sharply as numbers increase. If population number is plotted against time on a graph a characteristic J-shaped curve

Graph showing exponential growth of the human population

results (see graph). In animal and plant populations, such factors as overcrowding, lack of nutrients, and disease limit population increase beyond a certain point and the J-shaped exponential curve tails off giving an S-shaped (sigmoid) curve.

extensor Any muscle that causes a limb to extend. *See* **voluntary muscle.** *Compare* **flexor.**

exteroceptor Any *receptor that detects external stimuli. Examples of exteroceptors are the thermoreceptors in the skin, which monitor the temperature of the external environment. *Compare* **interoceptor.**

extinction 1. The irreversible condition of a species or other group of organisms of having no living representatives in the wild, which follows the death of the last surviving individual of that species or group. Extinction may occur on a local or global level; it can result from various human activities, including the destruction of habitats or the overexploitation of species that are hunted or harvested as a resource. Species at the top of a *food chain (e.g. large birds of prey) will be more prone to extinction since they exist in relatively small numbers and will be affected by a deleterious change at any of the levels in the food chain. **2.** The termination of a behaviour pattern that is no longer appropriate. For example, dogs can be conditioned to salivate when they hear a bell ring in the absence of a food stimulus (*see* **conditioning**). However, if the bell continues to be rung in the absence of food the dogs will gradually stop salivating on hearing the bell.

extracellular Located or occurring outside the cell. *Cuticularization is an example of an extracellular process.

extraembryonic membranes (embryonic membranes) The tissues produced by an animal *embryo for protection and nutrition but otherwise taking no part in its development. The four membranes, which are called *fetal membranes* in man, are the *chorion, *amnion, *allantois, and *yolk sac.

extrafusal *See* **intrafusal.**

extranuclear genes Genes included in the DNA present in organelles other than the nucleus, such as the mitochondria and chloroplasts, some of which code for the synthesis of proteins. The DNA of these organelles is inherited by the offspring via the cytoplasm of the gametes (*see* **cytoplasmic inheritance**). In organisms in which one gamete is much larger than the other, so that the smaller gamete contains very little cytoplasm, one parent will contribute most or all of the extranuclear genes. For example, human *mitochondrial DNA is passed from generation to generation through the maternal line in the ovum.

eye The organ of sight. The most primitive eyes are the *eyespots of some unicellular organisms. More advanced eyes are the *ocelli and *compound eyes of arthropods (e.g. insects). The cephalopod molluscs (e.g. the octopus

Structure of the vertebrate eye

and squid) and vertebrates possess the most highly developed eyes (see illustration). These normally occur in pairs, are nearly spherical, and filled with fluid. Light is refracted by the *cornea through the pupil in the *iris and onto the *lens, which focuses images onto the retina. These images are received by light-sensitive cells in the retina (*see* **cone; rod**), which transmit impulses to the brain via the optic nerve.

eye muscle Any of the muscles associated with the functioning of the eye, which are divided into two groups. The *intrinsic muscles*, which are involuntary, are situated inside the eyeball and comprise the ciliary muscle (*see* **ciliary body**) and the iris. The *extrinsic muscles*, which comprise three pairs of voluntary muscles, are inserted on the sclera (outer surface) of the eyeball and control its movements.

eyepiece (ocular) The lens or system of lenses in an optical instrument that is nearest to the eye. It usually produces a magnified image of the previous image formed by the instrument.

eyespot (stigma) 1. A structure found in some free-swimming unicellular algae and in plant reproductive cells that contains orange or red pigments (carotenoids) and is sensitive to light. It enables the cell to move in relation to a light source (*see* **phototaxis**). **2.** A spot of pigment found in some lower animals, e.g. jellyfish.

eye tooth A *canine tooth in the upper jaw.

F

F₁ (first filial generation) The first generation of offspring resulting from an arranged cross between *homozygous parents in breeding experiments. *See* **monohybrid**.

F₂ (second filial generation) The second generation of offspring in breeding experiments, obtained by crosses between individuals of the *F₁ generation. *See* **monohybrid**.

facilitated diffusion The transport of molecules across the outer membrane of a living cell by a process that involves a specific carrier located within the cell membrane but does not require expenditure of energy by the cell. The carrier combines with a molecule on one side of the membrane, moves through the membrane, and releases the molecule on the other side. *Compare* **active transport**.

facilitation *(in neurophysiology)* The effect of successive stimuli on the postsynaptic membrane, which results in the generation of an *excitatory postsynaptic potential (EPSP). Although a single impulse may fail to cross the junction between adjacent nerve cells, the synapse becomes more responsive to the following impulse.

Factor VIII (antihaemophilic factor) One of the blood *clotting factors. Factor VIII is a soluble protein that stimulates the activation of Factor X by Factor IXa, which in turn converts *prothrombin to thrombin, thus causing the fibrin matrix of a blood clot to form. *Haemophilia is due to a deficiency or defect of Factor VIII and is treated by administration of blood plasma or plasma concentrate containing the factor. Factor VIII can now be obtained from genetically engineered cell cultures. Such preparations avoid the risk of contamination with viruses, notably HIV (the AIDS virus).

facultative anaerobes Organisms, such as certain bacteria, fungi, and some internal parasites of animals, that are able to alter their metabolism to grow in either the presence or absence of oxygen. The best-known facultative anaerobe is *Saccharomyces cerevisiae*, the yeast used in brewing. *See* **anaerobic respiration**.

FAD (flavin adenine dinucleotide) A *coenzyme important in various biochemical reactions. It comprises a phosphorylated vitamin B_2 (riboflavin) molecule linked to the nucleotide adenine monophosphate (AMP). FAD is usually tightly bound to the enzyme forming a *flavoprotein*. It functions as a hydrogen acceptor in dehydrogenation reactions, being reduced to $FADH_2$. This in turn is oxidized to FAD by the *electron transport chain, thereby generating ATP (two molecules of ATP per molecule of $FADH_2$).

faeces Waste material that is eliminated from the alimentary canal through the *anus. Faeces consist of the indigestible residue of food that remains after the processes of digestion and absorption of nutrients and water have taken place, together with bacteria and dead cells shed from the gut lining.

Fahrenheit scale A temperature scale in which (by modern definition) the temperature of boiling water is taken as 212 degrees and the temperature of melting ice as 32 degrees. It was invented in 1714 by the German scientist G. D. Fahrenheit (1686–1736), who set the zero at the lowest temperature he knew how to obtain in the laboratory (by mixing ice and common salt) and took his own body temperature as 96°F. The scale is no longer in scientific use. To convert to the *Celsius scale the formula is

$$C = 5(F - 32)/9.$$

fallopian tube (oviduct) The tube that carries egg cells from the *ovary to the womb in mammals. The eggs are carried by the action of muscles and cilia. It was named after the Italian anatomist Gabriel Fallopius (1523–62).

false fruit *See* pseudocarp.

family A category used in the *classification of organisms that consists of one or several similar or closely related genera. Similar families are grouped into an order. Family names end in *-aceae* or *-ae* in botany (e.g. Cactaceae) and *-idae* in zoology (e.g. Equidae). The names are usually derived from a type genus (*Cactus* and *Equus* in the examples above) that is characteristic of the whole family (*see* **type specimen**). In botany, families are sometimes called *natural orders*.

farming *See* agriculture.

fascia A sheet of fibrous connective tissue occurring beneath the skin and also enveloping glands, vessels, nerves, and forming muscle and tendon sheaths.

fascicle **1.** A small bundle of nerve or muscle fibres. **2.** *See* **vascular bundle**.

fascicular cambium *See* cambium.

fast green A green dye used in optical microscopy that stains cellulose, cytoplasm, collagen, and mucus green. It is frequently used to stain plant tissues, with *safranin as a counterstain. Unlike *light green*, a similar dye, it does not fade easily.

fat A mixture of lipids, chiefly *triglycerides, that is solid at normal body temperatures. Fats occur widely in plants and animals as a means of storing food energy, having twice the calorific value of carbohydrates. In mammals, fat is deposited in a layer beneath the skin (subcutaneous fat) and deep within the body as a specialized *adipose tissue (*see also* **brown fat**). The insulating properties of fat are also important, especially in animals lacking fur and those inhabiting cold climates (e.g. seals and whales).

Fats derived from plants and fish generally have a greater proportion of unsaturated *fatty acids than those from mammals. Their melting points thus tend to be lower, causing a softer consistency at room temperatures. Highly unsaturated fats are liquid at room temperatures and are therefore more properly called *oils.

fat body 1. An abdominal organ in amphibians attached to the anterior of each kidney. It contains a reserve of fat that nourishes the gonads during the winter hibernation in readiness for the spring breeding season. **2.** A mass of fatty tissue spreading throughout the body cavity of insects in which fats, proteins, and glycogen are stored as a reserve for hibernation or pupation.

fat cell (adipocyte) Any of the cells of *adipose tissue, in which fats (triglycerides) are stored. Fat cells contain enzymes (lipases) that can break down fat into glycerol and fatty acids, which can be transported in the blood to the liver, where they are used in *fatty-acid oxidation.

fatigue A decline in the level of response of tissues (such as muscle), cells, etc., to nervous stimulation, which occurs after prolonged and continued stimulation of these structures.

fatty acid An organic compound consisting of a hydrocarbon chain and a terminal carboxyl group (*see* **carboxylic acids**). Chain length ranges from one hydrogen atom (methanoic, or formic, acid, HCOOH) to nearly 30 carbon atoms. Ethanoic (acetic), propanoic (propionic), and butanoic (butyric) acids are important in metabolism. Long-chain fatty acids (more than 8–10 carbon atoms) most commonly occur as constituents of certain lipids, notably glycerides, phospholipids, sterols, and waxes, in which they are esterified with alcohols. These long-chain fatty acids generally have an even number of carbon atoms; unbranched chains predominate over branched chains. They may be *saturated* (e.g. *palmitic (hexadecanoic) acid and *stearic (octadecanoic) acid) or *unsaturated*, with one double bond (e.g. *oleic (cis-octodec-9-enoic) acid) or two or more double bonds, in which case they are called *polyunsaturated fatty acids* (e.g. *linoleic acid and *linolenic acid). *See also* **essential fatty acids**.

The physical properties of fatty acids are determined by chain length, degree of unsaturation, and chain branching. Short-chain acids are pungent liquids, soluble in water. As chain length increases, melting points are raised and water-solubility decreases. Unsaturation and chain branching tend to lower melting points.

fatty-acid oxidation (β-oxidation) The metabolic pathway in which fats are metabolized to release energy. Fatty-acid oxidation occurs continually but does not become a major source of energy until the animal's carbohydrate resources are exhausted, for example during starvation. Fatty-acid oxidation occurs in the mitochondria. A series of reactions cleave off two carbon atoms at a time from the hydrocarbon chain of the fatty acid. These two-carbon fragments are combined with *coenzyme A to form *acetyl coenzyme A (acetal CoA), which then enters the *Krebs cycle. The formation of acetyl CoA occurs repeatedly until all the hydrocarbon chain has been used up. *See also* **glyoxylate cycle**.

fauna All the animal life normally present in a given habitat at a given time. *See also* **macrofauna; microfauna**. *Compare* **flora**.

feathers The body covering of birds, formed as outgrowths of the epidermis and composed of the protein *keratin. Feathers provide heat insulation, they give the body its streamlined shape, and those of the wings and tail are important in flight. Basically a feather consists of a *quill*, which is embedded in the skin attached to a feather follicle and is continuous with the shaft (*rachis*) of the feather, which carries the *barbs. This basic structure is modified depending on the type of feather (*see* **contour feathers; down feathers; filoplumes**).

fecundity The fertility of an organism (in higher animals, generally of the female of the species). Normally all organisms, assuming they reach reproductive age, are sufficiently fecund to replace themselves several times over. Darwin noted this, together with the fact that population numbers nevertheless tended to remain fairly constant: these observations led him to formulate his theory of evolution by *natural selection.

feedback The use of part of the output of a system to control its performance. In *positive feedback*, the output is used to enhance the input; in *negative feedback*, the output is used to reduce the input. Many biological processes rely on negative feedback. As the population of a species expands, so its food supply per individual is diminished; the result is that the population then begins to fall. Many biochemical processes are controlled by feedback *inhibition. Feedback mechanisms play an important role in maintaining a state of equilibrium within an organism (*see* **homeostasis**).

feeding *See* **ingestion**.

Fehling's test A chemical test to detect *reducing sugars and aldehydes in solution, devised by the German chemist H. C. von Fehling (1812–85). *Fehling's solution* consists of Fehlings A (copper(II) sulphate solution) and Fehling's B (alkaline 2,3-dihydroxybutanedioate (sodium tartrate) solution), equal amounts of which are added to the test solution. After boiling, a positive result is indicated by the formation of a brick-red precipitate of copper(I) oxide. Methanal, being a strong reducing agent, also produces copper metal; ketones do not react. The test is now little used, having been replaced by *Benedict's test.

female 1. Denoting the gamete (sex cell) that, during *sexual reproduction, fuses with a *male gamete in the process of fertilization. Female gametes are generally larger than the male gametes and are usually immotile (*see* **oosphere; ovum**). **2.** (Denoting) an individual organism whose reproductive organs produce only female gametes. *Compare* **hermaphrodite**.

femoral Of or relating to the thigh or the femur (thigh bone). For example, the *femoral artery* runs down the front of the thigh.

femto- Symbol f. A prefix used in the metric system to denote 10^{-15}. For example, 10^{-15} second = 1 femtosecond (fs).

femur 1. The thigh bone of terrestrial vertebrates. It articulates at one end with the pelvic girdle at the hip joint and at the other (via two *condyles) with the *tibia. **2.** The third segment of an insect's leg, attached to the *trochanter. *See also* **coxa**.

fenestra Either of the two delicate membranes between the *middle ear and the *inner ear. The upper membrane is the *fenestra ovalis* (*see* **oval window**); the lower membrane is the *fenestra rotunda* (*see* **round window**).

fenestration The condition of tissues or other structures within an organism of being perforated by window-like slits. For example, the endothelium of the capillaries in the glomerulus within the kidneys is fenestrated in order to facilitate the filtration of substances.

feral animal Any domesticated animal, such as a dog or a cat, that has returned to live in wild conditions.

fermentation A form of *anaerobic respiration occurring in certain microorganisms, e.g. yeasts. *Alcoholic fermentation* comprises a series of biochemical reactions by which pyruvate (the end product of *glycolysis) is converted to ethanol and carbon dioxide. It is the basis of the baking and *brewing industries (*see* **bakers' yeast**). In *lactic-acid fermentation*, which occurs in many microorganisms and (when sugar is in short supply) in animal cells, the end product is lactic acid.

ferns *See* **Filicinophyta**.

ferredoxin A protein containing iron associated with sulphur that is a *carrier molecule in the electron transport chain in *photosynthesis. It accepts electrons from photosystem I and passes them on to NADP reductase, which donates them for the reduction of $NADP^+$. *See* **photophosphorylation; photosystems I and II**.

ferritin A protein, found predominantly in the tissues of the liver and spleen but also present in nearly all cells of the body, that is used for the storage of iron. A ferritin molecule, which is spherical and has a crystalline core, can store as many as 4000 iron atoms, which can be released when required for *haemoglobin formation. *Compare* **transferrin**.

fertilization (syngamy) The union of male and female gametes (reproductive cells) during the process of sexual reproduction to form a *zygote*. It involves the fusion of the gametic nuclei (*karyogamy*) and cytoplasm (*plasmogamy*). As each gamete contains only half the correct number of chromosomes, fertilization and zygote formation results in a cell with the full complement of chromosomes, half of which are derived from each of the parents. In animals the process involves fusion of the nuclei of a spermatozoan and an ovum. In most aquatic animals (e.g. fish) this takes place in the surrounding water, into which the gametes are shed. Among most terrestrial animals (e.g. insects, many mammals) fertilization occurs in the body of the female, into which the sperms are introduced. In flowering plants, after *pollination, the pollen grain produces a *pollen tube, which grows down into the female reproductive organ (carpel) to enable a male gamete nucleus to fuse with the egg nucleus (see **double fertilization**).

In *self-fertilization* the male and female gametes are derived from the same individual. Among plants, self-fertilization (also called *autogamy*) is common in many cultivated species, e.g. wheat and oats. However, self-fertilization is a form of *inbreeding and does not allow for the mixing of genetic material; if it occurs over a number of generations it will result in offspring being less vigorous and productive than those resulting from cross-fertilization. In *cross-fertilization* (also called *allogamy* in plants) the gametes are derived from different individuals. In plants the pollen comes either from another flower of the same plant or from a different plant (see also **incompatibility**).

fertilizer Any substance that is added to soil in order to increase its productivity. Fertilizers can be of natural origin, such as *composts, or they can be made up of synthetic chemicals, particularly nitrates and phosphates. Synthetic fertilizers can increase crop yields dramatically, but when leached from the soil by rain, which runs into lakes, they also increase the process of eutrophication (see **algal bloom; eutrophic**). Bacteria that can fix nitrogen are sometimes added to the soil to increase its fertility; for example, in tropical countries the cyanobacterium *Anabaena* is added to rice paddies to increase soil fertility.

fetal membranes *See* **extraembryonic membranes**.

fetus (foetus) The *embryo of a mammal, especially a human, when development has reached a stage at which the main features of the adult form are recognizable. In humans the embryo from eight weeks to birth is called a fetus.

Feulgen's test A histochemical test in which the distribution of DNA in the chromosomes of dividing cell nuclei can be observed. It was devised by the German chemist R. Feulgen (1884–1955). A tissue section is first treated with dilute hydrochloric acid to remove the purine bases of the DNA, thus exposing the aldehyde groups of the sugar deoxyribose. The section is then immersed in *Schiff's reagent, which combines with the aldehyde groups to form a magenta-coloured compound.

fibre 1. An elongated plant cell whose walls are extensively (usually completely) thickened with lignin (see **sclerenchyma**). Fibres are found in the vascular tissue, usually in the xylem, where they provide structural support. The term is often used loosely to mean any kind of xylem element. The fibres of many species, e.g. flax, are of commercial importance. **2.** Any of various threadlike structures in the animal body,

such as a muscle fibre, a nerve fibre, a collagen fibre, or an elastic fibre.
3. (*or* **dietary fibre; roughage**) The part of food that cannot be digested
and absorbed to produce energy. Dietary fibre falls into four groups:
cellulose, hemicelluloses, lignins, and *pectins.* Highly refined foods, such as
sucrose, contain no dietary fibre. Foods with a high fibre content include
wholemeal cereals and flour, root vegetables, nuts, and fruit. Dietary fibre
is considered by some to be helpful in the prevention of many of the
diseases of Western civilization, such as diverticulosis, constipation,
appendicitis, obesity, and diabetes mellitus.

fibre optics *See* **optical fibres**.

fibril Any small fibre or threadlike structure. *See also* **microfibril**.

fibrin The insoluble protein that forms fibres at the site of an injury and
is the foundation of a blood clot. *See* **blood clotting**.

fibrinogen The protein dissolved in the blood plasma that, when suitably
activated, is converted to insoluble *fibrin fibres. *See* **blood clotting**.

fibrinolysis The breakdown of the protein *fibrin by the enzyme
*plasmin (or *fibrinolysin*), which occurs when blood clots are removed from
the circulation.

fibroblast A cell that secretes fibres in the intercellular substance of
*connective tissue. The cells are long, flat, and star-shaped and lie close to
collagen fibres.

fibrocartilage *See* **cartilage**.

fibrous protein *See* **protein**.

fibula The smaller and outer of the two bones between the knee and the
ankle in terrestrial vertebrates. *Compare* **tibia**.

field-emission microscope (FEM) A type of electron microscope in
which a high negative voltage is applied to a metal tip placed in an
evacuated vessel some distance from a glass screen with a fluorescent
coating. The tip produces electrons by *field emission,* i.e. the emission of
electrons from an unheated sharp metal part as a result of a high electric
field. The emitted electrons form an enlarged pattern on the fluorescent
screen, related to the individual exposed planes of atoms. As the resolution
of the instrument is limited by the vibrations of the metal atoms, it is
helpful to cool the tip in liquid helium. Although the individual atoms
forming the point are not displayed, individual adsorbed atoms of other
substances can be, and their activity is observable.

field-ionization microscope (field-ion microscope; FIM) A type of
electron microscope that is similar in principle to the *field-emission
microscope, except that a high positive voltage is applied to the metal tip,
which is surrounded by low-pressure gas (usually helium) rather than a
vacuum. The image is formed in this case by *field ionization:* ionization at
the surface of an unheated solid as a result of a strong electric field

creating positive ions by electron transfer from surrounding atoms or molecules. The image is formed by ions striking the fluorescent screen. Individual atoms on the surface of the tip can be resolved and, in certain cases, adsorbed atoms may be detected.

filament 1. *(in zoology)* A long slender hairlike structure, such as any of the *barbs of a bird's feather. **2.** *(in botany)* The stalk of the *stamen in a flower. It bears the anther and consists mainly of conducting tissue.

Filicinophyta A phylum of mainly terrestrial vascular plants (*see* **tracheophyte**) – the ferns. Ferns are perennial plants bearing large conspicuous leaves (*fronds*: *see* **megaphyll**) usually arising from either a rhizome or a short erect stem. Bracken is a common example. Only the tree ferns have stems that reach an appreciable height. There is a characteristic uncurling of the young leaves as they expand into the adult form. Reproduction is by means of spores borne on the underside of specialized leaves (*sporophylls).

filoplumes Minute hairlike *feathers consisting of a shaft (*rachis*) bearing a few unattached barbs. They are found between the contour feathers.

filter feeding A method of feeding in which tiny food particles are strained from the surrounding water by various mechanisms. It is used by many aquatic invertebrates, especially members of the plankton, and by some vertebrates, notably baleen whales. *See also* **ciliary feeding**; **whalebone**.

filtrate The clear liquid obtained by filtration.

filtration The process of separating solid particles using a filter. In *vacuum filtration*, the liquid is drawn through the filter by a vacuum pump. *Ultrafiltration is filtration under pressure; for example, ultrafiltration of the blood occurs in the *nephrons of the vertebrate kidney.

finger domain A finger-shaped structure produced in a protein when a series of the constituent amino acids combines with a metal atom. Finger domains are often found repeated in *transcription factors. *See also* **domain**.

fingerprinting *See* **genetic fingerprinting**; **peptide mapping**.

fins The locomotory organs of aquatic vertebrates. In fish there are typically one or more *dorsal* and *ventral fins* (sometimes continuous), whose function is balance; a *caudal fin* around the tail, which is the main propulsive organ; and two paired fins: the *pectoral fins* attached to the pectoral (shoulder) girdle and the *pelvic fins* attached to the pelvic (hip) girdle, which are used in steering. These paired fins are homologous with the limbs of tetrapods. Fins are strengthened by a number of flexible fin rays, which may be cartilaginous, bony and jointed, horny, or fibrous and jointed.

first convoluted tubule *See* **proximal convoluted tubule**.

fish *See* **Chondrichthyes** (cartilaginous fish); **Osteichthyes** (bony fish); **Pisces**.

fission A type of asexual reproduction occurring in some unicellular organisms, e.g. diatoms, protozoans, and bacteria, in which the parent cell divides to form two (*binary fission*) or more (*multiple fission*) similar daughter cells.

fission-track dating A method of estimating the age of glass and other mineral objects by observing the tracks made in them by the fission fragments of the uranium nuclei that they contain. By irradiating the objects with neutrons to induce fission and comparing the density and number of the tracks before and after irradiation it is possible to estimate the time that has elapsed since the object solidified.

fitness *(in evolution)* The condition of an organism that is well adapted to its environment, as measured by its ability to reproduce itself. *See also* **inclusive fitness**.

fixation 1. The first stage in the preparation of a specimen for microscopical examination, in which the tissue is killed and preserved in as natural a state as possible by immersion in a chemical *fixative*. The fixative prevents the distortion of cell components by denaturing its constituent protein. Some commonly used fixatives are formaldehyde, ethanol, and Bouin's fluid (for light microscopy), and osmium tetroxide and gluteraldehyde (for electron microscopy). Fixation may also be brought about by heat. **2.** *See* **nitrogen fixation**.

fixed action pattern *See* **instinct**.

flaccid *(in botany)* Describing plant tissue that has become soft and less rigid than normal because the cytoplasm within its cells has shrunk and contracted away from the cell walls through loss of water (*see* **plasmolysis**).

flagellum (*pl.* **flagella**) A relatively long (up to 150 μm) fine whiplike structure present on the surface of certain cells, particularly motile reproductive cells (e.g. spermatozoa), bacteria, and certain protozoans. Flagella occur singly or in small groups. Beating of flagella produces movement of the cell, but some flagella (e.g. in *Hydra*) beat to cause movement of the surrounding fluid. *See also* **axoneme**. *Compare* **cilium**.

flame cells Ciliated cells that form part of the excretory and osmoregulatory system of platyhelminths, rotifers, and nemertine worms. This system, known as a *protonephridium*, consists of branching tubules that open to the exterior through excretory pores; flame cells occur at the ends of the tubules, into which their cilia project. Fluid, containing water and nitrogenous waste products, diffuses into the flame cells and is directed through the tubules to the exterior by movements of the cilia, which resemble the flickering of flames. Flame cells that possess only one cilium are known as *solenocytes*; these are found in the protonephridia of marine worms of the phylum Priapulida.

flatworms *See* Platyhelminthes.

flavin adenine dinucleotide *See* FAD.

flavonoid One of a group of naturally occurring phenolic compounds many of which are plant pigments. They include the *anthocyanins, flavonols*, and *flavones*. Patterns of flavonoid distribution have been used in taxonomic studies of plant species.

flavoprotein *See* FAD.

fleas *See* Siphonaptera.

flexor A muscle that causes a limb to bend by bringing the two parts of the limb together. An example is the *biceps. Flexors work antagonistically with *extensors. *See* **voluntary muscle**.

flies *See* Diptera.

flight 1. Any form of *locomotion in air, which can be active or passive (*gliding*). Mechanisms of flight have evolved mainly in birds, bats, and insects: these animals are adapted for flight by the presence of *wings*, which increases the ratio of surface area to body weight. Birds possess powerful flight muscles: the *depressor* muscle runs from the underside of the humerus to the sternum and is responsible for the downstroke of the wing; the *levator* muscle works antagonistically, producing the upstroke. Flight in insects works in a similar fashion but the muscles that control the wing movement are attached to the thorax. A few species of mammals, reptiles, and fish have developed flight to a lesser extent. For example, flying squirrels (order Dermoptera) possess a membrane attached to the limbs that can open and function as a parachute, allowing the animals to glide. **2.** Part of a survival mechanism in an animal that is generated in response to a threatening situation. A potentially dangerous situation can induce the release of *adrenaline, which prepares the animal for 'fight or flight' by increasing the blood pressure and heart rate and diverting the blood flow to the muscles and heart. *See* **alarm response**.

flip-flop The movement (*transverse diffusion*) of a lipid molecule from one surface of a *lipid bilayer membrane to the other, which occurs at a very slow rate. This contrasts with the much faster rate at which lipid molecules exchange places with neighbouring molecules on the same surface of the membrane (*lateral diffusion*).

flocculation The process in which particles in a colloid aggregate into larger clumps. Flocculation of clay particles in soil can be induced by the addition of calcium salts. Clay particles have an overall negative charge and therefore attract positive ions, such as Ca^{2+}, which form bridges holding the particles together. Flocculation is also often observed in cultures of bacterial and yeast cells.

flora All the plant life normally present in a given habitat at a given time. *See also* **microflora**. *Compare* **fauna**.

floral formula A summary of the structure and components of a flower using symbols and numbers. The symbols representing symmetry include ⊕ (for actinomorphy) and ·|·, ↓, or ↑ (for zygomorphy). The parts of the flower are represented by K (for calyx), C (for corolla), P (for perianth), A (for androecium), and G (for gynoecium). These are followed by numbers indicating the number of parts in each whorl (e.g. K5 indicates a calyx of five sepals); ∞ indicates an indefinite number of parts (more than 12). If the parts are fused the number is in parentheses; if they are in separate groups or whorls the number is split (e.g. C2 + 2 indicates a corolla of two whorls each of two petals). The symbol G has a line above it to indicate an inferior ovary and a line beneath it to indicate a superior ovary. For example, the buttercup flower has the following floral formula:

⊕ K5 C5 A∞ G∞,

i.e. it has a calyx of five sepals, a corolla of five petals, an androecium of many stamens, and a superior ovary consisting of many carpels.

florigen A hypothetical plant growth substance that is postulated to transmit the stimulus for flowering, which is a response to photoperiod (*see* **photoperiodism**), from the leaves to the apex of the plant. However, florigen has never been isolated and some plant physiologists question its existence.

Section through a monocarpellary flower at the time of pollination

flower The structure in angiosperms (flowering plants) that bears the organs for sexual reproduction. Flowers are very variable in form, ranging from the small green insignificant wind-pollinated flowers of many grasses to spectacular brightly coloured insect-pollinated flowers. Flowers are often grouped together into *inflorescences, some of which (e.g. that of dandelion) are so compacted as to resemble a single flower. Typically flowers consist of a receptacle that bears sepals, petals, stamens, and carpels (see illustration). The flower parts are adapted to bring about pollination and fertilization resulting in the formation of seeds and fruits. The sepals are usually green and leaflike and protect the flower bud. The petals of insect-pollinated flowers are adapted in many ingenious ways to attract insects and, in some instances, other animals. For example, some flowers are adapted to attract short-tongued insects by having an open shallow *corolla tube and nectar situated in an exposed position. Flowers adapted for pollination by long-tongued insects have a long corolla tube of fused petals with nectar in a concealed position. The tongue of the insect brushes against the anthers and stigma before reaching the nectar. Wind-pollinated flowers, in contrast, are inconspicuous. The anthers dangle outside the corolla and the stigmas have a feathery surface to catch the pollen grains.

Some species are adapted for self-pollination and have small flowers, no nectar, and stamens and carpels that mature simultaneously.

flowering plants *See* **Angiospermophyta**.

Fluid mosaic model of cell membranes

fluid mosaic model A widely accepted model, proposed by S. J. Singer and G. L. Nicholson in the 1970s, for the structure of biological membranes. It is based on a *lipid bilayer forming the framework of the membrane, with proteins randomly embedded in it; these proteins may be restricted to either lipid layer (*extrinsic proteins*) or they may span both (*intrinsic proteins*), producing an asymmetric structure (see illustration). This structure allows the lateral movement of the constituent phospholipid and protein molecules of the membrane.

flukes *See* **Trematoda**.

fluoridation The process of adding very small amounts of fluorine salts (e.g. sodium fluoride, NaF) to drinking water to prevent tooth decay. The fluoride becomes incorporated into the fluoroapatite (*see* **apatite**) of the growing teeth and reduces the incidence of *dental caries.

focusing *(in animal physiology)* The process of directing and concentrating light from a source onto the *retina of the eye, by means of the lens, in order to obtain a clear image of objects at a range of distances. *See* **accommodation**.

foetus *See* **fetus**.

folacin *See* **folic acid**.

folic acid **(folacin)** A vitamin of the *vitamin B complex. In its active form, tetrahydrofolic acid, it is a *coenzyme in various reactions involved in the metabolism of amino acids, purines, and pyrimidines. It is synthesized by intestinal bacteria and is widespread in food, especially green leafy vegetables. Deficiency causes poor growth and nutritional anaemia.

follicle **1.** *(in animal anatomy)* Any enclosing cluster of cells that protects and nourishes a cell or structure within. For example, follicles in the *ovary contain developing egg cells, while *hair follicles envelop the roots of hairs. **2.** *(in botany)* A dry fruit that, when ripe, splits along one side to release its seeds. It is formed from a single carpel containing one or more seeds. Follicles do not occur singly but are grouped to form clusters (*etaerios*). Examples include larkspur, columbine, and monk's hood.

follicle-stimulating hormone **(FSH)** A hormone, secreted by the anterior pituitary gland in mammals, that stimulates, in female mammals, ripening of specialized structures in the ovary (*Graafian follicles) that produce ova and, in males, the formation of sperm in the testis. It is a major constituent of fertility drugs, used to treat failure of ovulation and decreased sperm production. *See also* **gonadotrophin**.

follicular phase The phase of the *oestrous cycle (menstrual cycle of women) during which the Graafian follicles mature and the lining of the uterus thickens under the influence of oestrogen.

fontanelle A gap in the bones of the skull. Newborn babies are born with a fontanelle, which disappears as the bones in the skull fuse; the fusion is complete at around 18 months of age.

food Any material containing *nutrients, such as carbohydrates, proteins and fats, which are required by living organisms in order to obtain energy and to grow. Heterotrophic organisms, such as animals, ingest their food (*see also* **diet**); autotrophic organisms, such as plants, manufacture their food materials.

food additive A substance added to a food during its manufacture or processing in order to improve its keeping qualities, texture, appearance, or stability or to enhance its taste or colour. Additives are usually present in minute quantities; they include colouring materials, sweeteners, preservatives (*see* **food preservation**), *antioxidants, emulsifiers, and stabilizers. In most countries the additives used must be selected from an approved list of such compounds, which have been tested for safety, and they must be listed on the food labels of individual products.

food chain The transfer of energy from green plants (the primary producers) through a sequence of organisms in which each eats the one below it in the chain and is eaten by the one above. Thus plants are eaten by herbivores, which are then eaten by carnivores. These may in turn be eaten by different carnivores. The position an organism occupies in a food chain is known as its *trophic level. In practice, many animals feed at several different trophic levels, resulting in a more complex set of feeding relationships known as a *food web. *See* **bioenergetics; consumer; producer; pyramid of biomass; pyramid of energy; pyramid of numbers**.

food poisoning An acute illness caused by food that may be naturally poisonous or contaminated by certain types of pathogenic microorganisms. The most common type of food poisoning in the UK is that caused by the bacteria belonging to the genus *Salmonella*, which inhabit the alimentary canal of livestock. Freezing and other types of *food preservation can prevent the growth of the bacteria and thorough cooking will kill the microorganisms before the meat is eaten. However, food poisoning can result if frozen meat is not completely thawed at its centre before cooking, as it may not reach sufficiently high temperatures to kill the bacteria during cooking. Another type of food poisoning, known as botulism, is caused by toxins produced by the bacterium *Clostridium botulinum*, which can grow in badly preserved canned foods.

food preservation Prevention of the spoilage of food, which is achieved by a variety of techniques. These aim to prevent bacterial and fungal decay and contamination of food, which can cause *food poisoning. For example, dehydration removes the water from food, which prevents microorganisms from growing. Treating food with salt (salting) causes the microorganisms to lose water due to osmosis. Pickling involves treatment with vinegar (ethanoic acid), which reduces the pH and prevents bacteria from growing. Heating food (*blanching*) to temperatures of 90°C denatures the enzymes that cause the breakdown of food and kills many bacteria. The food is then packed in air-tight containers, such as cans or bottles. Heating milk to high temperatures to kill the bacteria is the basis of *pasteurization. Freezing food prevents the growth of bacteria but does not necessarily kill them; thorough cooking is therefore essential. In *freeze drying, food is rapidly frozen and then dehydrated, usually in a vacuum. Preprepared food can be preserved by the addition of chemicals (*see* **food additive**), such as sodium benzenecarboxylate, proprionates, and sulphur dioxide, but some of these may have adverse side-effects. *Irradiation* is a recently developed method of food preservation in which the bacteria are killed by irradiating the food with gamma rays.

food production *See* **agriculture**.

food reserves Reserves of fat, carbohydrate, or (rarely) protein in cells and tissues that function as an important store of energy that can be released and used in ATP production when required by the organism. For example, in animals *fat is stored in adipose tissue, and carbohydrate – in the form of the *storage compound* *glycogen – is stored in liver and muscle cells. In plants *starch is a major storage compound, being found in perennating organs (*see* **perennation**) and seeds (in which it is mobilized at germination), and oils are important storage materials in some species (e.g. in the seeds of the castor-oil plant).

food supply *(in human ecology)* The production of food for human consumption. *See* **agriculture**.

food web A system of *food chains that are linked with one another. In a food web a particular organism may feed at more than one trophic level. For example, in a pond food web a freshwater mussel may feed directly on green algae, in which case it is a primary consumer. However, it can also feed on protozoa, which are themselves primary consumers, in which case the mussel is the secondary consumer. A food web does not usually include the decomposers, but these organisms are very important in the flow of energy through a food web (*see* **energy flow**).

foramen An aperture in an animal part or organ, especially one in a bone or cartilage. For example, the *foramen magnum* is the opening at the base of the skull through which the *spinal cord passes.

forebrain (prosencephalon) One of the three sections of the brain of a vertebrate embryo. The forebrain develops to form the *cerebrum, *hypothalamus, and *thalamus in the adult. *Compare* **hindbrain; midbrain**.

foregut 1. The anterior region of the alimentary canal of vertebrates, up to the anterior part of the duodenum. **2.** The anterior part of the alimentary canal of arthropods. *See also* **hindgut; midgut**.

forest An area of vegetation in which the dominant plants are trees; forests constitute major *biomes. Temperate forests have adequate or abundant rainfall and moderate temperatures. They may be dominated by deciduous trees (such as oak, ash, elm, beech, or maple), often growing together to form mixed deciduous forest, as in temperate regions of Europe, Asia, and North America; or by broad-leaved evergreens (such as southern beech, *Nothofagus*), as in Chile. Cold forests, of northern regions, are dominated by evergreen conifers (*see* **taiga**). Tropical forests include *rainforest, characterized by regular heavy rainfall; monsoon forest, found in SE Asia and having heavy rainfall interspersed with periods of drought; and thorn forest, as in SW North America, SW Africa, and parts of Central and South America and Australia, which has sparse rainfall, is dominated by small thorny trees, and grades into *savanna woodland and semidesert.

form **1.** A category used in the *classification of organisms into which different types of a variety may be placed. **2.** Any distinct variant within a species. Seasonal variants, e.g. the tawny brown (summer) and blue-white (winter) forms of the blue hare, may be called forms, as may the different types that constitute a *polymorphism.

fossil The remains or traces of any organism that lived in the geological past. In general only the hard parts of organisms become fossilized (e.g. bones, teeth, shells, and wood) but under certain circumstances the entire organism is preserved. For example, virtually unaltered fossils of extinct mammals, such as the woolly mammoth and woolly rhinoceros, have been found preserved in ice in the Arctic. Small organisms or parts of organisms (e.g. insects, leaves, flowers) have been preserved in *amber.

In the majority of fossils the organism has been turned to stone – a process known as *petrification*. This may take one of three forms. In *permineralization*, solutions originating underground fill the microscopic cavities in the organism. Minerals in these solutions (e.g. silica or calcite) may actually replace the original material of the organism so that even microscopic structures may be preserved; this process is known as *replacement* (or *mineralization*). A third form of petrification – *carbonization* (or *distillation*) – occurs in certain soft tissues that are composed chiefly of compounds of carbon, hydrogen, and oxygen (e.g. cellulose). After the organism has been buried, and in the absence of oxygen, carbon dioxide and water are liberated until only free carbon remains. This forms a black carbon film in the rock outlining the original organism. *Moulds* are formed when the original fossil is dissolved away leaving a mould of its outline in the solid rock. The deposition of mineral matter from underground solutions in a mould forms a *cast*. Palaeontologists often produce casts from moulds using such substances as dental wax. Moulds of thin organisms (e.g. leaves) are commonly known as *imprints*. *Trace fossils* are the fossilized remnants of the evidence of animal life, such as tracks, trails, footprints, burrows, and *coprolites* (fossilized faeces).

The ideal conditions for the formation of fossils occur in areas of rapid sedimentation, especially those parts of the seabed that lie below the zone of wave disturbance. *See also* **chemical fossil; index fossil; microfossil**.

fossil fuel Coal, oil, and natural gas, the fuels used by man as a source of energy. They are formed from the remains of living organisms and all have a high carbon or hydrogen content. Their value as fuels relies on the exothermic oxidation of carbon to form carbon dioxide ($C + O_2 \rightarrow CO_2$) and the oxidation of hydrogen to form water ($H_2 + \frac{1}{2}O_2 \rightarrow H_2O$).

fossil hominid *See* **hominid**.

fovea (fovea centralis) A shallow depression in the *retina of the eye, opposite the lens, that is present in some vertebrates. This area contains a large concentration of *cones with only a thin layer of overlying nerves. It is therefore specialized for the perception of colour and sharp intense images. The clarity is enhanced when light is focused on the foveae of both retinas simultaneously. *See* **binocular vision**.

fragmentation A method of asexual reproduction, occurring in some invertebrate animals, in which parts of the organism break off and subsequently differentiate and develop into new individuals. It occurs especially in certain coelenterates and annelids. In some, regeneration may occur before separation, producing chains of individuals budding from the parent.

frameshift An alteration in the sequence of DNA bases read as base triplets during *transcription, caused by the deletion (or addition) of a single nucleotide in the DNA sequence. The missing (or additional) base results in an abnormal triplet and also causes each subsequent triplet group to be altered. A frameshift mutation therefore produces messenger RNA with corresponding changed *codons and results in the synthesis of an abnormal protein. For example, the deletion of uracil (U) in the second codon of the following mRNA sequence:

AUU   CAU   CGG   UAG   ACC   UGU   AUG

results in the frameshifted sequence:

AUU   CAC   GGU   AGA   CCU   GUA   UG

fraternal twins (dizygotic twins) Two individuals that result from a single pregnancy, each having developed from a separate fertilized egg. The two egg cells contain different combinations of *alleles and so do the two sperm that fertilize them. The twins therefore have no more genetic similarity than brothers or sisters from single births. *Compare* **identical twins**.

free energy A measure of a system's ability to do work. The *Gibbs free energy* (or *Gibbs function*), G, is defined by $G = H - TS$, where G is the energy liberated or absorbed in a reversible process at constant pressure and constant temperature (T), H is the *enthalpy and S the *entropy of the system. Changes in Gibbs free energy, ΔG, are useful in indicating the conditions under which a chemical reaction will occur. If ΔG is positive the reaction will only occur if energy is supplied to force it away from the equilibrium position (i.e. when $\Delta G = 0$). If ΔG is negative the reaction will proceed spontaneously to equilibrium.

free radical An atom or group of atoms with an unpaired valence electron. Free radicals are formed when a bond is broken without forming ions. Because of their unpaired electrons, most free radicals are extremely reactive. For example, superoxide free radicals, formed in the body during normal metabolic processes and as a reaction to toxins and infections, have damaging effects on cells and tissues. *See also* **antioxidants; superoxide dismutase**.

freeze drying (lyophilization) The removal of liquid from heat-sensitive materials. The material is frozen, placed under a high vacuum, and maintained at a low temperature (−40°C or below). The pressure generated by the vacuum causes the ice to turn from a solid to a gaseous form without passing through a liquid state. This allows the removal of water from the material without otherwise disturbing its composition. Freeze drying is used to preserve tissues (e.g. blood plasma) and foods and to concentrate solutions.

freeze fracture A method of preparing material for electron microscopy that allows the visualization of the interior of cell membranes and organelles. Cells are frozen at −196°C and cracked so that the plane of fracture runs through the middle of *lipid bilayers, separating the two halves. The exposed surfaces are then coated with carbon and platinum and the organic material is digested with enzymes (*freeze etching*), leaving a carbon–platinum replica of the fractured surface, which can be examined using the microscope.

frogs *See* **Amphibia**.

frond *See* **megaphyll**.

frontal lobe The anterior part of each cerebral hemisphere, which is associated with the higher mental functions, such as abstract thought.

fructification *See* **sporophore**.

fructose (fruit sugar; laevulose) A simple sugar, $C_6H_{12}O_6$, stereoisomeric with glucose (*see* **monosaccharide**). (Although natural fructose is the D-form, it is in fact laevorotatory.) Fructose occurs in green plants, fruits, and honey and tastes sweeter than sucrose (cane sugar), of which it is a constituent. Derivatives of fructose are important in the energy metabolism of living organisms. Some polysaccharide derivatives (*fructans*) are carbohydrate energy stores in certain plants.

fructose 1,6-bisphosphate An intermediate formed in the initial stage of *glycolysis by the phosphorylation of glucose using ATP.

fruit The structure formed from the ovary of a flower, usually after the ovules have been fertilized (*see also* **parthenocarpy**). It consists of the *fruit wall* (*see* **pericarp**) enclosing the seed(s). Other parts of the flower, such as the receptacle, may develop and contribute to the structure, resulting in a *false fruit* (*see* **pseudocarp**). The fruit may retain the seeds and be dispersed whole (an *indehiscent fruit*), or it may open (dehisce) to release the seeds (a *dehiscent fruit*). Fruits are divided into two main groups depending on whether the ovary wall remains dry or becomes fleshy (*succulent*). Succulent fruits are generally dispersed by animals and dry fruits by wind, water, or by some mechanical means. See illustration. *See also* **composite fruit**.

fruit sugar *See* **fructose**.

frustule The cell wall of a diatom (*see* **Bacillariophyta**), which is

Different types of fruit and methods of seed dispersal

impregnated with silica. The wall consists of two halves that overlap one another.

FSH *See* **follicle-stimulating hormone.**

fucoxanthin The major *carotenoid pigment present, with chlorophyll, in the brown algae (*see* **Phaeophyta**).

fumaric acid A carboxylic acid, HCOOHC:CHCOOH, that is an intermediate in the *Krebs cycle, being formed by the dehydrogenation of succinic acid.

fungi A group of organisms formerly regarded as simple plants lacking chlorophyll but now classified in a separate kingdom, Fungi. They can

either exist as single cells or make up a multicellular body called a
*mycelium, which consists of filaments known as *hyphae. Most fungal
cells are multinucleate and have cell walls composed chiefly of *chitin.
Fungi exist primarily in damp situations on land and, because of the
absence of chlorophyll, are either parasites or saprotrophs on other
organisms. The principal criteria used in classification are the nature of the
spores produced and the presence or absence of cross walls within the
hyphae (*see* **Ascomycota; Basidiomycota; Deuteromycota; Zygomycota**).
See also **lichens**.

fungicide *See* pesticide.

Fungi Imperfecti *See* Deuteromycota.

funicle The stalk that attaches an ovule to the placenta in the ovary of a
flowering plant. It contains a strand of conducting tissue leading from the
placenta into the chalaza.

furanose A *sugar having a five-membered ring containing four carbon
atoms and one oxygen atom.

fusion The combining together of cells, nuclei, or cytoplasm. *See* **cell
fusion; fertilization**.

G

GABA *See* **gamma-aminobutyric acid.**

Gaia hypothesis The theory, based on an idea put forward by the British scientist James Ephraim Lovelock (1919–), that the whole earth, including both its biotic (living) and abiotic (nonliving) components, functions as a single self-regulating system. Named after the Greek earth goddess, it proposes that the responses of living organisms to environmental conditions ultimately bring about changes that make the earth better adapted to support life; the system would rid itself of any species that adversely affects the environment. The theory has found favour with many conservationists.

galactose A simple sugar, $C_6H_{12}O_6$, stereoisomeric with glucose, that occurs naturally as one of the products of the enzymic digestion of milk sugar (lactose) and as a constituent of gum arabic.

β-galactosidase *See* **lactase.**

gall bladder A small pouch attached to the *bile duct, present in most vertebrates. *Bile, produced in the *liver, is stored in the gall bladder and released when food (especially fatty substances) enters the duodenum.

gallstone A hard ball of material formed in the gall bladder from surplus cholesterol, which precipitates in the *bile. Gallstones may lodge in and obstruct the bile duct.

GALP *See* **glyceraldehyde 3-phosphate.**

gametangium An organ that produces gametes. The term is usually restricted to the sex organs of algae, fungi, mosses, and ferns. *See* **antheridium; archegonium; oogonium.**

gamete A reproductive cell that fuses with another gamete to form a zygote. Examples of gametes are ova and spermatozoa. Gametes are *haploid, i.e. they contain half the normal (diploid) number of chromosomes; thus when two fuse, the diploid number is restored (*see* **fertilization**). Gametes are formed by *meiosis. Gametes often differ in size, the smaller (usually male) gamete being known as the *microgamete* and the larger (usually female) as the *macrogamete. See also* **sexual reproduction.**

gametogenesis The processes involved in the formation of gametes. Gametes are normally formed by *meiosis but sometimes by *mitosis (as in the gametophyte generation of the ferns). In mammals gametogenesis in the female is known as *oogenesis and occurs in the ovaries; in the male it is known as *spermatogenesis and occurs in the testes.

gametophyte The generation in the life cycle of a plant that bears the gamete-producing sex organs. The gametophyte is *haploid. It is the dominant phase in the life cycle of mosses and other bryophytes, the *sporophyte generation depending on it either partially or completely. In clubmosses, horsetails, and ferns it is the *prothallus. In seed plants it is very much reduced. For example, in angiosperms the pollen grain is the male gametophyte and the embryo sac is the female gametophyte. *See also* **alternation of generations**.

gamma-aminobutyric acid (GABA) An inhibitory *neurotransmitter in the central nervous system (principally the brain) that is capable of increasing the permeability of postsynaptic membranes (*see* **inhibitory postsynaptic potential**). GABA is synthesized by *decarboxylation of the amino acid glutamate.

gamma globulin *See* globulin.

gamopetalous Describing a flower in which the petals are fused to form a *corolla tube. *Compare* **polypetalous**.

gamosepalous Describing a flower that possesses a calyx consisting of fused sepals. *Compare* **polysepalous**.

ganglion A mass of nervous tissue containing many *cell bodies and *synapses, usually enclosed in a connective-tissue sheath. In vertebrates most ganglia occur outside the central nervous system; exceptions are the *basal ganglia in the brain. In invertebrates ganglia occur along the nerve cords and the most anterior pair (*cerebral ganglia*) are analogous to the vertebrate brain; invertebrate ganglia constitute a part of the central nervous system.

ganglion cell *See* retina.

ganoid scale *See* scales.

gap junction A passage through the lipid bilayers of adjacent cell membranes that mediates the transfer of small molecules or ions between interacting cells. Gap junctions are abundant in epithelial tissues and cardiac muscle. They consist of hexagonally packed tubes, approximately 7 nm in diameter, through which small molecules or ions may directly pass from the interior of one cell to the interior of the other. Gap junctions, together with chemical *synapses (which function through *neurotransmitters), are *communicating junctions* and comprise one of several types of intercellular junction (*compare* **desmosome; tight junction**).

gas bladder *See* swim bladder.

gaseous exchange The transfer of gases between an organism and the external environment in either direction. It occurs by diffusion across a *concentration gradient and includes the exchange of oxygen and carbon dioxide in respiration and photosynthesis. Successful gaseous exchange requires a large surface area, as is provided by the alveoli of the lungs and the leaves of plants.

gas–liquid chromatography A technique for separating or analysing mixtures of gases by *chromatography. The apparatus consists of a very long tube containing the stationary phase, a nonvolatile liquid, such as a hydrocarbon oil coated on a solid support. The sample is often a volatile liquid mixture (e.g. of fatty acids), which is vaporized and swept through the column by a carrier gas (e.g. hydrogen). The components of the mixture pass through the column at different rates and are detected as they leave, either by measuring the thermal conductivity of the gas or by a flame detector.

Gas chromatography is usually used for analysis; components can be identified by the time they take to pass through the column. It is also used for separating mixtures into their components, which are then directly injected into a mass spectrometer in the technique of gas chromatography–*mass spectroscopy.

gasohol Alcohol that has been produced by fermentation of sugar cane by the yeast *Saccharomyces cerevisiae. Gasohol has been produced on a large scale in Brazil, where it is manufactured in large distilleries and provides an alternative source of energy for motor vehicles. In Brazil all cars are adapted so as to be able to use gasohol or a mixture of gasohol and petrol. Gasohol is an example of a *biofuel.

gastric Of or relating to the stomach.

gastric gland *See* gastric juice.

gastric juice An acidic mixture of inorganic salts, *hydrochloric acid, and digestive enzymes (e.g. *pepsin) produced by *gastric glands* in the stomach lining.

gastric mill (proventriculus) A type of *gizzard occurring in many crustaceans. It is situated in the anterior region of the stomach and consists of a set of bones (ossicles) and muscles that grind food particles. The food particles are then filtered by bristles in the posterior section of the stomach.

gastrin A hormone, produced in the stomach, that controls the release of gastric juice. The secretion of gastrin is stimulated by the presence of food in the stomach. It is one of the hormones that integrates and controls digestive processes (*see also* **secretin**).

gastrodermis The epithelial lining of the digestive tract of certain invertebrates, including the nematode worms and coelenterates.

Gastropoda A class of molluscs that includes the snails, whelks, limpets, land and sea slugs, and conches. Molluscs have a well-developed head with tentacles, a large flattened foot, and a coiled twisted shell. They occupy marine, freshwater, and terrestrial habitats; in the terrestrial and some freshwater gastropods the *mantle cavity acts as a lung instead of enclosing gills.

gastrula The stage in the *development of an animal embryo that succeeds the *blastula. It begins with the production of the primary *germ layers and the embryo becomes converted to a cup-shaped structure containing a cavity (the *archenteron).

gel A lyophilic *colloid that has coagulated to a rigid or jelly-like solid. In a gel, the disperse medium has formed a loosely-held network of linked molecules through the dispersion medium. Examples of gels are silica gel and gelatin.

gelatin(e) A colourless or pale yellow water-soluble protein obtained by boiling collagen with water and evaporating the solution. It swells when water is added and dissolves in hot water to form a solution that sets to a gel on cooling. It is used in bacteriology for preparing culture media, in pharmacy for preparing capsules and suppositories, and in jellies and other foodstuffs.

gel filtration A type of column *chromatography in which a mixture of liquids is passed down a column containing a gel. Small molecules in the mixture can enter pores in the gel and move slowly down the column; large molecules, which cannot enter the pores, move more quickly. Thus, mixtures of molecules can be separated on the basis of their size. The technique is used particularly for separating proteins but it can also be applied to cell nuclei, viruses, etc.

gemmation A type of *vegetative propagation in which small clumps of undifferentiated cells (*gemmae*) develop on the surface of a plant. These are shed and dispersed to other areas, where they grow to produce new individuals. Gemmation is found only in certain lower plants, such as mosses and liverworts.

gene A unit of heredity composed of DNA. In classical genetics (*see* **Mendelism; Mendel's laws**) a gene is visualized as a discrete particle, forming part of a *chromosome, that determines a particular characteristic. It can exist in different forms called *alleles, which determine which aspect of the characteristic is shown (e.g. tallness or shortness for the characteristic of height).

A gene occupies a specific position (*locus) on a chromosome. It may be defined as the shortest length of chromosome that cannot be broken by recombination (*see* **meiosis**) or that can undergo *mutation, both units being recognized as a single base pair in the nucleotide sequence that makes up the DNA. The gene as a unit of function may be defined as the sequence of nucleotides concerned with a specific function, such as the synthesis of a single polypeptide chain or of a messenger RNA molecule, corresponding to a particular sequence of the *genetic code. One or more of these *structural genes*, coding for protein, may be associated with other genes controlling their expression (*see* **operon**).

gene amplification The multiple replication of a section of the
*genome, which occurs during a single cell cycle and results in the
production of many copies of a specific sequence of the DNA molecule. For
example, in the oocytes of amphibians and other animals, in which large
numbers of ribosomes are needed, the genes encoding ribosomal RNA are
greatly amplified. Viral genes that cause the formation of tumours (*see*
oncogene) are amplified in tumour cells.

gene bank *See* **gene library**.

gene cloning (DNA cloning) The production of exact copies (*clones*) of a
particular gene or genes using genetic engineering techniques. The DNA
containing the target gene(s) is split into fragments using *restriction
enzymes. These fragments are then inserted into cloning *vectors, such as
bacterial plasmids or bacteriophages, which transfer the recombinant DNA
to suitable host cells, such as the bacterium *E. coli*. Alternatively,
*complementary DNA is inserted into the vectors, or 'naked' DNA
fragments can be taken up directly by a host bacterium from its medium
(this is less efficient than vector transfer).

Inside the host cell the recombinant DNA undergoes replication; thus, a
bacterial host will give rise to a colony of cells containing the cloned target
gene. Various screening methods may be used to identify such colonies,
enabling them to be selected and cultured. Gene cloning facilitates *gene
sequencing; it also enables large quantities of a desired protein product to
be produced: human insulin, for example, is now produced by bacteria
containing the cloned insulin gene.

gene expression The manifestation of the effects of a gene by the
production of the particular protein, polypeptide, or type of RNA whose
synthesis it controls. Individual genes can be 'switched on' (exert their
effects) or 'switched off' according to the needs and circumstances of the
cell at a particular time. A number of mechanisms are thought to be
responsible for the control of gene expression; the *Jacob–Monod
hypothesis postulates the mechanism operating in prokaryotes (*see* **operon**).
Control of gene expression is known to be more complicated in eukaryotes,
which possess various control mechanisms not seen in prokaryotes. For
example, the methylation of cytosine bases of specific genes in eukaryotic
DNA (*DNA methylation*) is observed in cells in which the gene is not
expressed; if DNA methylation is prevented by the use of inhibitory
chemicals, this can cause certain genes to be expressed.

gene flow The exchange of genetic material by interbreeding between
populations of the same species or between individuals within a population.
Gene flow increases the variation in the genetic composition of a
population.

gene frequency *See* **allele frequency**.

gene imprinting The differential expression of a single gene according
to its parental origin. For example, the symptoms of Huntingdon's disease,

an inherited neuromuscular disorder caused by a dominant mutant allele, first occur during adolescence if the mutant allele is inherited from the male parent but symptoms generally do not arise until middle age if the allele derives from the female parent.

gene library (gene bank) A collection of cloned DNA fragments representing the entire genetic material of an organism. This facilitates screening and isolation of any particular gene. Gene libraries are created by fractionating the genomic DNA into fragments using *restriction enzymes and/or physical methods. These fragments are cloned (*see* **gene cloning**) and the host cells containing the recombinant fragments are centrifuged and frozen; alternatively, the phage *vectors are maintained in culture. Individual genes in the library are identified using specific *gene probes with the *Southern blotting technique or, via their protein products, using *Western blotting. Gene libraries are thus repositories of raw material for use in genetic engineering. A human gene library may contain some 500 000 separate clones.

gene manipulation *See* **biotechnology; genetic engineering.**

gene mutation *See* **point mutation.**

gene pool All the *genes and their different alleles that are present in a population of a plant or animal species. *See also* **population genetics.**

gene probe (DNA probe) A single-stranded DNA or RNA fragment used in genetic engineering to search for a particular gene or other DNA sequence. The probe has a base sequence complementary to the target sequence and will thus attach to it by *base pairing. By labelling the probe with a radioactive isotope it can be identified on subsequent separation and purification. Probes of varying lengths, up to about 100 nucleotides, can be constructed in the laboratory. They are used in the *Southern blotting technique to identify particular DNA fragments, for instance in conjunction with *restriction mapping to diagnose gene abnormalities or to map certain sequences.

generation A group of organisms of approximately the same age within a population. Organisms that are crossed to produce offspring in a genetics study are referred to as the *parental generation* and their offspring as the first *filial generation*. *See also* F_1; F_2; **P**.

generation time The interval between the beginnings of consecutive cell divisions. It may be as short as 20 minutes in bacteria. *See also* **interphase.**

generative nucleus One of the two nuclei in an angiosperm *pollen grain (*compare* **tube nucleus**). It divides to produce two male gamete nuclei (*see* **double fertilization**).

generator potential The local electrical charge that develops in the sensitive part of a receptor cell, known as the *generator region*. The generator potential is caused by *depolarization of the membrane surrounding this part of the cell, brought about by an exchange of ions in response to a stimulus. The strength of the potential is proportional to the strength of the stimulus: when the potential exceeds a certain threshold, it fires an action potential. More action potentials are triggered as long as the generator potential exists.

gene sequencing (DNA sequencing) The process of elucidating the nucleotide sequence of a gene. Two techniques are used. The Maxam–Gilbert method (named after Allan Maxam and Walter Gilbert) involves cleaving the DNA with a *restriction enzyme and labelling each of the resulting fragments with ^{32}P-phosphate at one end. The fragments are subjected to four different sets of reactions, each set specifically cleaving DNA at a particular base or bases. The cleaved fragments are separated by electrophoresis according to their chain length and identified by autoradiography. The base (nucleotide) sequence is deduced from the position of bands in each of the four lanes in the gel.

The Sanger method (named after Frederick Sanger), also called the dideoxy method, involves synthesizing a new DNA strand using as template single-stranded DNA from the gene being sequenced. Synthesis of the new strand can be stopped at any of the four bases by adding the corresponding dideoxy (dd) derivative of the deoxyribonucleoside phosphates; for example, by adding ddATP the synthesis terminates at an adenosine; by adding ddGTP it terminates at a guanosine, etc. As in the first method, the fragments, which comprise radiolabelled nucleotides, are finally subjected to electrophoresis and autoradiography. Gene sequencing is now employed on a major scale, for example in determining the nucleotide sequence of the entire human genome (see **Human Genome Project**).

gene splicing A stage in the processing of messenger *RNA (see **transcription**), occurring only in eukaryote cells, in which noncoding *introns are removed from the primary mRNA transcript and the coding *exons are spliced together to form the functional mRNA molecule. Splicing is catalysed by a complex of small RNA molecules and proteins called a *spliceosome*. In some organisms, self-splicing of mRNA is known to occur.

gene therapy The application of genetic engineering techniques to alter or replace defective genes. A defective gene may result from an incorrect sequence of bases in the DNA molecule or an inability of the gene to code for the expression of a particular polypeptide. Techniques currently being investigated involve the transfer of normal genes into the genetic material of the cell to replace the defective gene. *Retroviruses are often used as *vectors for transferring genes into cells as part of the natural retrovirus life cycle involves the insertion of their own genetic material into the chromosomes of their host. Alternatively *liposomes may be used. Gene therapy is being developed to cure such genetic diseases as cystic fibrosis.

genetic code The means by which genetic information in *DNA controls the manufacture of specific proteins by the cell. The code takes the form of a series of triplets of bases in DNA, from which is transcribed a

complementary sequence of *codons in messenger *RNA (*see*
transcription). The sequence of these codons determines the sequence of
amino acids during *protein synthesis. There are 64 possible codes from
the combinations of the four bases present in DNA and messenger RNA and
20 amino acids present in body proteins: some of the amino acids are coded
by more than one codon, and some codons have other functions (*see* **start
codon; stop codon**). See illustration.

First base in codon						Second base in codon				Third base in codon
	U		**C**		**A**		**G**			
U	UUU	Phe	UCU	Ser	UAU	Tyr	UGU	Cys		U
	UUC	Phe	UCC	Ser	UAC	Tyr	UGC	Cys		C
	UUA	Leu	UCA	Ser	UAA	(stop codon)	UGA	(stop codon)		A
	UUG	Leu	UCG	Ser	UAG	(stop codon)	UGG	Trp		G
C	CUU	Leu	CCU	Pro	CAU	His	CGU	Arg		U
	CUC	Leu	CCC	Pro	CAC	His	CGC	Arg		C
	CUA	Leu	CCA	Pro	CAA	Gln	CGA	Arg		A
	CUG	Leu	CCG	Pro	CAG	Gln	CGG	Arg		G
A	AUU	Ile	ACU	Thr	AAU	Asn	AGU	Ser		U
	AUC	Ile	ACC	Thr	AAC	Asn	AGC	Ser		C
	AUA	Ile	ACA	Thr	AAA	Lys	AGA	Arg		A
	AUG	Met	ACG	Thr	AAG	Lys	AGG	Arg		G
	(start codon)									
G	GUU	Val	GCU	Ala	GAU	Asp	GGU	Gly		U
	GUC	Val	GCC	Ala	GAC	Asp	GGC	Gly		C
	GUA	Val	GCA	Ala	GAA	Glu	GGA	Gly		A
	GUG	Val	GCG	Ala	GAG	Glu	GGG	Gly		G

The genetic code

genetic drift (Sewall Wright effect) Evolutionary changes in small
populations produced by random variation in *allele frequencies and not
by *natural selection. The resulting *mutant genes generally confer no
significant benefit on the organisms in which they occur, and they may be
disadvantageous.

genetic engineering (recombinant DNA technology) The techniques
involved in altering the characters of an organism by inserting genes from
another organism into its DNA. This altered DNA (known as *recombinant
DNA*) is usually produced by *gene cloning. Genetic engineering has many
applications, ranging from the commercial production of insulin and other
hormones, vaccines, etc., to the creation of *transgenic animals and crop
plants in agriculture. *See also* **gene library; gene probe; gene therapy;
monoclonal antibody**.

genetic fingerprinting (DNA fingerprinting) A technique in which an individual's DNA is analysed to reveal the pattern of repetition of particular nucleotide sequences (the marker sequences) throughout the genome. This pattern is claimed to be unique to the individual concerned, and the technique is therefore used for identification purposes in forensic science and paternity disputes, and in veterinary science. Sufficient DNA can be obtained from very small samples of body tissue, such as blood, semen, or hair. *Restriction enzymes are used to cleave the DNA and the marker sequences are revealed by specific *gene probes and *Southern blotting.

genetic load The concentration of deleterious genes carried within a population.

genetic mapping *See* **chromosome map; restriction mapping.**

genetic marker *See* **marker gene.**

genetic polymorphism *See* **polymorphism.**

genetics The branch of biology concerned with the study of heredity and variation. *Classical genetics* is based on the work of Gregor Mendel (*see* **Mendelism**). During the 20th century genetics has expanded to overlap with the fields of ecology and animal behaviour (*see* **behavioural genetics; population genetics**), and important advances in biochemistry and microbiology have led to clarification of the chemical nature of *genes and the ways in which they can replicate and be transmitted. *See also* **genetic engineering**.

genetic variation *See* **variation.**

gene tracking A method for determining the inheritance of a particular gene in a family. It is used in the diagnosis of genetic diseases, such as cystic fibrosis and Huntington's chorea. *Restriction fragment length polymorphisms (RFLPs) situated in or near the locus of interest are identified using *gene probes, and suitable marker RFLPs selected. These can then be traced through members of the family and used to detect the presence or absence of the disease locus prenatally in future at-risk pregnancies.

genome All the genes contained in a single set of chromosomes, i.e. in a *haploid nucleus. Each parent, through its reproductive cells, contributes its genome to its offspring.

genotoxicity The condition resulting from the interaction of toxic agents (*genotoxins*) with DNA molecules in genes. Since the genes are passed down to the next generation, the toxicity induced by genotoxins is heritable. Genotoxins can induce mutations in chromosomes (*clastogenesis*) or in a small number of base pairs (*mutagenesis*). Genotoxic agents include X-rays, natural *carcinogens, some man-made products (e.g. acridine and vinyl chloride), and viruses.

genotype The genetic composition of an organism, i.e. the combination of *alleles it possesses. *Compare* **phenotype**.

genotype frequency The percentage of individuals in a population that possess a specific genotype. The genotype frequency can be calculated using the Hardy–Weinberg equation (*see* **Hardy–Weinberg equilibrium**).

genus (*pl.* **genera**) A category used in the *classification of organisms that consists of a number of similar or closely related species. The common name of an organism (especially a plant) is sometimes similar or identical to that of the genus, e.g. *Lilium* (lily), *Crocus, Antirrhinum.* Similar genera are grouped into families. *See also* **binomial nomenclature**.

geochronology *See* varve dating.

geographical isolation The separation of two populations of the same species or breeding group by a physical barrier, such as a mountain or body of water. Geographical isolation may ultimately lead to the populations becoming separate species by adaptive radiation. *See also* **allopatric; speciation**.

geological time scale A time scale that covers the earth's history from its origin, estimated to be about 4600 million years ago, to the present. The chronology is divided into a hierarchy of time intervals: eras, periods, epochs, ages, and chrons (see Appendix).

geotaxis The movement of a cell or microorganism in response to gravity. For example, certain cnidarian larvae that swim towards the seabed exhibit positive geotaxis. *See* **taxis**.

geotropism The growth of plant organs in response to gravity. A main root is positively geotropic and a main stem negatively geotropic, growing downwards and upwards respectively, irrespective of the positions in which they are placed. For example, if a stem is placed in a horizontal position it will still grow upwards. Geotropism is thought to be controlled by *auxin, a plant growth substance. *See* **tropism**.

germ cell Any cell in the series of cells (the *germ line*) that eventually produces *gametes, especially the first cell in such a series. In mammals the germ cells are the oogonia (in the ovaries) and the spermatogonia (in the testes). *See* **oogenesis; spermatogenesis**.

germinal epithelium 1. A layer of epithelial cells on the surface of the ovary that are continuous with the *mesothelium. These cells do not give rise to the ova (they were formerly thought to do this). **2.** The layer of epithelial cells lining the seminiferous tubules of the testis, which gives rise to spermatogonia (*see* **spermatogenesis**).

germination 1. The initial stages in the growth of a seed to form a seedling. The embryonic shoot (plumule) and embryonic root (radicle) emerge and grow upwards and downwards respectively. Food reserves for germination come from *endosperm tissue within the seed and/or from the seed leaves (cotyledons). *See also* **epigeal; hypogeal. 2.** The first signs of growth of spores and pollen grains.

germ layers (primary germ layers) The layers of cells in an animal embryo at the *gastrula stage, from which are derived the various organs of the animal's body. There are two or three germ layers: an outer layer (*see* **ectoderm**), an inner layer (*see* **endoderm**), and in most animal groups a middle layer (*see* **mesoderm**). *See also* **development**.

germ plasm *See* Weismannism.

gestation The period in animals bearing live young (especially mammals) from the fertilization of the egg to birth of the young (parturition). In man gestation is known as *pregnancy* and takes about nine months (40 weeks).

giant chromosome *See* polyteny.

giant fibre A nerve fibre with a very large diameter, found in many types of invertebrate (e.g. earthworms and squids). Its function is to allow extremely rapid transmission of nervous impulses and hence rapid escape movements in emergencies.

Gibberellic acid

gibberellic acid (GA₃) A plant *growth substance, abundant in young actively growing areas of the plant, that is involved in stem elongation. A *terpene, it was discovered in 1954 (see formula). Gibberellic acid and related growth substances are called *gibberellins.

gibberellin Any of a group of plant *growth substances chemically related to *gibberellic acid. Gibberellins promote shoot elongation in certain plants, often overcoming genetic dwarfism. They are also involved in the release of buds from dormancy and in promoting seed germination.

giga- Symbol G. A prefix used in the metric system to denote one thousand million times. For example, 10^9 joules = 1 gigajoule (GJ).

gigantism *See* growth hormone.

gill 1. *(in zoology)* A respiratory organ used by aquatic animals to obtain oxygen from the surrounding water. A gill consists essentially of a membrane or outgrowth from the body, with a large surface area and a plentiful blood supply, through which diffusion of oxygen and carbon dioxide between the water and blood occurs. Fishes have *internal gills*, formed as outgrowths from the pharynx wall and contained within *gill slits. Water entering the mouth is pumped out through these slits and over the gills. The gills of most aquatic invertebrates and amphibian larvae are *external gills*, which project from the body so that water passes over them as the animal moves. **2. (in botany)** One of the ridges of tissue that radiate from the centre of the underside of the cap of mushrooms. The spores are produced on these gills.

gill bar A cartilaginous support for the tissue between the gill slits in lower chordates, such as lancelets.

gill slit An opening leading from the pharynx to the exterior in aquatic vertebrates and lancelets. In lancelets they function in *filter feeding. In fish they contain the *gills and are usually in the form of a series of long slits. They are absent in adult tetrapod vertebrates (except for some amphibians) but their presence in some form in the embryos of all vertebrates is a characteristic of the phylum *Chordata.

gingiva (gum) The part of the epithelial tissue lining the mouth that covers the jaw bones. It is continuous with the sockets surrounding the roots of the teeth.

Ginkgophyta *See* **gymnosperm**.

gizzard A muscular compartment of the alimentary canal of many animals that is specialized for breaking up food. In birds the gizzard lies between the *proventriculus and the duodenum and contains small stones and grit, which assist in breaking up the food when the gizzard contracts. *See also* **gastric mill**.

gland A group of cells or a single cell in animals or plants that is specialized to secrete a specific substance. In animals there are two types of glands, both of which synthesize their secretions. *Endocrine glands discharge their products directly into the blood vessels; *exocrine glands secrete through a duct or network of ducts into a body cavity or onto the body surface. Secretory cells are characterized by having droplets (*vesicles*) containing their products. *See also* **secretion**.

In plants glands are specialized to secrete certain substances produced by the plant. The secretions may be retained within a single cell, secreted into a special cavity or duct, or secreted to the outside. Examples are the water glands (*hydathodes) of certain leaves, nectaries (*see* **nectar**), and the digestive glands of certain carnivorous plants.

glenoid cavity The socket-shaped cavity in the *scapula (shoulder blade) that holds the head of the *humerus in a ball-and-socket joint.

glia (glial cells; neuroglia) Cells of the nervous system that support the neurones. There are four classes of glial cells: *astrocytes, oligodendrocytes, ependymal cells*, and *microglia*. Oligodendrocytes form insulating sheaths of *myelin round neurones in the central nervous system, preventing impulses from travelling between adjacent neurones. Other functions of glial cells include providing nutrients for neurones and controlling the biochemical composition of the fluid surrounding the neurones.

global warming *See* greenhouse effect.

globin *See* haemoglobin.

globular protein *See* protein.

globulin Any of a group of globular proteins that are generally insoluble in water and present in blood, eggs, milk, and as a reserve protein in seeds. Blood serum globulins comprise four types: α_1-, α_2-, and β-globulins, which serve as carrier proteins; and γ-globulins (*gamma globulins*), which include the *immunoglobulins responsible for immune responses.

glomerular filtrate The fluid in the lumen of the Bowman's capsule of the *nephron that has been filtered from the capillaries of the glomerulus (*see* **ultrafiltration**). The glomerular filtrate has the same composition as the plasma except that it does not contain any of the larger components, such as plasma proteins or cells.

glomerulus (*pl.* **glomeruli**) A tangled mass of blood capillaries enclosed by the cup-shaped end (*Bowman's capsule) of a kidney tubule (*see* **nephron**). Fluid is filtered from these capillaries into the Bowman's capsule and down the nephron (*see* **glomerular filtrate**).

glottis The opening from the pharynx to the trachea (windpipe). In mammals it also serves as the space for the *vocal cords. *See also* **epiglottis**; **larynx**.

glucagon A hormone, secreted by the *islets of Langerhans in the pancreas, that increases the concentration of glucose in the blood by stimulating the metabolic breakdown of glycogen. It thus antagonizes the effects of *insulin (*see* **antagonism**).

glucocorticoid *See* corticosteroid.

gluconeogenesis The synthesis of glucose from noncarbohydrate sources, such as fat and protein. This occurs when the glycogen supplies in the liver are exhausted. The pathway is essentially a reversal of *glycolysis from pyruvate to glucose and it can utilize many sources, including amino acids, glycerol, and *Krebs cycle intermediates. Large-scale protein and fat catabolism normally occurs only in those suffering from starvation or certain endocrine disorders.

gluconic acid An optically active hydroxycarboxylic acid, $CH_2(OH)(CHOH)_4COOH$. It is the carboxylic acid corresponding to the aldose sugar glucose, and can be made by the action of certain moulds.

glucosamine *See* amino sugar.

glucosan Any one of a class of *polysaccharide compounds that can be converted to glucose by hydrolysis. Glucosans include dextrin, starch, and cellulose.

glucose (dextrose; grape sugar) A white crystalline sugar, $C_6H_{12}O_6$, occurring widely in nature. Like other *monosaccharides, glucose is optically active: most naturally occurring glucose is dextrorotatory. Glucose and its derivatives are crucially important in the energy metabolism of living organisms. It is a major energy source, being transported around the body in blood, lymph, and cerebrospinal fluid to the cells, where energy is released in the process of *glycolysis. Glucose is present in the sap of plants, in fruits, and in honey and is also a constituent of many polysaccharides, most notably of starch and cellulose. These yield glucose when broken down, for example by enzymes during digestion.

glucuronic acid A compound, $OC_6H_9O_6$, derived from the oxidation of glucose. It is an important constituent of *gums and *mucilages. Glucuronic acid can combine with hydroxyl (–OH), carboxyl (–COOH), or amino (–NH$_2$) groups to form a *glucuronide*. The addition of a glucuronide group to a molecule (*glucuronidation*) generally increases the solubility of a compound; hence glucuronidation plays an important role in the excretion of foreign substances (*see* **phase II metabolism**).

glucuronide *See* glucuronic acid.

glume Either member of a pair of bracts that are found at the base of a spikelet (*see* **spike**) of the grasses. *Compare* **lemma**.

glutamic acid *See* amino acid.

glutamine *See* amino acid.

glutathione A *peptide comprising the amino acids glutamic acid, cysteine, and glycine. It occurs widely in plants, animals, and microorganisms, serving chiefly as an *antioxidant. Reduced glutathione reacts with potentially harmful oxidizing agents and is itself oxidized. This is important in ensuring the proper functioning of proteins, haemoglobin, membrane lipids, etc. Glutathione is also involved in amino acid transport across cell membranes.

gluten A mixture of two proteins, gliadin and glutenin, occurring in the endosperm of wheat grain. Their amino acid composition varies but glutamic acid (33%) and proline (12%) predominate. The composition of wheat glutens determines the 'strength' of the flour and whether or not it is suitable for biscuit or bread making. Sensitivity of the lining of the intestine to gluten occurs in *coeliac disease*, a condition that must be treated by a gluten-free diet.

glyceraldehyde 3-phosphate (GALP) A triose phosphate, $CHOCH(OH)CH_2OPO_3H_2$, that is an intermediate in the *Calvin cycle (*see also* **photosynthesis**) and glycolysis.

glycerate 3-phosphate A phosphorylated three-carbon monosaccharide

that is an intermediate in the *Calvin cycle of photosynthesis and also in
*glycolysis. It was formerly known as *3-phosphoglycerate* or *phosphoglyceric
acid (PGA)*.

glyceride (acylglycerol) A fatty-acid ester of glycerol. Esterification can
occur at one, two, or all three hydroxyl groups of the glycerol molecule
producing *monoglycerides*, *diglycerides*, and *triglycerides respectively.
Triglycerides are the major constituent of fats and oils found in living
organisms. Alternatively, one of the hydroxyl groups may be esterified with
a phosphate group forming a phosphoglyceride (*see* **phospholipid**) or a
sugar forming a *glycolipid.

glycerine *See* glycerol.

glycerol (glycerine; propane-1,2,3,-triol) A trihydric alcohol,
$HOCH_2CH(OH)CH_2OH$. Glycerol is a colourless sweet-tasting viscous liquid,
miscible with water but insoluble in ether. It is widely distributed in all
living organisms as a constituent of the *glycerides, which yield glycerol
when hydrolysed.

glycine *See* amino acid.

glycobiology The study of carbohydrates and carbohydrate complexes,
especially *glycoproteins.

glycocalyx (cell coat) A layer of carbohydrate on the surface of the cell
membrane of most eukaryotic cells. It is made up of the oligosaccharide
side-chains of the glycolipid and glycoprotein components of the
membrane and may include oligosaccharides secreted by the cell. The
function of the glycocalyx is not known.

glycogen (animal starch) A *polysaccharide consisting of a highly
branched polymer of glucose occurring in animal tissues, especially in liver
and muscle cells. It is the major store of carbohydrate energy in animal
cells.

glycogenesis The conversion of glucose to glycogen, which is stimulated
by insulin from the pancreas. Glycogenesis occurs in skeletal muscles and
to a lesser extent in the liver. Glucose that is taken up by cells is
phosphorylated to glucose 6-phosphate; this is converted successively to
glucose 1-phosphate, uridine diphosphate glucose, and finally to glycogen.
See also **gluconeogenesis**. *Compare* **glycogenolysis**.

glycogenolysis The conversion of glycogen to glucose, which occurs in
the liver and is stimulated by glucagon from the pancreas and adrenaline
from the adrenal medulla. These hormones activate an enzyme that
phosphorylates glucose molecules in the glycogen chain to form glucose 1-
phosphate, which is converted to glucose 6-phosphate. This is then
converted to glucose by a phosphatase enzyme. In skeletal muscle glycogen
is degraded to glucose 6-phosphate, which is then converted into pyruvate
and used in ATP production during glycolysis and the Krebs cycle. However,

pyruvate can also be converted, in the liver, to glucose; thus muscle glycogen is indirectly a source of blood glucose. *Compare* **glycogenesis**.

glycolipid Any of a group of sugar-containing lipids, in which the lipid portion of the molecule is usually based on glycerol (*see* **glyceride**) or sphingosine and the sugar is typically galactose, glucose, or inositol. Glycolipids are components of biological membranes. In animal plasma membranes they are found in the outer layer of the lipid bilayer; the simplest animal glycolipids are the *cerebrosides. Plant glycolipids are glycerides in which the sugar group is most commonly galactose. They are the principal lipid constituents of chloroplasts.

The principal stages of glycolysis

glycolysis (Embden–Meyerhof pathway) The series of biochemical reactions in which glucose is broken down to pyruvate with the release of usable energy in the form of *ATP (see illustration). One molecule of glucose undergoes two phosphorylation reactions and is then split to form two triose-phosphate molecules. Each of these is converted to pyruvate. The net energy yield is two ATP molecules per glucose molecule. In *aerobic respiration pyruvate then enters the *Krebs cycle. Alternatively, when oxygen is in short supply or absent, the pyruvate is converted to various products by *anaerobic respiration. Other simple sugars, e.g. fructose and galactose, and glycerol (from fats) enter the glycolysis pathway at intermediate stages. *Compare* **gluconeogenesis**.

glycoprotein A carbohydrate linked covalently to a protein. Formed in the Golgi apparatus in the process of *glycosylation, glycoproteins are important components of cell membranes, in which they extend throughout the *lipid bilayer. They are also constituents of body fluids, such as mucus, that are involved in lubrication. Many of the hormone receptors on the surfaces of cells have been identified as glycoproteins. Glycoproteins produced by viruses attach themselves to the surface of the host cell, where they act as markers for the receptors of leucocytes. Viral glycoproteins can also act as target molecules and help viruses to detect certain types of host cell; for example, a glycoprotein on the surface of *HIV enables the virus to find and infect white blood cells.

glycosaminoglycan Any one of a group of polysaccharides that contain *amino sugars (such as glucosamine). Formerly known as *mucopolysaccharides*, they include *hyaluronic acid and chondroitin (*see* **cartilage**), which provide lubrication in joints and form part of the matrix of cartilage. The three-dimensional structure of these molecules enables them to trap water, which forms a gel and gives glycosaminoglycans their elastic properties.

glycoside Any one of a group of compounds consisting of a pyranose sugar residue, such as glucose, linked to a noncarbohydrate residue (R) by a *glycosidic bond: the hydroxyl group (−OH) on carbon-1 of the sugar is replaced by −OR. Glycosides are widely distributed in plants; examples are the *anthocyanin pigments and the *cardiac glycosides*, such as digoxin (*see* **digitalis**) and ouabain, which are used medicinally for their stimulant effects on the heart.

Formation of a glycosidic bond

glycosidic bond (glycosidic link) The type of chemical linkage between the monosaccharide units of disaccharides, oligosaccharides, and polysaccharides, which is formed by the removal of a molecule of water (i.e. a *condensation reaction). The bond is normally formed between the carbon-1 on one sugar and the carbon-4 on the other (see illustration). An α-glycosidic bond is formed when the −OH group on carbon-1 is below the plane of the glucose ring and a β-glycosidic bond is formed when it is above the plane. *Cellulose is formed of glucose molecules linked by 1-4 β-glycosidic bonds, whereas starch is composed of 1-4 α-glycosidic bonds.

glycosuria *See* **insulin**.

glycosylation The process in which a carbohydrate is joined to another molecule, such as a protein to form a *glycoprotein or to a lipid to form a *glycolipid. Glycosylation occurs in the rough endoplasmic reticulum and the Golgi apparatus of cells.

Glyoxylate cycle

glyoxylate cycle A metabolic pathway in plants and microorganisms that is a modified form of the *Krebs cycle. It utilizes fats as a source of carbon and enables the synthesis of carbohydrate from fatty acids by avoiding the stages of the Krebs cycle in which carbon dioxide is released. It occurs in tissues rich in fats, such as those of germinating seeds; the enzymes involved in the cycle, which have not been found in mammals, are contained in organelles called *glyoxysomes*. The glyoxylate cycle differs from the Krebs cycle in that it utilizes two molecules of acetyl coenzyme A (rather than one), which are derived from *fatty-acid oxidation. Isocitrate is converted to succinate (from which glucose can be synthesized in *gluconeogenesis) and glyoxylate (see illustration).

glyoxysome *See* **glyoxylate cycle**.

Gnathostomata A subphylum of the *Chordata consisting of all vertebrates that possess jaws. It is divided into two superclasses (*see* **Pisces**; **Tetrapoda**) and contains six extant classes: *Chondrichthyes (cartilaginous fishes), *Osteichthyes (bony fishes), *Amphibia, *Reptilia, *Aves (birds), and *Mammalia. *Compare* **Agnatha**.

Gnetophyta *See* **gymnosperm**.

gnotobiotic Designating germ-free conditions, especially those in which experimental animals are inoculated with known strains of microorganisms.

goblet cell A goblet-shaped cell, found in the epithelium of the intestine and respiratory system in mammals and in the epidermis of fish, that secretes *mucus. Goblet cells have a wide top and constricted base and possess glycoprotein-containing vesicles.

goitre Enlargement of the thyroid gland resulting in a swelling of the neck. This may be caused by lack of iodine in the diet (which is required for the production of thyroid hormones), as the gland enlarges to compensate for this deficiency; or by *hyperthyroidism.

Golgi apparatus An assembly of vesicles and folded membranes within the cytoplasm of plant and animal *cells that stores and transports secretory products (such as enzymes and hormones) and plays a role in formation of a cell wall (when this is present). It is named after its discoverer, the Italian cytologist Camillo Golgi (1843–1926). *See also* **endoplasmic reticulum**.

gonad Any of the usually paired organs in animals that produce reproductive cells (gametes). The most important gonads are the male *testis, which produces spermatozoa, and the female *ovary, which produces ova (egg cells). The gonads also produce hormones that control secondary sexual characteristics.

gonadotrophin (gonadotrophic hormone) Any of several hormones, secreted by the mammalian anterior *pituitary gland, that stimulate reproductive activity of the testes or ovaries (the gonads). Pituitary gonadotrophins include *follicle-stimulating hormone and *luteinizing hormone. *Chorionic gonadotrophin* is a hormone produced by the placenta of higher mammals that maintains the *corpus luteum. The presence of large amounts of *human chorionic gonadotrophin (hCG)* in the urine of women is an indication of pregnancy.

Gondwanaland *See* **continental drift**.

gonorrhoea *See* **sexually transmitted disease**.

G-protein Any one of a group of proteins that have an important role in relaying signals in mammalian cells. They occur on the inner surface of the plasma membrane and transmit signals from hormone receptors on the outer surface of the cell to *adenylate cyclase, which catalyses the formation of the *second messenger, cyclic AMP, inside the cell. G-proteins bind to GTP and GDP (*see* **guanosine**); the GTP–protein complex is able to activate adenylate cyclase, whereas the GDP–protein complex does not. Normally, deactivation of a G-protein is controlled by the hormone receptor on the cell surface and it also occurs as the GTP of a GTP–protein complex gradually hydrolyses to GDP. The cholera toxin exerts its effects by changing the G-protein in the epithelial cells of the intestine so that it is continually activated, which causes an abnormal increase in cellular adenylate cyclase levels. One consequence of this is that sodium ions are actively pumped into the intestine, causing water to follow by osmosis: the result is diarrhoea and dehydration.

Graafian follicle (ovarian follicle) The fluid-filled cavity that surrounds and protects the developing egg cell (oocyte) in the ovary of a mammal. The oocyte is surrounded by *granulosa cells, which secrete fluid into the cavity of the follicle. After the release of the ovum the follicle develops into a *corpus luteum. It is named after the Dutch anatomist Reinier de Graaf (1641–73).

grade A group of organisms that share certain morphological characteristics but are not necessarily related in evolutionary terms. For example, all organisms that possess a coelom can be regarded as a grade. A grade can be contrasted with a *clade, which represents phylogenetic lineage.

graft An isolated portion of living tissue that is joined to another tissue, either in the same or a different organism, the consequent growth resulting in fusion of the tissues. (The word is also used for the process of joining the tissues.)

Grafting of plant tissues is a horticultural practice used to propagate plants, especially certain bushes and fruit trees, artificially. A shoot or bud of the desired variety (the *scion*) is grafted onto a rootstock of either a common or a wild related species (the *stock*). The scion retains its desirable characteristics (e.g. flower form or fruit yield) and supplies the stock with food made by photosynthesis. The stock supplies the scion with water and mineral salts and affects only the size and vigour of the scion.

Animal and human grafts are used to replace faulty or damaged parts of the body. An *autograft* is taken from one part of the body and transferred to another part of the same individual, e.g. a skin graft used for severe burns. A *homograft* (*allograft*) is taken from one individual (the *donor*) and implanted in another (the *recipient*), the process being known as *transplantation*, e.g. a heart or kidney transplant. In such cases the graft may be regarded by the body as foreign (a state of *incompatibility*): an *immune response follows and the graft is rejected (*see also* **histocompatibility**).

graft hybrid A type of plant *chimaera that may be produced when a part of one plant (the scion) is grafted onto another plant of a different genetic constitution (the stock). Shoots growing from the point of union of the graft contain tissues from both the stock and the scion.

gram Symbol g. One thousandth of a kilogram. The gram is the fundamental unit of mass in *c.g.s. units and was formerly used in such units as the *gram-atom*, *gram-molecule*, and *gram-equivalent*, which have now been replaced by the *mole.

Gram's stain A staining method used to differentiate bacteria. The bacterial sample is smeared on a microscope slide, stained with a violet dye, treated with acetone-alcohol (a decolourizer), and finally counterstained with a red dye. *Gram-positive* bacteria retain the first dye, appearing blue-black under the microscope; such bacteria have a thick layer of *peptidoglycan in their cell walls. In *Gram-negative* bacteria, the acetone-alcohol washes out the violet dye and the counterstain is taken up, the cells appearing red. The cell walls of these bacteria have an outer layer of lipoprotein overlying a thin layer of peptidoglycan. The stain is named after the Danish bacteriologist H. C. J. Gram (1853–1938), who first described the technique (since modified) in 1884.

granulocyte Any white blood cell (*see* **leucocyte**) that contains granular material (*secretory vessels*) and *lysosomes in its cytoplasm. *Neutrophils, *basophils, and *eosinophils are examples of granulocytes. *Compare* **agranulocyte**.

granulosa cells (granular cells) Secretory cells in a *Graafian follicle that surround the oocyte. They have a function analogous to that of the *Sertoli cells in the testis, supplying nutrients to the developing oocyte. After ovulation they form the *corpus luteum.

granum (*pl.* **grana**) A stack of platelike bodies (*thylakoids), many of which are found in plant *chloroplasts (each chloroplast contains about 50 grana). Grana bear the light-receptive pigment chlorophyll and contain the enzymes responsible for the light reaction of *photosynthesis.

grape sugar *See* glucose.

graptolites A group of extinct marine colonial animals that were common in the Palaeozoic era. Graptolites are generally regarded as being related to the *Cnidaria. They had chitinous outer skeletons in the form of simple or branched stems, the individual polyps occupying minute cups (*thecae*) along these stems. Fossils of these skeletons are found in Palaeozoic rocks of all continents; they are particularly abundant in Ordovician and Silurian rock strata, for which they are used as *index fossils. At the end of the Silurian many graptolites became extinct but a few groups continued into the early Carboniferous.

grassland A major terrestrial *biome in which the dominant plants are species of grass; the rainfall is insufficient to support extensive growth of trees, which are also suppressed by grazing animals. Tropical grassland (*savanna*), which covers much of Africa south of the Sahara, has widely spaced trees, such as acacias and baobabs, and supports large herds of grazing animals and their predators. Temperate grasslands, such as the *steppes* of Asia, the *prairies* of North America, and the *pampas* of South America, have few trees and are largely used for agriculture.

gray Symbol Gy. The derived SI unit of absorbed *dose of ionizing radiation (*see* **radiation units**). It is named after the British radiobiologist L. H. Gray (1905–65).

grazing The consumption of vegetation, usually on *grassland, by animals, particularly cattle and sheep. Overgrazing can lead to *desertification.

green algae *See* Chlorophyta.

greenhouse effect An effect occurring in the atmosphere because of the presence of certain gases (*greenhouse gases*) that absorb infrared radiation. Light and ultraviolet radiation from the sun are able to penetrate the atmosphere and warm the earth's surface. This energy is re-radiated as infrared radiation, which, because of its longer wavelength, is absorbed by such substances as carbon dioxide. The overall effect is that the average temperature of the earth and its atmosphere is increasing (so-called *global warming*). The effect is similar to that occurring in a greenhouse, where light and long-wavelength ultraviolet radiation can pass through the glass into the greenhouse but the infrared radiation is absorbed by the glass and part of it is re-radiated into the greenhouse.

The greenhouse effect is seen as a major environmental hazard. Average increases in temperature could change weather patterns and agricultural output. It might also lead to melting of the polar ice caps and a corresponding rise in sea level. Carbon dioxide, from coal-fired power stations and car exhausts, is the main greenhouse gas. Other contributory pollutants are nitrogen oxides, ozone, methane, and *chlorofluorocarbons. *See also* **pollution**.

grey matter Part of the tissue that makes up the central nervous system of vertebrates. It is brown-grey in colour, consisting largely of nerve *cell bodies, *synapses, and *dendrites. The grey matter is the site of coordination between nerves of the central nervous system. *Compare* **white matter**.

grooming The actions of an animal of rearranging fur or feathers and cleaning the body surface by biting, scratching, licking, etc., which is important for removing parasites and spreading oils over the body surface. In many mammals, especially primates, grooming between individuals (*allogrooming*) has an important role in maintaining social cohesion.

ground meristem The meristem in plant shoots and roots, derived from the *apical meristem, that gives rise to the cortex and pith (the *ground tissues) in stems and the cortex and endodermis in roots.

ground substance The matrix of connective tissue, in which various cells and fibres are embedded. The ground substance of cartilage consists of *chondrin.

ground tissues All the plant tissues formed by the *apical meristems except the epidermis and vascular tissue. The principal ground tissues are the *cortex, *pith, and primary *medullary rays, and they consist chiefly of *parenchyma. *See also* **collenchyma; sclerenchyma**.

growth An increase in the dry weight or volume of an organism through cell division and cell enlargement. Growth may continue throughout the life of the organism, as occurs in woody plants, or it may cease at maturity, as in man and other mammals. *See also* **allometric growth; exponential growth**.

growth factor Any of various chemicals, particularly polypeptides, that have a variety of important roles in the stimulation of new cell growth and cell maintenance. They bind to the cell surface on receptors. Specific growth factors can cause new cell proliferation (*epidermal growth factor, haemopoietic growth factor – see* **haemopoietic tissue**) and the migration of cells (*fibroblast growth factor*) and play a role in wound healing (*platelet-derived growth factor; PDGF*). Some growth factors act in the embryonic stage of development; for example, *nerve growth factor stimulates the growth of axons and dendrites from developing sensory and sympathetic neurones. It is thought that some growth factors are involved in the abnormal regulation of growth seen in cancer when produced in excessive amounts.

growth hormone (GH; somatotrophin) A hormone, secreted by the mammalian pituitary gland, that stimulates protein synthesis and growth of the long bones in the legs and arms. It also promotes the breakdown and use of fats as an energy source, rather than glucose. Production of growth hormone is greatest during early life. Its secretion is controlled by the opposing actions of two hormones from the hypothalamus: *growth hormone releasing hormone (somatocrinin)*, which promotes its release; and *somatostatin, which inhibits it. Overproduction of *human growth hormone* (*hGH*) results in gigantism in childhood and *acromegaly in adults; underproduction results in dwarfism. *Bovine somatotrophin (BST)* has been used to increase milk and meat production in cattle.

Transverse section through a three-year-old woody stem to show the growth rings

growth ring (annual ring) Any of the rings that can be seen in a cross-section of a woody stem (e.g. a tree trunk). It represents the *xylem formed in one year as a result of fluctuating activity of the vascular *cambium. In temperate climates pale soft *spring wood*, characterized by large xylem vessels, is formed in spring and early summer. Growth slows down in late summer and a darker dense *autumn wood* with smaller xylem vessels is formed (see illustration). The age of a tree can be determined by counting the rings. Under certain circumstances two or more growth rings may form in one year, giving rise to false annual rings.

growth substance (phytohormone; plant hormone) Any of a number of organic chemicals that are synthesized by plants and regulate growth and development. They are usually made in a particular region, such as the shoot tip, and transported to other regions, where they take effect. *See* **abscisic acid; auxin; cytokinin; ethylene; gibberellin.**

guanine A *purine derivative. It is one of the major component bases of *nucleotides and the nucleic acids *DNA and *RNA.

guanosine A nucleoside consisting of one guanine molecule linked to a D-ribose sugar molecule. The derived nucleotides, guanosine mono-, di-, and triphosphate (GMP, GDP, and GTP, respectively), participate in various metabolic reactions.

guard cell *See* **stoma.**

gullet *See* **oesophagus.**

gum 1. Any of a variety of substances obtained from plants. Typically they are insoluble in organic solvents but form gelatinous or sticky solutions with water. Gum resins are mixtures of gums and natural resins. Gums are produced by the young xylem vessels of some plants (mainly trees) in response to wounding or pruning. The exudate hardens when it reaches the plant surface and thus provides a temporary protective seal while the cells below divide to form a permanent repair. Excessive gum formation is a symptom of some plant diseases. *See also* **mucilage. 2.** *See* **gingiva.**

gustatory receptors *See* **taste buds.**

gut *See* **alimentary canal.**

guttation *See* **hydathode.**

gymnosperm Any plant whose ovules and the seeds into which they develop are borne unprotected, rather than enclosed in ovaries, as are those of the flowering plants (the term gymnosperm means naked seed). In traditional systems of classification such plants were classified as the Gymnospermae, a class of the Spermatophyta, but they are now divided into separate phyla: *Coniferophyta (conifers), *Cycadophyta (cycads), Ginkgophyta (ginkgo), and Gnetophyta (e.g. *Welwitschia*).

gynandromorph An animal that possesses both male and female characteristics because it is genetically a *mosaic, i.e. some of its cells are genetically male and others are female. This phenomenon is found particularly in the insects but also appears in the birds and mammals; it is often due to the loss of an X chromosome in a *stem cell of a female (XX), so that all tissues derived from that cell are phenotypically male. *Compare* **intersex.**

gynoecium (gynaecium) The female sex organs (*carpels) of a flower. *Compare* **androecium.**

H

habitat The place in which an organism lives, which is characterized by its physical features or by the dominant plant types. Freshwater habitats, for example, include streams, ponds, rivers, and lakes. *See also* **microhabitat**.

habituation 1. A simple type of learning consisting of a gradual waning in the response of an animal to a continuous or repeated stimulus that is not associated with *reinforcement. **2.** The condition of being psychologically, but not physically, dependent on a drug, with a desire to continue its use but not to increase the dosage.

Haem

haem (heme) An iron-containing molecule (see formula) that binds with proteins as a *cofactor or *prosthetic group to form the *haemoproteins*. These are *haemoglobin, *myoglobin, and the *cytochromes. Essentially, haem comprises a *porphyrin with its four nitrogen atoms holding the iron(II) atom as a chelate. This iron can reversibly bind oxygen (as in haemoglobin and myoglobin) or (as in the cytochromes) conduct electrons by conversion between the iron(II) and iron(III) series.

haemagglutination *See* **agglutination**.

haematophagous Feeding on blood. For example, horseflies and certain leeches are haematophagous.

haematoxylin A compound used in its oxidized form (*haematein*) as a blue dye in optical microscopy, particularly for staining smears and sections of animal tissue. It stains nuclei blue and is frequently used with *eosin as a counterstain for cytoplasm. Haematoxylin requires a mordant, such as iron alum, which links the dye to the tissue. Different types of haematoxylin can be made up depending on the mordant used, the method of oxidation, and the pH. Examples are *Delafield's haematoxylin* and *Ehrlich's haematoxylin*.

haemocoel The body cavity of arthropods and molluscs, which is filled with blood. The haemocoel is an enlarged blastocoel (*see* **blastula**), which greatly reduces the coelom (this is restricted to the cavities of the gonads and excretory organs). The haemocoel can act as a *hydrostatic skeleton.

haemocyanin (hemocyanin) Any of a group of copper-containing respiratory proteins found in solution in the blood of certain arthropods and molluscs. Haemocyanins contain two copper atoms that reversibly bind oxygen, changing between the colourless deoxygenated form (CuI) and the blue oxygenated form (CuII). In some species, haemocyanin molecules form giant polymers with molecular weights of several million.

haemocytometer A device used to estimate the concentration of cells in blood samples. Typically a haemocytometer consists of a glass slide with a shallow depression divided into a fine-meshed grid; each of the grid compartments is about 0.05 mm square. The slide is placed under a microscope and the number of cells in each square of the grid can be counted, which enables the concentration of blood cells in the sample to be assessed. Haemocytometers are also used in estimating the number of microorganisms present in a sample of water or other fluid.

haemoerythrin A red iron-containing *respiratory pigment that occurs in the blood of annelids and some other invertebrates. Its structure is essentially the same as that of *haemoglobin except the prosthetic group has a different chemical composition.

haemoglobin One of a group of globular proteins occurring widely in animals as oxygen carriers in blood. Vertebrate haemoglobin comprises two pairs of polypeptide chains, known as α-chains and β-chains (forming the *globin* protein), with each chain folded to provide a binding site for a *haem group. Each of the four haem groups binds one oxygen molecule to form *oxyhaemoglobin*. Dissociation occurs in oxygen-depleted tissues: oxygen is released and haemoglobin is reformed (*see* **Bohr effect; haemoglobinic acid; oxygen dissociation curve**). The haem groups also bind other inorganic molecules, including carbon monoxide (to form *carboxyhaemoglobin). In vertebrates, haemoglobin is contained in the red blood cells (*erythrocytes).

haemoglobinic acid A very weak acid formed inside red blood cells when hydrogen ions combine with haemoglobin. The presence of the hydrogen ions, which are produced by the dissociation of carbonic acid (*see* **carbonic anhydrase**), encourages oxyhaemoglobin to dissociate into haemoglobin and oxygen (*see* **Bohr effect**). The oxygen diffuses into the tissue cells and the haemoglobin acts as a *buffer for the excess hydrogen ions, which it takes up to form haemoglobinic acid.

haemoglobin S An abnormal form of haemoglobin produced in sickle-cell disease. *See* **polymorphism**.

haemolysis The breakdown of red blood cells. It may be due to the action of disease-causing microorganisms, poisons, antibodies in mismatched blood transfusions, or certain allergic reactions. It produces anaemia.

haemophilia A hereditary sex-linked disease (*see* **sex linkage**) in which there is a deficiency or defect of *Factor VIII, causing the blood to clot very slowly. There may be prolonged bleeding following injury, and in severe cases spontaneous bleeding into the joints and muscles. The disorder is due to a defective recessive allele of the Factor VIII gene, which is located on the X chromosome. Female carriers of the defective allele are unaffected, whereas all males who inherit a defective allele exhibit the disease.

haemopoietic tissue The tissue that gives rise to blood cells in the process of *haemopoiesis*. The haemopoietic tissue of the embryo and fetal stage of vertebrates is the bone marrow, lymph nodes, yolk sac, liver, spleen, and thymus but after birth haemopoiesis occurs in the red bone marrow (*see* **myeloid tissue**). The different types of *stem cells in haemopoietic tissue that rise to erythrocytes and leucocytes are all originally derived from *haemopoietic stem cells* (or *haemocytoblasts*). The formation of the different types of blood cell is under the control of *haemopoietic growth factors*, which include hormones and *cytokines. *See also* **erythropoiesis**.

haemostasis The prevention of blood loss following rupture of blood vessels, which is effected by several physiological processes. Initially, bleeding is restricted by constriction of the damaged vessels, whose endothelial surfaces also stick together. This is followed by the formation of a plug of *platelets and then by the various stages of *blood clotting.

hair 1. A multicellular threadlike structure, consisting of many dead keratinized cells, that is produced by the epidermis in mammalian *skin. The section of a hair below the skin surface (the *root*) is contained within a *hair follicle, the base of which produces the hair cells. Hair assists in maintaining body temperature by reducing heat loss from the skin. Bristles and whiskers are specialized types of hair. **2.** Any of various threadlike structures on plants, such as a *trichome.

hair follicle A narrow tubular depression in mammalian skin containing the root of a *hair. It is lined with epidermal cells and extends down through the epidermis and dermis to its base in the subcutaneous tissue. The ducts of *sebaceous glands empty into hair follicles.

hallucinogen A drug or chemical that causes alterations in perception (usually visual), mood, and thought. Common hallucinogenic drugs include *lysergic acid diethylamide (LSD) and mescaline. There is no common mechanism of action for this class of compounds although many hallucinogens are structurally similar to *neurotransmitters in the central nervous system, such as serotonin and the catecholamines.

hallux The innermost digit on the hindlimb of a tetrapod vertebrate. In man it is the big toe and contains two phalanges. The hallux is absent in some mammals and in many birds it is directed backwards as an adaptation to perching. *Compare* **pollex**.

halophyte A plant that can tolerate a high concentration of salt in the soil. Such conditions occur in salt marshes and mudflats. Halophytes possess some of the structural modifications of *xerophytes; for example, many of them are *succulents. In addition, they are physiologically adapted to withstand the high salinity of the soil water: their root cells have a higher than normal concentration of solutes, which enables them to take up water by osmosis from the surrounding soil. Examples of halophytes are mangrove trees (*see* **mangrove swamp**), thrift (*Armeria*), sea lavender (*Limonium*), and rice grass (*Spartina*). *Compare* **hydrophyte; mesophyte**.

haploid Describing a nucleus, cell, or organism with a single set of unpaired chromosomes. The haploid number is designated as *n*. Reproductive cells, formed as a result of *meiosis, are haploid. Fusion of two such cells (*see* **fertilization**) restores the normal (*diploid) number.

haplont An organism that is at the *haploid stage of its life cycle. *Compare* **diplont**.

hapten A foreign molecule that can bind to an antibody but does not evoke an *immune response unless combined with a carrier protein. In *hypersensitivity reactions haptens bound to endogenous proteins cause the body's immune system to attack both the hapten and the endogenous protein.

haptonasty *See* **nastic movements**.

haptotropism *See* **thigmotropism**.

hardwood *See* **wood**.

Hardy–Weinberg equilibrium The balance in the relative numbers of *alleles (*see* **allele frequency**) that is maintained within a large population over a period of time assuming that: (1) mating is random; (2) there is no natural selection; (3) there is no migration; (4) there is no mutation. In such a stable population, for a gene with two alleles, *A* (dominant) and *a* (recessive), if the frequency of *A* is *p* and the frequency of *a* is *q*, then the frequencies of the three possible genotypes (*AA*, *Aa*, and *aa*) can be expressed by the *Hardy–Weinberg equation*:

$$p^2 + 2pq + q^2 = 1,$$

where p^2 = frequency of *AA* (homozygous dominant) individuals, $2pq$ = frequency of *Aa* (heterozygous) individuals, and q^2 = frequency of *aa* (homozygous recessive) individuals. The equation can be used to calculate allele frequencies if the numbers of homozygous recessive individuals in the population is known. The equation and the equilibrium are named after British mathematician G. H. Hardy (1877–1947) and German physician W. Weinberg (1862–1937).

harvesting 1. The processes involved in gathering in ripened crops (*see*

agriculture). **2.** The collection of cells from cell cultures or of organs from donors for the purpose of transplantation (*see* **graft**).

Hatch–Slack pathway *See* C_4 **pathway**. Is is named after M. D. Hatch and R. Slack.

haustorium (*pl.* **haustoria**) A specialized structure of certain parasitic plants and fungi that penetrates the cells of the host plant to absorb nutrients. In parasitic fungi haustoria are formed from enlarged hyphae and in parasitic flowering plants, such as the dodder (*Cuscuta*), they are outgrowths of the stem.

Haversian canals Narrow tubes within compact *bone containing blood vessels and nerves. They generally run parallel to the bone surface. Each canal surrounded by a series of rings of bone (*lamellae*) is known as a *Haversian system*. Haversian systems are joined to each other by bone material.

health physics The branch of medical physics concerned with the protection of medical, scientific, and industrial workers from the hazards of ionizing radiation and other dangers associated with atomic physics. Establishing the maximum permissible *dose of radiation, the disposal of radioactive waste, and the shielding of dangerous equipment are the principal activities in this field.

hearing The sense by which sound is detected. In vertebrates the organ of hearing is the *ear. In higher vertebrates variation in air pressure caused by sound waves are amplified in the outer and middle ears and transmitted to the inner ear, where sensory cells in the cochlea (*see* **organ of Corti**) register the vibrations. The resulting information is transmitted to the brain via the auditory nerve. The ear can distinguish between sounds of different intensity (*loudness*) and frequency (*pitch*).

heart A hollow muscular organ that, by means of regular contractions, pumps blood through the circulatory system (*see* **circulation**). The vertebrate heart has a thick wall (*see* **myocardium**) composed of a specialized muscle (*see* **cardiac muscle**); it is surrounded by the *pericardium. Mammals have a four-chambered heart consisting of two atria and two ventricles; the right and left sides are completely separate from each other so there is no mixing of oxygenated and deoxygenated blood (see illustration). Oxygenated blood from the pulmonary veins enters the heart through the left atrium, passes to the left ventricle, and leaves the heart through the *aorta. Deoxygenated blood from the *venae cavae enters the right atrium and is pumped through the right ventricle to the pulmonary artery, which conveys it to the lungs for oxygenation. The tricuspid and bicuspid valves ensure that there is no backflow of blood. The contractions of the heart are initiated and controlled by the sinoatrial node (*see* **pacemaker**); in an average adult human the heart contracts about 70 times per minute. *See also* **cardiac cycle; cardiac output**.

The hearts of other vertebrates are similar except in the number of atria and ventricles (there may be one or two) and in the degree of separation of oxygenated and deoxygenated blood. Invertebrates, however, show great variation in the form and functioning of the heart.

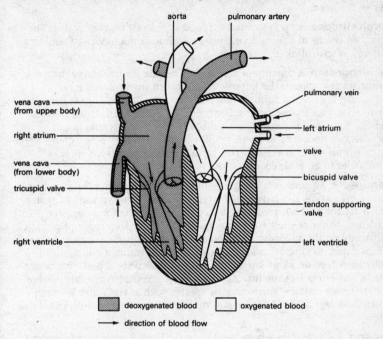

Structure of the mammalian heart

heartwood (duramen) The wood at the centre of a tree trunk or branch. It consists of dead *xylem cells heavily thickened with lignin and provides structural support. Many heartwood cells contain oils, gums, and resins, which darken the wood. *Compare* **sapwood**.

heavy-metal pollution Environmental *pollution by metals with a high *relative atomic mass, such as lead and mercury. These metals derive from a number of sources, including lead in petrol, industrial effluents, and leaching of metal ions from the soil into lakes and rivers by *acid rain. They are easily incorporated into biological molecules and exert their toxic effects by displacing essential metals of a lower binding power in biologically active molecules or by acting as noncompetitive inhibitors of enzymes (*see* **inhibition**).

hecto- Symbol h. A prefix used in the metric system to denote 100 times. For example, 100 coulombs = 1 hectocoulomb (hC).

heliotropism *See* phototropism.

helper T cell *See* T cell.

heme *See* haem.

hemicellulose A *polysaccharide found in the cell walls of plants. The branched chains of this molecule bind to cellulose microfibrils, forming a network of cross-linked fibres.

Hemichordata A phylum of soft-bodied marine invertebrates that comprises two classes: the Enteropneusta (acorn worms), which are cylindrical burrowing animals; and the Pterobranchia, which are colonial forms with vase-shaped bodies. The coelom is divided into three regions and the body into three sections: proboscis, collar, and trunk; like chordates, hemichordates possess *gill slits but they lack a notochord. Some acorn worms develop via a ciliated larva (*tornaria*), which has some similarities to an echinoderm larva.

Hemiptera An order of *exopterygote insects comprising the true bugs. Hemipterans typically have oval flattened bodies with two pairs of wings, which are folded back across the abdomen at rest. The forewings are hardened, either at their bases only (in the suborder Heteroptera) or uniformly (in the suborder Homoptera). The mouthparts are modified for piercing and sucking, with long slender stylets forming a double tube. Many bugs feed on plant sap and are serious agricultural pests, including aphids, leaf-hoppers, scale insects, mealy bugs, etc. Others are carnivorous, and the order contains many aquatic species, such as the water boatmen, which have legs adapted for swimming and the exchange of respiratory gases.

heparin A glycosaminoglycan (mucopolysaccharide) with *anticoagulant properties, occurring in vertebrate tissues, especially the lungs and blood vessels.

hepatic Of or relating to the liver. For example, the *hepatic portal vein* (*see* **hepatic portal system**) and the *hepatic artery* supply blood to the liver, and the *hepatic vein* carries blood away from the liver.

Hepaticae A class of plants within the *Bryophyta comprising the liverworts, which occur in moist situations (including fresh water) and as epiphytes on other plants. The plant body may be a thallus, growing closely pressed to the ground (thallose liverworts, e.g. *Pellia*), or it may bear many leaflike lobes (leafy liverworts). It gives rise to leafless stalks bearing capsules. Spores formed in the capsules are released and grow to produce new plants.

hepatic portal system The vein (*hepatic portal vein*) or veins that transport blood containing the absorbed products of digestion from the intestine directly to the liver.

hepatocyte A specialized epithelial cell that is the most abundant type of cell in the *liver. Hepatocytes are arranged around a central vein in units called *lobules*, in close association with branches of the hepatic portal vein, hepatic artery, and bile canaliculi. They are involved with the various functions of the liver, including metabolism, *detoxification, and the production of bile. *See also* **Kupffer cell**.

hepatotoxin Any chemical that has adverse effects on the liver. Alcohol (ethanol) is one of the most common hepatotoxins.

herb 1. A *herbaceous plant, i.e. a seed-bearing plant that does not form hard woody tissue. **2.** A plant with medicinal or culinary uses. Culinary herbs are usually plants whose leaves are used for flavouring food.

herbaceous Describing a plant that contains little permanent woody tissue. The aerial parts of the plant die back after the growing season. In *annuals the whole plant dies; in *biennials and herbaceous *perennials the plant has organs (e.g. bulbs or corms) that are modified to survive beneath the soil in unfavourable conditions.

herbicide *See* pesticide.

herbivore An animal that eats vegetation, especially any of the plant-eating mammals, such as ungulates (cows, horses, etc.). Herbivores are characterized by having teeth adapted for grinding plants and alimentary canals specialized for digesting cellulose (*see* **caecum**).

heredity The transmission of characteristics from parents to offspring via the chromosomes. The study of heredity (*genetics) was first undertaken by Gregor Mendel, who derived a series of laws that govern heredity (*see* **Mendel's laws**).

hermaphrodite (bisexual) 1. An animal, such as the earthworm, that has both male and female reproductive organs. **2.** A plant whose flowers contain both stamens and carpels. This is the usual arrangement in most plants. *Compare* **monoecious; dioecious**.

heroin (diacetylmorphine) A *narcotic compound that is a synthetic derivative of morphine (*see* **opiate**). The compound is easily absorbed by the brain, due to its lipid-like nature, and is used as a sedative and powerful *analgesic. Highly addictive, it is abused by drug users.

herpesvirus One of a group of complex DNA-containing viruses causing infections in man and most other vertebrates that tend to recur. The group includes *herpes simplex*, the agent of cold sores; *herpes varicella/zoster*, the virus causing chickenpox and shingles; *Epstein–Barr (EB) virus*, the causal agent of glandular fever and also implicated in the cancer Burkitt's lymphoma; and the *cytomegalovirus.

hertz Symbol Hz. The *SI unit of frequency equal to one cycle per second. It is named after Heinrich Hertz (1857–94), a German physicist.

hesperidium *See* berry.

heterochromatin *See* **chromatin**.

heterocyst A specialized cell found in the nitrogen-fixing cyanobacteria. Heterocysts are enlarged cells with thick cell walls and they lack chlorophyll, giving them a colourless appearance. They are the site of nitrogen fixation, for which they produce the enzyme nitrogenase. The lack of photosynthetic activity (which would produce oxygen), together with the thick cell wall, are thought to maintain the anoxic conditions in heterocysts that are essential for the activity of nitrogenase. Heterocysts are connected by *plasmodesmata to surrounding cells, on which they depend for nutrients.

heterodont Describing animals that possess teeth of more than one type (i.e. *incisors, *canine teeth, *premolars, and *molars), each with a particular function. Most mammals are heterodont. *See* **permanent teeth**. *Compare* **homodont**.

heteroduplex DNA Double-stranded DNA in which the two strands are derived from different DNA molecules. Heteroduplex DNA is formed during genetic recombination (*see* **Holliday intermediate**) and can be produced *in vitro* in *DNA hybridization.

heteroecious Describing rust fungi (*see* **rusts**) that pass different phases of their life cycle on different host plants. *Compare* **autoecious**.

heterogametic sex The sex that is determined by possession of two dissimilar *sex chromosomes (i.e. XY). In humans and many other mammals this is the male sex. The heterogametic sex produces reproductive cells (gametes) of two kinds, half containing an X chromosome and half a Y chromosome. *Compare* **homogametic sex**.

heterogamy *Alternation of generations in which parthenogenesis alternates with bisexual reproduction in the life cycle, as seen in certain aphids.

heterokaryon *See* **heterokaryosis**.

heterokaryosis The presence in the same cell of two or more genetically different nuclei. Heterokaryosis occurs naturally in certain fungi, in which it results from the fusion of the cytoplasm of cells from different strains without the fusion of their nuclei. The cell, and the hypha or mycelium containing it, is known as a *heterokaryon*; the most common type of heterokaryon is a *dikaryon. Heterokaryosis can also be induced *in vitro*, to study the interaction between the cellular components from different species (*see* **cell fusion**).

heterosis *See* **hybrid vigour**.

heterospory The condition of producing two types of spore, megaspores and microspores. Heterospory occurs in all seed-bearing plants and in some mosses and ferns.

heterostyly The condition of flowering plants in which flowers of the same species have styles of different lengths, so that the stigma is positioned below the anthers in some flowers and above them in others. This ensures that pollinating insects are more likely to transfer pollen from the anther of one flower to the stigma of another, thus promoting cross-pollination. Species showing heterostyly include the primrose (see illustration).

Heterostyly in the primrose (Primula vulgaris)

heterothallic Describing species of algae and fungi that reproduce sexually by the *conjugation of cells, thalli, or mycelia from different strains (*mating types), normally referred to as + and − strains. *Compare* **homothallic**.

heterotrichy The condition occurring in certain filamentous algae in which the algal body is composed of both prostrate filaments and protruding upright filaments.

heterotrophic nutrition A type of nutrition in which energy is derived from the intake and digestion of organic substances, normally plant or animal tissues. The breakdown products of digestion are used to synthesize the organic materials required by the organism. All animals obtain their food this way: they are *heterotrophs*. *See also* **ingestion**. *Compare* **autotrophic nutrition**.

heterozygous Describing an organism or cell in which the *alleles at a given locus on *homologous chromosomes are different. The aspect of the feature displayed by the organism will be that determined by the *dominant allele. Heterozygous organisms, called *heterozygotes*, do not breed true. *Compare* **homozygous**.

hexose A *monosaccharide that has six carbon atoms in its molecules.

hibernation A sleeplike state in which some animals pass the winter months as a way of surviving food scarcity and cold weather. Various physiological changes occur, such as lowering of the body temperature and slowing of the pulse rate and other vital processes, and the animal lives on its reserve of body fat. Animals that hibernate include bats, hedgehogs, and many fish, amphibians, and reptiles. *See also* **dormancy**. *Compare* **aestivation**.

high-performance liquid chromatography (HPLC) A sensitive technique for separating or analysing mixtures, in which the sample is forced through the chromatography column under pressure.

Hill reaction The release of oxygen from isolated illuminated chloroplasts when suitable electron acceptors (e.g. potassium ferricyanide) are added to the surrounding water. The reaction was discovered by Robert Hill (1899–1991) in 1939; the electron acceptors substitute for NADP$^+$, the natural acceptor for the light reaction of *photosynthesis.

hilum A scar on the seed coat of a plant marking the point at which the seed was attached to the fruit wall by the *funicle. It is a feature that distinguishes seeds from fruits.

hindbrain (rhombencephalon) One of the three sections of the brain of a vertebrate embryo. It develops to form the *cerebellum, *pons, and *medulla oblongata, which control and coordinate fundamental physiological processes (including respiration and circulation of blood). *Compare* **forebrain; midbrain**.

hindgut 1. The posterior part of the alimentary canal of vertebrates, comprising the posterior section of the colon. **2.** The posterior section of the alimentary canal of arthropods. *See also* **foregut; midgut**.

hip girdle *See* **pelvic girdle**.

hippocampus A part of the vertebrate brain consisting of two ridges, one over each of the two lateral *ventricles. It is highly developed in advanced mammals (primates and whales) and its function appears to be related to the expression of responses that generate emotion (such as fear and anger).

Hirudinea A class of freshwater and terrestrial annelid worms that comprises the leeches. They have suckers at both anterior and posterior ends but no bristles. Some are blood-sucking parasites of vertebrates and invertebrates but the majority are predators.

histamine A substance that is released during allergic reactions, e.g. hay fever. Formed from the amino acid histidine, histamine can occur in various tissues but is concentrated in connective tissue. It causes dilation and increased permeability of small blood vessels, which results in such symptoms as localized swelling, itching, sneezing, and runny eyes and nose. The effects of histamine can be countered by the administration of *antihistamine drugs.

histidine *See* **amino acid**.

histochemistry The study of the distribution of the chemical constituents of tissues by means of their chemical reactions. It utilizes such techniques as *staining, light and electron microscopy, *autoradiography, and *chromatography.

histocompatibility The degree to which tissue from one organism will be tolerated by the immune system of another organism. The body recognizes a tissue *graft or invading organism as foreign because of the *histocompatibility antigens* present on its surface. Each individual has a unique set of such antigens, consequently a foreign tissue graft will be rejected by the recipient's immune system. In mice and humans the set of antigens is determined by the *major histocompatibility complex (MHC) – a complex cluster of genes. This encodes two distinct classes of histocompatibility antigens, both of which are glycoproteins. Class I antigens determine graft rejection and help cytotoxic *T cells recognize virus-infected cells. They are found on the surface of most cells in the body. Class II antigens act as receptors for presenting foreign antigen to helper T cells (*see* **antigen-presenting cell**). Their distribution is restricted to certain cell types of the immune system, including macrophages, B cells, and activated T cells.

histocompatibility antigen *See* histocompatibility; HLA system.

histology The microscopic study of the tissues of living organisms. The study of cells, a specialized branch of histology, is known as *cytology.

histone Any of a group of water-soluble proteins found in association with the *DNA of plant and animal chromosomes. They contain a large proportion of the basic (positively charged) amino acids lysine, arginine, and histidine. They are believed to be involved in the condensation and coiling of chromosomes during cell division and have also been implicated in nonspecific suppression of gene activity. Histones do not occur in vertebrate sperm cells (*see* **protamine**) or in bacteria.

HIV (human immunodeficiency virus) The *retrovirus that causes *AIDS in humans. It has a specific affinity for the helper *T cells of its host. The membrane envelope glycoproteins encasing the virus show great variability in their amino-acid sequences, hence the difficulty of preparing an effective AIDS vaccine.

HLA system (human leucocyte antigen system) A series of four gene loci (*A, B, C*, and *D*) in humans that code for a group of antigens, present on the surface of cell membranes, that are important in determining the acceptance or rejection by the body of a tissue or organ transplant (*see* **graft**). These antigens are one group of the so-called *histocompatibility antigens* (*see* **histocompatibility**). Two individuals with identical HLA types are said to be *histocompatible*. Successful transplantation requires a minimum number of HLA differences between the donor's and recipient's tissues. *See also* **major histocompatibility complex**.

Holliday intermediate An intermediate structure in genetic recombination, proposed by Robin Holliday in 1964, in which two double-stranded homologous DNA molecules are joined by means of a reciprocal crossover involving one DNA strand of each molecule. It is formed when a single strand of DNA from each chromosome is broken and joined to the other strand at the point of crossover. The region in which strands from different DNA molecules are paired (a *heteroduplex* DNA sequence) is extended and two strands of the Holliday intermediate are cleaved. The breaks in the DNA sequences are then repaired to form the recombinant products (see illustration).

Holliday intermediate

holocarpic Denoting a fungus in which the entire thallus is differentiated into a reproductive sporangium when mature. *Compare* **eucarpic**.

Holocene (Recent) The most recent geological epoch of the *Quaternary period, comprising roughly the past 10 000 years since the end of the *Pleistocene up to the present. It follows the final glacial of the Pleistocene and thus is sometimes known as the *Postglacial* epoch. Some geologists consider the Holocene to be an interglacial phase of the Pleistocene that will be followed by another glacial.

holocrine secretion *See* **secretion**.

holoenzyme A complex comprising an enzyme molecule and its *cofactor. Only in this state is an enzyme catalytically active. *Compare* **apoenzyme**.

holophytic Describing organisms that feed like plants, i.e. that are photoautotrophic. *See* **autotrophic nutrition**.

holotype *See* **type specimen**.

holozoic Describing organisms that feed by ingesting complex organic matter, which is subsequently digested and absorbed. *See* **heterotrophic nutrition; ingestion**.

homeobox A nucleotide sequence containing about 180 base pairs, which are identical or very similar in many eukaryotic organisms, that encodes a series of amino acids known as a *homeodomain*. Present in many eukaryotic *regulatory proteins, this *domain is thought to be an important region involved in the binding of *regulatory proteins to the DNA molecule. A homeobox was first identified in the *homeotic genes of *Drosophila* and has since been found in the *homeogenes of many animals (including humans) and in plants.

homeodomain *See* **homeobox**.

homeogenes Genes in which a *homeobox has been identified. Homeogenes are very important in the early stages of embryonic development of many eukaryotic organisms as they play a role in the complex series of interactions between genes during early development. *Homeotic genes are examples of homeogenes.

homeostasis The regulation by an organism of the chemical composition of its body fluids and other aspects of its *internal environment so that physiological processes can proceed at optimum rates. It involves monitoring changes in the external and internal environment by means of *receptors and adjusting the composition of the body fluids accordingly; *excretion and *osmoregulation are important in this process. Examples of homeostatic regulation are the maintenance of the *acid–base balance and body temperature (*see* **homoiothermy; poikilothermy**).

homeotic genes Genes that have an important role in the control of early development and differentiation in eukaryotic organisms. By being either 'switched on' or 'switched off', they determine the direction of development of particular organs or regions of the body. Mutations in homeotic genes cause one body part to be substituted for another. For example in *Drosophila* fruit flies, in which these genes were first identified, a *bithorax* mutation causes the segment of the thorax that normally bears halteres (balancing organs; *see* **Diptera**) to be transformed into a segment bearing a second pair of wings. Homeotic genes occur on one chromosome in a cluster that specifies the entire body plan of the organism. They contain specific nucleotide sequences that are the same or very similar in many different organisms (*see* **homeobox**).

hominid Any member of the primate family Hominidae, which includes man and his fossil ancestors (*fossil hominids*) in the genus *Homo*.

Homo The genus of primates that includes modern man (*H. sapiens*, the only living representative) and various extinct species. The oldest *Homo* fossils are those of *H. habilis*, found in the Olduvai Gorge in Tanzania in

1960 and estimated to be about 1.75 million years old. The find was significant as it had been thought that man emerged in Asia rather than Africa. *H. habilis* appears to have been 1–1.5 m tall and had more manlike features and a larger brain than *Australopithecus*. However, the direct ancestor of modern man is thought to be *H. erectus*, which includes Java man and Peking man and appeared about 1.5 million years ago. Fossils of *H. erectus*, which are sometimes called *Pithecanthropus* (ape man), are similar to present-day man except that there was a prominent ridge above the eyes and no forehead or chin. They had crude stone tools and used fire. *H. sapiens* appeared about 100 000 years ago as *Neanderthal man and subsequently (about 35 000 years ago) as *Cromagnon man.

homodont Describing animals whose teeth are all of the same type. Most vertebrates except mammals are homodont. *Compare* **heterodont**.

homogametic sex The sex that is determined by possession of two similar *sex chromosomes (i.e. XX). In humans and many other mammals this is the female sex. All the reproductive cells (gametes) produced by the homogametic sex have the same kind of sex chromosome (i.e. an X chromosome). *Compare* **heterogametic sex**.

homogamy The condition in a flower in which the male and female reproductive organs mature at the same time, thereby allowing self-fertilization. *Compare* **dichogamy**.

homoiothermy The maintenance by an animal of its internal body temperature at a relatively constant value by using metabolic processes to counteract fluctuations in the temperature of the environment. Homoiothermy occurs in birds and mammals, which are described as *endotherms. The heat produced by their tissue metabolism and the heat lost to the environment are balanced by various means to keep body temperature constant: 36–38°C in mammals and 38–40°C in birds. The *hypothalamus in the brain monitors blood temperature and controls thermoregulation by both nervous and hormonal means. This produces both short-term responses, such as shivering or sweating, and long-term adjustments to metabolism according to seasonal changes in climate (acclimatization). Endotherms generally possess insulating feathers or fur. Their relatively high internal temperature permits fast action of muscles and nerves and enables them to lead highly active lives even in cold climates. However, in certain animals, homoiothermy is abandoned during periods of *hibernation. *Compare* **poikilothermy**.

homologous Describing features of organisms that have the same evolutionary origin but have developed different functions. For example the wings of a bat, the flippers of a dolphin, and the arms of a man are homologous organs, having evolved from the paired pectoral fins of a fish ancestor. *Compare* **analogous**.

homologous chromosomes Chromosomes having the same structural features. In *diploid nuclei, pairs of homologous chromosomes can be identified at the start of meiosis. One member of each pair comes from the female parent and the other from the male. Homologous chromosomes have the same pattern of genes along the chromosome but the nature of the genes may differ (*see* **allele**).

homoplastic Denoting similarity resulting from *convergent evolution. *Compare* **patristic**.

homothallic Describing species of algae and fungi that reproduce sexually by the *conjugation of cells, thalli, or mycelia from the same strain. *Compare* **heterothallic**.

homozygous Describing an organism or cell in which the *alleles at a given locus on homologous chromosomes are identical (they may be either dominant or recessive). Homozygous organisms, which are called *homozygotes*, breed true when crossed with genetically identical organisms. *Compare* **heterozygous**.

horizontal cell *See* **retina**.

hormogonium A filament produced by filamentous cyanobacteria that becomes detached from the parent organism and acts as a means of vegetative propagation.

hormone 1. A substance that is manufactured and secreted in very small quantities into the bloodstream by an *endocrine gland or a specialized nerve cell (*see* **neurohormone**) and regulates the growth or functioning of a specific tissue or organ in a distant part of the body. For example, the hormone *insulin controls the rate and manner in which glucose is used by the body. Other hormones include the *sex hormones, *corticosteroids, *adrenaline, *thyroxine, and *growth hormone. **2.** A plant *growth substance.

horsetails *See* **Sphenophyta**.

host An organism whose body provides nourishment and shelter for a parasite (*see* **parasitism**). A *definitive* (or *primary*) *host* is one in which an animal parasite becomes sexually mature; an *intermediate* (or *secondary*) *host* is one in which the parasite passes the larval or asexual stages of its life cycle.

HPLC *See* **high-performance liquid chromatography**.

human chorionic gonadotrophin (hCG) *See* **gonadotrophin**.

Human Genome Project A coordinated international project, begun in 1988, to map the entire human *genome so that the genes can be isolated and eventually sequenced (*see* **gene sequencing**). It involves the production of a *gene library. The human genome contains about 3×10^9 nucleotide base pairs, making up an unknown number of genes (about 100 000 according to most estimates, although some scientists suggest that more than 2 million genes may be present). It is hoped that the project will be completed within the first ten years of the 21st century.

human growth hormone (hGH) *See* growth hormone.

human immunodeficiency virus *See* HIV.

humerus The long bone of the upper arm of tetrapod vertebrates. It articulates with the *scapula (shoulder blade) at the *glenoid cavity and with the *ulna and *radius (via a *condyle) at the elbow.

humidity The concentration of water vapour in the atmosphere. The *absolute humidity* is the mass of water vapour per unit volume of air, usually expressed in $kg\,m^{-3}$. A useful measure is the *relative humidity*, the ratio, expressed as a percentage, of the moisture in the air to the moisture it would contain if it were saturated at the same temperature and pressure.

humus The dark-coloured amorphous colloidal material that constitutes the organic component of soil. It is formed by the decomposition of plant and animal remains and excrement (*see* **litter**) and has a complex and variable chemical composition. Being a colloid, it can hold water and therefore improves the water-retaining properties of soil; it also enhances soil fertility and workability. Acidic humus (*mor*) is found in regions of coniferous forest, where the decay is brought about mainly by fungi. Alkaline humus (*mull*) is typically found in grassland and deciduous forest: it supports an abundance of microorganisms and small animals (e.g. earthworms).

hyaline cartilage *See* cartilage.

hyaluronic acid A *glycosaminoglycan that is part of the *ground substance of connective tissue. Hyaluronic acid binds cells together and helps to lubricate joints. It may play a role in the migration of cells at wounds; this activity ceases when *hyaluronidase breaks down hyaluronic acid.

hyaluronidase The enzyme that breaks down *hyaluronic acid, thereby decreasing its viscosity and increasing the permeability of connective tissue. It is used in medicine to increase the absorption and diffusion of drugs administered by injection or application.

hybrid The offspring of a mating in which the parents differ in at least one characteristic. The term is usually used of offspring of widely different parents, e.g. different varieties or species. Hybrids between different animal species are usually sterile, as is the mule (a cross between a horse and a donkey). *See also* **hybrid vigour**.

hybrid dysgenesis Infertility and an increased incidence of chromosome mutations that is observed in the progeny of laboratory-reared organisms crossed with wild organisms. Hybrid dysgenesis has been best studied in strains of *Drosophila* fruit flies, when laboratory females are crossed with wild males; it is thought to be caused by transposable elements (*see* **transposon**) in the wild-type males, which are activated in eggs from certain strains of laboratory females.

hybridization 1. The production of one or more *hybrid organisms by the mating of genetically different parents. **2.** The production of hybrid cells. *See* **cell fusion (somatic cell hybridization)**. **3.** *See* DNA **hybridization**.

hybridoma A type of hybrid cell that is produced by the fusion of a tumour cell (a myeloma cell) with a normal antibody-producing *B cell (*see* **cell fusion**). The resulting hybrid cell line is able to produce large amounts of normal antibody, which is described as monoclonal (*see* **monoclonal antibody**) as it results from a cloned cell line.

hybrid vigour (heterosis) The increased vigour displayed by the offspring from a cross between genetically different parents. Hybrids from crosses between different crop varieties (F_1 hybrids) are often stronger and produce better yields than the original varieties. Mules, the offspring of mares crossed with donkeys, have greater strength and resistance to disease and a longer lifespan than either parent.

hybrid zone A zone that exists between the populations of two related species, whose geographical ranges overlap, which contain hybrids resulting from matings between the two species. The hybrids, which have a reduced fertility, will eventually act as a barrier between further interspecific mating and hence will restrict gene flow between the two species.

hydathode A pore found in the *epidermis of the leaves of certain plants. Like *stomata, hydathodes are surrounded by two crescent-shaped cells but these, unlike guard cells, do not regulate the size of the aperture. Hydathodes are used by the plant to secrete water under conditions in which *transpiration is inhibited; for example, when the atmosphere is very humid. This process of water loss is called *guttation*.

hydrochloric acid A strong acid, HCl, that is produced by *oxyntic cells in the wall of the stomach and forms a constituent of *gastric juice. Hydrochloric acid is required for the conversion of pepsinogen to pepsin in the lumen of the stomach and kills various microorganisms that enter with food.

hydrocortisone *See* **cortisol**.

hydrogen acceptor *See* **hydrogen carrier**.

hydrogen bond A type of electrostatic interaction between electronegative (fluorine, nitrogen, or oxygen) atoms in one molecule and hydrogen atoms bound to electronegative atoms in another molecule. It is a strong dipole–dipole attraction caused by the electron-withdrawing properties of the electronegative atom. Thus, in the water molecule the oxygen atom attracts the electrons in the O–H bonds. The hydrogen atom has no inner shells of electrons to shield the nucleus, and there is an electrostatic interaction between the hydrogen proton and a lone pair of electrons on an oxygen atom in a neighbouring molecule. Each oxygen atom has two lone pairs and can make hydrogen bonds to two different hydrogen atoms. The strengths of hydrogen bonds are about one tenth of the strengths of normal covalent bonds. Hydrogen bonding does, however, have significant effects on physical properties. Thus it accounts for the unusual properties of water and for its relatively high boiling point. It is also of great importance in living organisms. Hydrogen bonding occurs between bases in the chains of DNA (*see* **base pairing**). It also occurs between the C=O and N–H groups in proteins, and is responsible for maintaining the secondary structure.

hydrogencarbonate (bicarbonate) A salt of *carbonic acid in which one hydrogen atom has been replaced; it thus contains the hydrogencarbonate ion HCO_3^-. *See also* **buffer**.

hydrogen carrier (hydrogen acceptor) A molecule that accepts hydrogen atoms or ions, becoming reduced in the process (*see* **oxidation–reduction**). The electron transport system, whose function is to generate energy in the form of ATP during respiration, involves a series of hydrogen carriers, including *NAD and *FAD, which pass on the hydrogen (derived from the breakdown of glucose) to the next carrier in the chain.

hydrolase Any of a class of enzymes that catalyse the addition of water to, or the removal of water from, a molecule. Hydrolases play an important role in the construction and breakdown of storage materials, such as starch.

hydrological cycle (water cycle) The circulation of water between the atmosphere, land, and oceans on the earth. Water evaporates from the oceans and other water bodies on earth to form water vapour in the atmosphere. This may condense to form clouds and be returned to the earth's surface as precipitation (e.g. rainfall, hail, and snow). Some of this precipitation is returned to the atmosphere directly through evaporation or transpiration by plants; some flows off the land surface as overland flow, eventually to be returned to the oceans via rivers; and some infiltrates the ground to flow underground forming groundwater storage.

hydrophilic Having an affinity for water.

hydrophily A rare form of pollination in which pollen is carried to a flower by water. It occurs by one of two methods. In Canadian pondweed (*Elodea canadensis*) the male flowers break off and float downstream until they contact the female flowers. In *Zostera*, a marine species, the filamentous pollen grains are themselves carried in the water. *Compare* **anemophily; entomophily**.

hydrophobic Lacking affinity for water.

hydrophyte Any plant that lives either in very wet soil or completely or partially submerged in water. Structural modifications of hydrophytes include the reduction of mechanical and supporting tissues and vascular tissue, the absence or reduction of a root system, and specialized leaves that may be either floating or finely divided, with little or no cuticle. Examples of hydrophytes are waterlilies and certain pondweeds. *Compare* **halophyte; mesophyte; xerophyte.**

hydroponics A commercial technique for growing certain crop plants in culture solutions rather than in soil. The roots are immersed in an aerated solution containing the correct proportions of essential mineral salts. The technique is based on various water culture methods used in the laboratory to assess the effects of the absence of certain mineral elements on plant growth.

hydrosphere The water on the surface of the earth. Some 74% of the earth's surface is covered with water, 97% (or some 10^{21} kilograms) of which is in the oceans. Icecaps and glaciers contain about 3×10^{19} kg, rivers about 10^{15} kg, lakes and inland seas about 2×10^{17} kg, and groundwater (down to 4000 metres) about 8×10^{19} kg. Water in the atmosphere contains only about 10^{16} kg.

hydrostatic skeleton The system of support found in soft-bodied invertebrates, which relies on the incompressibility of fluids contained within the body cavity. For example, in earthworms the coelomic fluid is under pressure within the coelom and therefore provides support for internal organs.

hydrotropism The growth of a plant part in response to water. Roots, for example, grow towards water in the soil. *See* **tropism.**

hydroxonium ion *See* **acid.**

hydroxyl Denoting the group –OH in a chemical compound or the ion OH^-.

5-hydroxytryptamine *See* **serotonin.**

Hydrozoa *See* **Cnidaria.**

hygroscopic Describing a substance that can take up water from the atmosphere. *See also* **deliquescence.**

hymen A fold of mucous membrane that covers the opening of the vagina at birth. It normally perforates at puberty, to allow the flow of menstrual blood, but if the opening is small it may be ruptured during the first occasion of sexual intercourse.

Hymenoptera An order of insects that includes the ants, bees, wasps, ichneumon flies, and sawflies. Hymenopterans generally have a narrow waist between thorax and abdomen. The smaller hindwings are interlocked with the larger forewings by a row of tiny hooks on the leading edges of the hindwings. Some species are wingless. The mouthparts are typically adapted for biting, although some advanced forms (e.g. bees) possess a tubelike proboscis for sucking liquid food, such as nectar. The long slender *ovipositor can serve for sawing, piercing, or stinging. Metamorphosis occurs via a pupal stage to the adult form. *Parthenogenesis is common in the group.

Ants and some bees and wasps live in colonies, often comprising numerous individuals divided into *castes and organized into a coordinated and complex society. The colony of the honeybee (*Apis mellifera*), for example, consists of workers (sterile females), *drones (fertile males), and usually a single fertile female – the queen. The sole concern of the queen is egg laying. She determines the gender of the egg by either withholding or releasing stored sperm. Unfertilized eggs become males; fertilized eggs become females. The workers fulfil a variety of tasks, including nursing the developing larvae, building the wax cells (combs) of the hive, guarding the colony, and foraging for nectar and pollen. The single function of the larger drones is to mate with the young queen on her nuptial flight.

hyoid arch The second of seven bony V-shaped arches that support the gills of fish. The dorsal part is specialized as the *hyomandibular*, which is involved in jaw suspension (*see* **amphistylic jaw suspension; hyostylic jaw suspension**). In tetrapods the dorsal part of the hyoid arch has evolved to form the stapes (one of the *ear ossicles). The ventral part forms the *hyoid bone*, which supports the tongue.

hyomandibular *See* hyoid arch.

Hyostylic jaw suspension

hyostylic jaw suspension A type of jaw suspension seen in most fishes, in which the upper jaw is not directly connected to the cranium. The attachment between cranium and upper jaw is made by a ligament at the front end and by the hyomandibular (*see* **hyoid arch**) at the rear end (see illustration). *Compare* **amphistylic jaw suspension; autostylic jaw suspension**.

hyper- A prefix denoting over, above, high; e.g. hyperpolarization.

hyperglycaemia Excessive amounts of glucose in the bloodstream, which occurs in diabetes mellitus due to underproduction of *insulin. *Compare* hypoglycaemia.

hypermetropia (hyperopia) Long-sightedness. A vision defect in which the lens of the eye is unable to accommodate sufficiently to throw the image of near objects onto the retina. It is caused usually by shortness of the eyeball rather than any fault in the lens system. The subject requires

Hypermetropia

spectacles with converging lenses to bring the image from behind the retina back on to its surface (see illustration).

hyperplasia Increase in the size of a tissue or organ due to an increase in the number of its component cells. *Compare* **hypertrophy**.

hyperpolarization *See* **inhibitory postsynaptic potential**.

hypersensitivity 1. *(in immunology)* Increased sensitivity to compounds, which can elicit a specific *immune response accompanied by tissue damage. Hypersensitive reactions include *allergies and *anaphylaxis. *See also* **hapten**. **2.** *(in botany)* Localized cell death in plants following invasion by a pathogen. This is a protective mechanism preventing the spread of plant diseases, such as powdery mildew, throughout the rest of the plant tissues.

hypertension *See* **blood pressure**.

hyperthyroidism Overactivity of the *thyroid gland. This condition causes an increase in *basal metabolic rate, with such symptoms as rapid pulse and weight loss. It is sometimes associated with a *goitre. *Compare* **hypothyroidism**.

hypertonic solution A solution that has a lower *water potential and a correspondingly higher osmotic pressure than another solution. *See* **osmosis**.

hypertrophy An increase in the size of a tissue or organ due to an increase in the size of its component cells. Hypertrophy often occurs in response to an increased workload in an organ, which may result from malfunction or disease. *Compare* **hyperplasia**.

hyperventilation An increase in the amount of air taken into the lungs caused by an increase in the depth or rate of breathing. *See also* **ventilation**.

hypha (*pl.* **hyphae**) A delicate filament in fungi many of which may form either a loose network (*mycelium) or a tightly packed interwoven mass of *pseudoparenchyma, as in the fruiting body of mushrooms. Hyphae may be branched or unbranched and may or may not possess cross walls. The cell wall consists of either fungal cellulose or a nitrogenous compound called *chitin. The cell wall is lined with cytoplasm, which often contains oil globules and glycogen, and there is a central vacuole. The hyphae produce enzymes that in parasitic fungi digest the host tissue, and in saprotrophic fungi digest dead organic matter.

hypo- A prefix denoting under, below, low; e.g. hypogyny, hyponasty.

hypocotyl The region of the stem beneath the stalks of the seed leaves (*cotyledons) and directly above the young root of an embryo plant. It grows rapidly in seedlings showing *epigeal germination and lifts the cotyledons above the soil surface. In this region (the *transition zone*) the arrangement of vascular bundles in the root changes to that of the stem. *Compare* **epicotyl**.

hypodermis (exodermis) The outermost layer of cells in the plant *cortex, lying immediately below the epidermis. These cells are sometimes modified to give additional structural support or to store food materials or water. After the loss of the *piliferous layer of the root the hypodermis takes over the protective functions of the epidermis.

hypogeal 1. Describing seed germination in which the seed leaves (cotyledons) remain below ground. Examples of hypogeal germination are seen in oak and runner bean. *Compare* **epigeal. 2.** Describing fruiting bodies that develop underground, such as truffles and peanuts.

hypoglycaemia Below-normal levels of glucose in the blood plasma, causing weakness, dizziness, and sweating. *Compare* **hyperglycaemia**.

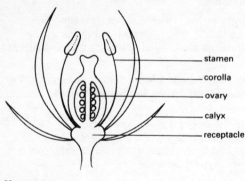

stamen
corolla
ovary
calyx
receptacle

Hypogyny

hypogyny A floral arrangement in which the ovary is *superior, i.e. it arises from the receptacle above the sepals, petals, and stamens (see illustration). The perianth and stamens are said to be *hypogynous* with respect to the ovary, as seen in the tulip. *Compare* **epigyny; perigyny**.

hypolimnion The lower layer of water in a lake. *See* **thermocline**. *Compare* **epilimnion**.

hypophysis *See* **pituitary gland**.

hypothalamus Part of the vertebrate brain that is derived from the *forebrain and located on the ventral surface below the *thalamus and the *cerebrum. The hypothalamus regulates a wide variety of physiological processes, including maintenance of body temperature, water balance, sleeping, and feeding, via both the *autonomic nervous system (which it controls) and the *neuroendocrine system. Its endocrine functions are largely mediated by the *pituitary gland. The pituitary responds to releasing hormones produced by the hypothalamus, which thereby indirectly controls hormone production in other glands.

hypothyroidism Underactivity of the *thyroid gland, which is generally due to a deficiency in *thyroid-stimulating hormone. In adults hypothyroidism causes a decrease in *basal metabolic rate, with tissue swelling and weight gain (*myxoedema*). In children thyroid hormone is required for development of the bones and nervous system, and hypothyroidism can result in *cretinism. *Compare* **hyperthyroidism**.

hypotonic solution A solution that has a higher *water potential and a correspondingly lower osmotic pressure than another solution. *See* **osmosis**.

hypoxia A deficiency of oxygen in body tissues, which can result from living in an oxygen-deficient environment, inadequate *inspiration, or deficiency of red blood cells or haemoglobin (required for oxygen transport in the body).

H zone The region of a striated muscle fibre that contains only thick (*myosin) filaments. The H zone appears as a lighter band in the middle of the dark *A band at the centre of a *sarcomere.

IAA (indoleacetic acid) *See* **auxin**.

I band The region of a striated muscle fibre that contains only thin (*actin) filaments. I bands are visible as light bands at either end of a *sarcomere.

ice age A period in the earth's history during which ice advanced towards the equator and a general lowering of temperatures occurred. The last major ice age, that of the Pleistocene period (sometimes known as the *Ice Age*), ended about 10 000 years ago. At least four major ice advances (glacials) occurred during the Pleistocene period; these were separated by interglacials during which the ice retreated and temperatures rose. At present it is not known if the earth is between ice ages or is in an interglacial of the Pleistocene Ice Age. It has been established that ice ages also occurred during the Precambrian (over 500 million years ago) and during the Permo-Carboniferous (about 250 million years ago).

ICSH *See* **luteinizing hormone**.

identical twins (monozygotic twins) Two individuals that develop from a single fertilized egg cell by its division into two genetically identical parts. Each part eventually gives rise to a separate individual and these twins are therefore identical in every respect. *Compare* **fraternal twins**.

identification key *See* **key**.

idiogram *See* **karyogram**.

idiosyncrasy An abnormal reaction to a drug or other foreign substance shown by an individual, which is usually genetically determined. An individual that shows immunological idiosyncrasy is said to be *hypersensitive* to a particular substance, agent, etc.

ileum The portion of the mammalian *small intestine that follows the *jejunum and precedes the *large intestine. It is a site of digestion and absorption. The internal lining of the ileum bears numerous small outgrowths (*see* **villus**), which increase its absorptive surface area.

ilium The largest of the three bones that make up each half of the *pelvic girdle. The ilium bears a flattened wing of bone that is attached by ligaments to the sacrum (*see* **sacral vertebrae**). *See also* **ischium; pubis**.

imaginal disc A group of undifferentiated cells in an insect larva that develops into a specific adult structure. Imaginal discs are derived from the *blastoderm. They do not undergo further development until the pupal stage, when they start to differentiate into adult epidermal structures, such as eyes, antennae, and wings, under the control of the hormone *ecdysone.

imago The adult sexually mature stage in the life cycle of an insect after metamorphosis.

imbibition The uptake of water by substances that do not dissolve in water, so that the process results in swelling of the substance. Imbibition is a property of many biological substances, including cellulose (and other constituents of plant cell walls), starch, and some proteins. It occurs in dry seeds before they germinate and – together with osmosis – is responsible for the uptake of water by growing plant cells.

immersion objective An optical microscope objective in which the front surface of the lens is immersed in a liquid on the cover slip of the microscope specimen slide. Cedar-wood oil (for an *oil-immersion lens*) or sugar solution is frequently used. It has the same refractive index as the glass of the cover slip, so that the object is effectively immersed in it. The presence of the liquid increases the effective aperture of the objective, thus increasing the resolution.

immune response The reaction of the body to foreign or potentially dangerous substances (*antigens), particularly disease-producing microorganisms. The response involves the production by specialized white blood cells (*lymphocytes) of proteins known as *antibodies, which react with the antigens to render them harmless. The antibody–antigen reaction is highly specific. *See also* **anaphylaxis; immunity**.

immunity The state of relative insusceptibility of an animal to infection by disease-producing organisms or to the harmful effects of their poisons (toxins). Immunity depends on the presence in the blood of *antibodies and white blood cells (*lymphocytes), which produce an *immune response. *Inherited (natural or innate) immunity* is that with which an individual is born. *Acquired immunity* is of two types, active and passive. *Active immunity* arises when the body produces antibodies against an invading foreign substance (*antigen), either through infection or *immunization; this type of immunity may be *humoral*, in which B lymphocytes produce free antibodies that circulate in the bloodstream (*see* **B cell**), or *cell-mediated*, caused by the action of T lymphocytes (*see* **T cell**). *Passive immunity* is induced by injection of serum taken from an individual already immune to a particular antigen; it can also be acquired by the transfer of maternal antibodies to offspring via the placenta or breast milk (*see* **colostrum**). Active immunity tends to be long-lasting; passive immunity is short-lived. *See also* **autoimmunity**.

immunization The production of *immunity in an individual by artificial means. *Active immunization (vaccination)* involves the introduction, either orally or by injection (*inoculation*), of specially treated bacteria, viruses, or their toxins to stimulate the production of *antibodies (*see* vaccine). *Passive immunization* is induced by the injection of preformed antibodies.

immunoassay Any of various techniques that measure the amount of a particular substance by virtue of its binding antigenically to a specific antibody. In *solid-phase immunoassay* the specific antibody is attached to a solid supporting medium, such as a PVC sheet. The sample is added and any test antigens will bind to the antibody. A second antibody, specific for a different site on the antigen, is added. This carries a radioactive or fluorescent label, enabling its concentration, and thus that of the test antigen, to be determined by comparison with known standards. Variations on this technique include *ELISA and *Western blotting.

immunoelectrophoresis An analytical technique used to identify antigens that have been separated by *electrophoresis. The gel medium used for the electrophoresis is irrigated with specific antibodies for the required antigens: when an antigen meets its corresponding antibody a precipitate forms, which can be seen and isolated.

immunofluorescence A technique used to identify specific proteins in tissue samples by applying antibodies that have been labelled with a fluorescent compound. The labelled antibody combines with the protein under investigation and the resulting complex can be identified under a microscope by its fluorescence. Proteins and enzymes in tissue samples can be identified and located in this way.

immunogenicity The degree to which an antigen can produce an *immune response.

immunoglobulin One of a group of proteins (*globulins) in the body that act as *antibodies. They are produced by specialized white blood cells (*lymphocytes) and are present in blood serum and other body fluids. There are several classes (e.g. IgE, IgG, and IgM) having different functions.

immunosuppression The suppression of an *immune response. Immunosuppression is necessary following organ transplants in order to prevent the host rejecting the grafted organ (*see* **graft**); it is artificially induced by radiation or chemical agents that inhibit cell division of lymphocytes. Immunosuppression occurs naturally in certain diseases, notably *AIDS.

imperfect fungi *See* **Deuteromycota**.

Imperial units The British system of units based on the pound and the yard. The former f.p.s. system was used in engineering and was loosely based on Imperial units; for all scientific purposes *SI units are now used. Imperial units are also being replaced for general purposes by metric units.

implantation (nidation) *(in embryology)* The embedding of a fertilized mammalian egg into the wall of the uterus (womb) where it will continue its development. After fertilization in the fallopian tube the egg passes into the womb in the form of a ball of cells (*blastocyst*). Its outer cells destroy cells of the uterine wall, forming a cavity into which the blastocyst sinks.

imprinting A specialized form of *learning in which young animals, during a critical period in their early development, learn to recognize and approach some large moving object nearby. In nature this is usually the mother, though simple models or individuals of a different species (including man) may suffice. Imprinting was first described by Konrad Lorenz (1903–89), working with young ducks and geese.

impulse (nerve impulse) The signal that travels along the length of a *nerve fibre and is the means by which information is transmitted through the nervous system. It is marked by the flow of ions across the membrane of the *axon caused by changes in the permeability of the membrane, producing a reduction in potential difference that can be detected as the *action potential. The strength of the impulse produced in any nerve fibre is constant (*see* **all-or-none response**).

inbreeding Mating between closely related individuals, the extreme condition being self-fertilization, which occurs in many plants and some primitive animals. A population of inbreeding individuals generally shows less variation than an *outbreeding population. Continued inbreeding among a normally outbreeding population leads to *inbreeding depression* (the opposite of *hybrid vigour) and an increased incidence of harmful characteristics. For example, in humans, certain mental and other defects tend to occur more often in families with a history of cousin marriages.

incisor A sharp flattened chisel-shaped *tooth in mammals that is adapted for biting food and – in rodents – for gnawing. In humans there are normally two pairs of incisors (central and lateral) in each jaw. *See* **permanent teeth**.

inclusion body Any of various particulate structures formed in plant and animal cells due to viral infection. Inclusion bodies sometimes become embedded in cellular proteins.

inclusive fitness The quality that organisms attempt (unconsciously) to maximize as the result of natural selection acting on genes that are influential in controlling their behaviour and physiology. It includes the individual's own *reproductive success* (usually taken as the number of its offspring that survive to adulthood) and also the effects of the individual's actions on the reproductive success of its relatives, because relatives have a higher probability of sharing some identical genes with the individual than do other members of the population. When interactions between relatives are likely to occur (which happens during the lives of many animals and plants) *kin selection will operate.

incompatibility 1. The condition that exists when foreign grafts or blood transfusions evoke a marked *immune response and are rejected. **2.** The phenomenon in which pollen from one flower fails to fertilize other flowers on the same plant (*self-incompatibility*) or on other genetically similar plants. This genetically determined mechanism prevents self-fertilization (breeding between likes) and promotes cross-fertilization (breeding between individuals with different genetic compositions). *See also* **fertilization; pollination**.

incomplete dominance The condition that arises when neither *allele controlling a characteristic is dominant and the aspect displayed by the organism results from the partial influence of both alleles. For example, a snapdragon plant with alleles for red and for white flowers produces pink flowers. *Compare* **co-dominance**.

incubation 1. The process of maintaining the fertilized eggs of birds and of some reptiles and egg-laying mammals at the optimum temperature for the successful development of the embryos. A period of incubation follows the laying of the eggs and precedes their hatching. **2.** The process of maintaining a *culture of bacteria or other microorganisms at the optimum temperature for growth of the culture. **3.** The phase in the development of an infectious disease between initial infection and the appearance of the first symptoms.

incus (anvil) The middle of the three *ear ossicles of the mammalian middle ear.

indefinite inflorescence *See* **racemose inflorescence**.

indehiscent Describing a fruit or fruiting body that does not open to release its seeds or spores when ripe. Instead, release occurs when the fruit wall decays or, if eaten by an animal, is digested. *Compare* **dehiscence**.

independent assortment The separation of the alleles of one gene into the reproductive cells (gametes) independently of the way in which the alleles of other genes have segregated. By this process all possible combinations of alleles should occur equally frequently in the gametes. In practice this does not happen because alleles situated on the same chromosome tend to be inherited together. However, if the allele pairs *Aa* and *Bb* are on different chromosomes, the combinations *AB*, *Ab*, *aB*, and *ab* will normally be equally likely to occur in the gametes. *See* **meiosis; Mendel's laws**.

index fossil (zone fossil) An animal *fossil of a group that existed continuously during a particular span of geological time and can therefore be used to date the rock in which it is found. Index fossils are found chiefly in sedimentary rocks. They are an essential tool in stratigraphy for comparing the geological ages of sedimentary rock formations. Examples are *ammonites and *graptolites.

indicator species A plant or animal species that is very sensitive to a particular environmental factor, so that its presence (or absence) in an area can provide information about the levels of that factor. For example, some lichens are very sensitive to the concentration of sulphur dioxide (a major pollutant) in the atmosphere. Examination of the lichens present in an area can provide a good indication of the prevailing levels of sulphur dioxide.

indigenous Describing a species that occurs naturally in a certain area, as distinct from one introduced by man; native.

indoleacetic acid (IAA) *See* auxin.

induced-fit model A proposed mechanism of interaction between an enzyme and a substrate. It postulates that exposure of an enzyme to a substrate causes the *active site of the enzyme to change shape in order to allow the enzyme and substrate to bind (*see* **enzyme–substrate complex**). This hypothesis is generally preferred to the *lock-and-key mechanism.

induction 1. *(in embryology)* The ability of natural stimuli to cause unspecialized embryonic tissue to develop into specialized tissue. **2.** *(in obstetrics)* The initiation of childbirth by artificial means; for example, by injection of the hormone *oxytocin.

indusium The kidney-shaped covering of the *sorus of certain ferns that protects the developing sporangia. It withers when the sorus ripens to expose the sporangia.

industrial melanism The increase of melanic (dark) forms of an animal in areas darkened by industrial pollution. The example most often quoted is that of the peppered moth (*Biston vetularia*), melanic forms of which markedly increased in the industrial north of England during the 19th century. Experiments have shown that the dark forms increase in polluted regions because they are less easily seen by birds against a dark background; conversely the paler forms survive better in unpolluted areas. *See also* **directional selection**.

infection The invasion of any living organism by disease-causing microorganisms (*see* **pathogen**), which proceed to establish themselves, multiply, and produce various symptoms in their host. Pathogens may invade via a wound or (in animals) through the mucous membranes lining the alimentary, respiratory, and reproductive tracts, and may be transmitted by an infected individual, a *carrier, or an arthropod *vector. Symptoms in animals appear after an initial symptomless *incubation period* and typically consist of localized *inflammation, often with pain and fever. Infections are combated by the body's natural defences (*see* **immune response**). Treatment with drugs (*see* **antibiotic; antiseptic**) is effective against most bacterial, fungal, and protozoan infections but not against viral infections. *See also* **immunization**.

inferior Describing a structure that is positioned below or lower than another structure in the body. For example, in flowering plants the ovary is described as inferior when it is located below the other organs of the flower (*see* **epigyny**). *Compare* **superior**.

inflammation The defence reaction of tissue to injury, infection, or irritation by chemicals or physical agents. Cells in the affected tissue release a chemical (*histamine) that causes localized dilatation of blood vessels so that fluid leaks out and blood flow is increased. This leads to swelling, redness, heat, and often pain. White blood cells, particularly *phagocytes, enter the tissue and an *immune response is stimulated. A gradual healing process usually follows.

inflorescence A particular arrangement of flowers on a single main stalk of a plant. There are many different types of inflorescence, which are classified into two main groups depending on whether the tip of the flower axis goes on producing new flower buds during growth (*see* **racemose inflorescence**) or loses this ability (*see* **cymose inflorescence**).

infraspecific Occurring within a species. For example, *infraspecific variation* is the *variation occurring between individuals of the same species. *Compare* **interspecific; intraspecific**.

ingestion (feeding) A method of *heterotrophic nutrition in which bulk food is taken into an organism and subsequently digested (*see* **digestion**). Ingestion is the principal mechanism of animal nutrition. *See also* **macrophagous; microphagous**.

inguinal Of, relating to, or situated in the groin. For example, the *inguinal canal*, which contains the spermatic cord, lies in the abdominal cavity of a male fetus.

inhalation *See* **inspiration**.

inheritance The transmission of particular characteristics from generation to generation by means of the *genetic code, which is transferred to offspring in the gametes. *See also* **Mendel's laws**.

inhibition 1. (in biochemistry) (enzyme inhibition) A reduction in the rate of an enzyme-catalysed reaction by substances called *inhibitors. Competitive inhibition* occurs when the inhibitor molecules resemble the substrate molecules and bind to the *active site of the enzyme, so preventing normal enzymatic activity. Competitive inhibition can be reversed by increasing the concentration of the substrate. In *noncompetitive inhibition* the inhibitor binds to a part of the enzyme or *enzyme–substrate complex other than the active site, known as an *allosteric site*. This deforms the active site so that the enzyme cannot catalyse the reaction. Noncompetitive inhibition cannot be reversed by increasing the concentration of the substrate. The toxic effects of many substances are produced in this way. Inhibition by reaction products (*feedback inhibition*) is important in the control of enzyme activity. *See also* **allosteric enzyme**.
2. (in physiology) The prevention or reduction of the activity of effectors (such as muscles) by means of certain nerve impulses. Inhibitory activity often provides a balance to stimulation of a process; for example, the impulse to stimulate contraction of a voluntary muscle may be accompanied by an inhibitory impulse to prevent contraction of its antagonist.

inhibitory postsynaptic potential (IPSP) The electric potential that is generated in a postsynaptic neurone when an inhibitory neurotransmitter (such as *gamma-aminobutyric acid) is released into the synapse and causes a slight increase in the potential difference across the postsynaptic membrane (*hyperpolarization*). This makes the neurone less likely to transmit an impulse. *Compare* **excitatory postsynaptic potential**.

initial One of a group of cells (or, in lower plants, a single cell) that divides to produce the cells of a plant tissue or organ. The cells of the apical meristem, cambium, and cork cambium are initials.

initiation codon *See* start codon.

initiation factor (IF) Any of a group of proteins that are required for initiating the *translation stage of protein synthesis. Each ribosome is assembled on the messenger RNA (mRNA) chain in two subunits, one smaller than the other. IFs catalyse the binding of an *initiator tRNA* molecule, which recognizes the *start codon of the mRNA, to the smaller subunit of the ribosome. This subunit then binds to the mRNA and the IFs dissociate, allowing the larger ribosomal subunit to bind to the smaller one. IFs also control the rate of translation in some cells.

initiation site *See* promoter.

innate behaviour An inherited pattern of behaviour that appears in a similar form in all normally reared individuals of the same sex and species. *See* **instinct**.

innate immunity *See* immunity.

inner ear The structure in vertebrates, surrounded by the temporal bone of the skull, that contains the organs of balance and hearing. It consists of soft hollow sensory structures (the *membranous labyrinth*), containing fluid (*endolymph*), surrounded by fluid (*perilymph*), and encased in a bony cavity (the *bony labyrinth*). It consists of two chambers, the *sacculus and *utriculus, from which arise the *cochlea and *semicircular canals respectively.

innervation The supply of nerve fibres to and from an organ.

innominate artery A short artery that branches from the aorta to divide into the *subclavian artery (the main artery to the arm) and the right *carotid artery (which supplies blood to the head).

innominate bone One of the two bones that form each half of the *pelvic girdle in adult vertebrates. This bone is formed by the fusion of the *ilium, *ischium, and *pubis.

inoculation 1. *See* vaccine 2. The placing of a small sample of microorganisms or any other type of cell into a *culture medium so that the cells can grow and proliferate.

inositol A cyclic alcohol, $C_6H_{12}O_6$, that is a constituent of certain cell

phosphoglycerides; the most important isomer is *myo*-inositol (see formula).
It is sometimes classified as a member of the vitamin B complex but it can
be synthesized by many animals and it is not regarded as an essential
nutrient in humans. *Phosphatidyl inositol*, a constituent of cell membranes,
is a precursor of the intracellular *second messenger molecules, *inositol
1,4,5-trisphosphate* (IP_3) and *diacylglycerol*; these are produced in response to
the binding of a substance, e.g. serotonin, to receptors on the cell surface.
They mediate such cellular events as smooth muscle contraction, adrenaline
secretion, and histamine secretion.

Myo-*inositol*

Insecta (Hexapoda) A class of arthropods of the subphylum *Uniramia
comprising about a million known species (many more are thought to
exist). They are distributed worldwide in nearly all terrestrial habitats.
Ranging in length from 0.5 to over 300 mm, an insect's body consists of a
head, a thorax of three segments and usually bearing three pairs of legs
and one or two pairs of wings, and an abdomen of eleven segments. The
head possesses a pair of sensory *antennae and a pair of large *compound
eyes, between which are three simple eyes (*ocelli). The *mouthparts are
variously adapted for either chewing or sucking, enabling insects to feed
on a wide range of plant and animal material. Insects owe much of their
success to having a highly waterproof *cuticle (to resist desiccation) and, in
most species, wings – outgrowths of the body wall that confer the greater
mobility of *flight. Breathing occurs through a network of tubes (*see*
trachea).

Most insect species have separate sexes and undergo sexual reproduction.
In some, this may alternate with asexual *parthenogenesis and in a few,
males are unknown and reproduction is entirely asexual. In the wingless
insects (subclass *Apterygota) *metamorphosis is slight or absent. In the
winged insects (subclass *Pterygota) the newly hatched young grow by
undergoing a series of moults. In the more primitive *exopterygotes
(including the orders *Dermaptera, *Orthoptera, *Dictyoptera, and
*Hemiptera) the young (called a *nymph) resembles the adult. The more
advanced *endopterygotes (e.g. *Coleoptera, *Diptera, *Lepidoptera, and
*Hymenoptera) undergo metamorphosis, in which the young (called a
*larva) is transformed into a quiescent *pupa from which the fully formed
adult emerges. Insects are of vital importance in many ecosystems and
many are of economic significance to man – as animal or plant pests or

disease vectors or beneficially as crop pollinators or producers of silk, honey, etc.

insecticide *See* **pesticide**.

Insectivora An order of small, mainly nocturnal, mammals that includes the hedgehogs, moles, and shrews. They have long snouts covered with stiff tactile hairs and their teeth are specialized for seizing and crushing insects and other small prey. The insectivores have changed very little since they evolved in the Cretaceous period, 130 million years ago.

insectivore An animal that eats insects, especially a mammal of the order Insectivora (hedgehogs, shrews, etc.).

insectivorous plant *See* **carnivorous plant**.

insertion **1.** *(of muscles)* *See* **voluntary muscle**. **2.** *(in genetics)* A *point mutation in which an extra nucleotide base is added to the DNA sequence. This results in the misreading of the base sequence during the *translation stage of protein synthesis.

insertion sequence *See* **transposon**.

insight learning A form of learning in which an animal responds to new situations by adapting experiences gained in other contexts. Insight learning requires an animal to solve problems by viewing a situation as a whole instead of relying wholly on trial-and-error learning. Chimpanzees are capable of insight learning. For example, in one experiment they were able to reach food placed at some distance outside their cage by joining poles together to make a longer pole.

inspiration (inhalation) The process by which gas is drawn into the lungs through the trachea (*see* **respiratory movement**). In mammals the rib cage is raised by contraction of the external *intercostal muscles and the muscles of the diaphragm. These actions enlarge the thorax, so that pressure in the lung cavity is reduced below atmospheric pressure, which causes an influx of air until the pressures are equalized. *Compare* **expiration**.

inspiratory centre The most important subcentre of the *ventilation centre in the brain, which controls the regularity of breathing. The inspiratory centre is situated in the medulla oblongata and is composed of a group of neurones (the *dorsal respiratory group*). These neurones receive information from stretch receptors in the bronchial tubes and chemoreceptors in the *carotid bodies and generate impulses that pass down the phrenic nerve to the diaphragm, causing it to contract (*see* **inspiration**).

instar A stage in the larval development of an insect between two moults (ecdyses). There are usually a number of larval instars before metamorphosis. *See* **exopterygote**.

instinct An innate tendency to behave in a particular way, which does not depend critically on particular learning experiences for its development and therefore is seen in a similar form in all normally reared individuals of the same sex and species. Much instinctive behaviour takes the form of *fixed action patterns*. These are movements that – once started – are performed in a stereotyped way unaffected by external stimuli. For example, a frog's prey-catching tongue flick is performed in the same way whether or not anything is caught. Some complex instinctive behaviour, however, requires some learning by the animal before it is perfected. Birdsong, for example, consists of an innate component that is modified and made more complex by the influence of other birds, the habitat, etc.

insulin A protein hormone, secreted by the β cells of the *islets of Langerhans in the pancreas, that promotes the uptake of glucose by body cells, particularly in the liver and muscles, and thereby controls its concentration in the blood. Insulin was the first protein whose amino-acid sequence was fully determined (in 1955). Underproduction of insulin results in the accumulation of large amounts of glucose in the blood (*hyperglycaemia*) and its subsequent excretion in the urine in abnormally high concentrations (*glycosuria*). This condition, known as *diabetes mellitus*, can be treated successfully by insulin injections.

integration *(in neurophysiology)* The coordination within the brain of separate but related nervous processes. For example, sensory information from the inner ear and the eye are both necessary for the sense of balance. These stimuli must be integrated by the brain not only with each other but also with various motor nerves, which coordinate the muscles that control posture.

integument 1. The outermost body layer of an animal, characteristically comprising a layer of living cells – the *epidermis – together with a superficial protective coat, which may be a secreted hardened *cuticle, as in arthropods, or dead keratinized cells, as in vertebrates (*see* **skin**). **2.** The outer protective covering of a plant *ovule. It is perforated by a small pore, the *micropyle. Usually two integuments are present in angiosperms and one in gymnosperms. After fertilization the integuments form the *testa of the seed.

intelligence The coordination of *memory, *learning, and reasoning in animals. Intelligence has also been defined as the ability of an animal to form associative links between events or objects of which it has had no previous experience (*see* **insight learning**). In humans intelligence is generally expressed as an *intelligence quotient* (*IQ*): the mental age of the subject (as measured by standard tests) divided by his or her real age × 100.

intercalary Occurring between differentiated tissues. For example, *intercalary meristem*, which is not part of the *apical meristem, occurs in the internodes of grasses (between leaf nodes) and enables longitudinal growth of the stem.

intercellular Located or occurring between cells. *Compare* **intracellular**.

intercostal muscles The muscles located between the *ribs, surrounding the lungs. Comprising the superficial *external intercostal muscles* and the deep *internal intercostal muscles*, they play an essential role in breathing (*see* **expiration; inspiration**).

interfascicular cambium *See* **cambium**.

interferon Any of a number of proteins (*see* **cytokine**) that increase the resistance of a cell to attack by viruses by unmasking genes that synthesize antiviral proteins. In humans, three groups of interferons have been discovered: α-interferons from white blood cells; β-interferons from connective tissue fibroblasts; and γ-interferons from lymphocytes (*see* **interleukin**). Interferons are also produced by lymphocyte *killer cells, which attack altered tissue cells, such as cancer cells. This converts other normal lymphocytes to killer cells and effects other changes in the immune system. Interferon is therefore of great potential use in treating both acute viral infections and cancer and attempts are underway to produce it on a large scale by means of *gene cloning in bacterial hosts.

intergenic suppressor A *point mutation that negates or lessens the effects of another mutation in the same gene. For example, the effects of a primary *frameshift mutation may be suppressed by a second mutation in which the *deletion or *insertion of a nucleotide occurs.

interleukin Any of several *cytokines that act specifically as mediators between leucocytes. *Interleukin-1 (IL-1)* is secreted by antigen-activated macrophages; it induces activated helper *T cells to secrete *interleukin-2 (IL-2)*. IL-2 stimulates the production of other cytokines, including B-cell growth factor, B-cell differentiation factor, *colony-stimulating factor, and γ-*interferon. *Interleukin-3* is involved in regulating mast-cell proliferation, and *interleukin-4* is thought to be a B-cell growth factor.

internal environment The conditions that prevail within the body of an organism, particularly with respect to the composition of the *tissue fluid. The concept of an internal environment was first proposed by the French physiologist Claude Bernard (1813–78), who stated that maintenance of a constant internal environment was necessary for the survival of an organism in a varying external environment. Selective absorption of materials across cell membranes plays a large part in controlling the internal environment of both animals and plants. Animals in addition can regulate their body fluids by the action of hormones and the nervous system. *See* **homeostasis**.

interneurone A relatively short neurone, which in vertebrates is confined to the grey matter of the central nervous system, that provides a link between a sensory neurone and a motor neurone in a *polysynaptic reflex.

internode 1. (*in botany*) The part of a plant stem between two *nodes. **2.** (*in neurology*) The myelinated region of a nerve fibre between two nodes of Ranvier. *See* **myelin sheath**.

interoceptor A *receptor that detects stimuli from the internal environment of an organism. *Chemoreceptors that detect changes in the levels of oxygen concentration in the blood are examples. *Compare* **exteroceptor**.

interphase The period following the completion of *cell division, when the nucleus is not dividing. During this period changes in both the nucleus and the cytoplasm result in the complete development of the daughter cells. *See* **cell cycle**.

intersex An organism displaying characteristics that are intermediate between those of the typical male and typical female of its species. For example, a human intersex may have testes that fail to develop, so that although he is technically a man he has the external appearance of a woman. Intersexes may be produced in various ways; for example, by malfunctioning of the sex hormones. *See also* **hermaphrodite**. *Compare* **gynandromorph**.

interspecific Occurring between members of different species. The term is applied, for example, to some types of *competition and *communication. *Compare* **infraspecific; intraspecific**.

interspecific competition *See* **competition**.

interstitial cell A cell that forms part of the connective tissue (the *interstitium*) between other tissues and structures, especially any of the cells of the *testis that lie between the seminiferous tubules and secrete androgens in response to stimulation by interstitial-cell-stimulating hormone (*see* **luteinizing hormone**).

interstitial-cell-stimulating hormone *See* **luteinizing hormone**.

intervertebral disc Any of the discs of cartilage that separate the bones of the *vertebral column. The intervertebral discs allow the vertebral column a certain degree of flexibility and they also absorb shock.

intestinal juice A slightly alkaline liquid, containing digestive enzymes and mucus, that is secreted into the lumen of the small intestine from the cells that line the *crypts of Lieberkühn. Together with pancreatic juice, the intestinal juice provides an alkaline environment for the enzymes of the small intestine and helps in the absorption of digested food molecules that enter the small intestine in chyme from the stomach.

intestine The portion of the *alimentary canal posterior to the stomach. Its major functions are the final digestion of food matter from the stomach, the absorption of soluble food matter, the absorption of water, and the production of *faeces. *See* **large intestine; small intestine**.

intine *See* **pollen**.

intracellular Located or occurring within cells. *Compare* **intercellular**.

intrafusal Describing a type of muscle fibre that occurs in the *muscle spindles in voluntary muscles. Two to twelve intrafusal fibres enclosed in a capsule are found in each muscle spindle, lying parallel to the main (*extrafusal*) fibres. Each intrafusal fibre consists of a noncontractile (*equatorial*) region connected to a contractile (*polar*) region at each end of the fibre. The equatorial region is connected to stretch receptors.

intramembranous ossification *See* ossification.

intraspecific Occurring between members of the same species. The term is applied, for example, to some types of *competition and *communication. *Compare* **infraspecific**; **interspecific**.

intraspecific competition *See* competition.

intrinsic factor *See* vitamin B complex.

introgression (**introgressive hybridization**) The insertion of the genes of one species into the gene pool of another. This can occur when two species interbreed to produce fertile hybrids. These can then back-cross with individuals of one of the parent species. *See also* **hybrid zone**.

intron (**intervening sequence**) A nucleotide sequence in a gene that does not code for the gene product (*compare* **exon**). Introns, which are confined to eukaryotes, are transcribed into messenger *RNA but are subsequently removed from the transcript before translation (*see* **gene splicing**). Their functional significance is still subject to controversy.

intussusception The insertion of new cellulose fibres in the space between existing fibres in the wall of an elongating plant cell. This type of growth results in increasing the surface area of the cell wall. *Compare* **apposition**.

inulin A polysaccharide, made up from fructose molecules, that is stored as a food reserve in the roots or tubers of many plants, such as the dahlia.

in utero Describing any event occurring in the uterus of a mammal during pregnancy.

inversion (*in genetics*) **1.** A *chromosome mutation caused by reversal of part of a chromosome, so that the genes within that part are in inverse order. Inversion mutations usually occur during *crossing over in meiosis. **2.** A *point mutation caused by the reversal of two or more bases in the DNA sequence within a gene.

invertebrate Any animal that lacks a vertebral column (backbone). Invertebrates include all nonchordate animals as well as the more primitive chordates (*see* **Chordata**).

in vitro Describing biological processes that are made to occur outside the living body, in laboratory apparatus (literally 'in glass', i.e. in a test tube). In *in vitro* fertilization, mature egg cells are removed from the ovary of a woman unable to conceive normally and fertilized externally; the resultant *blastocyst is implanted into her uterus. *Compare* **in vivo**.

in vivo Describing biological processes as they are observed to occur in their natural environment, i.e. within living organisms. *Compare* **in vitro**.

involucre A protective structure in some flowering plants and bryophytes. In flowering plants it consists of a ring of *bracts arising beneath the flower cluster of those species with a *capitulum (i.e. members of the dandelion family) or an *umbel (i.e. members of the carrot family). In bryophytes the involucre is a projection of tissue from the thallus that arches over the developing *archegonium.

involuntary Not under the control of the will of an individual. Involuntary responses by muscles, glands, etc., occur automatically when required; many such responses, such as gland secretion, heartbeat, and peristalsis, are controlled by the *autonomic nervous system and effected by *involuntary muscle.

involuntary muscle (smooth muscle) Muscle whose activity is not under the control of the will; it is supplied by the *autonomic nervous system. Involuntary muscle comprises long spindle-shaped cells without striations. These cells occur singly, in groups, or as sheets in the skin, around hair follicles, and in the digestive tract, respiratory tract, urinogenital tract, and the circulatory system. The cells contract slowly in spontaneous rhythms or when stretched; they may show sustained contraction (tonus) for long periods without fatigue. *Compare* **voluntary muscle**.

involution 1. A decrease in the size of an organ or the body. It may be associated with functional decline, as occurs in the ageing process, or follow enlargement, as when the uterus returns to its normal size after pregnancy. 2. The turning or rolling inwards of cells that occurs during the development of some vertebrate embryos.

iodine Symbol I. A dark violet nonmetallic element that is required as a trace element (*see* **essential element**) by living organisms; in animals it is concentrated in the thyroid gland as a constituent of thyroid hormones.

iodopsin *See* **colour vision**.

ion An atom or group of atoms that has either lost one or more electrons, making it positively charged (a cation), or gained one or more electrons, making it negatively charged (an anion).

ion exchange The exchange of ions of the same charge between a solution (usually aqueous) and a solid in contact with it. The process occurs widely in nature, especially in the absorption and retention of water-soluble fertilizers by soil. For example, if a potassium salt is dissolved in water and applied to soil, potassium ions are absorbed by the soil and sodium and calcium ions are released from it. The soil, in this case, is acting as an ion exchanger. Synthetic *ion-exchange resins* consist of various copolymers having a cross-linked three-dimensional structure to which ionic groups have been attached. An *anionic resin* has negative ions built into its structure and therefore exchanges positive ions. A *cationic resin* has positive ions built in and exchanges negative ions. Ion-exchange resins are used as the stationary phase in ion-exchange *chromatography.

ion-exchange chromatography *See* chromatography.

ionizing radiation Radiation of sufficiently high energy to cause ionization in the medium through which it passes. It may consist of a stream of high-energy particles (e.g. electrons, protons, alpha-particles) or short-wavelength electromagnetic radiation (ultraviolet, X-rays, gamma-rays). This type of radiation can cause extensive damage to the molecular structure of a substance either as a result of the direct transfer of energy to its atoms or molecules or as a result of the secondary electrons released by ionization. In biological tissue the effect of ionizing radiation can be very serious, usually as a consequence of the ejection of an electron from a water molecule and the oxidizing or reducing effects of the resulting highly reactive species:

$$2H_2O \rightarrow e^- + H_2O^* + H_2O^+$$

$$H_2O^* \rightarrow .OH + .H$$

$$H_2O^+ + H_2O \rightarrow .OH + H_3O^+$$

where the dot before a radical indicates an unpaired electron and an * denotes an excited species.

ionophore A relatively small hydrophobic molecule that facilitates the transport of ions across lipid membranes. Most ionophores are produced by microorganisms. There are two types of ionophore: *channel formers*, which combine to form a *channel in the membrane through which ions can flow; and *mobile ion carriers*, which transport ions across a membrane by forming a complex with the ion. Examples of ionophores include vancomycin and nigericin.

iridium anomaly The occurrence of unusually high concentrations of the relatively scarce metal iridium at the boundaries of certain geological strata. Two such layers have been discovered, one at the end of the Cretaceous, 65 million years ago, and the second at the end of the Eocene, 34 million years ago. One theory to account for these suggests that on each occasion a huge iridium-containing meteorite may have collided with the earth, producing a cloud of dust that settled out to form an iridium-rich layer. The environmental consequences of such an impact, notably in causing a general warming of the earth by the *greenhouse effect, may have led to the extinction of the dinosaurs at the end of the Cretaceous and the extinction of many radiolarians at the end of the Eocene.

iris The pigmented ring of muscular tissue, lying between the cornea and the lens, in the eyes of vertebrates and some cephalopod molluscs. It has a central hole (the *pupil*) through which light enters the eye and it contains both circular and radial muscles. Reflex contraction of the former occurs in bright light to reduce the diameter of the pupil (*see* **pupillary reflex**); contraction of the radial muscles in dim light increases the pupil diameter and therefore the amount of light entering the eye. Colour is determined by the amount of the pigment melanin in the iris. Blue eyes result from relatively little melanin; grey and brown eyes from increasingly larger amounts.

iron Symbol Fe. A silvery malleable and ductile metallic element that is the fourth most abundant element in the earth's crust. It is required as a trace element (*see* **essential element**) by living organisms. Iron is an important constituent of *haemoglobin and the *cytochromes, being stored in the liver in the form of *ferritin. In animals deficiency of iron results in a form of *anaemia.

irradiation Exposure to any form of radiation; often exposure to *ionizing radiation is implied. *See also* **food preservation**.

irrigation The provision of water for crops by artificial methods; for example by constructing ditches, pipe systems, and canals. Irrigation can lead to problems when the water leaches trace elements from the soil; selenium, for example, can be toxic to both local fauna and flora. Irrigation can also increase the salinity of the soil, if diverted rivers are used to provide the water. Evaporation of surface water leaves a crust of salt, which can drain down to deeper layers of the soil.

irritability *See* sensitivity.

ischium The most posterior of the three bones that make up each half of the *pelvic girdle. *See also* **ilium; pubis**.

isidium (*pl.* **isidia**) An outgrowth from the thallus of a *lichen, 0.01–0.03 mm in diameter, that includes both algal and fungal cells. Isidia can detach from the main body of the lichen and serve as structures of vegetative propagation. *Compare* **soredium**.

islets of Langerhans Small groups of cells in the pancreas that function as an endocrine gland. The α *cells* secrete the hormone *glucagon and the β *cells* secrete *insulin. The islets are named after their discoverer, the German anatomist and microscopist Paul Langerhans (1847–88).

isoelectric point The pH of a medium at which a protein carries no net charge and therefore will not migrate in an electric field. Proteins precipitate most readily at their isoelectric points; this property can be utilized to separate mixtures of proteins or amino acids.

isoenzyme *See* isozyme.

isogamy Sexual reproduction involving the production and fusion of gametes that are similar in size and structure. It occurs in some protoctists, e.g. certain protozoans and algae. *Compare* **anisogamy**.

isolating mechanism Any of the biological properties of organisms that prevent interbreeding (and therefore exchange of genetic material) between members of different species that inhabit the same geographical area. These mechanisms include *seasonal isolation*, in which the *breeding seasons of the different populations do not overlap; and *behavioural isolation*, in which different *courtship behaviour in the populations ensures that mating takes place only between members of the same species. Both these are examples of *premating mechanisms*. *Postmating mechanisms* include hybrid infertility and inviability.

isoleucine *See* **amino acid.**

isomerase Any of a class of *enzymes that catalyse the rearrangement of the atoms within a molecule, thereby converting one *isomer into another.

isomers Chemical compounds that have the same molecular formulae but different molecular structures or different arrangements of atoms in space. *Structural isomers* have different molecular structures, i.e. they may be different types of compound or they may simply differ in the position of the functional group in the molecule. Structural isomers generally have different physical and chemical properties. *Stereoisomers* have the same formula and functional groups, but differ in the arrangement of groups in space. Optical isomers are examples of stereoisomers (*see* **optical activity**).

isoprene A colourless liquid diene, $CH_2:C(CH_3)CH:CH_2$. The systematic name is *2-methylbuta-1,3-diene*. It is the structural unit in *terpenes and natural rubber, and is used in making synthetic rubbers.

Isoptera An order of social exopterygote insects that comprises the termites. These mainly tropical insects have a complex system of *castes, including wingless workers and soldiers and primary and secondary winged reproductive members. A termite colony is founded by a single reproductive pair, the nest comprising an elaborate system of tunnels in wood or soil. Termites rely on gut microflora for cellulose digestion, causing considerable damage to wooden structures if they invade houses.

isotonic Describing solutions that have the same osmotic pressure (*see* **osmosis**).

isotope One of two or more atoms of the same element that have the same number of protons in their nucleus but different numbers of neutrons. Hydrogen (1 proton, no neutrons), deuterium (1 proton, 1 neutron), and tritium (1 proton, 2 neutrons) are isotopes of hydrogen. Most elements in nature consist of a mixture of isotopes.

isozyme (isoenzyme) One of several forms of an enzyme in an individual or population that catalyse the same reaction but differ from each other in such properties as substrate affinity and maximum rates of enzyme–substrate reaction (*see* **Michaelis–Menton curve**).

itaconic acid A product of the fermentation of the filamentous fungus *Aspergillus niger*. Itaconic acid is used commercially in the production of adhesives and paints.

J

Jacob–Monod hypothesis The theory postulated by the French biologists F. Jacob (1920–   ) and J. Monod (1910–76) in 1961 to explain the control of *gene expression in bacteria (*see* **operon**). Jacob and Monod investigated the expression of the gene that codes for the enzyme β-galactosidase, which breaks down lactose; the operon that regulates lactose metabolism is called the *lac operon.

jaw The part of the vertebrate skeleton that provides a support for the mouth and holds the teeth. It consists of the upper jaw (maxilla) and the lower jaw (mandible; *see also* **dentary**). Members of the Agnatha lack jaws.

jejunum The portion of the mammalian *small intestine that follows the *duodenum and precedes the *ileum. The surface area of the lining of the jejunum is greatly increased by numerous small outgrowths (*see* **villus**). This facilitates the absorption of digested material, which is the prime function of the jejunum.

jellyfish *See* **Cnidaria**.

Types of freely movable joints

joint The point of contact between two (or more) bones, together with the tissues that surround it. Joints fall into three classes that differ in the degree of freedom of movement they allow: (1) *immovable joints*, e.g. the *sutures between the bones that form the cranium; (2) *slightly movable joints*, e.g. the *symphyses between the vertebrae of the spinal column; and (3) *freely movable* or *synovial joints*, e.g. those that occur between the limb bones. Synovial joints include the *ball-and-socket joints* (between the limbs and the hip and shoulder girdles), which allow movement in all directions; and the *hinge joints* (e.g. at the knee and elbow), which allow movement in one plane only (see illustration). A synovial joint is bound by ligaments and lined with *synovial membrane.

joule Symbol J. The *SI unit of work and energy equal to the work done when the point of application of a force of one newton moves, in the direction of the force, a distance of one metre. 1 joule = 10^7 ergs = 0.2388 calorie. It is named after James Prescott Joule (1818–89).

jugular vein A paired vein in the neck of mammals that returns blood from the head to the heart. It joins the subclavian vein at the base of the neck.

junk DNA *See* **selfish DNA.**

Jurassic The second geological period of the Mesozoic era. It followed the Triassic, which ended about 190 million years ago, and extended until the beginning of the Cretaceous period, about 139 million years ago. It was named in 1829 by A. Brongniart after the Jura Mountains on the borders of France and Switzerland. Jurassic rocks include clays and limestones in which fossil flora and fauna are abundant. Plants included ferns, cycads, ginkgos, rushes, and conifers. Important invertebrates included *ammonites (on which the Jurassic is zoned), corals, brachiopods, bivalves, and echinoids. Reptiles dominated the vertebrates and the first flying reptiles – the pterosaurs – appeared. The first primitive bird, *Archaeopteryx*, also made its appearance.

juvenile hormone A hormone secreted by insects from a pair of endocrine glands (*corpora allata*) close to the brain. It inhibits metamorphosis and maintains the presence of larval features.

Kainozoic *See* **Cenozoic.**

kairomone A chemical given off by organisms of one species that acts as an olfactory signal to individuals of a different species, which thereby benefit. For example, the cabbage white butterfly is attracted to cabbages by means of a kairomone produced by the cabbage, and certain insect predators detect their prey by means of kairomones.

kallidin *See* **kinin.**

karyogamy The fusion of nuclei or nuclear material that occurs during sexual reproduction. *See* **fertilization.**

karyogram (idiogram) A diagram representing the characteristic features of the *chromosomes of a species.

karyokinesis The division of a cell nucleus. *See* **meiosis; mitosis.**

karyolysis *See* **karyorrhexis.**

karyoplasm *See* **nucleoplasm.**

karyorrhexis A stage of cell death (*see* **necrosis**) that involves fragmentation of a cell nucleus. The nucleus breaks down into small dark beads of damaged *chromatin. This usually leads to further dissolution of the nucleus (*karyolysis*), whereupon the damaged chromatin gradually fades.

karyotype The number and structure of the *chromosomes in the nucleus of a cell. The karyotype is identical in all the *diploid cells of an organism.

keel (carina) The projection of bone from the sternum (breastbone) of a bird or bat, to which the powerful flight muscles are attached. The sterna of flightless birds (e.g. ostrich and emu) lack keels.

kelp Any large brown seaweed (*see* **Phaeophyta**) or its ash, used as a source of iodine.

kelvin Symbol K. The *SI unit of thermodynamic temperature equal to the fraction 1/273.16 of the thermodynamic temperature of the triple point of water. The magnitude of the kelvin is equal to that of the degree Celsius (centigrade), but a temperature expressed in degrees Celsius is numerically equal to the temperature in kelvins less 273.15 (i.e. °C = K − 273.15). The unit is named after Lord Kelvin (1824–1907).

keratin Any of a group of fibrous *proteins occurring in hair, feathers, hooves, and horns. Keratins have coiled polypeptide chains that combine to form supercoils of several polypeptides linked by disulphide bonds between adjacent cysteine amino acids. Aggregates of these supercoils form microfibrils, which are embedded in a protein matrix. This produces a strong but elastic structure.

keratinization (cornification) The process in which the cytoplasm of the outermost cells of the mammalian *epidermis is replaced by *keratin. Keratinization occurs in the *stratum corneum, feathers, hair, claws, nails, hooves, and horns.

ketohexose See monosaccharide.

ketone Any one of a group of organic compounds that contain the carbonyl group (>C=O) linked to two hydrocarbon groups. The ketone group is a carbonyl group with two single bonds to other carbon atoms (−CO−). Examples are acetone (propanone), CH_3COCH_3, and methyl ethyl ketone (butanone), $CH_3COC_2H_5$. See also **ketone body**.

ketone body Any of three compounds, acetoacetic acid (3-oxobutanoic acid, CH_3COCH_2COOH), β-hydroxybutyric acid (3-hydroxybutanoic acid, $CH_3CH(OH)CH_2COOH$), and acetone (propanone, CH_3COCH_3), all produced by the liver as a result of the metabolism of body fat deposits. Ketone bodies are normally used as energy sources by peripheral tissues. However, if carbohydrate supply is limited (e.g. during starvation or in diabetics), the blood level of ketone bodies rises and they may be present in urine, giving it a characteristic 'pear drops' odour. This condition is called ketosis.

ketopentose See monosaccharide.

ketose See monosaccharide.

key (identification key) A means of identifying a specimen organism by selecting from a series of choices relating to external features. Each stage in the key presents several character descriptions (two in a dichotomous key); each description is followed by a direction to another stage in the key. Selection of the character that matches the specimen leads to another stage, with another two choices, and so on until the specimen is identified.

kidney The main organ of *excretion and *osmoregulation in vertebrates, through which nitrogenous waste material (usually in the form of *urine) and surplus water, ions, etc., are eliminated from the body. In mammals there is a pair of kidneys situated in the abdomen (see illustration). Each has an outer cortex and an inner medulla and is made up of tubular units called *nephrons, through which nitrogenous waste is filtered from the blood, with the formation of urine. The nephrons drain into a basin-like cavity in the kidney (the renal pelvis), which leads to the *ureter and *bladder.

killer cell Either of two types of *lymphocyte that destroy infected or cancerous host cells. Natural killer cells are distinct from both T cells and B cells. They can act without needing the presence on the surface of the target cell of foreign antigen combined with host histocompatibility proteins. This contrasts with the other type of killer cells − cytotoxic *T cells − which do require such antigen presentation by the target cell. Killer

vena cava — aorta
right kidney — renal artery — left kidney
— collecting duct
— cortex
— medulla
renal vein
pelvis
— ureter
→ direction of blood flow
bladder —
urethra — sphincter muscle

The kidneys of a mammal

cells destroy their target cells by releasing proteins (perforins) that create pores through the target cell membrane, thus causing lysis and cell death.

kilo- Symbol k. A prefix used in the metric system to denote 1000 times. For example, 1000 volts = 1 kilovolt (kV).

kilobase (kb) A unit used at the molecular level for measuring distances along nucleic acids, chromosomes, or genes, equal to 1000 bases (equivalent to 1000 nucleotides or base pairs).

kilogram Symbol kg. The *SI unit of mass defined as a mass equal to that of the international platinum–iridium prototype kept by the International Bureau of Weights and Measures at Sèvres, near Paris.

kinase (phosphokinase) An enzyme that can transfer a phosphate group from a high-energy phosphate, such as ATP, to an organic molecule. *Phosphorylation is normally required to activate the molecule, which is often an enzyme. For example, kinases activate the precursors of enzymes secreted in pancreatic juice (*see* **chymotrypsin; trypsin**). *See also* **protein kinase**.

kinesis The movement of a cell or organism in response to a stimulus in which the rate of movement depends on the intensity (rather than the direction) of the stimulus. For example, a woodlouse moves slowly in a damp atmosphere and quickly in a dry one.

kinetochore Two parallel brushlike filaments situated in the *centromere of a chromosome. Kinetochores attach chromatids to the *spindle fibres during nuclear division.

kingdom The highest category into which organisms are classified. Traditionally two kingdoms were recognized, the Plantae (*see* **plant**) and the

Animalia (*see* **animal**). More recently a number of other kingdoms have been suggested, and most modern classification systems now recognize five kingdoms: Prokaryotae (bacteria; *see* **prokaryote**), *Protoctista (including protozoa and algae), Fungi (*see* **fungi**), Plantae, and Animalia.

kinin 1. One of a group of peptides, occurring in blood, that are involved in inflammation. Kinins are formed by the enzymatic splitting of blood plasma globulins (*kininogens*) at the site of inflammation. Kinins so far identified include *bradykinin* and *kallidin*. They cause local increases in the permeability of small blood vessels. **2.** *See* **cytokinin**.

kinomere *See* **centromere**.

kin selection Natural selection of genes that tend to cause the individuals bearing them to be altruistic to close relatives. These relatives therefore have a higher probability of bearing identical copies of those same genes than do other members of the population. Thus kin selection for a gene that tends to cause an animal to share food with a close relative will result in the gene being spread through the population because it (unconsciously) benefits itself. The more closely two animals are related, the higher the probability that they share some identical genes and therefore the more closely their interests coincide. Parental care is a special case of kin selection. *See* **inclusive fitness**.

Klinefelter's syndrome A genetic disorder affecting men in which an individual gains an extra X chromosome, so that the usual karyotype of XY (*see* **sex chromosome**) is replaced by one of XXY (*see* **nondisjunction**). Symptoms of Klinefelter's syndrome, named after US physician H. F. Klinefelter (1912–), include underdeveloped testes, infertility, and some female characteristics (such as breast enlargement).

klinostat A device used in experiments to test the influence of gravity on the growth movements of plants (*see* **geotropism**). It consists of a motor that slowly rotates a drum inside which seedlings are attached. This prevents any single part of the seedlings from receiving uninterrupted gravitational stimulation and results in horizontal growth of the seedlings.

knee-jerk reflex *See* **stretch reflex**.

Kranz anatomy The arrangement of palisade mesophyll cells in a circle around the vascular bundle of C_4 plants. *See* **bundle sheath cells**.

Krebs cycle (citric acid cycle; tricarboxylic acid cycle; TCA cycle) A cyclical series of biochemical reactions that is fundamental to the metabolism of aerobic organisms, i.e. animals, plants, and many microorganisms (see illustration). The enzymes of the Krebs cycle are located in the *mitochondria and are in close association with the components of the *electron transport chain. The two-carbon *acetyl coenzyme A (acetyl CoA) reacts with the four-carbon oxaloacetate to form the six-carbon citrate. In a series of seven reactions, this is reconverted to oxaloacetate and produces two molecules of carbon dioxide. Most importantly, the cycle generates one molecule of guanosine triphosphate (GTP – equivalent to 1 ATP) and reduces three molecules of the coenzyme *NAD to NADH and one molecule of the coenzyme *FAD to FADH$_2$. NADH and FADH$_2$ are then oxidized by the electron transport chain to generate three and two molecules of ATP respectively. This gives a net yield of 12 molecules of ATP per molecule of acetyl CoA.

Acetyl CoA can be derived from carbohydrates (via *glycolysis), fats, or certain amino acids. (Other amino acids may enter the cycle at different stages.) Thus the Krebs cycle is the central 'crossroads' in the complex system of metabolic pathways and is involved not only in degradation and energy production but also in the synthesis of biomolecules. It is named after its principal discoverer, Sir Hans Adolf Krebs (1900–81).

Krebs cycle

K selection A type of selection that favours organisms with a low rate of reproduction but whose populations expand to the maximum number of individuals that the habitat can support (the *carrying capacity* or *K value* of the habitat). K-selected species (or *K strategists*) tend to be highly adapted to their environment and are able to compete successfully for food and other resources. They also tend to inhabit stable environments and have relatively long life spans. *See* **survival curve**. *Compare* **r selection**.

Kupffer cells Specialized *macrophages that dispose of old blood cells and particulate matter. Kupffer cells, named after Karl Wilhelm von Kupffer (1829–1902), are found in the bloodstream and in the liver, attached to the walls of the *sinusoids.

kwashiorkor *See* **malnutrition**.

labelling The process of replacing a stable atom in a compound with a radioisotope of the same element to enable its path through a biological or mechanical system to be traced by the radiation it emits. In some cases a different stable isotope is used and the path is detected by means of a mass spectrometer. A compound containing either a radioactive or stable isotope is called a *labelled compound* and the atom used is a *label*. If a hydrogen atom in each molecule of the compound has been replaced by a tritium atom, the compound is called a *tritiated compound*. A radioactive labelled compound will behave chemically and physically in the same way as an otherwise identical stable compound, and its presence can easily be detected. This process of *radioactive tracing* is widely used in biology and medicine.

labia *See* labium.

labium (*pl.* **labia**) **1.** The lower lip in the *mouthparts of an insect, which is used in feeding and is formed by the fusion of a pair of appendages (the second *maxillae). **2.** Either member of two pairs of fleshy folds that form part of the *vulva. The outer and larger pair, the *labia majora*, are covered by pubic hair and contain adipose tissue; the smaller *labia minora* lack adipose tissue and pubic hair. Both pairs of labia contain sebaceous glands.

labrum The upper lip in the *mouthparts of an insect. It is formed from a plate of cuticle hinged to the head above the mouth and is used in feeding.

labyrinth The system of cavities and tubes that comprises the *inner ear of vertebrates. It consists of a system of membranous structures (*membranous labyrinth*) housed in a similar shaped bony cavity (*bony labyrinth*).

***lac* operon** The *operon that regulates lactose metabolism in the bacterium *Escherichia coli*. Its form was first postulated in 1961 by François Jacob (1920–) and Jacques Monod (1910–76) to explain control of β-galactosidase synthesis, and it is used as a model for the structure of all other operons. *See* **Jacob–Monod hypothesis.**

lacrimal gland (lachrymal gland) The tear gland, present in the eyelids of some vertebrates. The fluid (tears) produced by this gland cleanses and lubricates the exposed surface of the eye; it drains into the nose through the lacrimal duct.

lactase (β-galactosidase) The enzyme that breaks down the milk sugar, lactose, to glucose and galactose by catalysing the hydrolysis of galactose residues.

lactation The discharge of milk from the *mammary glands. This generally only occurs after birth of the young and is stimulated by the sucking action of the infants. Lactation is under the control of hormones, notably *prolactin and *oxytocin.

lacteal A minute blind-ended lymph vessel that occurs in each *villus of the small intestine. Digested fats are absorbed into the lacteals (*see* **chyle**) and transported to the bloodstream through the *thoracic duct.

lactic acid (2-hydroxypropanoic acid) An alpha hydroxy carboxylic acid, $CH_3CH(OH)COOH$, with a sour taste. Lactic acid is produced from pyruvic acid in active muscle tissue when oxygen is limited (*see* **oxygen debt**) and subsequently removed for conversion to glucose by the liver. During strenuous exercise it may build up in the muscles, causing cramplike pains. It is also produced by fermentation in certain bacteria and is characteristic of sour milk.

lactobacillus (*pl.* **lactobacilli**) Any one of a group of rod-shaped Gram-positive anaerobic bacteria of the genus *Lactobacillus*, in which lactic acid is an end product of fermentation. Lactobacilli are important in food technology for the production of cheese, yoghurt, and other foods from milk.

lactogenic hormone *See* prolactin.

lactose (milk sugar) A sugar comprising one glucose molecule linked to a galactose molecule. Lactose is manufactured by the mammary gland and occurs only in milk. For example, cows' milk contains about 4.7% lactose. It is less sweet than sucrose (cane sugar).

lacuna A gap or cavity in the tissues of an organism; for example, the hollow centre of certain plant stems or any of the small cavities in bone in which the bone-forming cells are found.

laevorotatory Designating a chemical compound that rotates the plane of plane-polarized light to the left (anticlockwise for someone facing the oncoming radiation). *See* **optical activity**.

laevulose *See* fructose.

lag phase *See* bacterial growth curve.

Lamarckism One of the earliest superficially plausible theories of *evolution, proposed by the French biologist Jean-Baptiste de Lamarck (1744–1829) in 1809. He suggested that changes in an individual are acquired during its lifetime, chiefly by increased use or disuse of organs in response to "a need that continues to make itself felt", and that these changes are inherited by its offspring. Thus the long neck and limbs of a giraffe are explained as having evolved by the animal stretching its neck to browse on the foliage of trees. This so-called inheritance of acquired characteristics has never unquestionably been demonstrated to occur and the theory was largely displaced by *Darwinism. Lamarckism is also incompatible with the *Central Dogma of molecular biology. *See also* **Lysenkoism**.

lamella (*pl.* **lamellae**) **1.** *(in botany)* **a.** Any of the paired folds of membranes seen between the *grana in a plant chloroplast. **b.** Any of the spore-bearing gills on the underside of the cap of many mushrooms and toadstools. *See also* **middle lamella**. **2.** *(in zoology)* Any of various thin layers of membranes, especially any of the thin layers of tissue of which compact bone is formed.

Lamellibranchia *See* **Pelecypoda**.

lamina 1. The thin and usually flattened blade of a leaf, in which photosynthesis and transpiration occurs. The bulk of the lamina is made up of *mesophyll cells interspersed by a network of veins (*vascular bundles). The mesophyll is enclosed by a protective epidermis that produces a waxy cuticle. **2.** The leaflike part of the thallus of certain algae, notably kelps. *See also* **stipe**.

lancelet *See* **Cephalochordata**.

large intestine The portion of the alimentary canal of vertebrates between the *small intestine and the *anus. It consists of the *caecum, *colon, and *rectum and its principal function is the absorption of water and formation of faeces.

larva (*pl.* **larvae**) The juvenile stage in the life cycle of most invertebrates, amphibians, and fish, which hatches from the egg, is unlike the adult in form, and is usually incapable of sexual reproduction (*see* **paedogenesis**). It develops into the adult by undergoing *metamorphosis. Larvae can feed themselves and are otherwise self-supporting. Examples are the tadpoles of frogs, the caterpillars of butterflies, and the ciliated planktonic larvae of many marine animals. *Compare* **nymph**.

larynx The anterior portion of the *trachea (windpipe) of tetrapod vertebrates, which in amphibians, reptiles, and mammals contains the *vocal cords. Movement of the cartilage in the walls of the larynx (by means of the laryngeal muscles) alters the tension of the vocal cords. This changes the pitch of the sound emitted by the vocal cords when they vibrate. The final voiced sound is modified by resonance within the oral and nasal cavities.

lasso cell (colloblast) A type of cell in comb jellies (phylum *Ctenophora) that is similar to the *thread cells of cnidarians. Lasso cells are embedded in the tentacles of the organism. Each has two protruding filaments, one straight and one spiralled, with adhesive heads used to catch prey.

latent period The short time that elapses between the reception of a stimulus and the start of the response in an irritable tissue. For a contracting muscle the latent period lasts about 0.02 seconds, during which time calcium ions are released from the *sarcoplasmic reticulum.

latent virus Any virus that remains in its host organism without undergoing replication. Latent viruses, which include herpes simplex virus (the agent of cold sores), may be induced to replicate and cause cell lysis some time after the initial infection, for example when the host's immunity is reduced.

lateral-line canal A sense organ in fish consisting of a groove on the head and along each side of the body that contains sensory hairs. These hairs are sensitive to variations in water pressure and are used to detect moving objects, such as predators, and to monitor progress through water during swimming. The canals and the hairs within them collectively form the *lateral-line system*.

lateral root *See* root.

latex A milky fluid of mixed composition found in some herbaceous plants and trees. Its function is not clear but it may assist in protecting wounds (*compare* **gum**) and it may be involved in the nutrition of the plant. The latex of some species, notably rubber trees, is collected for commercial purposes.

Laurasia *See* continental drift.

lauric acid (dodecanoic acid) A white crystalline *fatty acid, $CH_3(CH_2)_{10}COOH$. Glycerides of the acid are present in natural fats and oils (e.g. coconut and palm-kernel oil).

Law of Independent Assortment *See* Mendel's laws; independent assortment.

Law of Segregation *See* Mendel's laws.

LD$_{50}$ Lethal dose 50, or median lethal dose: the amount of a pharmacological or toxic substance (such as ionizing radiation) that causes death in 50% of a group of experimental animals. For each LD$_{50}$ the species and weight of the animal and the route of administration of the substance is specified. LD$_{50}$s are used both in toxicology and in the *bioassay of therapeutic compounds.

L-dopa *See* dopa.

leaching Extraction of the soluble components of a solid mixture by percolating a solvent through it.

leaf A flattened structure that develops from a superficial group of tissues, the leaf buttress, on the side of the stem apex. Each leaf has a lateral bud in its axil. Leaves are arranged in a definite pattern (*see* **phyllotaxis**) and usually show limited growth. Each consists of a broad flat *lamina (leaf blade) and a leaf base, which attaches the leaf to the stem; a leaf stalk (*petiole*) may also be present. The leaves of bryophytes are simple appendages, which are not homologous with the leaves of vascular plants as they develop on the gametophyte generation.

Leaves show considerable variation in size, shape, arrangement of veins,

Simple leaves   *Compound leaves*

type of attachment to the stem, and texture. They may be *simple* or divided into *leaflets*, i.e. *compound* (see illustration). Types of leaf include: *cotyledons (seed leaves); *scale leaves*, which lack chlorophyll and develop on rhizomes or protect the inner leaves of a bud; *foliage leaves*, which are the main organs for photosynthesis and transpiration; and *bracts and *floral leaves*, such as sepals, petals, stamens, and carpels, which are specialized for reproduction.

Leaves may be modified for special purposes. For example the leaf bases of bulbs are swollen with food to survive the winter. In some plants leaves are reduced to spines for protection and their photosynthetic function is carried out by another organ, such as a *cladode.

Transverse section through a leaf blade

leaf area index (LAI) The total surface area of the leaves of plants in a given area divided by the area of ground covered by the plants. In an area of dense vegetation, such as a forest, the LAI will be high.

leaf buttress *See* primordium.

leaf litter *See* litter.

learning A process by which an animal's response to a particular situation may be permanently altered, usually in a beneficial way, as a result of its experience. Learning allows an animal to respond more flexibly to the situations it encounters: learning abilities in different species vary widely and are adapted to the species' environment. Numerous different categories of learning have been proposed, but there is no general agreement on how many different processes are involved. These categories include *habituation, associative learning (through *conditioning), *insight learning, and *imprinting.

lecithin (phosphatidylcholine) A phosphoglyceride (*see* **phospholipid**) containing the amino alcohol *choline esterified to the phosphate group. It is the most abundant animal phospholipid (being a component of cell membranes) and also occurs in higher plants, but rarely in microorganisms.

lectin Any of a group of proteins, derived from plants, that can bind to specific oligosaccharides on the surface of cells, causing the cells to clump together (*see* **agglutination**). Lectins can be used to identify mutant cells in cell cultures and to determine blood groups as they can cause the agglutination of red blood cells. Lectins are found in seeds of legumes and in other tissues, in which they are thought to act as a toxin.

leeches *See* Hirudinea.

legume (pod) A dry fruit formed from a single carpel and containing one or more seeds, which are shed when mature. It is the characteristic fruit of the Leguminosae (pea family). It splits, often explosively, along both sides and the two halves of the fruit move apart to expose the seeds. A special form of the legume is the *lomentum.

Leishman's stain A neutral stain for blood smears devised by the British surgeon W. B. Leishman (1865–1926). It consists of a mixture of *eosin (an acidic stain), and *methylene blue (a basic stain) in alcohol and is usually diluted and buffered before use. It stains the different components of blood in a range of shades between red and blue. The similar *Wright's stain* is favoured by American workers.

lemma One of two bracts that protects the floret of grasses. The lemma is situated beneath the second (smaller) bract, known as the *palea*, which it surrounds. *See* spike.

lens A transparent biconvex structure in the eyes or analogous organs of many animals, responsible for directing light onto light-sensitive cells. In vertebrates it is a flexible structure centred behind the iris and attached by *suspensory ligaments* to the *ciliary body. In terrestrial species its main function is to focus images onto the retina. To focus on near objects, the circular muscles in the ciliary body contract and the lens becomes more convex; contraction of the radial muscles in the ciliary body flattens the lens for focusing on distant objects (*see also* **accommodation**).

lenticel Any of the raised pores in the stems of woody plants that allow gas exchange between the atmosphere and the internal tissues. The pore is formed by the *cork cambium, which, at certain points, produces a loose bulky form of cork that pushes through the outer tissues to create the lenticel.

Lepidoptera An order of insects comprising the butterflies and moths, found mainly in tropical regions. Adults possess two pairs of membranous wings, often brightly coloured and usually coupled together. The wings, body, and legs are covered with minute scales. Adult mouthparts are generally modified to form a long proboscis for sucking nectar, fruit juices, etc. Butterflies are typically small-bodied, active during daylight, and rest with their wings folded vertically; moths have larger bodies, are nocturnal, and rest with their wings in various positions. The larvae (caterpillars) have a prominent head and a segmented wormlike body, most segments bearing a pair of legs. They chew leaves and stems, sometimes causing considerable damage to crop plants. The larvae undergo metamorphosis via a *pupa (chrysalis) to the adult form. In some groups, the pupa is enclosed in a cocoon of silk derived from silk glands (modified salivary glands); others use leaves, etc. to build a cocoon.

leptotene The beginning of the first prophase of *meiosis, when the chromatids can be seen and *pairing begins.

lethal allele (lethal gene) A mutant form of a gene that eventually results in the death of an organism if expressed in the phenotype. Most lethal genes are recessive; for example, sickle-cell anaemia (*see* **polymorphism**) results from a recessive lethal gene that causes the production of abnormal and inefficient haemoglobin.

lethal dose 50 *See* LD_{50}.

leucine *See* **amino acid**.

leucocyte (white blood cell) A colourless cell with a nucleus, found in blood and lymph. Leucocytes are formed in lymph nodes and red bone marrow and are capable of amoeboid movement. They can produce *antibodies and move through the walls of vessels to migrate to the sites of injuries, where they surround and isolate dead tissue, foreign bodies, and bacteria. There are two major types: those without granules in the cytoplasm, such as *lymphocytes and *monocytes (*see* **agranulocyte**), and those with granular cytoplasm (*granulocytes), which include *basophils, *eosinophils, and *neutrophils.

leucoplast Any of various organelles in plant cells that contain no pigment and are therefore colourless. Leucoplasts are usually found in tissues not normally exposed to light and frequently contain reserves of starch (in *amyloplasts*), protein, or oil. *See* **plastid**. *Compare* **chromoplast**.

leukaemia *See* cancer.

leukotrienes (LTs) A class of compounds synthesized from *arachidonic acid in a variety of tissues, including the lungs and leucocytes. There are at least six different types of leukotrienes, all of which play an active role in regulating *inflammation. In the *immune response, LTs attract neutrophils and eosinophils to sites of infection and they cause *vasoconstriction.

***l*-form** *See* optical activity.

LH *See* luteinizing hormone.

lice *See* **Mallophaga** (bird lice); **Siphunculata** (sucking lice).

lichens A group of organisms that are symbiotic associations (*see* **symbiosis**) between a fungus (usually one of the *Ascomycota) and a green alga or a blue-green bacterium. The fungal partner (*mycobiont*) usually makes up most of the lichen body and the cells of the alga or bacterium (*phycobiont*) are distributed within it. The phycobiont photosynthesizes and passes most of its food to the fungus and the fungus protects its partner's cells. The lichen reproduces by means of *soredia, *isidia, or by fungal spores, which must find a suitable partner on germination. Lichens are slow growing but can live in regions that are too cold or exposed for other plants. They may form a flattened crust or be erect and branching. Many grow as *epiphytes, especially on tree trunks. Some species are very sensitive to air pollution and have been used as *indicator species. Formerly placed together in the taxon Lichenes, lichens are now classified as a phylum of fungi, the Mycophycophyta.

life cycle The complete sequence of events undergone by organisms of a particular species from the fusion of gametes in one generation to the same stage in the following generation. In most animals gametes are formed by *meiosis of germ cells in the reproductive organs of the parents. The zygote, formed by the fusion of two gametes, eventually develops into an organism essentially similar to the parents. In plants, however, the products of meiosis are spores, which develop into plants (the *gametophyte generation) often very different in form from the spore-forming (*sporophyte) generation. The sporophyte generation is restored when gametes, formed by the gametophyte generation, fuse. *See* **alternation of generations**.

ligament A resilient but flexible band of tissue (chiefly *collagen) that holds two or more bones together at a movable *joint. Ligaments restrain the movement of bones at a joint and are therefore important in preventing dislocation.

ligase Any of a class of enzymes that catalyse the formation of covalent bonds using the energy released by the cleavage of ATP. Ligases are important in the synthesis and repair of many biological molecules, including DNA (*see* **DNA ligase**), and are used in genetic engineering to insert foreign DNA into cloning *vectors.

light green *See* **fast green**.

light reaction (light-dependent reaction) *See* **photosynthesis**.

lignin A complex organic polymer that is deposited within the cellulose of plant cell walls during secondary thickening. Lignification makes the walls woody and therefore rigid. *See* **sclerenchyma**.

lignite *See* **coal**.

ligule **1.** A membranous scalelike outgrowth from the leaves of certain flowering plants. Many grasses have a ligule at the base of the leaf blade. **2.** A small membranous structure that develops on the upper surface of a young leaf base in certain lycopods, for example *Selaginella*. It withers as the plant matures. **3.** A strap-shaped extension from the corolla tube in certain florets of a *capitulum. They are termed *ligulate* (or *ray*) *florets*.

limb **1.** An appendage of a vertebrate animal, such as the leg or arm of a mammal or the wing of a bird. *See also* **pentadactyl limb**. **2.** The expanded upper part of a sepal, petal, or leaf. **3.** The widened upper section of a gamopetalous *corolla.

limbic system A group of regions in the brain that is involved in the expression and control of mood and instinct and plays a major role in long-term memory. The limbic system includes the *hippocampus and the *hypothalamus.

liming The application of lime (calcium hydroxide) to soils to increase levels of calcium and decrease acidity.

limiting factor Any environmental factor that – by its decrease, increase, absence, or presence – limits the growth, metabolic processes, or distribution of organisms or populations. In a desert ecosystem, for example, low rainfall and high temperature will be factors limiting colonization. When a metabolic process is affected by more than one factor, the *law of limiting factors* states that its rate is limited by the factor that is nearest its minimum value. For example, photosynthesis is affected by many factors, such as light, temperature, and carbon dioxide concentration, but on a warm sunny day carbon dioxide concentration will be the limiting factor as light and temperature will be at optimum levels.

limnology The study of freshwater habitats and ecosystems.

linear energy transfer (LET) The energy transferred per unit path length by a moving high-energy charged particle (such as an electron or a proton) to the atoms and molecules along its path. It is of particular importance when the particles pass through living tissue as the LET modifies the effect of a specific dose of radiation. LET is proportional to the square of the charge on the particle and increases as the velocity of the particle decreases.

linkage The tendency for two different genes on the same chromosome to remain together during the separation of *homologous chromosomes at meiosis. Linkage can be broken by *crossing over or by a *chromosome mutation, when sections of chromosomes are exchanged and new combinations of genes are produced. *See also* **sex linkage**.

linkage map *See* **chromosome map**.

Linnaean system *See* **binomial nomenclature**.

linoleic acid A liquid polyunsaturated *fatty acid with two double bonds, $CH_3(CH_2)_4CH:CHCH_2CH:CH(CH_2)_7COOH$. Linoleic acid is abundant in many plant fats and oils, e.g. linseed oil, groundnut oil, and soya-bean oil. It is an *essential fatty acid.

linolenic acid A liquid polyunsaturated *fatty acid with three double bonds in its structure: $CH_3CH_2CH:CHCH_2CH:CHCH_2CH:CH(CH_2)_7COOH$. Linolenic acid occurs in certain plant oils, e.g. linseed and soya-bean oil, and in algae. It is one of the *essential fatty acids.

lipase An enzyme secreted by the pancreas and the glands of the small intestine of vertebrates that catalyses the breakdown of fats into fatty acids and glycerol.

lipid Any of a diverse group of organic compounds, occurring in living organisms, that are insoluble in water but soluble in organic solvents, such as chloroform, benzene, etc. Lipids are broadly classified into two categories: *complex lipids*, which are esters of long-chain fatty acids and include the *glycerides (which constitute the *fats and *oils of animals and plants), *glycolipids, *phospholipids, and *waxes; and *simple lipids*, which do not contain fatty acids and include the *steroids and *terpenes.

Lipids have a variety of functions in living organisms. Fats and oils are a convenient and concentrated means of storing food energy in plants and animals. Phospholipids and *sterols, such as cholesterol, are major components of cell membranes (*see* **lipid bilayer**). Waxes provide vital waterproofing for body surfaces. Terpenes include vitamins A, E, and K, and phytol (a component of chlorophyll) and occur in essential oils, such as menthol and camphor. Steroids include the adrenal hormones, sex hormones, and bile acids.

Lipids can combine with proteins to form *lipoproteins (e.g. in cell membranes). In bacterial cell walls, lipids may associate with polysaccharides to form *lipopolysaccharides*.

lipid bilayer The arrangement of lipid molecules in biological *membranes, which takes the form of a double sheet. Each lipid molecule comprises a hydrophilic 'head' (having a high affinity for water) and a hydrophobic 'tail' (having a low affinity for water). In the lipid bilayer the molecules are aligned so that their hydrophilic heads face outwards, forming the outer and inner surfaces of the membrane, while the hydrophobic tails face inwards, away from the external aqueous environment.

Structure of a lipid bilayer

lipoic acid A vitamin of the *vitamin B complex. It is one of the *coenzymes involved in the decarboxylation of pyruvate by the enzyme pyruvate dehydrogenase. This reaction has to take place before carbohydrates can enter the *Krebs cycle during aerobic respiration. Good sources of lipoic acid include liver and yeast.

lipolysis The breakdown of storage lipids in living organisms. Most long-term energy reserves are in the form of triglycerides in fats and oils. When these are needed, e.g. during starvation, lipase enzymes convert the triglycerides into glycerol and the component fatty acids. These are then transported to tissues and oxidized to provide energy.

lipoprotein One of a group of compounds consisting of a lipid combined with a protein. Lipoproteins are the main structural materials of the membranes of cells and cell organelles. They also occur in blood and lymph, being the form in which lipids are transported in these media. *Cholesterol is transported in the bloodstream in the form of *low-density lipoproteins (LDLs)* and is removed by means of LDL receptors in cell membranes; the LDLs are bound to the receptors, which are then taken into the cells. Lack of LDL receptors, occurring as a genetic defect in some individuals, is believed to be a cause of high levels of cholesterol in the blood, predisposing to atherosclerosis.

liposome A microscopic spherical membrane-enclosed vesicle or sac (20–30 nm in diameter) made artificially in the laboratory by the addition of an aqueous solution to a phospholipid gel. The membrane resembles a cell membrane and the whole vesicle is similar to a cell organelle. Liposomes can be incorporated into living cells and are used to transport relatively toxic drugs into diseased cells, where they can exert their maximum effects. For example, liposomes containing the drug methotrexate, used in the treatment of cancer, can be injected into the patient's blood. The cancerous organ is heated to a temperature higher than body temperature, so that when the liposome passes through its blood vessels, the membrane melts and the drug is released. Liposomes can also be used as vectors in *gene therapy. The study of the behaviour of liposome membranes is used in research into membrane function, particularly to observe the behaviour of membranes during anaesthesia with respect to permeability changes.

lipotrophin Either of two peptide hormones produced in the anterior pituitary gland that trigger the mobilization of fat deposits and the transfer of lipid components to the bloodstream. β-lipotrophin is formed by cleavage of the precursor pro-opiocortin, and is itself cleaved to form γ-lipotrophin and β-endorphin.

Listeria A genus of rod-shaped aerobic motile Gram-positive bacteria. Only one species, *L. monocytogenes*, causes disease (*listeriosis*). It is resistant to physical and chemical treatments and can occur as a contaminant in certain foods, in faeces, etc. Listeriosis can take various forms, depending on the site of infection: localization in the central nervous system causes meningoencephalitis, while uterine infection can result in abortion or congenital handicap in the fetus.

litre Symbol l. A unit of volume in the metric system regarded as a special name for the cubic decimetre. It was formerly defined as the volume of 1 kilogram of pure water at 4°C at standard pressure, which is equivalent to 1.000 028 dm^3.

litter Dead organic matter in the soil that has not yet decomposed. It consists of fallen leaves and other plant remains (*leaf litter*), animal excrement, etc. After decomposition by *decomposers and *detritivores litter becomes *humus.

littoral Designating or occurring in the marginal shallow-water zone of a sea or lake, especially (in the sea) between high and low tide lines. In this zone enough light penetrates to the bottom to support rooted aquatic plants. *Compare* **profundal; sublittoral**.

liver A large lobed organ in the abdomen of vertebrates that plays an essential role in many metabolic processes by regulating the composition and concentration of nutrients and toxic materials in the blood. It is made up of units called *lobules*, each of which is a roughly hexagonal structure consisting largely of *hepatocytes arranged around a central vein. The liver receives the products of digestion dissolved in the blood via the *hepatic portal vein and its most important functions are to convert excess glucose to the storage product *glycogen, which serves as a food reserve; to break down excess amino acids to ammonia, which is converted to *urea or *uric acid and excreted via the kidneys; and to store and break down fats (*see* **lipolysis**). Other functions of the liver are (1) the production of *bile; (2) the breakdown (*detoxification) of poisonous substances in the blood; (3) the removal of damaged red blood cells; (4) the synthesis of vitamin A and the blood-clotting substances prothrombin and fibrinogen; and (5) the storage of iron (*see* **ferritin**).

liverworts *See* **Hepaticae**.

living fossil Any organism whose closest relatives are extinct and that was once itself thought to be extinct. An example is the coelacanth, a primitive fish that was common in the Devonian era, the first recent living specimen of which was discovered in 1938.

lizards *See* **Squamata**.

loam A fertile *soil that is made up of organic matter mixed with clay, sand, and silt. Loams differ in their ratios of clay, sand, and silt, which influences which types of plants they can support.

lock-and-key mechanism A mechanism proposed in 1890 by Emil Fischer (1852–1919) to explain binding between the active site of an enzyme and a substrate molecule. The active site was thought to have a fixed structure (the lock), which exactly matched the structure of a specific substrate (the key). Thus the enzyme and substrate interact to form an *enzyme–substrate complex. The substrate is converted to products that no longer fit the active site and are therefore released, liberating the enzyme. Recent observations made by X-ray diffraction studies have shown that the active site of an enzyme is more flexible than the lock-and-key theory would suggest. *Compare* **induced-fit model**.

locomotion The ability of an organism to move in a particular direction in its environment, which requires a propulsive force acting against a supporting structure. Most animals and many single-celled organisms have powers of locomotion. Some protozoa possess contractile fibres that exert force on the cell membrane to change the shape of the cell; this may be combined with *cytoplasmic streaming to bring about locomotion (*see* **amoeboid movement**). In many other protozoa and bacteria the propulsive force is provided by the action of *cilia or *flagella. In animals the force required to initiate locomotion is generated by *voluntary muscles, which act against a supporting framework provided by a *skeleton. *See also* **fins; flight**.

locule (loculus) A small cavity in a plant or animal body. In plants the locule of the ovary is the cavity containing the ovules and the locules of the anther contain the developing pollen grains.

locus (*pl.* **loci**) The position of a gene on a chromosome or within a DNA molecule. The two alleles of a gene occupy the same locus on *homologous chromosomes.

lodicule Either of two small scales situated at the base of the floret of grasses. They may represent the reduced perianth and be involved in the opening of the floret at the time of pollination.

logarithmic scale A scale of measurement in which an increase or decrease of one unit represents a tenfold increase or decrease in the quantity measured. Decibels and pH measurements are common examples of logarithmic scales of measurement.

log phase *See* **bacterial growth curve**.

lomentum A type of dry dehiscent fruit formed from a single carpel but divided into one-seeded compartments by constrictions between the seeds. *Legumes (e.g. those of *Acacia*) and *siliquas (e.g. those of wild radish) can be divided in this way.

long-day plant A plant in which flowering can be induced or enhanced by long days, usually of more than 12 hours of daylight. Examples are spinach and spring barley. *See* **photoperiodism**. *Compare* **day-neutral plant; short-day plant**.

long-sightedness *See* **hypermetropia**.

loop of Henle The hairpin-shaped section of a kidney tubule situated between the proximal and distal tubules in the *nephron. The loop of Henle extends from the cortex into the medulla; it consists of a thin descending limb, which is permeable to water, and a thick ascending limb, which is impermeable to water. Complex movements of ions and water across the walls of the loop enable it to function as a *countercurrent multiplier, resulting in the production of concentrated urine in the *collecting duct. It is named after German anatomist F. G. J. Henle (1809–85).

lophophore An organ characteristic of aquatic invertebrates of the phyla Ectoprocta, Phoronida, and Brachiopoda that functions in filter feeding. It consists of a ridge of hollow tentacles being cilia, which waft food particles into the mouth.

LSD *See* **lysergic acid diethylamide**.

luciferase *See* **bioluminescence**.

luciferin *See* **bioluminescence**.

lumbar vertebrae The *vertebrae in the region of the lower back. They occur below the *thoracic vertebrae and above the *sacral vertebrae. In mammals they bear processes for the attachment of back muscles.

lumen 1. The space enclosed by a vessel, duct, or other tubular or saclike organ. The central cavity of blood vessels and of the digestive tract are examples. **2.** Symbol lm. The SI unit of luminous flux equal to the flux emitted by a uniform point source of 1 candela in a solid angle of 1 steradian.

lung The *respiratory organ of air-breathing vertebrates. A pair of lungs is situated in the thorax, within the ribcage. Each consists essentially of a thin moist membrane that is folded to increase its surface area. Exchange of oxygen and carbon dioxide takes place between blood capillaries on one side of the membrane and air on the other. The lung is supplied with air through a *bronchus. In mammals and reptiles the membrane of the lung takes the form of numerous sacs (*see* **alveolus**) that are connected to the bronchus via *bronchioles (see illustration). The lungs themselves contain no muscular tissue and are ventilated by *respiratory movements, the mechanisms of which vary with the species.

The lungs and air passages of a mammal (right lung cut open to show internal structure)

lung book The respiratory organ of some arachnids. Lung books occur in pairs on the ventral side of the abdomen. They consist of leaflike folds of ectoderm sunk into pockets having slitlike openings at the abdominal surface. Gaseous exchange occurs by diffusion through the ectodermal folds.

lungfish *See* Dipnoi.

luteal phase The phase of the *oestrous cycle (menstrual cycle in women) that occurs after ovulation. During this phase the corpus luteum secretes oestrogens and progesterone, which prepare the uterine wall to receive a fertilized egg.

luteinizing hormone (LH; interstitial-cell-stimulating hormone; ICSH) A hormone, secreted by the anterior pituitary gland in mammals, that stimulates in males the production of sex hormones (*androgens) by the *interstitial cells of the testes and in females ovulation, *progesterone synthesis, and *corpus luteum formation.

luteotrophic hormone *See* prolactin.

lyase Any of a class of enzymes that catalyse either the cleavage of a double bond and the addition of new groups to a substrate, or the formation of a double bond.

Lycopodophyta A phylum of *tracheophyte plants containing the clubmosses (genus *Lycopodium*) and related genera (including *Selaginella*) as well as numerous extinct forms, which reached their peak in the Carboniferous period with giant coal-forming tree species. Lycopods have roots and their stems are covered with numerous small leaves. Reproduction is by means of spores; the sporangia are usually grouped into cones.

lymph The colourless liquid found within the *lymphatic system, into which it drains from the spaces between the cells. Lymph (called *tissue fluid in the intercellular spaces) resembles ^blood plasma, consisting mostly of water with dissolved salts and proteins. Fats are found in suspension and their presence varies wth food intake. The lymph eventually enters the bloodstream near the heart.

lymphatic system The network of vessels that conveys *lymph from the tissue fluids to the bloodstream. Tiny *lacteals (in the small intestine) and *lymph capillaries* (in other tissues) drain into larger tubular vessels that converge to form the right lymphatic duct and the *thoracic duct, which connect with the venous blood supply to the heart. Associated with the lymphatic vessels at intervals along the system are the *lymph nodes. The lymph capillary walls are very permeable, so lymph bathing the body's tissues can drain away molecules that are too large to pass through blood capillary walls. In most vertebrates lymph is pumped by cycles of contraction and relaxation of the lymphatic vessels and also by the action of adjoining muscles. *See also* **lymph heart**.

lymph capillary *See* **lymphatic system**.

lymph heart Any of a series of muscular pumping structures that occur in the major lymphatic vessels of certain lower vertebrates and maintain a flow of lymph around the body. Many animals (including all mammals) do not possess lymph hearts; in these the flow of lymph is maintained by a series of muscle contractions and valves (*see* **lymphatic system**).

lymph node A mass of *lymphoid tissue, many of which occur at intervals along the *lymphatic system. Lymph in the lymphatic vessels flows through the lymph nodes, which filter out bacteria and other foreign particles, so preventing them from entering the bloodstream and causing infection. The lymph nodes also produce *lymphocytes. In humans, major lymph nodes occur in the neck, under the arms, and in the groin.

lymphocyte A type of white blood cell (*leucocyte) that has a large nucleus and little cytoplasm. Lymphocytes are formed in the *lymph nodes and provide about a quarter of all leucocytes. They are important in the body's defence and are responsible for immune reactions as the presence of *antigens stimulates them to produce *antibodies. There are two populations of lymphocytes: *B lymphocytes* (*see* **B cell**), which produce circulating antibodies and are responsible for humoral *immunity; and *T lymphocytes* (*see* **T cell**), which are responsible for cell-mediated immunity. *See also* **killer cell**.

lymphoid tissue The type of tissue found in the *lymph nodes, *tonsils, *spleen, and *thymus. It is responsible for producing lymphocytes and antibodies and therefore contributes to the body's defence against infection.

lymphokine Any soluble factor secreted by lymphocytes that mediates the activities of lymphoid cells. Examples include *interleukin-2, B-cell growth factor (BCGF), B-cell differentiation factor (BCDF), and macrophage migration inhibition factor (MIF). *See also* **cytokine**.

lyophilization *See* freeze drying.

Lysenkoism The official Soviet science policy governing the work of geneticists in the USSR from about 1940 to 1960. It was named after its chief promoter, the agriculturalist Trofim Lysenko (1898–1976). Lysenkoism dismissed all the advances that had been made in classical genetics, denying the existence of genes, and held that the variability of organisms was produced solely by environmental changes. There was also a return to a belief in the inheritance of acquired characteristics (*see* **Lamarckism**). This state of affairs continued, despite overwhelming conflicting evidence from Western scientists, because it provided support for communist theory.

lysergic acid diethylamide (LSD) A chemical derivative of lysergic acid that has potent hallucinogenic properties (*see* **hallucinogen**). It occurs in the cereal fungus ergot and was first synthesized in 1943. LSD acts as an *antagonist at *serotonin receptors.

lysigeny The localized disruption of plant cells to form a cavity (surrounded by remnants of the broken cells) in which secretions accumulate. Examples are the oil cavities in the leaves of citrus trees. *Compare* **schizogeny**.

lysimeter An instrument used to measure the loss of water from an area of land covered with vegetation. Water that evaporates from both the soil and from the plants can be determined.

lysine *See* amino acid.

lysis The destruction of a living cell. This may be effected by *lysosomes or *lymphocytes, either as part of the normal metabolic process (as when cells are damaged or worn out) or as a reaction against invading cells (e.g. bacteria). *Bacteriophages eventually cause lysis of their host cells.

lysogeny The relationship between a temperate phage (*see* **bacteriophage**) and a bacterium. A bacterium whose chromosome has been integrated with the DNA of a temperate phage is called a *lysogen*; the viral DNA is known as a *prophage.

lysosome A membrane-bound sac within the cytoplasm of animal *cells that contains enzymes responsible for the digestion of material in food vacuoles, the dissolution of foreign particles entering the cell (such as bacteria) and, on the death of the cell, the breakdown of all cell structures. A similar structure is present in plant cells.

lysozyme An antibacterial enzyme widely distributed in body fluids and secretions, including tears and saliva. It disrupts the polysaccharide components of bacterial cell walls, leaving them susceptible to destruction.

M

macrofauna The larger animals, collectively, which can be observed without the aid of a microscope (*compare* **microfauna**). The macrofauna sometimes includes small soil-dwelling invertebrates, such as annelids and nematodes, but these may be separated into an intermediate category, the *mesofauna*.

macromolecule A very large molecule. Natural and synthetic polymers have macromolecules, as do such substances as haemoglobin. *See also* **colloids**.

macronutrient A chemical element required by plants in relatively large amounts. Macronutrients include carbon, hydrogen, oxygen, nitrogen, phosphorus, potassium, sulphur, magnesium, calcium, and iron. *See also* **essential element**. *Compare* **micronutrient**.

macrophage (histiocyte; clasmatocyte) A large phagocytic cell (*see* **phagocyte**) found in the spleen, lymph nodes, liver, and many other tissues. Macrophages ingest bacteria and cell debris and are responsible for the breakdown of worn-out red blood cells. They develop from *monocytes and are characterized by a horseshoe-shaped nucleus and nongranular cytoplasm. Tissues containing a high concentration of macrophages are known collectively as the *reticuloendothelial system*.

macrophagous Describing a method of feeding in heterotrophic organisms in which food is ingested in the form of relatively large chunks. *Compare* **microphagous**.

macrophyll *See* **megaphyll**.

macula 1. A patch of sensory hair cells in the *utriculus and *sacculus of

Macula (when the head is positioned at 90° to the direction of the force of gravity)

the inner ear that provides information about the position of the body in relation to gravity. The hairs of the cells are embedded in an *otolith*, a gelatinous cap containing particles of calcium carbonate. Movement of the particles in response to gravity pulls the gelatinous mass downwards, which bends the hairs and triggers a nerve impulse to the brain (see illustration). **2.** An area of the *retina of the vertebrate eye with increased *visual acuity. Maculae occur in some animals that lack *foveae and often surround foveae in those animals that possess them.

magnesium Symbol Mg. A silvery metallic element that is an *essential element for living organisms. It functions as a cofactor for various enzymes and is a constituent of *chlorophyll in plants. In animals it is involved in the transmission of nerve impulses.

magnetic resonance imaging (MRI) *See* nuclear magnetic resonance.

major histocompatibility complex (MHC) A large gene cluster that encodes various components of the immune system, including the *histocompatibility antigens and components of the *complement system. In humans the MHC is located on chromosome 6 and includes the *HLA system. Other vertebrate species have similar MHC regions. Certain MHC genes can have many variant alleles; this produces an enormous diversity of antigens in a population, each individual possessing a unique set.

malaria A disease caused by the parasitic protozoan *Plasmodium*, which requires two hosts, the bloodsucking female *Anopheles* mosquito and man, in order to complete its complex life cycle. Symptoms of fever and anaemia in man are caused by invasion and destruction of the red blood cells during an asexual phase of the life cycle. *See* **Apicomplexa**.

male 1. Denoting the gamete (sex cell) that, during *sexual reproduction, fuses with a *female gamete in the process of fertilization. Male gametes are generally smaller than the female gametes and are usually motile (*see* **spermatozoon**). **2.** (Denoting) an individual whose reproductive organs produce only male gametes. *Compare* **hermaphrodite**.

malic acid (2-hydroxybutanedioic acid) A crystalline solid, $HOOCCH(OH)CH_2COOH$. L-malic acid occurs in living organisms as an intermediate metabolite in the *Krebs cycle and also (in certain plants) in photosynthesis. It is found especially in the juice of unripe fruits, e.g. green apples.

malignant Describing a mutant cell or group of cells that proliferates at a faster rate than normal cells and has the capacity to spread to other sites in the body. *See* **cancer**.

malleus (hammer) The first of the three *ear ossicles of the mammalian *middle ear.

Mallophaga An order of secondarily wingless insects comprising the bird lice. Bird lice are minute with dorsoventrally flattened ovoid bodies, reduced eyes, and biting mouthparts. They are ectoparasites of birds, feeding on particles of dead skin, feather fragments, and sometimes blood. The eggs hatch to form nymphs resembling the adults.

malnutrition The condition arising due to the lack of one or more of the *nutrients that are required in the *diet to maintain health. Malnutrition can result from a reduced intake of nutrients (*undernourishment*), an inability to use absorbed nutrients, failure to meet a required increase in nutrient intake, or nutrient losses. There are three stages in the process of malnutrition: first, the carbohydrate stores in the body are depleted; secondly, the fat reserves are metabolized (*see* **fatty-acid oxidation**); and finally, proteins are broken down to provide energy. Death may result after protein levels have been reduced to half their normal value. *Kwashiorkor* is a type of malnutrition that develops when the diet lacks proteins and hence *essential amino acids. Malnutrition due to reduced absorption of nutrients in the intestine can develop with a cereal-based diet, due to sensitivity of the intestinal lining to gluten, a protein found in cereals. *See also* **mineral deficiency**.

Malpighian body (Malpighian corpuscle) The part of a *nephron in the kidney that consists of its cup-shaped end together with the *glomerulus that it encloses. It is named after its discoverer, the Italian anatomist M. Malpighi (1628–94).

Malpighian layer (stratum germinativum) The innermost layer of the *epidermis of mammalian *skin, separated from the underlying dermis by a fibrous *basement membrane. It is only in this layer of the epidermis that active cell division (*mitosis) occurs. As the cells produced by these divisions age and mature, they migrate upwards through the layers of the epidermis to replace the cells being continuously worn away at the surface.

Malpighian tubule Any of the organs that are involved in the excretion of nitrogenous waste in insects, arachnids, and centipedes. They lie in the haemocoel. In insects Malpighian tubules open into the intestine; they selectively extract from the blood uric acid, which – together with water and salts – is deposited into the hindgut and excreted in the faeces. Some reabsorption of water and salts can occur in the tubule itself but most reabsorption occurs in the hindgut.

malt The product of the hydrolysis of starch by β-*amylase that occurs during the germination of barley in *brewing. *See also* **maltose**.

maltase A membrane-bound enzyme in the small intestine that hydrolyses the disaccharide maltose into glucose.

maltose (malt sugar) A sugar consisting of two linked glucose molecules that results from the action of the enzyme *amylase on starch. Maltose occurs in high concentrations in germinating seeds; malt, used in the manufacture of beer and malt whisky, is produced by allowing barley seeds to germinate and then slowly drying them.

malt sugar *See* **maltose**.

Mammalia A class of vertebrates containing some 4250 species. Mammals are warm-blooded animals (*see* **homoiothermy**), typically having sweat glands whose secretion cools the skin and an insulating body covering of hair. All mammals have *mammary glands, which secrete milk to nourish the young. Mammalian teeth are differentiated into incisors, canines, premolars, and molars and the middle ear contains three sound-conducting *ear ossicles. The four-chambered heart enables complete separation of oxygenated and deoxygenated blood and a muscular *diaphragm takes part in breathing movements, both of which ensure that the tissues are well supplied with oxygen. This, together with well-developed sense organs and brain, have enabled mammals to pursue an active life and to colonize a wide variety of habitats.

Mammals evolved from carnivorous reptiles in the Triassic period about 225 million years ago. There are two subclasses: the primitive egg-laying *Prototheria (monotremes) and the Theria, which includes all other mammals and consists of the infraclasses *Metatheria (marsupials) and *Eutheria (placental mammals).

mammary glands The milk-producing organs (possibly modified sweat glands) of female mammals, which provide food for the young (*see* **milk; colostrum**). Their number (2 to 20) and position (on the chest or abdomen) vary according to the species. In most mammals the gland openings project as a *nipple* or *teat*. Nipples have a number of milk-duct openings; teats have one duct leading from a storage cavity.

mandible 1. One of a pair of horny *mouthparts in insects, crustaceans, centipedes, and millipedes. The mandibles lie in front of the weaker *maxillae and their lateral movements assist in biting and crushing the food. **2.** The lower jaw of vertebrates. **3.** Either of the two parts of a bird's beak.

manganese Symbol Mn. A grey brittle metallic element that is a trace element (*see* **essential element**) for living organisms. It functions as a cofactor for several enzymes.

mangrove swamp A region of vegetation, found along tropical coasts, in which mangrove trees (*Rhizophora* species) predominate. The waterlogged soil is highly saline, and – like other *halophytes – mangroves are adapted to withstand these conditions; they also possess aerial roots (*pneumatophores*) through which gaseous exchange occurs, to counteract effects of the badly aerated soil.

mannan *See* **mannose.**

mannitol A polyhydric alcohol, $CH_2OH(CHOH)_4CH_2OH$, derived from mannose or fructose. It is the main soluble sugar in fungi and an important carbohydrate reserve in brown algae. Mannitol is used as a sweetener in certain foodstuffs.

mannose A *monosaccharide, $C_6H_{12}O_6$, stereoisomeric with glucose, that occurs naturally only in polymerized forms called *mannans*. These are found in plants, fungi, and bacteria, serving as food energy stores.

manometer A device for measuring pressure differences, usually by the difference in height of two liquid columns. The simplest type is the U-tube manometer, which consists of a glass tube bent into the shape of a U. If a pressure to be measured is fed to one side of the U-tube and the other is open to the atmosphere, the difference in level of the liquid in the two limbs gives a measure of the unknown pressure.

mantle The fold of skin covering the dorsal surface of the body of molluscs, which extends into lateral flaps that protect the gills in the *mantle cavity* (the space between the body and mantle). The outer surface of the mantle secretes the shell (in species that have shells).

marker gene 1. A gene used to identify a particular bacterial colony or bacteriophage plaque. Such genes are incorporated into cloning *vectors to enable the isolation and replication of colonies containing a desired vector. Typically, marker genes confer resistance to specific antibiotics or produce colour changes. **2.** (*or* **genetic marker**) A gene or short sequence of DNA that acts as a tag for another, closely linked, gene. Such markers are used in mapping the order of genes along chromosomes and in following the inheritance of particular genes: genes closely linked to the marker will generally be inherited with it. Markers must be readily identifiable in the phenotype, for instance by controlling an easily observable feature (such as eye colour). *Restriction fragment length polymorphisms (RFLPs) are now widely used as markers, for example in tracking the inheritance of genes associated with diseases (*see* **gene tracking**).

marsupials *See* **Metatheria**.

mass flow A hypothesis to explain the movement of sugars in the phloem tissue of plants. At a *source* (site of production) sugars are actively secreted from phloem *companion cells into the *sieve elements, causing water to follow by osmosis. The pressure of water in the tubes (the hydrostatic pressure) causes it to move along the tubes to a *sink* (site of utilization), where the reverse process occurs. Here sugars are actively transported from the sieve elements into the companion cells and then into the surrounding tissues, establishing a concentration gradient from source to sink. However, the mass flow hypothesis does not explain how different solutes can be transported in the phloem in different directions at the same time.

mass spectroscopy A technique used to determine relative atomic masses and the relative abundance of isotopes and for chemical analysis, as in the identification of metabolites, drugs, and other molecules isolated from biological samples. In a *mass spectrometer* a sample (usually gaseous) is ionized and the positive ions produced are accelerated into a high-vacuum region containing electric and magnetic fields. These fields deflect and focus the ions onto a detector. The fields can be varied in a controlled way so that ions of different types can impinge on the detector. A *mass spectrum* is thus obtained consisting of a series of peaks of variable intensity to which mass/charge (m/e) values can be assigned. For organic molecules, the mass spectrum consists of a series of peaks, one corresponding to the parent ion and the others to fragment ions produced by the ionization process. Different molecules can be identified by their characteristic pattern of lines. Analysis of mixtures can be done by gas chromatography–mass spectroscopy (*see* **gas–liquid chromatography**).

mast cells Large amoeboid cells that are found in vertebrate *connective tissue. They may be connected with the deposition of fat. In certain situations they produce *histamine, responsible for allergy symptoms. They also produce *heparin, an anticoagulant, which may prevent fibrin from forming clots in the intercellular spaces.

mastication The process of chewing food, which involves movements of the jaws and teeth. Mastication breaks up the food into small particles, which provides a greater surface area for digestion and enables the formation of a *bolus, which is small enough to pass through the oesophagus.

mastoid process An outgrowth from the temporal bone of the skull containing air cavities that communicate with the cavity of the middle ear. In man it is a route through which infection may spread from the middle ear.

mating *See* **sexual intercourse**.

mating season *See* **breeding season**.

mating type The equivalent in microorganisms, fungi, and algae of the male and female individuals of higher organisms. Mating types are morphologically the same and therefore difficult to distinguish. For this reason different mating types, of the same species, are designated as + and –. The mating types are distinct strains that are compatible in sexual reproduction; they are able to give rise to a zygote, which can develop into a new individual.

matrix *(in histology)* The component of tissues (e.g. bone and cartilage) in which the cells of the tissue are embedded.

maturation 1. The process of becoming fully developed, especially the final phase in the development of *germ cells, which renders the egg or sperm capable of fertilization. **2.** Changes in the neuromuscular system as an animal develops that improve coordination regardless of experience.

maturation-promoting factor (MPF) A *kinase that induces mitosis in eukaryotic cells by promoting the transition from the G_2 stage to the M stage in the *cell cycle. MPF catalyses the phosphorylation of proteins that may in turn cause the breakdown of the nuclear membrane. It may also phosphorylate certain *histone proteins, possibly leading to the condensation of chromosomes before the M phase.

maturity 1. The stage in a life cycle that is reached when a developing organism has taken on the appearance of the adult form and is capable of reproduction. **2.** The stage reached in the formation of gametes (*gametogenesis) following meiotic division of precursor cells and their development into fully functional gametes.

maxicell A form of the bacterium *Escherichia coli* that is used to investigate recombinant DNA and other products produced by bacterial *plasmids.

maxilla (*pl.* **maxillae**) **1.** One of a pair of *mouthparts in insects, crustaceans, centipedes, and millipedes. They lie behind the *mandibles and their lateral movements assist in feeding. Crustaceans have two pairs of maxillae but in insects the second pair are fused together forming the *labium. **2.** One of a pair of large tooth-bearing bones in the upper jaw of vertebrates. In mammals they carry all the upper teeth except the incisors.

maxilliped A paired appendage in crustaceans that contributes to the *mouthparts. There may be up to three pairs of maxillipeds, which are located behind the mandibles and are used to move food to the mouth.

maxillula A paired appendage that contributes to the *mouthparts in certain crustaceans. Lying between the mandibles and the maxillae, they may bear hairs or spines designed to move food particles deeper into the mouth.

maximum permissible dose *See* dose.

M band (M line) *See* sarcomere.

mean The average value of a set of n numbers, i.e. the sum of the numbers divided by n.

meatus A small canal or passage in the body. An example is the *external auditory meatus* of the *outer ear in mammals, which connects the exterior opening to the eardrum.

mechanoreceptor A *receptor that responds to such mechanical stimuli as touch, sound, and pressure. The skin is rich in mechanoreceptors.

meconium The first faecal material passed by a newborn mammal, which is usually dark green in colour.

median The middle number or value in a series of numbers or values.

median eye (pineal eye) An eyelike structure, with a lens and retina, found on the top of the head of some lizards, *Sphenodon*, and the Cyclostomata (lampreys) as well as in many fossil vertebrates. Its function is unknown.

median lethal dose *See* LD_{50}.

mediastinum 1. A membrane in the midline of the *thorax of mammals that separates the lungs. **2.** The space between the two lungs, which is occupied by the heart and oesophagus.

medulla **1.** *(in zoology)* The central tissue of various organs, including the adrenal glands (*adrenal medulla*) and kidneys (*renal medulla*). **2.** *(in botany)* *See* **pith**.

medulla oblongata Part of the vertebrate *brainstem, derived from the *hindbrain, that is continuous with the spinal cord. Its function is to regulate the reflex responses controlling respiration, heart beat, blood pressure, and other involuntary processes. It gives rise to many of the *cranial nerves.

medullary ray (ray) Any of the vertical plates of *parenchyma cells running radially through the cylinder of vascular tissue in the stems and roots of plants. Each may be one to many cells in width. *Primary medullary rays* occur in young plants and in those not showing secondary thickening; they pass from the cortex through to the pith. *Secondary medullary rays* are produced by the vascular *cambium and terminate in xylem and phloem tissues. Medullary rays store and transport food materials.

medullated nerve fibre A nerve fibre that is characterized by a *myelin sheath, which insulates the axon.

medusa The free-swimming stage in the life cycle of the *Cnidaria. Medusae are umbrella-shaped, with tentacles round the edge and the mouth in the centre underneath. They swim by pulsations of the body and reproduce sexually. In the Hydrozoa (e.g. *Hydra*) they alternate in the life cycle with *polyps, from which they are produced by budding. In the Scyphozoa, which includes all the common jellyfish, the medusa is the dominant form and the polyp is reduced or absent.

mega- **1.** A prefix denoting large size; e.g. meganucleus, megasporangium. **2.** Symbol M. A prefix used in the metric system to denote one million times. For example, 10^6 volts = 1 megavolt (MV).

meganucleus *See* **nucleus**.

megaphyll A type of foliage leaf in ferns and seed plants that has branched or parallel vascular bundles running through the lamina. The megaphylls of ferns are large pinnate leaves called *fronds*. A megaphyll was formerly called a *macrophyll*. *Compare* **microphyll**.

megasporangium *See* **sporangium**.

megaspore *See* **megaspore mother cell**; **sporophyll**.

megaspore mother cell (megasporocyte) A diploid cell in plants that divides by meiosis to give rise to four haploid megaspores (*see* **sporophyll**). In flowering plants the megaspore mother cell (or *embryo mother cell*) is situated in the ovule. One of the megaspores it produces develops into the *embryo sac; the others abort.

megasporocyte *See* **megaspore mother cell.**

megasporophyll *See* **sporophyll.**

| 1st Prophase: Leptotene | 1st Prophase: Pachytene | 1st Prophase: Diplotene | 1st Prophase: Diakinesis |

centriole centromere

chiasmata

nuclear chromatids
membrane

| 1st Metaphase | 1st Anaphase | 1st Telophase | 2nd Metaphase |

pole equator

developing cell
membrane

pole spindle fibres

2nd Anaphase 2nd Telophase

The stages of meiosis in a cell containing two pairs of homologous chromosomes

meiosis (reduction division) A type of nuclear division that gives rise to four reproductive cells (gametes) each with half the chromosome number of the parent cell. Two consecutive divisions occur (see illustration). In the first, *homologous chromosomes become paired and may exchange genetic material (*see* **crossing over**) before moving away from each other into separate daughter nuclei. This is the actual reduction division because each of the two nuclei so formed contains only half of the original chromosomes. The daughter nuclei then divide by mitosis and four *haploid cells are produced. *See also* **prophase; metaphase; anaphase; telophase**.

Meissner's corpuscles Receptors for touch that are found in the papillary (upper) region of the dermis of the *skin. Named after German anatomist Georg Meissner (1829–1905), each consists of a mass of dendrites encapsulated by connective tissue. The corpuscles are found in abundance in the palms of the hands, fingertips, and soles of the feet as well as in the nipples and tip of the penis.

melanin Any of a group of polymers, derived from the amino acid tyrosine, that cause pigmentation of eyes, skin, and hair in vertebrates. Melanins are produced by specialized epidermal cells called *melanophores* (or *melanocytes*); their dispersion in these cells is controlled by *melanocyte-stimulating hormone and *melatonin. Certain invertebrates, fungi, and microorganisms also produce melanin pigments. The 'ink' of the octopus and squid is a notable example. Hereditary *albinism is caused by the absence of the enzyme tyrosinase, which is necessary for melanin production.

melanism Black coloration of the body caused by overproduction of the pigment melanin, often as a reaction to the environment. There are several species of melanic moths in industrially polluted areas (*see* **industrial melanism**) and the panther is a melanic form of leopard.

melanocyte-stimulating hormone (MSH) A hormone, secreted in the anterior region of the pituitary gland, that stimulates the concentration of melanin granules in the *chromatophores of the skin of lower vertebrates, such as amphibians. The role of MSH in humans is not clearly understood.

melanophore *See* **chromatophore**.

melatonin A hormone derived from *serotonin and secreted by the pineal gland and retinas of vertebrates. Melatonin secretion by the pineal is linked to the dark–light cycle of the organism's environment, being greatest at night and lowest by day. The hormone is involved in regulating certain diurnal and seasonal changes in the body, such as the reproductive cycle in seasonally breeding animals. Melatonin also controls pigmentation changes; it triggers aggregation of the pigment *melanin into melanophores in the skin, causing the skin to turn pale.

membrane A sheetlike tissue that covers, connects, or lines biological cells and their organelles, organs, and other structures. Membranes consist of polar lipids (predominantly the phospholipids lecithin and cephalin) and protein: the lipids form a double layer (*see* **lipid bilayer**) with protein molecules suspended within it. Membranes are permeable to water and fat-soluble substances but impermeable to polar substances (e.g. sugars). Movement of ions across a membrane occurs by carrier-mediated processes, including *active transport and *facilitated diffusion. *See also* **cell membrane; osmosis.**

membrane bone (dermal bone) *Bone formed directly in connective tissue, i.e. by intramembranous *ossification, rather than by replacing cartilage (*compare* **cartilage bone**). Some face bones, skull bones, and part of the clavicle are membrane bones. Small areas of membrane become jelly-like and attract calcium salts. Bone-forming cells break down these areas forming a bone lattice, which eventually fills in.

membranous labyrinth The soft tubular sensory structures that form the *inner ear of vertebrates and are housed within the bony labyrinth.

memory The means by which information is stored in the brain. The exact mechanism of processing and storing information is not known but is thought to involve the construction of circuits of neurones, which are strengthened by repeated use. Memory is essential to the processes of *learning and recognition of individuals and objects.

memory cell *See* B cell.

Mendelism The theory of heredity that forms the basis of classical *genetics, proposed by Gregor Mendel (1822–84) in 1866 and formulated in two laws (*see* **Mendel's laws; particulate inheritance**). Mendel suggested that individual characteristics were determined by inherited 'factors', and when improved microscopes revealed details of cell structure the behaviour of Mendel's factors could be related to the behaviour of chromosomes during *meiosis.

Mendel's laws Two laws summarizing Gregor Mendel's theory of inheritance (*see also* **Mendelism**). The *Law of Segregation* states that each hereditary characteristic is controlled by two 'factors' (now called *alleles), which segregate (separate) and pass into separate germ (reproductive) cells. The *Law of Independent Assortment* states that pairs of 'factors' segregate independently of each other when germ cells are formed (*see also* **independent assortment**). These laws are the foundation of genetics.

meninges The three membranes that surround the brain and spinal cord of vertebrates: the *pia mater, the *arachnoid membrane, and the outer *dura mater. The pia and arachnoid are separated by the *subarachnoid space*, which contains *cerebrospinal fluid.

meniscus *See* **surface tension.**

menopause The time in a woman's life when ovulation and menstruation cease (*see* **menstrual cycle**). It normally occurs between the ages of 45 and 55. The effects of the gonadotrophic hormones, *follicle-stimulating hormone and *luteinizing hormone, in the ovaries decrease so that the follicles do not develop properly. There is a change in the balance of the hormones oestrogen and progesterone, secreted by the ovaries, which may be associated with certain physical symptoms, such as weight gain and 'hot flushes', and there may also be mood changes.

menstrual cycle (sexual cycle) The approximately monthly cycle of events associated with *ovulation that replaces the *oestrous cycle in most primates (including humans). The lining of the uterus becomes progressively thicker with more blood vessels in preparation for the *implantation of a fertilized egg cell (blastocyst). Ovulation occurs during the middle of the cycle (the fertile period). If fertilization does not occur the uterine lining breaks down and is discharged from the body (*menstruation*); the discharge is known as a 'period'. In women the fertile period is 11–15 days after the end of the last menstruation.

menstruation *See* menstrual cycle.

mericarp *See* schizocarp.

meristem A plant tissue consisting of actively dividing cells that give rise to cells that differentiate into new tissues of the plant. The most important meristems are those occurring at the tip of the shoot and root (*see* apical meristem) and the lateral meristems in the older parts of the plant (*see* cambium; cork cambium).

meristoderm The outer meristematic layer of the thallus of certain brown algae (Phaeophyta). The cells of the meristoderm divide to increase the width of the thallus.

merocrine secretion *See* secretion.

meromictic lake *See* dimictic lake.

Merostomata *See* Chelicerata.

merozygote A bacterial cell that contains more than the *haploid number of chromosomes but less than the full *diploid number. Merozygotes arise when the genetic material from one bacterial cell is only partially transferred into another cell during *conjugation, *transduction, or *transformation.

mesencephalon *See* midbrain.

mesenteric artery The artery that delivers blood to the intestines.

mesentery A thin sheet of tissue, bounded on each side by *peritoneum, that supports the gut and other organs in the body cavities of animals. Vertebrates have a well-developed dorsal mesentery that anchors the stomach and intestine and contains blood vessels and nerves supplying the gut. The reproductive organs and their ducts are also supported by mesenteries.

mesocarp *See* pericarp.

mesoderm The layer of cells in the *gastrula that lies between the *ectoderm and *endoderm. It develops into the muscles, circulatory system, and sex organs and in vertebrates also into the excretory system and skeleton. *See also* germ layers.

mesofauna *See* macrofauna.

mesoglea The gelatinous noncellular layer between the endoderm and ectoderm in the body wall of coelenterates. It may be thin, as in *Hydra*, or tough and fibrous, as in the larger jellyfish and sea anemones. It often contains cells that have migrated from the two body layers but these do not form tissues and organs and the mesoglea is not homologous with the mesoderm of *triploblastic animals.

mesophyll The internal tissue of a leaf blade (lamina), consisting of *parenchyma cells. There are two distinct forms. *Palisade mesophyll* lies just beneath the upper epidermis and consists of cells elongated at right angles to the leaf surface. They contain a large number of *chloroplasts and their principal function is photosynthesis. *Spongy mesophyll* occupies most of the remainder of the lamina. It consists of spherical loosely arranged cells containing fewer chloroplasts than the palisade mesophyll. Between these cells are air spaces leading to the *stomata.

mesophyte Any plant adapted to grow in soil that is well supplied with water and mineral salts. Such plants wilt easily when exposed to drought conditions as they are not adapted to conserve water. The majority of flowering plants are mesophytes. *Compare* **halophyte**; **hydrophyte**; **xerophyte**.

mesothelium A single layer of thin platelike cells covering the surface of the inside of the abdominal cavity and thorax and surrounding the heart, forming part of the *peritoneum and *pleura (*see* **serous membrane**). It is derived from the *mesoderm. *Compare* **endothelium**; **epithelium**.

mesotrophic Describing a body of water, such as a lake, that is intermediate between a *eutrophic lake and an *oligotrophic lake in the amount of nutrients contained within it.

Mesozoic The geological era that extended from the end of the *Palaeozoic era, about 225 million years ago, to the beginning of the *Cenozoic era, about 65 million years ago. It comprises the *Triassic, *Jurassic, and *Cretaceous periods. The Mesozoic era is often known as the *Age of Reptiles* as these animals, which included the dinosaurs, pterosaurs, and ichthyosaurs, became the dominant lifeform; most became extinct before the end of the era.

messenger RNA *See* RNA.

metabolic pathway *See* metabolism.

metabolic rate A measure of the energy used by an animal in a given time period. The metabolic rate of an animal is affected by several interacting factors, including temperature and the level of activity. The metabolic rate of a resting animal is known as the *basal metabolic rate (BMR).

metabolic waste The *waste products, collectively, of metabolism.

metabolism The sum of the chemical reactions that occur within living organisms. The various compounds that take part in or are formed by these reactions are called *metabolites*. In animals many metabolites are obtained by the digestion of food, whereas in plants only the basic starting materials (carbon dioxide, water, and minerals) are externally derived. The synthesis (*anabolism) and breakdown (*catabolism) of most compounds occurs by a number of reaction steps, the reaction sequence being termed a *metabolic pathway*. Some pathways (e.g. *glycolysis) are linear; others (e.g. the *Krebs cycle) are cyclic. The changes at each step in a pathway are usually small and are promoted by efficient biological catalysts – the enzymes. In this way the amounts of energy required or released at any given stage are minimal, which helps in maintaining a constant *internal environment. Various *feedback mechanisms exist to govern *metabolic rates.

metabolite *See* metabolism.

metacarpal One of the bones in the *metacarpus.

metacarpus The hand (or corresponding part of the forelimb) in terrestrial vertebrates, consisting of a number of rod-shaped bones (*metacarpals*) that articulate with the bones of the wrist (*see* **carpus**) and those of the fingers (*see* **phalanges**). The number of metacarpals varies between species: in the basic *pentadactyl limb there are five, but this number is reduced in many species.

metacentric *See* centromere.

metameric segmentation (**metamerism; segmentation**) The division of an animal's body (except at the head region – *see* **cephalization**) into a number of compartments (*segments* or *metameres*) each containing the same organs. Metameric segmentation is most strongly marked in annelid worms (e.g. earthworms), in which the muscles, blood vessels, nerves, etc. are repeated in each segment. In these animals the segmentation is obvious both externally and internally. It also occurs internally in arthropods and in the embryonic development of all vertebrates, in which it is confined to parts of the muscular, skeletal, and nervous systems and does not show externally.

metamorphosis The rapid transformation from the larval to the adult form that occurs in the life cycle of many invertebrates and amphibians. Examples are the changes from a tadpole to an adult frog and from a pupa to an adult insect. Metamorphosis often involves considerable destruction of larval tissues by lysosomes, and in both insects and amphibians it is controlled by hormones.

metaphase The stage of cell division during which the membrane around the nucleus breaks down, the *spindle forms, and centromeres attach the chromosomes to the equator of the spindle. In the first metaphase of *meiosis pairs of chromosomes (bivalents) are attached, while in *mitosis and the second metaphase of meiosis, individual chromosomes are attached.

metaphloem The part of the primary *phloem that develops in a plant after the stem or other part has finished elongating. *Compare* **protophloem**.

metaplasia The transformation of a tissue into a different type. This is an abnormal process; for example, metaplasia of the epithelium of the bronchi may be an early sign of cancer.

metastasis *See* cancer.

metatarsal One of the bones in the *metatarsus.

metatarsus The foot (or corresponding part of the hindlimb) in terrestrial vertebrates, consisting of a number of rod-shaped bones (*metatarsals*) that articulate with the bones of the ankle (*see* **tarsus**) and those of the toes (*see* **phalanges**). The number of metatarsals varies between species: in the basic *pentadactyl limb there are five, but this number is reduced in some species.

Metatheria An infraclass of mammals containing the marsupials. The female bears an abdominal pouch (*marsupium*) into which the newly born young, which are in a very immature state, move to complete their development. They obtain nourishment from the mother's mammary teats. Modern marsupials are restricted to Australasia (where they include the kangaroos, koala bears, phalangers, and bandicoots) and America (the opossums). Marsupials evolved during the late Cretaceous period, 80 million years ago. In Australia, where the marsupials have been isolated for millions of years, they show the greatest diversity of form, having undergone *adaptive radiation to many of the niches occupied by placental mammals elsewhere. *Compare* **Eutheria**; **Prototheria**.

metaxylem The part of the primary *xylem that develops in a plant after the stem or other part has finished elongating (*compare* **protoxylem**). The walls of the metaxylem are more extensively lignified than those of the protoxylem.

Metazoa In traditional classification systems, a subkingdom comprising all multicellular animals. It excludes the *Porifera (sponges), which are traditionally placed in a separate subkingdom, Parazoa.

methanogen Any of various bacteria that produce methane; they are classified as a phylum (Methanocreatrices) of *Archaebacteria and include such genera as *Methanobacillus* and *Methanothrix*. Methanogens are *obligate anaerobes found in oxygen-deficient environments, such as marshes, swamps, sludge (formed during *sewage treatment), and the digestive systems of ruminants. They obtain their energy by reducing carbon dioxide and oxidizing hydrogen, with the production of methane:

$$CO_2 + 4H_2 \rightarrow CH_4 + 2H_2O$$

Formate, methanol, or acetate may be used as a source of electrons. Methanogenic bacteria are important in the production of *biogas.

methionine *See* amino acid.

methylene blue A blue dye used in optical microscopy to stain nuclei of animal tissues. It is also suitable as a vital stain and a bacterial stain.

metre Symbol m. The SI unit of length that is equal to 39.37 inches. It is formally defined as the length of the path travelled by light in vacuum during a time interval of $1/299\,792\,458$ of a second. This definition, adopted by the General Conference on Weights and Measures in October, 1983, replaced the 1967 definition based on the krypton lamp, i.e. $1\,650\,763.73$ wavelengths in a vacuum of the radiation corresponding to the transition between the levels $2p^{10}$ and $5d^5$ of the nuclide krypton-86. This definition (in 1958) replaced the older definition of a metre based on a platinum-iridium bar of standard length.

MHC *See* **major histocompatibility complex**.

micelle An aggregate of molecules in a *colloid. For example, phospholipids in aqueous solution form micelles – small clusters of molecules in which the nonpolar hydrocarbon groups are in the centre and the hydrophilic polar groups are on the outside. The products of fat digestion are dispersed into micelles by the action of bile salts, which facilitates their absorption in the small intestine.

Michaelis–Menten curve A graph that shows the relationship between the concentration of a substrate and the rate of the corresponding enzyme-controlled reaction. Named after L. Michaelis (1875–1949) and M. L. Menten (1879–1960), the curve only applies to enzyme reactions involving a single substrate. The graph can be used to calculate the *Michaelis constant* (K_m), which is the concentration of a substrate required in order for an enzyme to act at half of its maximum velocity (V_{max}). The Michaelis constant is a

Michaelis–Menten curve

measure of the affinity of an enzyme for a substrate. A low value corresponds to a high affinity, and vice versa. *See also* **enzyme kinetics**.

micro- 1. A prefix denoting very small size; e.g. microgamete, micronucleus. **2.** Symbol μ. A prefix used in the metric system to denote one millionth. For example, 10^{-6} metre = 1 micrometre (μm).

microbe *See* **microorganism**.

microbiology The scientific study of microorganisms (e.g. bacteria, viruses, and fungi). Originally this was directed towards their effects (e.g. in causing disease and decay), but during the 20th century the emphasis has shifted to their physiology, biochemistry, and genetics. Microbes are now recognized as important vehicles for the study of biochemical and genetic processes common to all living organisms, and their rapid growth enables their laboratory culture in large numbers for studies in genetics.

microclimate The local climate of a small area or of a particular *habitat, which is different from the *macroclimate* or the larger surrounding geographical area.

microdissection (micromanipulation) A technique used for the dissection of living cells under the high power of an optical microscope. It utilizes minute mechanically manipulated instruments, such as needles, scalpels, and pipettes. For example, the instruments may be used to remove a single nucleus from one cell and to implant it in another.

microfauna 1. Animals that cannot be seen with the naked eye. They are normally observed with the aid of a microscope. *Compare* **macrofauna**. **2.** The animals that live in a particular *microhabitat.

microfibril A microscopic fibre. Plant cell walls contain microfibrils, about 5 nm in diameter, each consisting of parallel cellulose chains that are associated together to form a flat ribbon. Cellulose microfibrils are arranged in layers at right angles to each other.

microfilament *See* **actin; cytoskeleton**.

microflora 1. Plants and algae that cannot be seen with the naked eye. They are normally observed with the aid of a microscope. **2.** The plants and algae that live in a particular *microhabitat.

microfossil A *fossil that is so small that it can only be studied under a microscope. Microfossils include bacteria, diatoms, and protozoa and parts of organisms, such as plant pollen and skeletal fragments. Microfossils are important in the correlation of rocks where only small samples are available. The study of microfossils, particularly pollen, is known as *palynology.

microhabitat The local habitat of a particular organism or microorganism. There are normally a number of different microhabitats within a large *habitat (*macrohabitat*), each with its distinct set of environmental conditions. For example, in a stream macrohabitat there will exist different microhabitats, depending on oxygen content, pH, speed of water flow, and other factors in localized areas of the stream.

micromanipulation *See* microdissection.

micronucleus *See* nucleus.

micronutrient A chemical element required by plants in relatively small quantities. Micronutrients are typically found in cofactors and coenzymes. They include copper, zinc, molybdenum, manganese, cobalt, and boron. *See* **essential element**. *Compare* **macronutrient**.

microorganism (microbe) Any organism that can be observed only with the aid of a microscope. Microorganisms include bacteria, viruses, protozoans, and some algae and fungi. *See* **microbiology**.

microphagous Describing the method of feeding of those heterotrophic organisms that take in their food in the form of tiny particles. *Filter feeding and *ciliary feeding are examples of this type of feeding. *Compare* **macrophagous**.

microphyll A type of foliage leaf in clubmosses and horsetails that has a single unbranched midrib. Such leaves are generally no more than a few millimetres long. *Compare* **megaphyll**.

micropropagation The *in vitro* propagation of plants by cloning. An explant, callus tissue, or suspension culture is grown on a medium containing the appropriate hormones and nutrients required for cell division and growth. Newly grown cells differentiate and roots and shoots are regenerated. Micropropagation is used in agriculture, horticulture, and forestry as special genotypes can be bred and maintained, the process is rapid, and plants can be kept disease-free.

micropyle 1. A small opening in the surface of a plant ovule through which the pollen tube passes prior to fertilization. It results from the incomplete covering of the nucellus by the integuments. It remains as an opening in the testa of most seeds through which water is absorbed. **2.** A small pore in some animal cells or tissues; for example, in insect eggs (*see* **chorion**).

microscope A device for forming a magnified image of a small object. The *simple microscope* consists of a biconvex magnifying glass or an equivalent system of lenses, either hand-held or in a simple frame. The *compound microscope* uses two lenses or systems of lenses, the second magnifying the real image formed by the first (see illustration). The lenses are usually mounted at the opposite ends of a tube that has mechanical controls to move it in relation to the object. An optical condenser and mirror, often with a separate light source, provide illumination of the object. The widely used *binocular microscope* consists of two separate instruments fastened together so that one eye looks through one while the other eye looks through the other. This gives stereoscopic vision. *See also* electron microscope; phase-contrast microscope; ultraviolet microscope.

Compound microscope

microsome A fragment of *endoplasmic reticulum formed when cells or tissues are disrupted. Microsomes can be isolated by centrifugation and are commonly used to investigate the functional properties of endoplasmic reticulum, such as enzymic activity and protein synthesis.

microsporangium *See* sporangium.

microspore *See* microspore mother cell; sporophyll.

microspore mother cell (microsporocyte) A diploid cell in plants that divides by meiosis to give rise to four haploid microspores (*see* **sporophyll**). In flowering plants microspore mother cells are formed within the pollen sacs of the anthers by mitosis; the microspores they produce develop into pollen grains.

microsporocyte *See* microspore mother cell.

microsporophyll *See* sporophyll.

microtome A machine used to cut thin sections (3–5 μm thick) of plant or animal tissue for microscopical observation. There are various designs of microtome, each basically consisting of a steel knife, a block for supporting the specimen, and a device for moving the specimen towards the knife. The specimen is usually supported by being embedded in wax; if a *freezing microtome* is used, the specimen is frozen. An *ultramicrotome* is used to cut much thinner sections (20–100 nm thick) for electron microscopy. The biological material is embedded in plastic or resin, sectioned with a glass or diamond knife, and the cut sections are allowed to float on the surface of water in an adjacent water bath.

microtubule A minute filament in living cells that is composed of the protein *tubulin* and occurs singly or in pairs, triplets, or bundles. Microtubules help cells to maintain their shape (*see* **cytoskeleton**); they also occur in cilia and flagella (*see* **axoneme**) and the *centrioles and form the *spindle during nuclear division.

microvillus One of a number of minute finger-like projections on the free surfaces of epithelial cells. Microvilli are covered with cell membrane and their cytoplasm is continuous with the main cell cytoplasm. Their purpose is probably to increase the absorptive or secretory surface area of the cell, and they are abundant on the villi of the intestine, where they form a *brush border.

micturition The discharge of urine from the bladder. It is brought about by reflex contraction of the *detrusor muscle after voluntary relaxation of the sphincter muscle at the junction of the bladder with the urethra.

midbrain (mesencephalon) One of the three sections of the brain of a vertebrate embryo. Unlike the *forebrain and the *hindbrain, the midbrain does not undergo further subdivision to form additional zones. In mammals it becomes part of the *brainstem, but in amphibians, reptiles, and birds the roof of the midbrain becomes enlarged as the *tectum*, a dominant centre for integration, and may include a pair of *optic lobes*.

middle ear (tympanic cavity) The air-filled cavity within the skull of vertebrates that lies between the *outer ear and the *inner ear. It is linked to the pharynx (and therefore to outside air) via the *Eustachian tube and in mammals contains the three *ear ossicles, which transmit auditory vibrations from the outer ear (via the *tympanum) to the inner ear (via the *oval window).

middle lamella A thin layer of material, consisting mainly of pectins, that binds together the walls of adjacent plant cells. *See also* **cell plate**.

midgut 1. The middle section of the alimentary canal of vertebrates, which is concerned with digestion and absorption. It comprises most of the small intestine. **2.** The middle section of the alimentary canal of arthropods. *See also* **foregut; hindgut**.

migration The seasonal movement of complete populations of animals to a more favourable environment. It is usually a response to lower temperatures resulting in a reduced food supply, and is often triggered by a change in day length (*see* **photoperiodism**). Migration is common in mammals (e.g. porpoises), fish (e.g. eels and salmon), and some insects but is most marked in birds. The Arctic tern, for example, migrates annually from its breeding ground in the Arctic circle to the Antarctic – a distance of some 17 600 km. Migrating animals possess considerable powers of orientation; birds seem to possess a compass sense, using the sun, pole stars, and (in cloud) the earth's magnetic lines of force as reference points (*see* **navigation**).

milk The fluid secreted by the *mammary glands of mammals. It provides a balanced and highly nutritious food for offspring. Cows' milk comprises about 87% water, 3.6% lipids (triglycerides, phospholipids, cholesterol, etc.), 3.3% protein (largely casein), 4.7% lactose (milk sugar), and, in much smaller amounts, vitamins (especially vitamin A and many B vitamins) and minerals (notably calcium, phosphorus, sodium, potassium, magnesium, and chlorine). Composition varies among species; human milk contains less protein and more lactose.

milk sugar *See* **lactose**.

milk teeth *See* **deciduous teeth**.

milli- Symbol m. A prefix used in the metric system to denote one thousandth. For example, 0.001 metre = 1 millimetre (mm).

millipedes *See* **Diplopoda**.

Millon's reagent A solution of mercuric nitrate and nitrous acid used to test for proteins. The sample is added to the reagent and heated for two minutes at 95°C; the formation of a red precipitate indicates the presence of protein in the sample. The reagent is named after French chemist Auguste Millon (1812–67).

mimicry The resemblance of one animal to another, which has evolved as a means of protection. In *Batesian mimicry*, named after British naturalist Henry Bates (1825–92), the markings of certain harmless insects closely resemble the *warning coloration of another insect (the *model*). Predators that have learnt to avoid the model will also avoid good mimics of it. This phenomenon is often found among butterflies. *Müllerian mimicry*, named after German zoologist J. F. T. Müller (1821–97), involves the mutual resemblance of a group of animals, all harmful, such as the wasp, bee, and hornet, so that a predator, having experienced one, will subsequently avoid them all.

mineral deficiency Lack of any essential mineral nutrient, such as nitrogen, phosphorus, or potassium, in living organisms, which can result in mineral deficiency diseases. For example, lack of calcium causes poor bone development in humans and stunted growth in plants; lack of nitrogen can cause the disease kwashiorkor, due to a deficiency in protein intake (*see* **malnutrition**); in plants it results in stunted growth and *chlorosis. Lack of phosphorus can cause reduced growth in plants. A deficiency of trace elements (*see* **essential element**) also leads to diseases; for example, a deficiency of iron can cause anaemia in humans and chlorosis in plants.

mineralocorticoid *See* **corticosteroid**.

mineral salts Inorganic salts that need to be ingested or absorbed by living organisms for healthy growth and maintenance. They comprise the salts of the trace elements in animals (*see* **essential element**) and the *micronutrients of plants.

minicell A fragment of a cell or a quantity of protoplasm that contains less than the haploid number of chromosomes. Minicells are produced, normally *in vitro*, by combining two unlike cells or protoplasms in a process that involves certain chromosomes being removed. Minicells are important in biotechnology as a means of introducing chromosomes into cells with which they are fused.

miracidium The first larval stage of trematode worms (flukes), which hatches from eggs excreted in the faeces of the primary host. Its leaflike body is covered with cilia, enabling the larva to swim towards a secondary host, in which it continues its development.

mitochondrial DNA (mtDNA) A circular ring of DNA found in mitochondria. In mammals mtDNA makes up less than 1% of the total cellular DNA, but in plants the amount is variable. It codes for ribosomal and transfer RNA but only some mitochondrial proteins (up to 30 proteins in animals), the nuclear DNA being required for encoding most of these. Human mtDNA codes for 13 proteins and some RNA. *See also* **mitochondrial Eve**.

mitochondrial Eve The hypothetical female claimed by some biologists to be the ancestor of all humankind. Analysis of *mitochondrial DNA (mtDNA) from groups of people throughout the world suggests that mitochondrial Eve lived 100 000–200 000 years ago, probably in Africa (hence she is also known as 'African Eve'). Mitochondrial DNA is particularly useful for investigating recent genetic history as it mutates quickly (ten times more rapidly than nuclear DNA) and is inherited solely through the female line (therefore it does not undergo recombination by *crossing over). The uniformity of the different samples of mtDNA indicates that modern humans evolved relatively recently from a single region in Africa. This view has been reinforced by studies of Y chromosomes from different groups around the world, which are transmitted only through the male line.

mitochondrion (*pl.* **mitochondria**) A structure within the cytoplasm of plant and animal *cells that carry out aerobic respiration: it is the site of the *Krebs cycle and *electron transport chain, and therefore the cell's energy production. Mitochondria vary greatly in shape, size, and number but are typically oval or sausage-shaped and bounded by two membranes, the inner one being folded into finger-like projections (*cristae*); they contain their own DNA (*see* **mitochondrial DNA**). They are most numerous in cells with a high level of metabolic activity.

The stages of mitosis in a cell containing two pairs of homologous chromosomes

mitosis A type of nuclear division that results in two daughter cells each having a nucleus containing the same number and kind of chromosomes as the mother cell. The changes during divisions are clearly visible with a light microscope. Each chromosome divides lengthwise into two *chromatids, which separate and form the chromosomes of the resulting daughter nuclei. The process is divided into four stages, *prophase, *metaphase, *anaphase, and *telophase, which merge into each other (see illustration). Mitotic divisions ensure that all the cells of an individual are genetically identical to each other and to the original fertilized egg. *See also* **cell cycle**.

mitral valve *See* **biscuspid valve**.

mixed function oxygenase (**mono-oxygenase**) An enzyme that introduces an oxygen atom into a substrate. Mixed function oxidases are essential components of many important metabolic pathways. An example of such an enzyme is *cytochrome oxidase.

mmHg A unit of pressure equal to that exerted under standard gravity by a height of one millimetre of mercury, or 133.322 pascals.

modern synthesis *See* neo-Darwinism.

modifier gene A gene that influences the expression of another gene. For example, one gene controls whether eye colour is blue or brown but other (modifier) genes can also influence the colour by affecting the amount or distribution of pigment in the iris.

molality *See* concentration.

molar 1. A broad ridged tooth in the adult dentition of mammals (*see* **permanent teeth**), found at the back of the jaws behind the premolars. There are two or more molars on each side of both jaws; their surfaces are raised into ridges or *cusps for grinding food during chewing. In man the third (and most posterior) molar does not appear until young adulthood: these molars are known as *wisdom teeth*. **2.** Denoting that an extensive physical property is being expressed per *amount of substance, usually per mole. For example, the molar heat capacity of a compound is the heat capacity of that compound per unit amount of substance, i.e. it is usually expressed in $J\ K^{-1}\ mol^{-1}$.

molarity *See* concentration.

mole Symbol mol. The SI unit of *amount of substance. It is equal to the amount of substance that contains as many elementary units as there are atoms in 0.012 kg of carbon–12. The elementary units may be atoms, molecules, ions, radicals, electrons, etc., and must be specified. 1 mole of a compound has a mass equal to its *relative molecular mass expressed in grams.

molecular biology The study of the structure and function of large molecules associated with living organisms, in particular proteins and the nucleic acids *DNA and *RNA. *Molecular genetics* is a specialized branch, concerned with the analysis of genes (*see* **gene sequencing**).

Molisch's test *See* alpha-naphthol test.

Mollusca A phylum of soft-bodied invertebrates characterized by an unsegmented body differentiated into a *head*, a ventral muscular *foot* used in locomotion, and a dorsal *visceral hump* covered by a fold of skin – the *mantle – which secretes a protective shell in many species. Respiration is by means of gills (*ctenidia) or a lunglike organ and the feeding organ is a *radula. Molluscs occur in marine, freshwater, and terrestrial habitats and there are six classes, including the *Gastropoda (snails, slugs, limpets, etc.), *Pelecypoda (bivalves, e.g. mussels, oysters), and *Cephalopoda (squids and octopuses).

molybdenum Symbol Mo. A silvery hard metallic element that is a trace element required by living organisms. *See* **essential element**.

Monera *See* prokaryote.

monoamine oxidase (MAO) An enzyme that breaks down monoamines (e.g. *adrenaline and *noradrenaline) in the body by oxidation. Drugs that inhibit this enzyme are used to treat forms of depression.

monochasium *See* **cymose inflorescence**.

monocistronic Describing a type of messenger *RNA that can encode only one polypeptide per RNA molecule. All eukaryotic cells are monocistronic. *Compare* **polycistronic**.

monoclonal antibody A specific *antibody produced by one of numerous identical cells derived from a single parent cell. (The population of these cells comprises a *clone and each cell is said to be *monoclonal*.) The parent cell is obtained by the fusion of a normal antibody-producing cell (a lymphocyte) with a cell derived from a malignant tumour of *lymphoid tissue of a mouse (*see* **cell fusion**). The resulting *hybridoma cell then multiplies rapidly and yields large amounts of antibody. Monoclonal antibodies are used to identify a particular antigen within a mixture and can therefore be used for identifying blood groups; they also enable the production of highly specific, and therefore effective, *vaccines.

Monocotyledonae One of the two classes of plants within the *Angiospermophyta, distinguished by having one seed leaf (*cotyledon) within the seed. The monocotyledons generally have parallel leaf veins, scattered vascular bundles within the stems, and flower parts in threes or multiples of three. Monocotyledon species include some crop plants (e.g. cereals, onions, fodder grasses), ornamentals (e.g. tulips, orchids, lilies), and a very limited number of trees (e.g. the palms). *Compare* **Dicotyledonae**.

monoculture *See* **agriculture**.

monocyte The largest form of white blood cell (*leucocyte) in vertebrates. It has a kidney-shaped nucleus and is actively phagocytic, ingesting bacteria and cell debris (*see* **phagocyte**).

monoecious Describing plant species that have separate male and female flowers on the same plant. Examples of monoecious plants are maize and birch. *Compare* **dioecious**.

monoglyceride *See* **glyceride**.

monohybrid cross A genetic cross between parents that differ in the alleles they possess for one particular gene, one parent having two dominant alleles and the other two recessives. All the offspring (called *monohybrids*) have one dominant and one recessive allele for that gene (i.e. they are hybrid at that one locus). Crossing between these offspring yields a characteristic 3:1 (monohybrid) ratio in the following generation of dominant:recessive phenotypes (see illustration). *Compare* **dihybrid cross**.

monokine *See* **cytokine**.

monolayer culture A type of *culture in which cells are grown in a single layer on a flask or Petri dish containing the culture medium. *Compare* **suspension culture**.

monomer A molecule (or compound) that consists of a single unit and can join with others in forming a dimer, trimer, or polymer.

monophyletic Denoting any group of organisms that are assumed to have originated from the same ancestor, i.e. any family, class, etc., of a natural classification. Sometimes the term has a more limited meaning and designates only those groups that include *all* the descendants of a common ancestor. In this restricted sense the birds are considered monophyletic because they are the sole descendants of a group of arboreal Triassic reptiles but the modern reptiles are not, because their common amphibian ancestor also gave rise to the birds and mammals. Such groups as the reptiles are described as *paraphyletic*. *Compare* **polyphyletic**.

monophyodont Describing a type of dentition that consists of a single set of teeth that last for the entire lifespan of an animal. *Compare* **diphyodont; polyphyodont**.

monopodium The primary axis of growth in such plants as pine trees. It consists of a single main stem that continues to grow from the tip and gives rise to lateral branches. *Compare* **sympodium**.

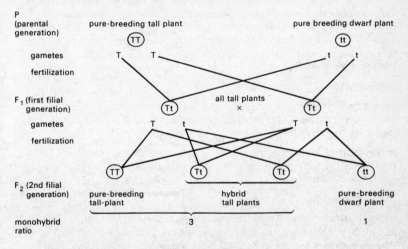

A monohybrid cross: the inheritance of stem lengths in garden peas

monosaccharide (simple sugar) A carbohydrate that cannot be split into smaller units by the action of dilute acids. Monosaccharides are classified according to the number of carbon atoms they possess: *trioses* have three carbon atoms; *tetroses*, four; *pentoses*, five; *hexoses*, six; etc. Each of these is further divided into *aldoses* and *ketoses*, depending on whether the molecule contains an aldehyde group (–CHO) or a ketone group (–CO–). For example glucose, having six carbon atoms and an aldehyde group, is an *aldohexose* whereas fructose is a *ketohexose*. These aldehyde and ketone groups confer reducing properties on monosaccharides: they can be oxidized to yield sugar acids. They also react with phosphoric acid to produce phosphate esters (e.g. in *ATP), which are important in cell metabolism. Monosaccharides can exist as either straight-chain or ring-shaped molecules (see illustration). They also exhibit *optical activity, giving rise to both dextrorotatory and laevorotatory forms.

Glucose (an aldohexose)

straight-chain form

α-glucose   β-glucose

ring forms

Fructose (a ketohexose)

straight-chain form

ring form

Monosaccharides

monosomy *See* aneuploid.

monosynaptic reflex A simple *reflex that involves transmission of information from a sensory neurone to the appropriate motor neurone across a single synapse in the spinal cord. The knee-jerk reflex action is an example of a monosynaptic reflex (*see* **stretch reflex**). *Compare* **polysynaptic reflex**.

monotremes *See* **Prototheria**.

monozygotic twins *See* **identical twins**.

morphine An alkaloid present in opium (*see* **opiate**). It is an analgesic and narcotic, used medically for the relief of severe pain.

morphogenesis The development, through growth and differentiation, of form and structure in an organism.

morphology The study of the form and structure of organisms, especially their external form. *Compare* **anatomy**.

mortality *See* **death rate**.

mosaic 1. An organism made up of cells that have different genotypes but have developed from the same zygote. *See also* **chimaera; gynandromorph. 2.** A viral disease in plants that causes yellow patches to develop on the leaves, giving these a variegated appearance (*see* **variegation**). An example is tobacco mosaic, caused by *tobacco mosaic virus.

mosses *See* **Musci**.

moths *See* **Lepidoptera**.

motivation The internal conditions responsible for temporary reversible changes in the responsiveness of an animal to external stimulation. Thus an animal that has been deprived of food will accept less palatable food than one that has not been deprived: the difference is attributed to a change in feeding motivation. Changes in responsiveness due to maturation, *learning, or injury are not usually readily reversible and are therefore not considered to be due to changes in motivation. Early attempts to describe motivation in terms of a number of separate *drives (e.g. food drive, sex drive) have not found general favour, partly because drives interact with one another; for example, water deprivation often affects an animal's willingness to feed.

motor neurone A *neurone that transmits nerve impulses from the central nervous system to an effector organ (such as a muscle or gland) and thereby initiates a physiological response (e.g. muscle contraction).

moulting 1. The seasonal loss of hair, fur, or feathers that occurs in mammals and birds. **2.** The periodic loss of the integument of arthropods and reptiles. *See* **ecdysis**.

mouth The opening of the *alimentary canal, which in most animals is used for the *ingestion of food. It leads to the *buccal cavity (mouth cavity).

mouth cavity *See* **buccal cavity**.

mouthparts Modified paired appendages on the head segments of arthropods, used for feeding. A typical insect has a *labium (lower lip), one pair each of *mandibles and *maxillae, and a *labrum (upper lip), although in many the mouthparts are modified to form piercing stylets or a sucking proboscis. Crustaceans, centipedes, and millipedes have one pair of mandibles and two pairs of maxillae used for cutting and holding the food. Crustaceans also have several pairs of *maxillipeds (see also **maxillula**). Arachnids have *chelicerae and *pedipalps.

MSH See **melanocyte-stimulating hormone**.

mtDNA See **mitochondrial DNA**.

mucilage A gumlike substance frequently present in the cell walls of aquatic plants and in the seed coats of certain other species. Mucilages are hard when dry and slimy when wet. Like *gums they probably have a general protective function or serve to anchor the plant. Some organisms (e.g. certain bacteria) are completely covered with mucilage and in such cases it probably prevents water loss.

mucin See **mucus**.

mucopolysaccharide See **glycosaminoglycan**.

mucosa A *mucous membrane, especially that forming the lining of the wall of the mammalian stomach and intestine. In these organs the mucosa includes an outer muscular layer, the *muscularis mucosae, which lies adjacent to the *submucosa.

mucous membrane (mucosa) A layer of tissue comprising an epithelium supported on connective tissue. Within the epithelium are *goblet cells, which secrete *mucus onto the surface, and the epithelium often bears cilia. Mucous membranes line body cavities communicating with the exterior, including the alimentary and respiratory tracts. Compare **serous membrane**.

mucus The slimy substance secreted by *goblet cells onto the surface of a *mucous membrane to protect and lubricate it and to trap bacteria, dust particles, etc. Mucus consists of water, the glycoprotein mucin, cells, and salts.

Müllerian mimicry See **mimicry**.

multicellular Describing tissues, organs, or organisms that are composed of a number of cells. Compare **unicellular**.

multienzyme system A complex of enzymes within a cell that form a reaction sequence of a biochemical pathway so that the product of the first enzyme reaction is transferred directly to the next enzyme and immediately undergoes a second reaction, and so on. The rate of an enzyme reaction often depends on the concentration of the enzyme and the substrate, both being required in relatively high amounts. Multienzyme systems, such as those involved in RNA and protein synthesis, help maintain a high rate of cellular metabolism since the intermediate products are transferred directly to the next enzyme and are therefore not required in large concentrations.

multifactorial inheritance The determination of a particular characteristic, e.g. height or skin colour, by many genes, each having a small effect individually. Characteristics controlled in this way show *continuous variation.

multiple alleles Three or more alternative forms of a gene (*alleles) that can occupy the same *locus. However, only two of the alleles can be present in a single organism. For example, the *ABO system of blood groups is controlled by three alleles, only two of which are present in an individual.

multipolar neurone A *neurone that has one axon and several dendrons extending from its cell body in different directions. Most vertebrate motor neurones and interneurones are multipolar neurones. *Compare* **bipolar neurone; unipolar neurone**.

Musci A class of plants within the *Bryophyta comprising the mosses, which are found in both damp (including freshwater) and drier situations. Mosses possess erect or prostrate leafy stems, which give rise to leafless stalks bearing capsules. Spores formed in the capsules are released and grow to produce new plants. A common moss is *Funaria*.

muscle A tissue consisting of sheets or bundles of cells (*muscle fibres*) that are capable of *contraction, so producing movement or tension in the body. There are three types of muscle. *Voluntary muscle produces voluntary movement (e.g. at joints); *involuntary muscle mainly effects the movements of hollow organs (e.g. intestine and bladder); and *cardiac muscle occurs only in the heart.

muscle spindle A *stretch receptor in vertebrate muscle. Muscle spindles run parallel to normal muscle fibres; each consists of a capsule containing small striated muscle fibres (*intrafusal fibres). Muscle spindles are responsible for the adjustment of muscle tone and play an important role in the subconscious maintenance of posture and movement. *See also* **stretch reflex**.

muscularis mucosae A thin layer of smooth muscle that forms the outermost part of the *mucosa of the mammalian stomach and intestine.

mutagen An agent that causes an increase in the number of mutants (*see* **mutation**) in a population. Mutagens operate either by causing changes in the DNA of the *genes, so interfering with the coding system, or by causing chromosome damage. Various chemicals (e.g. *colchicine) and forms of radiation (e.g. X-rays) have been identified as mutagens.

mutant A gene or an organism that has undergone a heritable change, especially one with visible effects (i.e. the change in *genotype is associated with a change in *phenotype). *See* **mutation**.

mutation A sudden random change in the genetic material of a cell that may cause it and all cells derived from it to differ in appearance or behaviour from the normal type. An organism affected by a mutation (especially one with visible effects) is described as a *mutant*. *Somatic mutations* affect the nonreproductive cells and are therefore restricted to the tissues of a single organism but *germ-line mutations*, which occur in the reproductive cells or their precursors, may be transmitted to the organism's descendants and cause abnormal development.

Mutations occur naturally at a low rate but this may be increased by radiation and by some chemicals (*see* **mutagen**). Most are *point (or gene) mutations, which consist of invisible changes in the DNA of the chromosomes, but some (the *chromosome mutations) affect the appearance or the number of the chromosomes. An example of a chromosome mutation is that giving rise to *Down's syndrome.

The majority of mutations are harmful, but a very small proportion may increase an organism's *fitness; these spread through the population over successive generations by natural selection. Mutation is therefore essential for evolution, being the ultimate source of genetic variation.

mutation frequency The average frequency with which a particular *mutation occurs in a population. This frequency can be increased by radiation or by exposure to chemicals, such as mustard gas or hydrogen peroxide.

mutualism An interaction between two species in which both species benefit. (The term *symbiosis is often used synonymously with mutualism.) A well-known example of mutualism is the association between termites and the specialized protozoans that inhabit their guts. The protozoans, unlike the termites, are able to digest the cellulose of the wood that the termites eat and release sugars that the termites absorb. The termites benefit by being able to use wood as a foodstuff, while the protozoans are supplied with food and a suitable environment. *See also* **mycorrhiza**.

mycelium (*pl.* **mycelia**) A network of *hyphae that forms the body of a fungus. It consists of feeding hyphae together with reproductive hyphae, which produce *sporangia and *gametangia.

mycobiont The fungal component of a *lichen.

mycology The scientific study of *fungi.

Mycophycophyta *See* **lichens**.

mycoplasmas A group of prokaryotes that lack a rigid cell wall and are considered to be the smallest living cells (diameter 0.1 μm–0.8 μm). They are either saprotrophic or parasitic and are found on animal mucous and synovial membranes, in insects, and in plants (in which they seem to inhabit sieve tubes). They cause a variety of diseases, including pleuropneumonia in cattle – hence they are also known as *pleuropneumonia-like organisms* (PPLO). Due to their small size and flexible cell wall they can pass through a 0.2-μm-diameter filter and they represent a major contaminant of biotechnological products, such as monoclonal antibodies and vaccines, and of other cell cultures, in which they may exist symbiotically with the cells. Six genera have been described (including *Mycoplasma*) with a total of 50–60 species.

mycorrhiza The mutually beneficial association (*see* **mutualism**) formed between fungi and the roots of plants. This is a very common form of mutualism; the absorption of mineral ions by the plant roots is enhanced by the presence of the fungus, which benefits by obtaining soluble organic nutrients from the root cells. *Ectotrophic* mycorrhizas form a network of hyphae around the root and grow into the air spaces between the cells of the root. The hyphae of *endotrophic* mycorrhizas are thought to actually enter the cells of the roots.

Mycota In older classification systems, a kingdom comprising the *fungi.

myelin A *phospholipid produced by the *Schwann cells of the nervous system. Myelin forms an insulating layer around the nerve fibres (*see* **myelin sheath**).

myelin sheath (medullary sheath) The layer of fatty material that surrounds and electrically insulates the axons of most vertebrate and some invertebrate neurones. The myelin sheath enables a more rapid transmission of nerve impulses (at speeds up to 120 m s^{-1}). It consists of layers of membrane derived from *Schwann cells. The sheath is interrupted at intervals along the axon by *nodes of Ranvier*; myelinated sections of axon are called *internodes*.

myeloid tissue Tissue within red *bone marrow that produces the blood cells. It is found around the blood vessels and contains various cells that are precursors of the blood cells. *See* **haemopoietic tissue**.

myocardium The muscular wall of the heart. The thickness of the myocardium varies, reflecting the magnitude of the pressure generated in the heart during contraction: the myocardium is thickest around the left ventricle, which does most of the work in the heart. The inner surface of the myocardium is lined by a layer of *endothelium (the *endocardium*), which is continuous with the endothelium lining the blood vessels.

myocyte A contractile cell, especially a muscle cell.

myofibril *See* **voluntary muscle**.

myogenic Originating in or produced by muscle cells. The contractions of *cardiac muscle fibres are described as myogenic, since they are produced spontaneously, without requiring stimulation from nerve cells (*see* **pacemaker**).

myoglobin A globular protein occurring widely in muscle tissue as an oxygen carrier. It comprises a single polypeptide chain and a *haem group, which reversibly binds a molecule of oxygen. This is only relinquished at relatively low external oxygen concentrations, e.g. during strenuous exercise when muscle oxygen demand outpaces supply from the blood. Myoglobin thus acts as an emergency oxygen store.

myometrium The thick layer of smooth muscle that makes up the bulk of the *uterus. The myometrium is lined by the *endometrium.

myopia Short-sightedness. It results from the lens of the eye refracting the parallel rays of light entering it to a focus in front of the retina generally because of an abnormally long eyeball. The condition is corrected by using diverging spectacle lenses to move the image back to the retina.

myosin A contractile protein that interacts with *actin to bring about contraction of muscle or cell movement. The type of myosin molecule found in muscle fibres consists of a tail, by which it aggregates with other myosin molecules to form so-called 'thick filaments'; and a globular head, which has sites for the attachment of actin and ATP molecules. *See* **sarcomere; sliding filament theory**.

myotome One of a series of segmented muscle blocks found in fishes and lancelets. Myotomes are arranged in pairs on either side of the body that work antagonistically (*see* **antagonism**) against the backbone (or notochord), providing a means of locomotion by causing the tail to sweep from side to side.

Myriapoda A former class of terrestrial arthropods that comprised the centipedes, millipedes, pauropods, and symphylids. In modern classifications myriapods, together with insects, are now regarded as a subphylum or superclass (*Uniramia) of the Arthropoda.

Myxomycota *See* **slime moulds**.

myxovirus One of a group of RNA-containing viruses associated with various diseases of man and other vertebrates. *Orthomyxoviruses* produce diseases of the respiratory tract, e.g. influenza; *paramyxoviruses* include the causal agents of mumps, measles, and fowl pest.

n *See* **haploid**.

NAD (nicotinamide adenine dinucleotide) A *coenzyme, derived from the B vitamin *nicotinic acid, that participates in many biological dehydrogenation reactions (see formula). NAD is characteristically loosely bound to the enzymes concerned. It normally carries a positive charge and can accept one hydrogen atom and two electrons to become the reduced form, *NADH*. NADH is generated during the oxidation of food, especially by the reactions of the *Krebs cycle. It then gives up its two electrons (and single proton) to the *electron transport chain, thereby reverting to NAD^+ and generating three molecules of ATP per molecule of NADH.

NADP (*nicotinamide adenine dinucleotide phosphate*) differs from NAD only in possessing an additional phosphate group. It functions in the same way as NAD although anabolic reactions (*see* **anabolism**) generally use NADPH (reduced NADP) as a hydrogen donor rather than NADH. Enzymes tend to be specific for either NAD or NADP as coenzyme.

NAD

nano- Symbol n. A prefix used in the metric system to denote 10^{-9}. For example, 10^{-9} metre = 1 nanometre (nm).

narcotic Any drug that induces stupor and relieves pain, especially morphine and other *opiates. Such narcotics are addictive and cause dependence, and their medical use is strictly controlled.

nares (nostrils) The paired openings of the *nasal cavity in vertebrates. All vertebrates have *external nares*, which open to the exterior; in some species these are situated on a *nose. *Internal nares* (or *choanae*) are present only in

air-breathing vertebrates (including lungfish) and open into the mouth
cavity. In mammals they open posteriorly, beyond the secondary *palate.

nasal cavity The cavity in the head of a vertebrate that is lined by a
membrane rich in sensitive olfactory receptors (*see* **olfaction**). It is
connected to the exterior by external nostrils and (in air-breathing
vertebrates) to the respiratory system by internal *nares.

nastic movements Movements of plant organs in response to external
stimuli that are independent of the direction of the stimuli. Examples are
the opening of crocus and tulip flowers in response to a rise in
temperature (*thermonasty*), the opening of evening primrose flowers at
night (*photonasty*), and the folding up and drooping of leaves of the
sensitive plant (*Mimosa pudica*) when lightly touched (*haptonasty*). *Compare*
tropism. *See also* **nyctinasty**.

natality *See* **birth rate**.

natural group A group of organisms of any taxonomic rank that are
believed to be descended from a common ancestor (*see* **monophyletic**). For
example, man and the apes are often regarded as a natural group,
descended from fossil ancestors, the dryopithecines (or their near relatives).
In an ideal natural classification all taxa should be natural groups. *See also*
cladistics.

natural history 1. The study of living organisms in their natural
habitats. **2.** The study of all natural phenomena.

natural order In the classification of plants, the former name for a
*family.

natural selection The process that, according to *Darwinism, brings
about the evolution of new species of animals and plants. Darwin noted
that the size of any population tends to remain constant despite the fact
that more offspring are produced than are needed to maintain it. He also
saw that variations existed between individuals of the population and
concluded that disease, competition, and other forces acting on the
population eliminated those individuals less well adapted to their
environment. The survivors would pass on any inheritable advantageous
characteristics (i.e. characteristics with survival value) to their offspring
and in time the composition of the population would change in adaptation
to a changing environment. Over a long period of time this process could
give rise to organisms so different from the original population that new
species are formed. *See also* **adaptive radiation**. *Compare* **punctuated
equilibrium**.

nature and nurture The combined effects of inherited factors (nature)
and environmental factors (nurture) on the development of an organism.
The genetic potential of an organism will only be realized under
appropriate environmental conditions. *See also* **phenotype**.

nauplius The free-swimming larva of many marine and freshwater crustaceans. It has an unsegmented body with a single eye at the front (the *nauplius eye*), mandibles, antennae, and three pairs of limbs.

navigation The complex process that enables animals to travel along a particular course in order to reach a specific destination. Navigation is an important aspect of behaviour in many animals, particularly those, such as birds, fish, and some insects, that undergo *migrations. Landmarks, such as coastlines and mountain ranges, are important reference points for navigation but many animals can navigate successfully without the aid of these, by using the sun, stars, magnetic fields, odours, and polarized light. For example, birds use the sun and stars as landmarks and are sensitive to the earth's magnetic fields, while salmon can identify the unique odour of their home river.

Neanderthal man A form of fossil man that appeared in Europe and Western Asia about 100 000 years ago. He was thought to be a different species (*Homo neanderthalensis*) from modern man but is now generally regarded as a subspecies of *H. sapiens*. The fossil remains indicate that Neanderthals were fairly short and had low brows but that the brain size was the same as modern man's. They were nomadic hunters and are thought to have had a language. *Cromagnon man began to replace Neanderthal man about 35 000 years ago. The name is derived from the site in the Neander valley, Germany, where fossils were found in 1856.

near point The nearest point at which the human eye can focus an object. As the lens becomes harder with age, the extent to which accommodation can bring a near object into focus decreases. Therefore with advancing age the near point recedes – a condition known as *presbyopia.

necrosis The *death of cells, which can be caused by a variety of chemicals and toxic substances. It often involves denaturation of proteins and may be preceded by a change in the appearance of cells and their contents (such as swelling of the mitochondria and *karyorrhexis) and the appearance of *lysosomes. *Compare* **apoptosis**.

nectar A sugary liquid produced in plants by *nectaries*, regions of secretory cells on the receptacle or other parts of a flower. It attracts pollinating insects or other animals.

nectary *See* nectar.

negative feedback *See* feedback.

nekton *Pelagic organisms that actively swim through the water. Examples are fish, jellyfish, turtles, and whales. *Compare* **benthos; neuston; plankton**.

nematoblast *See* thread cell.

nematocyst *See* thread cell.

Nematoda A phylum of *pseudocoelomate invertebrates comprising the roundworms. They are characterized by a smooth narrow cylindrical unsegmented body tapered at both ends. They shed their tough outer cuticle four times during life to allow growth. The microscopic free-living forms are found in all parts of the world, where they play an important role in the destruction and recycling of organic matter. The many parasitic nematodes are much larger; they include the filaria (*Wuchereria*) and Guinea worm (*Dracunculus*), which cause serious diseases in man.

neo-Darwinism (modern synthesis) The current theory of the process of *evolution, formulated between about 1920 and 1950, that combines evidence from classical genetics with the Darwinian theory of evolution by *natural selection (*see* **Darwinism**). It makes use of modern knowledge of genes and chromosomes to explain the source of the genetic variation upon which selection works. This aspect was unexplained by traditional Darwinism.

neo-Lamarckism Any of the comparatively modern theories of evolution based on Lamarck's theory of the inheritance of acquired characteristics (*see* **Lamarckism**). These include the unfounded dogma of *Lysenkoism and the recent controversial experiments on the inheritance of acquired immunological tolerance in mice.

Neolithic The New Stone Age, beginning in the Middle East approximately 9000 BC and lasting until 6000 BC, during which man first developed agriculture. Grinding and polishing of stone tools was also practised.

neopallium (neocortex) An area of the cerebral cortex of the vertebrate brain. The neopallium is most highly developed in mammals, in which it forms a surface layer covering most of the forebrain. The neopallium is the major centre for the coordination of sensory and motor information.

neoplasm (tumour) Any new abnormal growth of cells, forming either a harmless (benign) tumour or a malignant one (*see* **cancer**).

neoteny The retention of the juvenile body form, or particular features of it, in a mature animal. For example, the axolotl, a salamander, retains the gills of the larva in the adult. Neoteny is thought to have been an important mechanism in the evolution of certain groups, such as man, who is believed to have developed from the juvenile form of apes.

nephridiopore *See* **nephridium**.

nephridium (*pl.* **nephridia**) The excretory organ of many invertebrates, consisting of a simple or branched tube, formed by the ingrowth of ectoderm, with cilia at the inner end. Excretory products diffuse into the nephridium and are wafted to the exterior by ciliary action. The most primitive type is known as a *protonephridium* and consists of a system of *flame cells; it occurs in platyhelminths and rotifers. The *metanephridium*, which occurs in the earthworm and some other annelids, opens into the *coelom by combining with a coelomoduct; the internal opening is known

as the *nephrostome* and the external opening is the *nephridiopore*.
Metanephridia superficially resemble kidney tubules.

Structure of a single nephron

nephron The excretory unit of the vertebrate *kidney (see illustration).
Many constituents of the blood are filtered from the glomerulus into the
*Bowman's capsule at one end of the nephron. The *glomerular filtrate
passes along the length of the nephron and some of its water, plus some
salts, glucose, and amino acids, are reabsorbed into the surrounding blood
capillaries (*see* **proximal convoluted tubule; loop of Henle; distal
convoluted tubule**). More water is reabsorbed in the *collecting duct, and
the resulting concentrated solution of nitrogen-containing waste matter
(*urea in most mammals) plus inorganic salts drains from the collecting
ducts of the nephrons and is discharged as urine into the ureter.

nephrostome *See* **nephridium.**

nephrotoxin Any toxic substance that targets the kidneys. Common
examples of nephrotoxins are mercury salts and certain herbicides, such as
*Paraquat.

neritic zone The region of the sea over the continental shelf, which is less than 200 metres deep (approximately the maximum depth for organisms carrying out photosynthesis). *Compare* **oceanic zone**.

nerve A strand of tissue comprising many *nerve fibres plus supporting tissues (*see* glia), enclosed in a connective-tissue sheath. Nerves connect the central nervous system with the organs and tissues of the body. A nerve may carry only motor nerve fibres (*motor nerve*) or only sensory fibres (*sensory nerve*) or it may be mixed and carry both types (*mixed nerve*). Although the nerve fibres are in close proximity within the nerve, their physiological responses are independent of each other.

nerve cell *See* **neurone**.

nerve cord A large bundle of nerve fibres, running down the longitudinal axis of the body, that forms an important part of the *central nervous system. Most invertebrates have a pair of solid nerve cords, situated ventrally and bearing segmentally arranged *ganglia. All animals of the phylum *Chordata have a dorsal hollow nerve cord; in vertebrates this is the *spinal cord.

nerve fibre The *axon of a *neurone together with the tissues associated with it (such as a *myelin sheath). The length and diameter of nerve fibres are very variable, even within the same organism. *See also* **giant fibre**.

nerve growth factor (NGF) A protein (*see* **growth factor**), consisting of two polypeptide chains, that is required for the development of some neurones, especially those in the *sympathetic nervous system and sensory neurones. NGF is necessary for axon growth and also for initiating new neuronal connections with other cells.

nerve impulse *See* **impulse**.

nerve net A network of nerve cells connected with each other by synapses or fusion. The nervous system of certain invertebrates (e.g. coelenterates and echinoderms) consists exclusively of a nerve net in the body wall.

nervous system The system of cells and tissues in multicellular animals by which information is conveyed between sensory cells and organs and effectors (such as muscles and glands). It consists of the *central nervous system (in vertebrates the *brain and *spinal cord; in invertebrates the *nerve cord and *ganglia) and the *peripheral nervous system. Its function is to receive, transmit, and interpret information and then to formulate appropriate responses for the effector organs. It also serves to coordinate responses that require more than one physiological process. Nervous tissue consists of *neurones, which convey the information in the form of *impulses, and supporting tissue.

net primary productivity *See* **productivity**.

neural plate A strip of ectoderm, above the notochord, that lies along the

central axis of the early embryos (*see* **gastrula**) of chordates and develops
into the central nervous system. The neural plate folds and forms a tube
(the *neural tube*), which develops into the spinal cord and brain; this process
is called *neurulation*. During the stage when the neural tube is forming the
embryo is known as a *neurula*.

neuroendocrine system Any of the systems of dual control of certain
activities in the body of some higher animals by nervous and hormonal
stimulation. For example, the posterior *pituitary gland and the medulla of
the *adrenal gland receive direct nervous stimulation to secrete their
hormones, whereas the anterior pituitary gland is stimulated by releasing
hormones from the hypothalamus.

neurofibril Any of the fibres in the cytoplasm of a nerve *axon.
Neurofibrils include *neurofilaments and *neurotubules*, microtubules that
play a role in the transport of proteins and other substances within the
cytoplasm.

neurofilament A type of filament found in the axons of nerve cells.
Neurofilaments serve as elements of the *cytoskeleton supporting the axon
cytoplasm.

neurohormone Any hormone that is produced not by an endocrine
gland but by a specialized nerve cell and is secreted from nerve endings
into the bloodstream or directly to the tissue or organ whose growth or
function it controls (*see* **neurosecretion**). Examples of neurohormones are
*noradrenaline, *antidiuretic hormone, and hormones associated with
metamorphosis and moulting in insects (*see* **ecdysone; juvenile hormone**).

neurohypophysis *See* **pituitary gland**.

neuromuscular junction The point where a muscle fibre comes into
contact with a motor neurone carrying nerve impulses from the central
nervous system. The impulses travel from the neurone to the muscle fibre
by means of a neurotransmitter, in a similar way to the transmission of
impulses across a *synapse between two neurones. The neurotransmitter is
released from vesicles at the end of the motor neurone into a small gap
(the cleft), where it diffuses to the *end plate of the muscle fibre and
depolarizes the membrane. When depolarization has reached a certain
threshold an action potential is triggered in the muscle fibre.

neurone (neuron; nerve cell) An elongated branched cell that is the
fundamental unit of the *nervous system, being specialized for the
conduction of *impulses. A neurone consists of a *cell body, containing the
nucleus and *Nissl granules; *dendrites, which receive incoming impulses
and pass them towards the cell body; and an *axon, which conducts
impulses away from the cell body, sometimes over long distances. Impulses
are passed from one neurone to the next via *synapses. *Sensory neurones
transmit information from receptors to the central nervous system.
*Motor neurones conduct information from the central nervous system to

*effectors (e.g. muscles). (See illustration.) *See also* **bipolar neurone;
multipolar neurone; unipolar neurone**.

Sensory neurone

Motor neurone

neurone theory The hypothesis, now accepted, that the nervous system
consists of nerve cells (*neurones), which are functionally linked at
*synapses but are physically separate. This theory has superseded the idea
that the cytoplasm of the cells of the nervous system is continuous.

neurosecretion The secretion of *neurohormones by *neurosecretory cells*,
which possess characteristics of both nerve cells and endocrine cells. They
are found, for example, in the hypothalamus, where they receive nerve
impulses from other parts of the brain but transmit these signals to the
pituitary gland by neurohormones that are released into the blood.

neurotoxin A chemical that either physically damages a nerve or reduces
or alters the function of the nerve. An example of a neurotoxin is 6-
hydroxydopamine, which damages the nerve terminals of the sympathetic
neurones. This compound is structurally similar to dopamine and

noradrenaline and can enter the nerve terminal by existing transport systems.

neurotransmitter (transmitter) A chemical that mediates the transmission of a nerve impulse across a *synapse. Examples are *adrenaline, *noradrenaline, *dopamine, and *serotonin (in adrenergic nerves), *acetylcholine (in cholinergic nerves), and *gamma-aminobutyric acid. The neurotransmitter is released at the synaptic knob at the tip of the axon into the synaptic cleft. It diffuses across to the opposite membrane (the postsynaptic membrane), where it initiates the propagation of a nerve impulse in the next neurone.

neurotubule *See* **neurofibril.**

neurula *See* **neural plate.**

neuston The organisms that colonize the surface of an aquatic habitat. They are most abundant in freshwater habitats and include pond skaters, certain types of water beetles, and floating plants. *Compare* **benthos; nekton; plankton.**

neuter An organism that does not possess either male or female reproductive organs. Cultivated ornamental flowers that have neither pistils nor stamens are called neuters.

neutral Describing a compound or solution that is neither acidic nor basic. A neutral solution is one that contains equal numbers of both protonated and deprotonated forms of the solvent.

neutrophil A type of white blood cell (*leucocyte) that has a lobed nucleus and granular cytoplasm (*see* **granulocyte**). Neutrophils engulf bacteria (*see* **phagocytosis**) and release various substances, such as *lysozyme and oxidizing agents.

newton Symbol N. The *SI unit of force, being the force required to give a mass of one kilogram an acceleration of $1 \, m \, s^{-2}$. It is named after Sir Isaac Newton (1642–1727).

niacin *See* **nicotinic acid.**

niche *See* **ecological niche.**

nicotinamide *See* **nicotinic acid.**

nicotinamide adenine dinucleotide *See* **NAD.**

nicotinamide adenine dinucleotide phosphate (NADP) *See* **NAD.**

nicotine A colourless poisonous *alkaloid present in tobacco. It is used as an insecticide.

nicotinic acid (niacin) A vitamin of the *vitamin B complex. It can be manufactured by plants and animals from the amino acid tryptophan. The amide derivative, *nicotinamide*, is a component of the coenzymes *NAD and NADP. These take part in many metabolic reactions as hydrogen acceptors.

Deficiency of nicotinic acid causes the disease *pellagra in humans. Apart from tryptophan-rich protein, good sources are liver and groundnut and sunflower meals.

nictitating membrane A clear membrane forming a third eyelid in amphibians, reptiles, birds, and some mammals (but not man). It can be drawn across the cornea independently of the other eyelids, thus clearing the eye surface and giving added protection without interrupting the continuity of vision.

nidation *See* implantation.

nidicolous *See* altricial species.

nidifugous *See* precocial species.

ninhydrin A compound that reacts with amino acids to give a blue colour. Ninhydrin is commonly used in chromatography to analyse the amino-acid content of proteins.

nipple *See* mammary gland.

Nissl granules (Nissl bodies) Particles seen within the cell bodies of *neurones. They are rich in RNA and stain strongly with basic dyes. They are named after F. Nissl (1860–1919), the German neurologist who discovered them.

nitrate A salt or ester of nitric acid. The salts contain the ion NO_3^-.

nitrification A chemical process in which nitrogen (mostly in the form of ammonia) in plant and animal wastes and dead remains is oxidized at first to nitrites and then to nitrates. These reactions are effected mainly by the nitrifying bacteria *Nitrosomonas* and *Nitrobacter* respectively. Unlike ammonia, nitrates are readily taken up by plant roots; nitrification is therefore a crucial part of the *nitrogen cycle. Nitrogen-containing compounds are often applied to soils deficient in this element, as fertilizer. *Compare* denitrification.

nitrifying bacteria *See* nitrification.

nitrite A salt or ester of nitrous acid. The salts contain the ion NO_2^-.

nitrogen Symbol N. A colourless gaseous element that occurs in air (about 78% by volume) and is an essential constituent of proteins and nucleic acids in living organisms (*see* **nitrogen cycle**).

nitrogenase An important enzyme complex that is present in those microorganisms that are capable of fixing atmospheric nitrogen (*see* **nitrogen fixation**). Nitrogenase catalyses the conversion of atmospheric nitrogen into ammonia, which can then be used to synthesize nitrites, nitrates, or amino acids. The two main enzymes within the nitrogenase complex are *dinitrogenase reductase* and *dinitrogenase*.

nitrogen cycle One of the major cycles of chemical elements in the

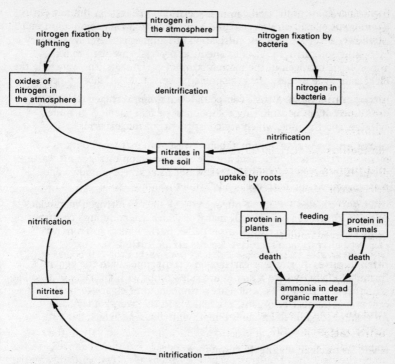

The nitrogen cycle

environment (*see* **biogeochemical cycle**). Nitrates in the soil are taken up
by plant roots and may then pass along *food chains into animals.
Decomposing bacteria convert nitrogen-containing compounds (especially
ammonia) in plant and animal wastes and dead remains back into nitrates,
which are released into the soil and can again be taken up by plants (*see*
nitrification). Though nitrogen is essential to all forms of life, the huge
amount present in the atmosphere is not directly available to most
organisms (*compare* **carbon cycle**). It can, however, be assimilated by some
specialized bacteria (*see* **nitrogen fixation**) and is thus made available to
other organisms indirectly. Lightning flashes also make some nitrogen
available to plants by causing the combination of atmospheric nitrogen and
oxygen to form oxides of nitrogen, which enter the soil and form nitrates.
Some nitrogen is returned from the soil to the atmosphere by denitrifying
bacteria (*see* **denitrification**). See illustration.

nitrogen fixation A chemical process in which atmospheric nitrogen is
assimilated into organic compounds in living organisms and hence into the
*nitrogen cycle. The ability to fix nitrogen, by means of *nitrogenase
enzymes, is limited to certain bacteria (e.g. *Azotobacter*, *Anabaena*). *Rhizobium*

bacteria are able to fix nitrogen in association with cells in the roots of leguminous plants, such as peas and beans, in which they form characteristic *root nodules; cultivation of legumes is therefore one way of increasing soil nitrogen. Various chemical processes are used to fix atmospheric nitrogen in the manufacture of *fertilizers. These include the Birkeland–Eyde process, the cyanamide process, and the Haber process.

nitrogenous base A basic compound containing nitrogen. The term is used especially of organic ring compounds, such as adenine, guanine, cytosine, and thymine, which are constituents of nucleic acids.

nitrogenous waste Any metabolic *waste product that contains nitrogen. *Urea and *uric acid are the most common nitrogenous waste products in terrestrial animals; freshwater fish excrete ammonia and marine fish excrete both urea and *trimethylamine oxide.

nitrogen oxides Oxides of nitrogen (NO_x), such as nitrogen monoxide (NO) and dinitrogen oxide (N_2O), many of which are pollutants contributing to *acid rain. Nitrogen oxides are expelled in the emissions from car exhausts, aircraft, and factories. *See also* **air pollution**.

nitrosamines A group of carcinogenic compounds with the general formula RR′NNO, where R and R′ are side groups with a variety of possible structures. Nitrosamines, which are a component of cigarette smoke, cause cancer in a number of organs, particularly in the liver, kidneys, and lungs. An example of a nitrosamine is dimethylnitrosamine, which has two methyl side groups (CH_3-).

NMR *See* **nuclear magnetic resonance**.

node 1. *(in botany)* The part of a plant stem from which one or more leaves arise. The nodes at the stem apex are very close together and remain so in species of monocotyledons that form bulbs. In older regions of the stem they are separated by areas of stem called *internodes*. **2. *(in anatomy)*** A natural thickening or bulge in an organ or part of the body. Examples are the *sinoatrial node* that controls the heartbeat (*see* **pacemaker**) and the *lymph nodes.

node of Ranvier *See* **myelin sheath**.

nodule *(in botany)* *See* **root nodule**.

nomad *(in cytology)* A cell that migrates or wanders from its site of formation. Certain types of *phagocytes are nomads.

noncompetitive inhibition *See* **inhibition**.

noncyclic phosphorylation (noncyclic photophosphorylation) *See* photophosphorylation.

nondisjunction The phenomenon that occurs when a pair of *homologous chromosomes do not separate in *meiosis but migrate to the same pole of the cell, resulting in an uneven number of chromosomes being present in the daughter cells. *Klinefelter's syndrome results from nondisjunction of the sex chromosomes.

nonreducing sugar A sugar that cannot donate electrons to other molecules and therefore cannot act as a reducing agent. Sucrose is the most common nonreducing sugar. The linkage between the glucose and fructose units in sucrose, which involves aldehyde and ketone groups, is responsible for the inability of sucrose to act as a *reducing sugar.

nonrenewable energy sources Sources of energy that use up the earth's finite mineral resources; these include *fossil fuels. Concern about the exhaustion of nonrenewable energy sources, together with the fact that burning fossil fuels contributes to *air pollution and the *greenhouse effect, is leading to increased use or investigation of *renewable energy resources*, which are not exhaustible. These include the sun (for solar heating and solar cells), wind power (for aerogenerators) and water (for hydroelectric generators).

nonsense mutation A mutation in one of the nucleotides in a DNA sequence that generates a *stop codon, resulting in the premature termination of synthesis of a protein.

noradrenaline (norepinephrine) A hormone produced by the *adrenal glands and also secreted from nerve endings in the *sympathetic nervous system as a chemical transmitter of nerve impulses (*see* **neurotransmitter**). Many of its general actions are similar to those of *adrenaline, but it is more concerned with maintaining normal body activity than with preparing the body for emergencies.

norepinephrine *See* noradrenaline.

Northern blotting *See* **Southern blotting**.

nose The protuberance on the face of some vertebrates that contains the nostrils (*see* **nares**) and part of the *nasal cavity. It therefore forms part of the olfactory system (*see* **olfaction**) and the external opening of the respiratory system.

nostrils *See* **nares**.

notochord An elastic skeletal rod lying lengthwise beneath the nerve cord and above the alimentary canal in the embryos or adults of all chordate animals (*see* **Chordata**). Its function is to strengthen and support the body and act as a protagonist for the muscles. It is found in both adult and larval lancelets but in adult vertebrates it is largely replaced by the *vertebral column.

nucellus The tissue that makes up the greater part of the ovule of seed plants. It contains the *embryo sac and nutritive tissue. It is enclosed by the integuments except for a small gap, the *micropyle. In certain flowering plants it may persist after fertilization and provide nutrients for the embryo.

nuclear–cytoplasmic ratio A measure of the size of a cell nucleus in relation to the cytoplasm. The nuclear–cytoplasmic ratio is often used as an index in the comparison of cells from normal and abnormal tissues. For example, cultured cancer cells show an increase in the nuclear–cytoplasmic ratio.

nuclear envelope The double membrane that separates the nucleoplasm (*see* **nucleus**) of a cell from the cytoplasm. The membranes consist of *lipid bilayers that are separated by a *perinuclear space* (or *compartment*). The outer membrane is continuous with the rough *endoplasmic reticulum and is structurally and functionally distinct from the inner membrane. The envelope is perforated at intervals by *nuclear pores*, which provide a channel for the selective transfer of water-soluble molecules between the nucleus and the cytoplasm. Each nuclear pore is surrounded by a disc-shaped structure (*nuclear pore complex*) consisting of an octagonal arrangement of eight protein granules.

nuclear magnetic resonance (NMR) The absorption of electromagnetic radiation (radio waves) by certain atomic nuclei placed in a strong and stable magnetic field. This results in a change of orientation of the nuclei, which respond to the magnetic field like miniature bar magnets. The main application of NMR is in a form of spectroscopy (*NMR spectroscopy*) used for chemical and biochemical analysis and structure determination. There are two methods of NMR spectroscopy. In *continuous wave (CW) NMR*, the sample is subjected to a strong magnetic field, which can be varied in a controlled way. As the field changes, absorption of radiation occurs at certain points; this produces oscillations in the field, which can be detected. *Fourier transform (FT) NMR* uses a fixed magnetic field and the sample is subjected to a high-intensity pulse of radiation covering a range of frequencies. The signal produced is analysed mathematically to give the NMR spectrum. The ^{1}H nucleus is the one commonly studied; other biochemically useful nuclei are ^{31}P, ^{13}C, ^{14}N, and ^{19}F, although these have lower natural abundance than hydrogen and produce weaker signals. The spectrum produced is characteristic of the molecule absorbing the radiation. In medicine, *magnetic resonance imaging (MRI)* has been developed, in which images of soft tissue are produced. This technique is useful for locating tumours and tissue abnormalities.

nuclear pore *See* **nuclear envelope**.

nuclease Any enzyme that breaks down nucleic acids to nucleotides. Nucleases are found in the small intestine. *See* **DNAase**.

nucleic acid A complex organic compound in living cells that consists of a chain of *nucleotides. There are two types: *DNA (deoxyribonucleic acid) and *RNA (ribonucleic acid).

nucleic acid hybridization The association, *in vitro*, of two complementary nucleic-acid strands to form a hybrid double strand. *See* **DNA hybridization**.

nucleoid (nuclear region) The part of a cell of a bacterium (i.e. a prokaryotic *cell) that contains the genetic material *DNA and therefore controls the activity of the cell. It corresponds to the nucleus of the more advanced eukaryotic cells but is not bounded by a membrane.

nucleolar organizer A group of genes situated on one of the chromosomes that code for ribosomal *RNA. The nucleolar organizer is involved with the collection of nucleolar material as it breaks down and is released into the nucleoplasm during prophase of mitosis and in the reconstruction of the *nucleolus during telophase.

nucleolus A small dense round body within the nondividing *nucleus of plant and animal cells that consists of protein, DNA, and ribosomal *RNA. It plays an important role in *ribosome manufacture (and therefore protein synthesis).

nucleoplasm (karyoplasm) The material contained within the *nucleus of a cell. The nucleoplasm is bound by the *nuclear envelope, which separates it from the cytoplasm.

nucleoprotein Any compound present in cells of living organisms that consists of a nucleic acid (DNA or RNA) combined with a protein. Chromosomes consist of nucleoprotein (DNA and proteins, mostly histones), as do ribosomes (RNA and protein). *See also* **ribonucleoprotein**.

nucleoside An organic compound consisting of a nitrogen-containing *purine or *pyrimidine base linked to a sugar (ribose or deoxyribose). An example is *adenosine. *Compare* **nucleotide**.

nucleosome The fundamental unit of *chromatin, the material of which eukaryotic chromosomes are made. It consists of a core of *histone proteins around which are coiled about 140 base pairs of DNA. Nucleosomes are packed in stacks in the chromatin threads.

nucleotidase An enzyme that catalyses the breakdown of nucleotides. It is present in the epithelial cells of the small intestine and plays an important role in the digestion of proteins.

nucleotide An organic compound consisting of a nitrogen-containing *purine or *pyrimidine base linked to a sugar (ribose or deoxyribose) and a phosphate group. *DNA and *RNA are made up of long chains of nucleotides (i.e. *polynucleotides*). *Compare* **nucleoside**.

nucleus The large body embedded in the cytoplasm of all eukaryote *cells that contains the genetic material *DNA. The nucleus functions as the control centre of the cell. It is bounded by a double membrane (the *nuclear envelope) and contains a viscous sap (*nucleoplasm*). When the cell is not dividing a *nucleolus is present in the nucleus and the genetic material is dispersed in the sap as *chromatin; in dividing cells the chromatin is organized into *chromosomes and the nucleolus disappears. In certain protozoans there are two nuclei per cell, a *macronucleus* (or *meganucleus*) concerned with vegetative functions and a *micronucleus* with sexual reproduction.

numerical taxonomy *See* taxonomy.

nut A dry single-seeded fruit that develops from more than one carpel and does not shed its seed when ripe. The fruit wall is woody or leathery. Many nuts are enclosed in a hard or membranous cup-shaped structure, the *cupule. The term nut is often loosely used of any hard fruit. For example, the walnut and coconut are in fact *drupes and the Brazil nut is a seed.

nutation The spiral movement of a plant organ during growth, also known as *circumnutation*. It is seen in climbing plants and helps the plant find a suitable support to twine around. Examples are the coiling movements of the shoot tips of runner beans and of the tendrils of sweet peas.

nutrient Any substance that is required for the nourishment of an organism, providing a source of energy or structural components. In animals nutrients form part of the *diet and include the *major nutrients*, i.e. carbohydrates, proteins (*see also* **essential amino acid**), and lipids (*see also* **essential fatty acid**), as well as vitamins and certain minerals (*see* **essential element**). Plant nutrients, derived from carbon dioxide in the atmosphere and water (containing minerals) absorbed from the soil by the roots, are divided into *macronutrients and *micronutrients.

nutrition The process by which organisms obtain energy (in the form of food) for growth, maintenance, and repair. There are two main types of nutrition: *heterotrophic nutrition, employed by animals, fungi, and certain bacteria; and *autotrophic nutrition, found in most plants and bacteria.

nyctinasty (sleep movements) *Nastic movements of plant organs in response to the changes in light and temperature that occur between day and night (and vice versa). Examples are the opening and closing of many flowers and the folding together of the leaflets of clover and other plants at night.

nymph The juvenile stage of *exopterygote insects, such as dragonflies, grasshoppers, and earwigs, which resembles the adult except that the wings and reproductive organs are undeveloped. There is no pupal stage, and the nymph develops directly into the adult. *Compare* **larva**.

objective The lens or system of lenses nearest to the object being examined through an optical instrument.

obligate anaerobes Organisms that cannot use free oxygen for respiration. Oxygen may inhibit the growth of obligate anaerobes or kill them. *See* **anaerobic respiration**.

occipital condyle A single or paired bony knob that protrudes from the occipital bone of the skull and articulates with the first cervical vertebra (the *atlas). In humans there is a pair of occipital condyles, one on each side of the *foramen magnum. Occipital condyles are absent in most fish, which cannot move their heads.

oceanic zone The region of the open sea beyond the edge of the continental shelf, where the depth is greater than 200 metres. *Compare* **neritic zone**.

ocellus A simple eye occurring in insects and other invertebrates. It typically consists of light-sensitive cells and a single cuticular lens.

ocular *See* **eyepiece**.

Odonata An order of *exopterygote insects containing the dragonflies and damselflies, most of which occur in tropical regions. Adult dragonflies have a pair of prominent *compound eyes, a compact thorax bearing two pairs of delicate membranous wings, and a long slender abdomen. They are strong fliers and prey on other insects, either in flight or at rest. The eggs are laid near or in water, and the newly hatched nymphs are aquatic and resemble the adults, with rudimentary wings. They breathe through gills and feed on small aquatic animals. The nymph leaves the water for its final moult into the terrestrial adult.

odontoblast A cell that is responsible for producing the *dentine of vertebrate teeth. Odontoblasts are found around the lining of the *pulp cavity and have processes that extend into the dentine.

oedema The accumulation of *tissue fluid in the tissues of the body, causing swelling of the affected part. Localized oedema occurs during *inflammation. Generalized oedema can result from a variety of pathological conditions, including kwashiorkor (*see* **malnutrition**) and heart or kidney failure; it can also occur as a side-effect of certain drugs and as a reaction to toxic chemicals.

oesophagus (gullet) The section of the *alimentary canal that lies between the *pharynx and the stomach. It is a muscular tube whose function is to transfer food to the stomach by means of wavelike contractions (*peristalsis) along its length.

oestrogen One of a group of female sex hormones, produced principally by the ovaries, that promote the onset of *secondary sexual characteristics (such as breast enlargement and development in women) and control the *oestrous cycle (*menstrual cycle in humans). *Oestradiol* is the most important. Oestrogens are secreted at particularly high levels during ovulation, stimulating the uterus to prepare for pregnancy. They are used in *oral contraceptives (with *progestogens) and as treatment for various disorders of the female reproductive organs. Small amounts of oestrogens are produced by the adrenal glands and testes.

oestrous cycle (sexual cycle) The cycle of reproductive activity shown by most sexually mature nonpregnant female mammals except most primates (*compare* **menstrual cycle**). There are four phases:
(1) *pro-oestrus* (*follicular phase*) – *Graafian follicles develop in the ovary and secrete oestrogens;
(2) *oestrus* (*heat*) – ovulation normally occurs, the female is ready to mate and becomes sexually attractive to the male;
(3) *metoestrus* (*luteal phase*) – *corpus luteum develops from ruptured follicle;
(4) *dioestrus* – *progesterone secreted by corpus luteum prepares uterus for implantation.
   The length of the cycle depends on the species: larger mammals typically have a single annual cycle with a well-defined breeding season (they are described as *monoestrous*). The males have a similar cycle of sexual activity. Other species may have many cycles per year (i.e. they are *polyoestrous*) and the male may be sexually active all the time.

oestrus (heat) *See* **oestrous cycle**.

offset *See* **runner**.

offspring (progeny) New individual organisms that result from the process of sexual or asexual reproduction. *See also* F_1; F_2.

oil Any of various viscous liquids that are generally immiscible with water. Natural plant and animal oils are either volatile mixtures of terpenes and simple esters (e.g. *essential oils) or are *glycerides of fatty acids.

oil-immersion lens *See* **immersion objective**.

oleaginous Producing or containing oil or lipids. Oleaginous microorganisms, which normally contain 20–25% oil, are of interest in biotechnology as alternative sources of conventional oils or as possible sources for novel oils. The majority of the oils produced by oleaginous eukaryotic microorganisms are similar to plant oils. One possibility under consideration is the production of oils and fats from waste material, to be used in animal feed.

olecranon process A bony process on the *ulna in the forelimb of vertebrates that projects beyond the joint between the humerus and the ulna.

oleic acid An unsaturated *fatty acid with one double bond, $CH_3(CH_2)_7CH:CH(CH_2)_7COOH$. Oleic acid is one of the most abundant constituent fatty acids of animal and plant fats, occurring in butterfat, lard, tallow, groundnut oil, soya-bean oil, etc. Its systematic chemical name is *cis-octadec-9-enoic acid*.

olfaction The sense of smell or the process of detecting smells. This is achieved by receptors in *olfactory organs* (such as the *nose) that are sensitive to air- or water-borne chemicals. Stimulation of these receptors results in the transmission of information to the brain via the *olfactory nerve*.

olfactory lobe Either member of a pair of lobes in the forebrain, at the anterior end of the cerebrum. They contain the endings of the olfactory nerves (the first pair of cranial nerves) and are concerned with the sense of smell, being prominent in the dogfish and other animals that depend on this sense.

Oligocene The third geological epoch of the *Tertiary period. It began about 38 million years ago, following the Eocene epoch, and extended for about 12 million years to the beginning of the Miocene epoch. The epoch was characterized by the continued rise of mammals; the first pigs, rhinoceroses, and tapirs made their appearance.

Oligochaeta A class of hermaphrodite annelid worms that bear only a few bristles (*chaetae). Oligochaetes are very abundant in freshwater and terrestrial habitats. The most familiar members of the class are the earthworm (*Lumbricus*) and the freshwater bloodworm (*Tubifex*).

oligonucleotide A short polymer of *nucleotides.

oligotrophic Describing a body of water (e.g. a lake) with a poor supply of nutrients and a low rate of formation of organic matter by photosynthesis. *Compare* **dystrophic; eutrophic; mesotrophic**.

omasum The third of four chambers that form the stomach of ruminants. *See* **Ruminantia**.

ommatidium (*pl.* **ommatidia**) *See* **compound eye**.

omnivore An animal that eats both animal and vegetable matter. Pigs, for example, are omnivorous. *Compare* **carnivore; herbivore**.

oncogene A gene that is capable of transforming a normal cell into a cancerous cell. Oncogenes result from the mutation of normal genes (*proto-oncogenes*); they occur in cancer cells and *oncogenic viruses. Viral oncogenes are derived from normal host genes that have become incorporated into the viral genome and subsequently undergo mutation.

oncogenic Describing a chemical, organism, or environmental factor that causes the development of cancer. Some viruses are oncogenic to vertebrates, notably the *retroviruses (including the Rous sarcoma virus of chickens), and some are suspected of being oncogenic (e.g. some of the *adenoviruses and *papovaviruses). Many of these viruses contain genes (known as *oncogenes) that are responsible for the transformation of a normal host cell into a cancerous cell. *See also* **growth factor**.

one gene–one polypeptide hypothesis The theory that each *gene is responsible for the synthesis of a single *polypeptide. It was originally stated as the *one gene–one enzyme hypothesis* by the US geneticist George Beadle (1903–89) in 1945 but later modified when it was realized that genes also encoded nonenzyme proteins and individual polypeptide chains. It is now known that some genes code for various types of RNA involved in protein synthesis.

ontogeny The developmental course of an organism from the fertilized egg through to maturity. It has been suggested that "ontogeny recapitulates *phylogeny", i.e. the stages of development, especially of the embryo, reflect the evolutionary history of the organism. This idea is now discredited. *See* **recapitulation**.

oocyte *See* **oogenesis**.

oogamy Sexual reproduction involving the formation and subsequent fusion of a large, usually stationary, female gamete and a small motile male gamete. The female gamete may contain nourishment for the development of the embryo, which is often retained and protected by the parent organism.

oogenesis The production and growth of the ova (egg cells) in the animal ovary. Special cells (*oogonia*) within the ovary divide repeatedly by mitosis to produce large numbers of prospective egg cells (*primary oocytes*). When mature, these undergo meiosis, which halves the number of chromosomes. During the first meiotic division a *polar body* and a *secondary oocyte* are produced. At the second meiotic division the secondary oocyte produces an ovum and a second polar body. Oocytes may be present in the ovaries at birth and may represent the total number of eggs to be produced.

oogonium (*pl.* **oogonia**) **1.** The female sex organ (*gametangium) of algae and fungi. **2.** Any of the immature sex cells in the animal ovary that give rise to oocytes by mitotic divisions (*see* **oogenesis**).

Oomycota A phylum of the Protoctista that includes the water moulds, downy mildews, and potato blight (*Phytophthora*), formerly classified as a class of fungi (Oomycetes). They are coenocytic and the cell wall is made of cellulose. Oomycotes are either saprotrophic or parasitic; they feed by extending hypha-like threads into the food source or host's body. Asexual reproduction is by means of flagellated *zoospores, which are released from a sporangium. Sexual reproduction involves the fusion of an antheridium and an oogonium and results in the production of a zygote, which can develop a wall of chitin and become a resistant oospore.

oosphere (ovum; egg cell) The nonmotile female gamete in plants and some algae. In angiosperms (flowering plants) it is a cell in the *embryo sac of the ovule. In other plants it is situated in an *archegonium. In algae, such as *Fucus*, the oosphere is protected by an *oogonium* until it is shed into the water prior to fertilization. Many oospheres store food in the form of starch or oil droplets.

oospore A zygote that is produced as a result of *oogamy in certain algae and fungi. It contains food reserves, develops a protective outer covering, and enters a resting phase before germination. *Compare* **zygospore**.

open circulation *See* circulation.

operant conditioning *See* conditioning.

operculum 1. *(in zoology)* A lid or flap of skin covering an aperture, such as the gill slit cover of fish and larval amphibians and the horny calcareous operculum secreted by many gastropod molluscs, which closes the opening of the shell when the animal is inside. **2. *(in botany)*** The cone-shaped lid of the *capsule of mosses, which is forcibly detached to release the spores.

operon A functionally integrated genetic unit for the control of gene expression in bacteria, as proposed in the *Jacob–Monod hypothesis. It comprises *structural genes*, coding for protein, and adjacent loci controlling their expression – an *operator site* and a *promoter site*. *Transcription of the structural genes is prevented by binding of a *repressor* molecule to the operator site. Another molecule, the *inducer*, can bind to the repressor molecule, preventing it from binding to the operator and thus allowing the promoter site to bind the enzyme RNA polymerase, thereby initiating transcription. The repressor molecule is encoded by a *regulator gene*, which may be close to or distant from the operon. Some operons also have an *attenuator site* preceding the first structural gene, where transcription may either stall or proceed according to the amount of end-product in the cell. *See also lac* operon.

opiate One of a group of drugs derived from *opium*, an extract of the poppy plant *Papaver somniferum* that depresses brain function (a *narcotic* action). Opiates include *morphine and its synthetic derivatives, such as *heroin and codeine. They are used in medicine chiefly to relieve pain, but the use of morphine and heroin is strictly controlled since they can cause drug dependence and tolerance.

opioid Any one of a group of substances that produce pharmacological and physiological effects similar to those of morphine. Opioids are not necessarily structurally similar to morphine, although a subgroup of opioids, the *opiates, are morphine-derived compounds.

opisthosoma The posterior section of the body (*see* **tagma**) of arachnids and other arthropods of the subphylum *Chelicerata, which consists of those body segments that do not bear legs. *Compare* **prosoma**.

opsin The lipoprotein component of *rhodopsin, the light-sensitive pigment that occurs in the rod cells of the retina.

opsonin *See* opsonization.

opsonization The process in which certain antibodies in the blood (known as *opsonins*) bind to the surface of an invading microorganism, which render it more susceptible to phagocytosis.

D-form   L-form   *meso*-form

Isomers of tartaric acid

optical activity The ability of certain substances to rotate the plane of plane-polarized light as it passes through a crystal, liquid, or solution. It occurs when the molecules of the substance are asymmetric, so that they can exist in two different structural forms each being a mirror image of the other. The two forms are *optical isomers* or *enantiomers* (see illustration). The existence of such forms is also known as *enantiomorphism* (the mirror images being *enantiomorphs*). One form will rotate the light in one direction and the other will rotate it by an equal amount in the other. The two possible forms are described as *dextrorotatory or *laevorotatory according to the direction of rotation, and prefixes are used to designate the isomer, as in *d*-tartaric and *l*-tartaric acids. An equimolar mixture of the two forms is not optically active. It is called a *racemic mixture* (or *racemate*) and designated by *dl*-. In addition, certain molecules can have a *meso form* in which one part of the molecule is a mirror image of the other. Such molecules are not optically active.

Molecules that show optical activity have no plane of symmetry. The commonest case of this is in organic compounds in which a carbon atom is linked to four different groups. An atom of this type is said to be a *chiral centre*. Many naturally occurring compounds show optical isomerism and usually only one isomer occurs naturally. For instance, glucose is found in the dextrorotatory form. The other isomer, *l*-glucose, can be synthesized in the laboratory, but cannot be synthesized by living organisms.

optical fibre A glass fibre through which light can be transmitted with very little leakage through the sidewalls. In the *step-index fibre* a pure glass core, with a diameter between 6 and 250 micrometres, is surrounded by a glass or plastic cladding of lower refractive index. The cladding is usually between 10 and 150 micrometres thick. The interface between core and cladding acts as a cylindrical mirror at which total internal reflection of the transmitted light takes place. This structure enables a beam of light to travel through many kilometres of fibre. In the *graded-index fibre*, each layer of glass, from the fibre axis to its outer wall, has a slightly lower refractive index than the layer inside it. This arrangement also prevents light from escaping through the fibre walls by a combination of refraction and total internal reflection, and can be made to give the same transit time for rays at different angles.

Fibre-optic systems use optical fibres to transmit information, in the form of coded pulses or fragmented images (using bundles of fibres), from a source to a receiver. They are used, for example, in medical instruments (*endoscopes* or *fibrescopes*) to examine internal body cavities, such as the stomach and bladder.

optical isomers *See* optical activity.

optical microscope *See* microscope.

optic chiasma The point where some of the fibres from both *optic nerves cross over to the opposite side, forming an X-shaped structure. Sensory information from the right side of both the right and left eye is transmitted to the right visual cortex in the brain (via the *right lateral geniculate nucleus*), while information from the left side of each eye is carried to the left visual cortex. Thus each half of the brain receives sensory information from both eyes.

optic lobes *See* midbrain.

optic nerve The second *cranial nerve: a paired sensory nerve that runs from each eye to the brain. It is responsible for conveying visual stimuli received by the rods and cones in the retina to the brain for interpretation. *See also* optic chiasma.

optic vesicle The outgrowth from the forebrain of a vertebrate embryo that develops into the retina of the eye. When the optic vesicle comes into contact with the ectoderm covering the head of the embryo, these ectodermal cells become thicker: the ectoderm in this region invaginates and eventually becomes detached from the adjacent ectodermal cells to form the lens of the eye.

oral cavity *See* buccal cavity.

oral contraceptive Any preparation taken in the form of a pill to prevent conception (*see* **birth control**). The most common form is the combination pill, which contains the hormones *oestrogen and a *progestogen. Both act to suppress ovulation, while the progestogen additionally causes changes in the viscosity of cervical mucus and alters the lining of the womb, both of which decrease the chances of fertilization should ovulation occur. The so-called 'minipill' contains only a progestogen.

oral groove A ciliated channel found in certain protozoa and aquatic invertebrates down which food is directed into the mouth.

orbit *(in anatomy)* Either of the two sockets in the skull of vertebrates that house the eyeballs.

order *(in taxonomy)* A category used in the *classification of organisms that consists of one or several similar or closely related families. Similar orders form a class. Order names typically end in -*ales* in botany, e.g. Rosales (roses and orchard fruits), and in -*a* in zoology, e.g. Carnivora (flesh eaters).

Ordovician The second geological period of the Palaeozoic era, following the Cambrian and preceding the Silurian periods. It began about 500 million years ago and lasted for about 60 million years. The period was named by the British geologist Charles Lapworth (1842–1920) in 1879. *Graptolites, in deep-water deposits, are the dominant fossils. Other fossils include *trilobites, brachiopods, ectoprocts, gastropods, bivalves, echinoids, crinoids, nautiloid cephalopods, and the first corals.

organ Any distinct part of an organism that is specialized to perform one or a number of functions. Examples are ears, eyes, lungs, and kidneys (in animals) and leaves, roots, and flowers (in plants). A given organ will contain many different *tissues.

organ culture The culture of complete living organs (*explants*) of animals and plants outside the body in a suitable culture medium. Animal organs must be small enough to allow the nutrients in the culture medium to penetrate all the cells. Whole plant roots and even root systems can be kept alive in such conditions for a considerable period of time. *See also* **explantation**.

organelle A minute structure within a plant or animal *cell that has a particular function. Examples of organelles are the nucleus, mitochondria, and lysosomes.

organic evolution The process by which changes in the genetic composition of populations of organisms occur in response to environmental changes. *See* **adaptation; evolution**. *Compare* **biochemical evolution**.

organism An individual living system, such as an animal, plant, or *microorganism, that is capable of reproduction, growth, and maintenance.

organizer An area of an animal embryo that causes adjacent areas of the embryo to develop in a certain way. The *primary organizer* (blastopore lip or archenteron roof) causes the *gastrula to develop as a complete organism.

organ of Corti The sensory part of the *cochlea in the inner ear, which responds to sound. This organ, named after Italian anatomist A. G. G. Corti (1822–88), projects into the cochlear duct and consists of two membranes that run parallel to each other: sensory hair cells, rooted in the *basilar membrane*, are in contact with the overlying *tectorial membrane* (see illustration). During the transmission of sound waves the basilar membrane vibrates, causing the sensory hairs to flex against the tectorial membrane; this results in the production of impulses, which are transmitted to the brain via the auditory nerve.

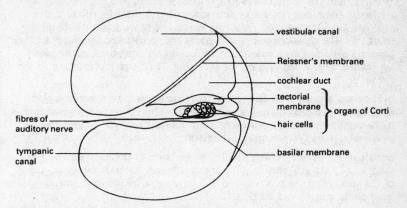

Organ of Corti (vertical section through one turn of the cochlea)

orgasm The climax of sexual excitement in humans, which – in males – coincides with *ejaculation. A sense of physiological and emotional release is accompanied by a feeling of extreme pleasure.

origin of life The process by which living organisms developed from inanimate matter, which is generally thought to have occurred on earth between 3500 and 4000 million years ago. It is supposed that the primordial atmosphere was like a chemical soup containing all the basic constituents of organic matter: ammonia, methane, hydrogen, and water vapour. These underwent a process of chemical evolution using energy from the sun and electric storms to combine into ever more complex molecules, such as amino acids, proteins, and vitamins. Eventually self-replicating nucleic acids, the basis of all life, could have developed. The very first organisms may have consisted of such molecules bounded by a simple membrane.

ornithine (Orn) An *amino acid, $H_2N(CH_2)_3CH(NH_2)COOH$, that is not a constituent of proteins but is important in living organisms as an intermediate in the reactions of the *urea cycle and in arginine synthesis.

ornithine cycle *See* **urea cycle**.

orthogenesis An early theory of the nature of evolutionary change, which proposed that organisms evolve along particular paths predetermined by some factor in their genetic make-up. More recent understanding of selection pressure and other external forces that can be shown experimentally to affect the survival of organisms has proved the improbability of the theory.

Orthoptera A large order of *exopterygote insects containing the grasshoppers, locusts, crickets, and – in some classification systems – the cockroaches (*see* **Dictyoptera**). They are characterized by enlarged hind legs modified for jumping and biting mouthparts and produce sounds by *stridulation. The crickets and long-horned grasshoppers (e.g. *Gryllus*, *Tettigonia*) have long threadlike antennae and stridulate by rubbing together modified veins on their forewings. The hearing organs are on the front legs. The short-horned grasshoppers and locusts (e.g. *Chorthippus*, *Locusta*) have short antennae and stridulate by rubbing pegs on the hind leg against a hardened vein on the forewing. The hearing organs are on the abdomen.

orthotropism The tendency for a *tropism (growth response of a plant) to be orientated directly in line with the stimulus concerned. An example is the vertical growth of main stems and roots in response to gravity (*orthogeotropism*). *Compare* **plagiotropism**.

osculum (*pl.* **oscula**) **1.** The mouthlike aperture in the body wall of a sponge (*see* **Porifera**) through which water leaves the body cavity. **2.** Any of the suckers on the head (scolex) of a tapeworm, by which it attaches itself to the gut wall of its host.

osmium tetroxide (osmium(IV) oxide) A yellow solid, OsO_4, made by heating osmium in air. It is used as a fixative in electron microscopy.

osmolyte Any compound that protects cells from desiccation by maintaining a high intracellular osmolality (osmotic *concentration). Osmolytes are active in the process of *osmoregulation and are of particular use in the kidneys, where cells are exposed to highly concentrated fluids. Compounds that are known osmolytes include *polyols, amines (e.g. *trimethylamine), certain amino acids, and urea.

osmoreceptor A receptor situated in the hypothalamus of the brain that responds to an increase in the concentration of the extracellular fluid. This results in the release of *antidiuretic hormone (ADH) and the subsequent conservation of water, thereby maintaining the *homeostasis of the body fluids.

osmoregulation The control of the water content and the concentration of salts in the body of an animal or protozoan. In freshwater species osmoregulation must counteract the tendency for water to pass into the animal by *osmosis. Various methods have been developed to eliminate the excess, such as *contractile vacuoles in protozoans and *kidneys with well-developed glomeruli in freshwater fish. Marine vertebrates have the opposite problem: they prevent excessive water loss and enhance the excretion of salts by having kidneys with few glomeruli and short tubules (*see also* **chloride secretory cell**). In terrestrial vertebrates the dangers of desiccation are reduced by the presence of long convoluted *renal tubules, which increase the reabsorption of water and salts.

osmosis The net movement of water molecules from a region where their concentration is high to a region where their concentration is low through a *partially permeable membrane. The distribution of water in living organisms is dependent to a large extent on osmosis, water entering the cells through their partially permeable plasma membranes. The pressure required to stop the flow of pure water into a solution across a partially permeable membrane is a characteristic of the solution, and is called the *osmotic pressure*. Thus water will move from a region of low osmotic pressure to a region of high osmotic pressure. In terms of *water potential, water moves from an area of high (less negative) water potential to an area of low (more negative) water potential. Both water potential and osmotic pressure can be used to explain osmosis but it is now recommended that only water potential be used in plant studies (*see also* **plasmolysis; turgor**). Animals have evolved various means to counteract the effects of osmosis (*see* **osmoregulation**); in animals solutions are still described in terms of osmotic pressure (*see* **hypertonic solution; hypotonic solution; isotonic**).

osmotic pressure *See* osmosis.

osmotroph Any heterotrophic organism that obtains its nutrients by absorbing organic matter in solution from its surroundings. *Compare* **phagotroph**.

ossicle Any small bony or chitinous structure found in various skeletal parts of animals. The term ossicle usually refers to any of the chain of bones in the mammalian ear (*see* **ear ossicles**).

ossification The process of *bone formation. It is brought about by the action of special cells called *osteoblasts, which deposit layers of bone in connective tissue. In *intramembranous ossification* bones are formed directly in connective tissue (*see* **membrane bone**); in *endochondral ossification* the bones are formed by the replacement of cartilage (*see* **cartilage bone**).

Osteichthyes The class of vertebrates comprising the bony fishes – marine and freshwater fish with a bony skeleton. All have gills covered with a bony operculum, and a layer of thin overlapping bony *scales covers the entire body surface. Bony fish have a *swim bladder, which acts as a hydrostatic organ enabling the animal to remain suspended in the water at any depth. In some fish this bladder acts as a lung. *See also* **Dipnoi; Teleostei**. *Compare* **Chondrichthyes**.

osteoblast Any of the cells, found in *bone, that secrete collagen and other substances that form the matrix of bone (*see* **osteoid**). Osteoblasts are derived from *osteoprogenitor cells* in the bone marrow; they eventually become *osteocytes. *See also* **ossification**.

osteoclast Any of the cells, found on the surface of bone, that are involved in the breakdown of the bone matrix to enable the further development and restructuring of bone during growth and repair. *See also* **parathyroid hormone; periosteum**.

osteocyte Any of the cells, found in bone, that perform the cellular activities, such as respiration and exchange of materials with the blood, that are required for maintenance of the bone tissue. They are derived from *osteoblasts.

osteoid A soft material, consisting mainly of *collagen, that is secreted by *osteoblasts and constitutes the uncalcified matrix of *bone. Osteoid is converted into hard bone matrix when it combines with calcium phosphate (hydroxyapatite) deposited from the blood (see **osteonectin**).

osteonectin A protein in bone that binds to collagen (see **osteoid**) and is involved in the formation of growth sites for calcium phosphate crystals, which are required for forming the hard bone matrix.

ostiole A pore in the fruiting body of certain fungi and algae, through which either spores or gametes are released. An ostiole occurs, for example, in the perithecium of ascomycete fungi (see **ascocarp**) and in the *conceptacle of brown algae.

ostium (*pl.* ostia) **1.** Any one of the pores in the body wall of a sponge (see **Porifera**) through which water enters the body cavity. **2.** Any one of the apertures in the arthropod heart through which blood enters from the haemocoel.

otolith A gelatinous mass containing a high concentration of particles of calcium carbonate, which forms part of the *macula of the inner ear.

outbreeding Mating between unrelated or distantly related individuals of a species. Outbreeding populations usually show more variation than *inbreeding ones and have a greater potential for adapting to environmental changes. Outbreeding increases the number of *heterozygous individuals, so disadvantageous recessive characteristics tend to be masked by dominant alleles.

outer ear (external ear) The part of the ear external to the *tympanum (eardrum). It is present in mammals, birds, and some reptiles and consists of a tube (the *external auditory meatus*) that directs sound waves onto the tympanum. In mammals it may include an external *pinna, which extends beyond the skull.

oval window (fenestra ovalis) A membrane-covered opening between the middle ear and the inner ear (see **ear**), situated above the *round window. Vibrations of the tympanum are transferred across the middle ear by the *ear ossicles and transmitted to the inner ear by the oval window, which is connected to the third ear ossicle (stapes). See also **vestibular canal**.

ovarian follicle See **Graafian follicle**.

ovary 1. The reproductive organ in female animals in which eggs (ova) are produced. In most vertebrates there are two ovaries (in some fish the ovaries fuse together to form a single structure and in birds the left ovary only is functional). As well as eggs, they produce steroid hormones (see oestrogen; progesterone). In mammals each ovary is situated close to the opening of a *fallopian tube; it contains numerous follicles in which the eggs develop and from which they are released in a regular cycle. *See also* **Graafian follicle; menstrual cycle; oogenesis; ovulation; reproductive system**. **2.** The hollow base of the *carpel of a flower, containing one or more *ovules. After fertilization, the ovary wall develops into the fruit enclosing the seeds. In some species, the carpels are fused together to form a complex ovary.

overpopulation The situation that arises when rapid growth of a population, usually a human population, results in numbers that cannot be supported by the available resources, such as space and food. This occurs when the birth rate exceeds the death rate, or when immigration exceeds emigration, or when a combination of these factors exists. *See* **population growth**.

oviduct The tube that conveys an animal egg cell from the ovary to other parts of the reproductive system or to the outside. Eggs are passed along the oviduct by the action of muscles and cilia. *See* **fallopian tube**.

oviparity Reproduction in which fertilized eggs are laid or spawned by the mother and hatch outside her body. It occurs in most animals except marsupial and placental mammals. *Compare* **ovoviviparity; viviparity**.

ovipositor An organ at the hind end of the abdomen of female insects through which eggs are laid. It consists of a pair of modified appendages and is often long and piercing, so that eggs can be laid in otherwise inaccessible places. The sting of bees and wasps is a modified ovipositor.

ovoviviparity Reproduction in which fertilized eggs develop and hatch in the oviduct of the mother. It occurs in many invertebrates and in some fish and reptiles (e.g. the viper). *Compare* **oviparity; viviparity**.

ovulation The release of an egg cell from the ovary, which in mammals is stimulated by *luteinizing hormone. The developing egg cell within its follicle migrates to the ovary surface; when mature, it is released from the follicle (which breaks open) into the body cavity, from where it passes into the oviduct. *See also* **menstrual cycle**.

ovule The part of the female reproductive organs of seed plants that consists of the *nucellus, *embryo sac, and *integuments. The ovules of gymnosperms are situated on ovuliferous scales of the female cones while those of angiosperms are enclosed in the carpel. After fertilization, the ovule becomes the seed.

ovuliferous scale One of a group of large woody specialized leaves that form the female *cone of conifers and related trees. It bears the ovules, which develop into seeds.

ovum (egg cell) (*pl.* **ova**) **1.** *(in zoology)* The mature reproductive cell (*see* **gamete**) of female animals, which is produced by the ovary (*see* **oogenesis**).

It is spherical, has a nucleus, is covered with an *egg membrane, and is not mobile. **2.** *(in botany)* The *oosphere of plants.

oxalic acid (ethanedioic acid) A crystalline solid, $(COOH)_2$, that is slightly soluble in water. Oxalic acid is strongly acidic and very poisonous. It occurs in certain plants, e.g. sorrel and the leaf blades of rhubarb.

oxaloacetic acid A compound, $HO_2CCH_2COCO_2H$, that plays an integral role in the *Krebs cycle. The anion, oxaloacetate, reacts with the acetyl group from acetyl coenzyme A to form citrate.

oxidase Any enzyme that catalyses *oxidation–reduction reactions that involve the transfer of electrons to molecular oxygen.

oxidation *See* oxidation–reduction.

oxidation–reduction (redox) Originally, *oxidation* was simply regarded as a chemical reaction with oxygen. The reverse process – loss of oxygen – was called *reduction*. Reaction with hydrogen also came to be regarded as reduction. Later, a more general idea of oxidation and reduction was developed in which oxidation was loss of electrons and reduction was gain of electrons. This wider definition covered the original one and also applies to reactions that do not involve oxygen. However, it applies only to reactions in which electron transfer occurs – i.e. to reactions involving ions. It can be extended to reactions between covalent compounds by using the concept of *oxidation number* (or *state*). This is a measure of the electron control that an atom has in a compound compared to the atom in the pure element. An oxidation number consists of two parts:
(1) its sign, which indicates whether the control has increased (negative) or decreased (positive);
(2) its value, which gives the number of electrons over which control has changed.

The change of electron control may be complete (in ionic compounds) or partial (in covalent compounds). Oxidation is a reaction involving an increase in oxidation number and reduction involves a decrease. Thus in

$$2H_2 + O_2 \rightarrow 2H_2O$$

the hydrogen in water is +1 and the oxygen −2. The hydrogen is oxidized and the oxygen is reduced. Compounds that tend to undergo reduction readily are *oxidizing agents*; those that undergo oxidation are *reducing agents*.

oxidative deamination A reaction involved in the catabolism of amino acids that assists their excretion from the body. An example of an oxidative deamination is the conversion of glutamate to α-ketoglutarate, a reaction catalysed by the enzyme glutamate dehydrogenase. *See* **deamination**.

oxidative decarboxylation The reaction in the *Krebs cycle in which oxygen, derived from two water molecules, is used to oxidize two carbon atoms to two molecules of carbon dioxide. The two carbon atoms result from the *decarboxylation reactions that occur during the Krebs cycle as the six-carbon compound citrate is converted to the four-carbon compound oxaloacetate.

oxidative phosphorylation A reaction occurring during the final stages of *aerobic respiration, in which ATP is formed from ADP and phosphate coupled to electron transport in the *electron transport chain. The reaction occurs in the mitochondria (*see* **chemiosmotic theory**). *See also* **phosphorylation**.

oxidoreductase Any of a class of enzymes that catalyse *oxidation–reduction reactions, i.e. they are involved in the transfer of hydrogen or electrons between molecules. They include the *oxidases and *dehydrogenases.

oxygen Symbol O. A colourless odourless gaseous element. It is the most abundant element in the earth's crust (49.2% by weight) and is present in the atmosphere (28% by volume). Atmospheric oxygen is of vital importance for all organisms that carry out *aerobic respiration.

oxygen cycle The cycling of oxygen between the biotic and abiotic components of the environment (*see* **biogeochemical cycle**). The oxygen cycle is closely linked to the *carbon cycle and the water cycle (*see* **hydrological cycle**). In the process of respiration oxygen is taken in by living organisms and released into the atmosphere, combined with carbon, in the form of carbon dioxide. Carbon dioxide enters the carbon cycle or is taken up by plants for *photosynthesis. During photosynthesis oxygen is evolved by the chemical splitting of water and returned to the atmosphere. In the upper atmosphere, ozone is formed from oxygen and dissociates to release oxygen (*see also* **ozone layer**).

oxygen debt The physiological state that exists in a normally aerobic animal when insufficient oxygen is available for metabolic requirements (e.g. during a period of strenuous physical activity). To meet the body's increased demand for energy, pyruvate is converted anaerobically (i.e. in the absence of oxygen) to lactic acid, which requires oxygen for its breakdown and accumulates in the tissues. When oxygen is available again lactic acid is oxidized in the liver, thus repaying the debt.

oxygen dissociation curve The S-shaped curve produced when the percentage saturation of haemoglobin with oxygen (i.e. the percentage of binding sites of haemoglobin that are occupied by oxygen molecules) is plotted against the partial pressure of oxygen (pO_2), which is a measure of the oxygen concentration in the surrounding medium. The steep rise of the curve indicates the high affinity of haemoglobin for oxygen: a small increase in pO_2 results in a relatively sharp increase in the percentage saturation of haemoglobin with oxygen. Therefore in the lungs, where the pO_2 is high, the blood is rapidly saturated with oxygen. Conversely, a small drop in pO_2 results in a large drop in percentage saturation of haemoglobin. Thus in tissues that utilize oxygen at a high rate, where the

Oxygen dissociation curve

pO_2 is low, oxygen readily dissociates from haemoglobin and is released for use by the tissues. *See also* **Bohr effect**.

oxygen sag curve The curve obtained when the concentration of dissolved oxygen in a river into which sewage or some other pollutant has been discharged is plotted against the distance downstream from the sewage outlet (see graph). Samples of water are taken at areas upstream and downstream from the sewage outlet. The presence of sewage reduces the oxygen content of the water and increases the *biochemical oxygen demand. This is due to the action of saprotrophic organisms that decompose the organic matter in the sewage and in the process use up the available oxygen.

Oxygen sag curve

oxyhaemoglobin *See* **haemoglobin**.

oxyntic cell (parietal cell) Any of the cells in the wall of the stomach that produce *hydrochloric acid, which forms part of the *gastric juice. The oxyntic cells also produce intrinsic factor, which is involved in the absorption of vitamin B_{12} in the small intestine (*see* **vitamin B complex**).

oxytocin A hormone, produced by birds and mammals, that in mammals causes both contraction of smooth muscle in the uterus during birth and expulsion of milk from the mammary glands during suckling. Like *antidiuretic hormone, oxytocin is produced in the neurosecretory cells of the hypothalamus (*see* **neurosecretion**) but is stored and secreted by the posterior pituitary gland.

ozonation The formation of ozone (O_3) in the earth's atmosphere. In the stratosphere, about 20–50 km above the surface of the earth, oxygen molecules (O_2) dissociate into their constituent atoms under the influence of *ultraviolet light of short wavelength (below about 240 nm). These atoms combine with oxygen molecules to form ozone (*see* **ozone layer**). Ozone is also formed in the lower atmosphere from nitrogen oxides and other pollutants by photochemical reactions (*see* **photochemical smog**).

ozone hole *See* ozone layer.

ozone layer (ozonosphere) A layer of the earth's atmosphere in which most of the atmosphere's ozone is concentrated. It occurs 15–50 km above the earth's surface and is virtually synonymous with the *stratosphere. In this layer most of the sun's ultraviolet radiation is absorbed by the ozone molecules, causing a rise in the temperature of the stratosphere and preventing vertical mixing so that the stratosphere forms a stable layer. By absorbing most of the solar ultraviolet radiation the ozone layer protects living organisms on earth. The fact that the ozone layer is thinnest at the equator is believed to account for the high equatorial incidence of skin cancer as a result of exposure to unabsorbed solar ultraviolet radiation. In the 1980s it was found that depletion of the ozone layer was occurring over both the poles, creating *ozone holes*. This is thought to have been caused by a series of complex photochemical reactions involving *nitrogen oxides produced from aircraft and, more seriously, *chlorofluorocarbons (CFCs) and *halons. CFCs rise to the stratosphere, where they react with ultraviolet light to release chlorine atoms; these atoms, which are highly reactive, catalyse the destruction of ozone. *See also* **air pollution**.

P

P (parental generation) The individuals that are selected to begin a breeding experiment, crosses between which yield the *F₁ generation. Only pure-breeding (homozygous) individuals are selected for the P generation.

pacemaker 1. (*or* **sinoatrial node**) A small mass of specialized muscle cells in the mammalian heart, found in the wall of the right atrium near the opening for the vena cava. The cells initiate and maintain the heart beat: by their rhythmic and spontaneous contractions they stimulate contraction of the atria (*see also* **atrioventricular node**). The cells themselves are controlled by the autonomic nervous system, which determines the heart rate. Similar pacemakers occur in the hearts of other vertebrates. **2.** An electronic or nuclear battery-charged device that can be implanted surgically into the chest to produce and maintain the heart beat. These devices are used when the heart's own pacemaker is defective or diseased.

pachytene The period in the first prophase of *meiosis when paired *homologous chromosomes are fully contracted and twisted around each other.

Pacinian corpuscle (lamellated corpuscle) Any of the sensory receptors in the subcutaneous layer of the skin that are sensitive to pressure. Each consists of a nerve ending surrounded by an oval capsule made of connective tissue. The receptors are named after Italian anatomist Filipo Pacini (1812–83).

paedogenesis Reproduction by an animal that is still in the larval or pre-adult form. Paedogenesis is a form of *neoteny and is particularly marked in the axolotl, a larval form of the salamander, which retains its larval features owing to a thyroid deficiency but can breed, producing individuals like itself. If the thyroid hormone thyroxine is given, metamorphosis occurs.

paedomorphosis The evolutionary process in which larval or juvenile features of an ancestral organism are displaced to the adult forms of its descendants. It can arise by *neoteny or *progenesis. Paedomorphosis is thought to have occurred in the evolution of higher chordates from free-swimming larval tunicates, in which metamorphosis was eventually lost and sexual development accelerated until larval forms were capable of breeding.

PAGE (polyacrylamide gel electrophoresis) A type of *electrophoresis used to determine the size and composition of proteins. Proteins are placed on a matrix of polyacrylamide gel and an electric field is applied. The protein molecules migrate towards the positive pole, the smaller molecules moving at a faster rate through the pores of the gel. The proteins are then detected by applying a stain, such as coomassie blue.

pairing (synapsis) The close association between *homologous chromosomes that develops during the first prophase of *meiosis. The two chromosomes move together and an exact pairing of corresponding points along their lengths occurs as they lie side by side. The resulting structure is called a *bivalent*.

palaeobotany The branch of *palaeontology concerned with the study of plants through geological time, as revealed by their *fossil remains (*see also* **palynology**). It overlaps with other aspects of plant study, including anatomy, ecology, evolution, and taxonomy.

Palaeocene The earliest geological epoch of the *Tertiary period. It began about 65 million years ago, following the Cretaceous period, and extended for about 11.5 million years to the beginning of the *Eocene (the Palaeocene is sometimes included in the Eocene). It was named by the palaeobotanist W. P. Schimper in 1874. A major floral and faunal discontinuity occurred between the end of the Cretaceous and the beginning of the Palaeocene: following the extinction of many reptiles the mammals became abundant on land. By the end of the epoch primates and rodents had evolved.

palaeoclimatology The study of climates of earlier geological periods. This is based largely on the study of sediments that were laid down during these periods and of fossils. The changes in the positions of the continents as a result of *continental drift and *plate tectonics complicate the study.

palaeoecology The study of the relationships of *fossil organisms to each other and to their environments. It involves the study both of the fossils and of the surrounding rocks in which they are found. Trace fossils may provide information on the behaviour of the organism.

Palaeolithic The Old Stone Age, lasting in Europe from about 2.5 million to 9000 years ago, during which man used primitive stone tools made by chipping stones and flints.

palaeontology The study of extinct organisms, including their structure, environment, evolution, and distribution, as revealed by their *fossil remains. Palaeontological work also makes important contributions to geology in revealing stratigraphic relationships between rock strata and determining the physical appearance and climate of the earth during past geological ages. *See also* **palaeobotany; palaeoecology; palaeozoology**.

Palaeozoic The first era of *Phanerozoic time. It follows the *Precambrian and is subdivided into the Lower Palaeozoic, comprising the *Cambrian, *Ordovician, and *Silurian periods, and the Upper Palaeozoic, comprising the *Devonian, *Carboniferous, and *Permian periods, It extended from about 570 million years ago to about 225 million years ago, when it was succeeded by the *Mesozoic era.

palaeozoology The branch of *palaeontology concerned with the study of animals throughout geological time, as revealed by their *fossil remains.

palate The roof of the mouth cavity of vertebrates, which separates the *buccal and nasal cavities. In mammals it is divided into two zones, the bony *hard palate* and the *soft palate*, and completely separates the buccal cavity from the air passage to enable simultaneous eating and breathing.

palea *See* **lemma.**

palindromic Describing a section of double-stranded DNA in which the sequence of bases on one strand is inverted and repeated on the other. Thus the following sequence is palindromic:

—ACTTGCAAGT—

—TGAACGTTCA—

Palindromic sequences are common in DNA and are the sites at which the DNA is cleaved by *restriction enzymes.

palisade (**palisade mesophyll**) *See* **mesophyll.**

pallium *See* **cerebral cortex.**

palmitic acid (**hexadecanoic acid**) A saturated fatty acid, $CH_3(CH_2)_{14}COOH$. Glycerides of palmitic acid occur widely in plant and animal oils and fats.

palp An elongated sensory organ, usually near the mouth, in many invertebrates. Examples are the tactile head appendages of polychaete worms, the ciliated flap of tissue that produces feeding currents in bivalve molluscs, the distal part of the *mandibles of crustaceans, and the olfactory parts of the first and second *maxillae of some insects.

palynology (**micropalaeontology**) The study of fossil pollen and spores (*pollen analysis*) and various other *microfossils, such as coccoliths and dinoflagellates. Palynology is used in stratigraphy, palaeoclimatology, and archaeology. Pollen and spores are very resistant to decay and therefore their fossils are found in sedimentary rocks. They may be extracted by various methods, including boiling with potassium hydroxide solution, washing with strong oxidizing mixtures, and centrifuging repeatedly. Spores and pollen are classified according to shape, form of aperture, and both internal and external details of the exine (outer coat). They indicate the nature of the dominant flora, and therefore the climate and conditions of the period in which they lived.

pancreas A gland in vertebrates lying between the duodenum and the spleen. Under the influence of the hormone *secretin it secretes *pancreatic juice* containing digestive enzymes (mainly *trypsin, *chymotrypsin, *amylase, and *lipase) into the duodenum via the pancreatic duct (*see* **acinus**). It also contains groups of cells – the *islets of Langerhans – that function as an *endocrine gland, producing the hormones *insulin and *glucagon, which regulate blood sugar levels.

pancreatin A mixture of digestive enzymes that have been extracted from the *pancreas.

pancreozymin *See* cholecystokinin.

Paneth cells *See* crypts of Lieberkühn.

panicle A type of flowering shoot common in the grass family. The primary axis bears groups of *racemes and is itself racemose, as the youngest groups of flowers are at the top (e.g. oat). The term may be used loosely for any form of branched *racemose inflorescence; for example, the horse chestnut is a raceme of cymes. Both these arrangements are seen in the family Polygonaceae (docks and sorrels).

pantothenic acid A vitamin of the *vitamin B complex. It is a constituent of *coenzyme A, which performs a crucial role in the oxidation of fats, carbohydrates, and certain amino acids. Deficiency rarely occurs because the vitamin occurs in many foods, especially cereal grains, peas, egg yolk, liver, and yeast.

papain A protein-digesting enzyme (*see* **protease**) occurring in the fruit of the West Indian papaya tree (*Carica papaya*). It is used as a digestant and in the manufacture of meat tenderizers.

paper chromatography A technique for analysing mixtures by *chromatography, in which the stationary phase is absorbent paper. A spot of the mixture to be investigated is placed near one edge of the paper and the sheet is suspended vertically in a solvent, which rises through the paper by capillary action carrying the components with it. The components move at different rates, partly because they absorb to different extents on the cellulose and partly because of partition between the solvent and the moisture in the paper. The paper is removed and dried, and the different components form a line of spots along the paper. Colourless substances are detected by using ultraviolet radiation or by spraying with a substance that reacts to give a coloured spot (e.g. ninhydrin gives a blue coloration with amino acids). The components can be identified by the distance they move in a given time.

papilla Any cone-shaped protuberance projecting from the surface of an organ or organism. Papillae occur, for example, on the tongue, in the kidneys, and, in plants, on the surface of many petals.

papovavirus One of a group of *DNA-containing viruses that produce tumours in their hosts. *Papillomaviruses* produce nonmalignant tumours (such as warts) in all vertebrates; *polyomaviruses* produce malignant tumours in only certain classes of vertebrates (not including man).

pappus A group of modified *sepals, often in the form of a ring of silky hairs. For example, when the fruit of the dandelion matures a pappus of hairs persists at the top of a thin stalk forming a parachute-like structure, which serves to disperse the fruit.

parallel evolution The development of related organisms along similar evolutionary paths due to strong selection pressures acting on all of them in the same way. It is debatable if the phenomenon really exists: many argue that all evolution is ultimately *convergent or divergent (*see* **adaptive radiation**).

paramorph A variant form within a species for which a more specific description does not exist because its taxonomic status cannot be determined.

parapatric speciation A type of *speciation in which there is free exchange of genes between two populations of organisms living in directly adjacent but environmentally different habitats. Although individuals from the two populations can interbreed, the offspring will not be very successful in either habitat, therefore natural selection tends to promote any mechanisms that decrease the number of crosses between such populations.

paraphyletic *See* **monophyletic**.

paraphysis (*pl.* **paraphyses**) Any of the sterile filaments of cells that are found around the sex organs of mosses, brown algae (Phaeophyta) and certain fungi (the Ascomycota and Basidiomycota).

parapodium (*pl.* **parapodia**) *See* **chaeta; Polychaeta**.

Paraquat Trade name for an organic herbicide used to control broadleaved weeds and grasses. It is poisonous to humans, having toxic effects on the liver, lungs, and kidneys if ingested. Paraquat is not easily broken down and can persist in the environment adsorbed to soil particles. *See also* **pesticide**.

parasexual cycle A type of reproduction in fungi that results in genetic recombination without the formation and fusion of gametes. It occurs in heterokaryotic mycelia, in which two haploid nuclei fuse and then divide by mitosis, during which a form of *crossing over occurs. During subsequent divisions chromosomes are progressively lost from the resultant diploid nucleus so that the haploid condition is restored. The asexual spores produced during the cycle will be genetically different from the original parent mycelium as a result of the crossing over that has occurred.

parasitism An association in which one organism (the *parasite*) lives on (*ectoparasitism*) or in (*endoparasitism*) the body of another (the *host), from which it obtains its nutrients. Some parasites inflict comparatively little damage on their host, but many cause characteristic diseases (these are, however, never immediately fatal, as killing the host would destroy the parasite's source of food). Parasites are usually highly specialized for their way of life, which may involve one host or several (if the *life cycle requires it). They typically produce vast numbers of eggs, very few of which survive to find their way to another suitable host. *Obligate parasites* can only survive and reproduce as parasites; *facultative parasites* can also live as *saprotrophs. The parasites of man include fleas and lice (which are ectoparasites), various bacteria, protozoans, and fungi (endoparasites causing characteristic diseases), and tapeworms (e.g. *Taenia solium*, which lives in the gut).

parasympathetic nervous system Part of the *autonomic nervous system. Its nerve endings release acetylcholine as a *neurotransmitter and its actions tend to antagonize those of the *sympathetic nervous system. For example, the parasympathetic nervous system increases salivary gland secretion, decreases heart rate, promotes digestion (by increasing *peristalsis), and dilates blood vessels, while the sympathetic nervous system has opposite effects.

parathyroid glands Two pairs of *endocrine glands situated behind, or embedded within, the thyroid gland in higher vertebrates. They produce *parathyroid hormone, which controls the amount of calcium in the blood. *See also* **C cell**.

parathyroid hormone A hormone secreted by the *parathyroid gland in response to low levels of calcium in the blood. It acts to maintain normal blood levels of calcium by (1) increasing the number of *osteoclasts, which break down the bone matrix and release calcium into the blood; (2) increasing the reabsorption of calcium and magnesium ions in the kidney tubules, so that their concentration is maintained in the blood; (3) converting *vitamin D to its active form, which increases calcium absorption in the intestine. Parathyroid hormone acts in opposition to *calcitonin.

Parazoa *See* **Metazoa**.

parenchyma 1. A plant tissue consisting of roughly spherical relatively undifferentiated cells, frequently with air spaces between them. The cortex and pith are composed of parenchyma cells (*see* **ground tissues**). **2.** Loose *connective tissue formed of large cells. Its function is to pack the spaces between organs in some simple acoelomate animals, such as flatworms (Platyhelminthes).

parent 1. Either male or female partner that together produce offspring in the process of sexual reproduction. *See also* **P. 2.** Denoting an organism or cell that gives rise to new organisms or cells, as by asexual reproduction or cell division.

parental care Any behaviour pattern in which a parent invests time or energy in feeding and protecting its offspring. Parental care is a form of *altruism since this type of behaviour involves increasing the *fitness of the offspring at the expense of the parents. The degree of parental care differs widely. For example, most species of fish show little or no parental care while humans and many other mammals care for their offspring until they reach adolescence.

parietal Of or relating to the wall of a body cavity or other structure.

parietal cell *See* oxyntic cell.

parthenocarpy The formation of fruit without prior fertilization of the flower by pollen. The resulting fruits are seedless and therefore do not contribute to the reproduction of the plant; examples are bananas and pineapples. Plant *growth substances may have a role in this phenomenon, which can be induced by auxins in the commercial production of tomatoes and other fruits.

parthenogenesis The development of an organism from an unfertilized egg. This occurs sporadically in many plants (e.g. dandelions and hawkweeds) and in a few animals, but in some species it is the main and sometimes only method of reproduction. For example, in some species of aphid, males are absent or very rare. The eggs formed by the females contain the full (diploid) number of chromosomes and are genetically identical. Variation is consequently very limited in species that reproduce parthenogenetically.

partially permeable membrane A membrane that is permeable to the small molecules of water and certain solutes but does not allow the passage of large solute molecules. This term is preferred to *semipermeable membrane* when describing membranes in living organisms. *See* osmosis.

particulate inheritance The transmission from parent to offspring of separate units that determine characteristics. Gregor Mendel observed that *recessive characteristics, absent in the offspring of a cross in which only one parent possessed them, reappeared repeatedly in the progeny of subsequent crosses. This led him to formulate his theory of inherited 'factors' (now called *alleles) that retain their identity through succeeding generations (*see* Mendel's laws). *Compare* blending inheritance.

parturition The act of giving birth to young at the end of the *gestation period. Hormones probably cause the process to start but the mechanism is not fully understood.

pascal The *SI unit of pressure equal to one newton per square metre.

passive immunity *See* immunity.

passive transport *See* diffusion.

pasteurization The treatment of milk to destroy disease-causing bacteria, such as those of tuberculosis, typhoid, and brucellosis. Milk is heated to 65°C for 30 minutes or to 72°C for 15 minutes followed by rapid cooling to below 10°C. The method was devised by the French microbiologist Louis Pasteur (1822–95). *See also* food preservation.

patch clamp technique A method for studying the activation of voltage-dependent transmembrane channels in neurones. A small piece of membrane containing only a few channels is isolated and microelectrodes are sealed to both sides of the membrane.

patella (kneecap) A small rounded movable bone that is situated in a tendon in front of the knee joint in most mammals (including humans). The function of the patella is to protect the knee.

pathogen Any disease-causing microorganism. Pathogens include viruses and many bacteria, fungi, and protozoans. *See* **infection**.

patristic Denoting similarity between organisms resulting from common ancestry. *Compare* **homoplastic**.

pattern recognition A branch of statistics used to classify data based upon several variables. This type of statistical analysis is particularly useful in interpreting biological data that consists of multiple observations on large numbers of individuals.

PCR *See* **polymerase chain reaction**.

PDGF *See* **growth factor**.

peat A mass of dark-brown or black fibrous plant debris produced by the partial disintegration of vegetation in wet places. It may accumulate in depressions. When subjected to burial and hence pressure and heat it may be converted to *coal. Peat is used to improve soil and as a fuel, especially in Ireland and Sweden.

peck order *See* **dominant**.

pecten Any of various comblike structures in animals. The pecten in the eyes of birds consists essentially of a network of blood vessels attached to the optic nerve and projecting into the vitreous humour. Its function is uncertain, but it may be involved in supplying the retina with nutrients and oxygen. A simple form of this structure is found in the eyes of reptiles.

pectic substances A group of polysaccharides made up primarily of sugar acids. They are important constituents of plant cell walls and the *middle lamella between adjacent cell walls. Normally they are present in an insoluble form, but in ripening fruits and in tissues affected by certain diseases they change into a soluble form, which is evidenced by softening of the tissues.

pectin A type of *pectic substance. It is used in making jam as it forms a gel with sucrose.

pectoral fins *See* **fins**.

pectoral girdle (shoulder girdle) The bony or cartilaginous structure in vertebrates to which the anterior limbs (pectoral fins, forelegs, or arms) are attached. In mammals it consists of two dorsal *scapulae (shoulder blades) attached to the backbone and two ventral *clavicles (collar bones) attached to the sternum (breastbone).

pedicel The stalk attaching a flower to the main floral axis (*see* **peduncle**).

Some flowers, described as *sessile*, do not have a pedicel and arise directly from the peduncle.

pedipalp Either member of the second pair of appendages in arachnids, immediately posterior to the *chelicerae. In species with large chelicerae the pedipalps are typically adapted for walking or sensory functions; otherwise they are variously adapted for defence, killing prey, or digging.

pedology The science of the study of soils, including their origin and characteristics and their utilization.

peduncle 1. The main stalk of a plant that bears the flowers, which may be solitary or grouped in an *inflorescence. *Compare* **pedicel. 2.** *See* **Brachiopoda**.

Pekilo process A process developed in Finland for the production of *Pekilo protein*, a *single-cell protein used as an animal feed. The fungus *Paecilomyces varioti* uses a substrate of sulphite waste liquor or wood hydrolysates for fermentation.

pelagic Describing organisms that swim or drift in a sea or a lake, as distinct from those that live on the bottom (*see* **benthos**). Pelagic organisms are divided into *plankton and *nekton.

Pelecypoda (Bivalvia; Lamellibranchia) A class of aquatic molluscs (the bivalves) that include the oysters, mussels, and clams. They are characterized by a laterally flattened body and a shell consisting of two hinged shells (i.e. a bivalved shell). The enlarged gills are covered with cilia and have the additional function of filtering microscopic food particles from the water flowing over them. Bivalves live on the sea bed or lake bottom and are sedentary, so the head and foot are reduced.

pellagra A disease resulting from a deficiency of *nicotinic acid, which is characterized by dermatitis and mental disorder.

pellicle The thin outer covering, composed of protein, that protects and maintains the shape of certain unicellular organisms, e.g. *Euglena*. It is transparent and in ciliated organisms, e.g. *Paramecium*, contains small pores through which the cilia emerge.

pelvic fins *See* **fins**.

pelvic girdle (pelvis; hip girdle) The bony or cartilaginous structure in vertebrates to which the posterior limbs (pelvic fins or legs) are attached. The pelvic girdle articulates dorsally with the backbone; it is made up of two halves, each produced by the fusion of the *ilium, *ischium, and *pubis.

pelvis 1. *See* **pelvic girdle. 2.** The lower part of the abdomen in the region of the pelvic girdle. **3.** A conical chamber in the *kidney into which urine drains from the kidney tubules before passing to the *ureter.

penicillin An *antibiotic derived from the mould *Penicillium notatum*;

specifically it is known as *penicillin G* and belongs to a class of similar substances called penicillins. They are all active against a wide variety of bacteria, producing their effects by disrupting synthesis of the bacterial cell wall, and are used to treat a variety of infections caused by these bacteria.

penis The male reproductive organ of mammals (and also of some birds and reptiles) used to introduce sperm into the female reproductive tract to ensure internal fertilization. It contains a duct (the *urethra) through which the sperms pass. The penis becomes erect during precopulatory activity, either by filling with blood or haemolymph or by the action of muscles, and can be inserted into the vagina (or cloaca). In mammals the urine also leaves the body through the penis.

A basic pentadactyl forelimb, as exemplified by the human arm

The modified pentadactyl forelimb of various vertebrates

pentadactyl limb A limb with five digits, characteristic of tetrapod vertebrates (amphibians, reptiles, birds, and mammals). It evolved from the paired fins of primitive fish as an adaptation to locomotion on land and is not found in modern fish. The limb has three parts (see illustration): the upper arm or thigh containing one long bone, the forearm or shank containing two long bones, and the hand or foot, which contains a number of small bones. This basic design is modified in many species, according to the function of the limb, particularly by the loss of fusion of the terminal bones.

pentose A sugar that has five carbon atoms per molecule. *See* **monosaccharide.**

pentose phosphate pathway (pentose shunt) A series of biochemical reactions that results in the conversion of glucose 6-phosphate to ribose 5-phosphate and generates NADPH, which provides reducing power for other metabolic reactions, such as synthesis of fatty acids. Ribose 5-phosphate and its derivatives are components of such molecules as ATP, coenzyme A, NAD, FAD, DNA, and RNA. In plants the pentose phosphate pathway also plays a role in the synthesis of sugars from carbon dioxide. In animals the pathway occurs at various sites, including the liver and adipose tissue.

PEP *See* **phosphoenolpyruvate.**

pepo *See* **berry.**

pepsin An enzyme that catalyses the breakdown of proteins to polypeptides in the vertebrate stomach. It is secreted as an inactive precursor, *pepsinogen.

pepsinogen The inactive precursor of the enzyme *pepsin. Pepsinogen is secreted by the lining of the vertebrate stomach into the lumen, where it is converted to pepsin by hydrochloric acid and also by the action of pepsin itself.

peptidase *See* **endopeptidase; exopeptidase; protease.**

Formation of a peptide bond

peptide Any of a group of organic compounds comprising two or more amino acids linked by *peptide bonds*. These bonds are formed by the reaction between adjacent carboxyl (−COOH) and amino (−NH₂) groups with the elimination of water (see illustration). *Dipeptides* contain two amino acids, *tripeptides* three, and so on. *Polypeptides contain more than ten and usually 100–300. Naturally occurring *oligopeptides* (of less than ten amino acids) include the tripeptide glutathione and the pituitary hormones antidiuretic hormone and oxytocin, which are octapeptides. Peptides also result from protein breakdown, e.g. during digestion.

peptide mapping (peptide fingerprinting) The technique of forming two-dimensional patterns of peptides (on paper or gel) by partial hydrolysis of a protein followed by electrophoresis and chromatography. The peptide pattern (or *fingerprint*) produced is characteristic for a particular protein and the technique can be used to separate a mixture of peptides.

peptidoglycan A macromolecule that is a component of the cell wall of bacteria; it is not found in eukaryotes. Consisting of chains of *amino sugars (N-acetylglucosamine and N-acetylmuramic acid) linked to a tripeptide (of alanine, glutamic acid, and lysine or diaminopimelic acid), it confers strength and shape to the cell wall.

perception The interpretation of sensory information using both the raw data detected by the senses and previous experience. *Compare* **sensation.**

perennation The survival of biennial or perennial plants from one year to the next by vegetative means. In biennials and herbaceous perennials the aerial parts of the plant die down and the plants survive by means of underground storage roots (e.g. carrot), *rhizomes (e.g. couch grass, Solomon's seal), *tubers (e.g. dahlia), *bulbs (e.g. daffodil, snowdrop), or *corms (e.g. crocus, gladiolus). These *perennating organs* are also frequently responsible for *vegetative propagation. Woody perennials survive the winter by reducing their metabolic activity (e.g. by leaf loss in deciduous trees and shrubs).

perennial A plant that lives for a number of years. Woody perennials (trees and shrubs) have a permanent aerial form, which continues to grow year after year. Herbaceous (i.e. nonwoody) perennials have aerial shoots that die down each autumn and are replaced in spring by new shoots from an underground structure (*see* **perennation**). Lupin and rhubarb are examples of herbaceous perennials. *Compare* **annual; biennial; ephemeral.**

perfusion techniques Methods of maintaining a live organ in isolation by circulating fluid containing essential nutrients and oxygen through and around the organ. Maintaining the circulation through an organ allows the delivery of nutrients to the tissues and the removal from them of toxins and waste products. Perfusion techniques are particularly useful in studying the metabolism of drugs in isolation in an intact organ, such as the liver or kidney.

perianth The part of a flower situated outside the stamens and carpels. In dicotyledons it consists of two distinct whorls, the outer of sepals (*see* **calyx**) and the inner of petals (*see* **corolla**). In monocotyledons the two whorls are similar and often brightly coloured. In wind-pollinated flowers both whorls may be reduced or absent. In many horticultural varieties the number of perianth parts is multiplied, but the resulting 'double' flowers are often sterile.

pericardial cavity The cavity in vertebrates that contains the heart and is bounded by a membrane (the *pericardium). It is part of the *coelom.

pericardium (pericardial membrane) The membrane that encloses the pericardial cavity, containing the vertebrate heart. The pericardium holds the heart in position while allowing it to relax and contract. It consists of two main parts: a tough outer fibrous layer (*fibrous pericardium*) and the more delicate *serous pericardium*, which consists of a double layer of *serous membrane, the inner layer being in close contact with the heart.

pericarp (fruit wall) The part of a fruit that develops from the ovary wall of a flower. The type of fruit that develops depends on whether the pericarp becomes dry and hard or soft and fleshy. The pericarp can be made up of three layers. The outer skin (*epicarp* or *exocarp*) may be tough and hard; the middle layer (*mesocarp*) may be succulent as in peach, hard as in almond, or fibrous as in coconut; and the inner layer (*endocarp*) may be hard and stony as in many *drupes, membranous as in citrus fruits, or indistinguishable from the mesocarp, as in many *berries.

perichondrium A dense layer of fibrous connective tissue that covers the surface of cartilage.

periclinal *(in botany)* Parallel to the surface of an organ or part. In periclinal cell division the plane of division is parallel to the surface of the plant body. *Compare* **anticlinal**.

pericycle A plant tissue comprising the outermost layer of the root vascular tissue, lying immediately beneath the *endodermis. Lateral roots originate from the pericycle.

periderm *See* **cork cambium**.

perigyny A floral arrangement in which the ovary is situated in a cup-shaped or flattened receptacle, from the margin of which the perianth and stamens arise (see illustration). The perianth and stamens are said to be

Perigyny

perigynous with respect to the ovary, as seen in cherry and plum flowers. *Compare* **epigyny**; **hypogyny**.

perilymph The fluid of the *inner ear that fills the space between the bony labyrinth and the membranous labyrinth. *See also* **cochlea**. *Compare* **endolymph**.

perinuclear space (perinuclear compartment) *See* **nuclear envelope**.

period *See* menstrual cycle.

periodontal membrane The membrane of connective tissue that surrounds the root of a *tooth and anchors it to its socket in the jawbone. Fibres of the periodontal membrane pass into the *cement covering the root, which provides a firm attachment.

periosteum The outer membrane that surrounds a bone. It contains connective tissue, capillaries, nerves, and a number of types of cell, including *osteoclasts. The periosteum plays an important role in bone repair and growth.

peripheral nervous system All parts of the nervous system excluding the *central nervous system. It consists of all the *cranial and *spinal nerves and their branches, which link the *receptors and *effectors with the central nervous system. *See also* **autonomic nervous system**.

periphyton The organisms, collectively, that live attached to the stems and leaves of freshwater plants.

periplasmic space The space between the cytoplasmic membrane and the outer membrane in Gram-negative bacteria. It contains a thin layer of *peptidoglycan and a number of enzymes considered to be the equivalent of extracellular enzymes of Gram-positive bacteria.

Perissodactyla An order of mammals having hoofed feet with an odd number of toes. They are all herbivores and include the tapirs, rhinoceros, and horse. The teeth are large and specialized for grinding. Cellulose digestion occurs in the caecum and large intestine. Fossils of the Eocene epoch, 60 million years ago, show that these animals were at that time already distinct from the cloven-hoofed *Artiodactyla.

peristalsis Waves of involuntary muscular contraction and relaxation that pass along the alimentary canal, forcing food contents along. It is brought about by contraction of the circular muscles of the gut wall in sequence.

peristome 1. A ring of toothlike structures around the opening of a moss *capsule. The teeth tend to bend and twist in dry weather, so opening the mouth of the capsule and allowing the spores to escape. In wet weather they close over the opening of the capsule. **2.** The area around the mouth in many invertebrates and some protozoa. It sometimes assists in food collecting. Examples are the spirally ciliated groove around the mouth of some ciliate protozoans and the first segment of the earthworm.

peritoneum The thin layer of tissue (*see* **serous membrane**) that lines the abdominal cavity of vertebrates and covers the abdominal organs. *See also* **mesentery**.

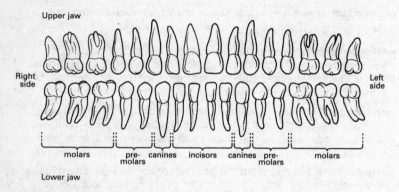

Permanent teeth of an adult human

permanent teeth The second and final set of teeth that mammals produce after shedding the *deciduous teeth. An adult human normally has 32 permanent teeth, consisting of incisors, canines, molars, and premolars (see illustration). These usually appear between the ages of approximately 6 and 21 years. *See also* **dental formula; diphyodont**.

Permian The last geological period in the Palaeozoic era. It extended from the end of the Carboniferous period, about 280 million years ago, to the beginning of the Mesozoic era, about 225 million years ago. It was named by the British geologist Roderick Murchison (1792–1871) in 1841 after the Perm province in Russia. In some areas continental conditions prevailed, which continued into the following period, the Triassic. These conditions resulted in the deposition of the New Red Sandstone. During the period a number of animal groups became extinct, including the trilobites, tabulate and rugose corals, and blastoids. Amphibians and reptiles continued to be the dominant land animals and gymnosperms replaced ferns, clubmosses, and horsetails as the dominant plants.

pernicious anaemia *See* **vitamin B complex**.

peroxisome An organelle, found in the cytoplasm of virtually all plant and animal cells, that contains several enzymes involved in oxidation processes. Peroxisomes are bound by a single membrane; the enzymes they contain include urate oxidase and *catalase. Peroxisomes are active in the process of *detoxification, being particularly abundant in liver parenchyma cells and renal *proximal convoluted tubule cells.

persistent Describing a *pesticide or other pollutant that is not readily broken down and can persist for long periods, causing damage in the environment. For example, the herbicides *Paraquat and *DDT can persist in the soil for many years after their application.

pest Any of various organisms, such as fungi, insects, rodents, and plants, that harm crops or livestock or otherwise interfere with the wellbeing of human beings. *Weeds* are plant pests that grow where they are not wanted – often on cultivated land, where they compete with crop plants for space, light, nutrients, etc. Pests are controlled by the use of *pesticides and *biological control methods.

pesticide Any chemical compound used to kill pests that destroy agricultural production or are in some way harmful to humans. Pesticides include *herbicides* (such as 2,4-*D and *Paraquat), which kill unwanted plants or weeds; *insecticides* (such as *pyrethrum), which kill insect pests; *fungicides*, which kill fungi; and *rodenticides* (such as *warfarin), which kill rodents. The problems associated with pesticides are that they are very often nonspecific and may therefore be toxic to organisms that are not pests; they may also be nonbiodegradable, so that they persist in the environment and may accumulate in living organisms (*see* **bioaccumulation**). Organophosphorus insecticides, such as malathion and parathion, are biodegradable but can also damage the respiratory and nervous systems in humans as well as killing useful insects, such as bees. They act by inhibiting the action of the enzyme *cholinesterase. Organochlorine insecticides, such as dieldrin, aldrin, and *DDT, are very persistent and not easily biodegradable.

peta- Symbol P. A prefix used in the metric system to denote one thousand million million times. For example, 10^{15} metres = 1 petametre (Pm).

petal One of the parts of the flower that make up the *corolla. Petals of insect-pollinated plants are usually brightly coloured and often scented. Those of wind-pollinated plants are usually reduced or absent. Petals are considered to be modified leaves but their structure is simpler. Epidermal hairs may be present and the cuticle is often covered by lines or dots known as *honey guides*, which direct insects to the *nectar.

petiole The stalk that attaches a *leaf blade to the stem. Leaves without petioles are described as *sessile*.

Petri dish A shallow circular flat-bottomed dish made of glass or plastic and having a fitting lid. It is used in laboratories chiefly for culturing bacteria and other microorganisms. It was invented by the German bacteriologist J. R. Petri (1852–1921).

petrification *See* **fossil**.

pH *See* **pH scale**.

PHA *See* **phytohaemagglutinin**.

Phaeophyta (brown algae) A phylum of *algae in which the green chlorophyll pigments are usually masked by the brown pigment fucoxanthin. Brown algae are usually marine (being abundant in cold water) and many species, such as the wracks (*Fucus*), inhabit intertidal zones. They vary in size from small branched filaments to ribbon-like bodies (known as kelps) many metres long.

phage *See* **bacteriophage**.

phagocyte A cell that is able to engulf and break down foreign particles, cell debris, and disease-producing microorganisms (*see* **phagocytosis**). Many protozoans and certain mammalian cells (e.g. *macrophages and *monocytes) are phagocytes. Phagocytes are important elements in the natural defence mechanism of most animals.

phagocytosis The process by which foreign particles invading the body or minute food particles are engulfed and broken down by certain animal cells (known as *phagocytes). The cell membrane of the phagocyte invaginates to capture the particle and then closes around it to form a sac or *vacuole. The vacuole coalesces with a *lysosome, which contains enzymes that break down the particle. *See* **endocytosis**. *Compare* **pinocytosis**.

phagotroph (macroconsumer) Any heterotrophic organism that feeds by ingesting organisms or organic particles, which are digested within its body. *Compare* **osmotroph**.

phalanges The bones that make up the *digits of the hand or foot in vertebrates. They articulate with the *metacarpals of the hand or with the *metatarsals of the foot. In the basic *pentadactyl limb there are two phalanges for the first digit (the thumb or big toe in humans) and three for each of the others.

Phanerozoic Geological time since the end of the Precambrian, represented by rock strata containing clearly recognizable fossils. It comprises the *Palaeozoic, *Mesozoic, and *Cenozoic eras and has extended for about 570 million years from the beginning of the Cambrian period. Fossils are extremely rare in Precambrian rocks.

pharmacodynamics The study of the action of drugs on the body. *Compare* **pharmacokinetics**.

pharmacokinetics The movement of foreign substances, particularly drugs, throughout the body of an animal. Processes that influence the pharmacokinetics of a compound include uptake, distribution throughout the body tissues, the length of time the compound remains in the body, and its rate of clearance (e.g. by metabolism or excretion).

pharmacology The study of the properties of drugs and their effects on living organisms. Clinical pharmacology is concerned with the effects of drugs in treating disease.

pharynx 1. The cavity in vertebrates between the mouth and the *oesophagus and windpipe (*trachea), which serves for the passage of both food and respiratory gases. The presence of food in the pharynx stimulates swallowing (see **deglutition**). In fish and aquatic amphibians the pharynx is perforated by *gill slits. **2.** The corresponding region in invertebrates.

phase-contrast microscope A type of *microscope that is widely used for examining cells and tissues. It makes visible the changes in phase that occur when nonuniformly transparent specimens are illuminated. In passing through an object the light is slowed down and becomes out of phase with the original light. With transparent specimens having some structure diffraction occurs, causing a larger phase change in light outside the central maximum of the pattern. The phase-contrast microscope provides a means of combining this light with that of the central maximum by means of an annular diaphram and a *phase-contrast plate*, which produces a matching phase change in the light of the central maximum only. This gives greater contrast to the final image, due to constructive interference between the two sets of light waves. This is *bright contrast*; in *dark contrast* a different phase-contrast plate is used to make the same structure appear dark, by destructive interference of the same waves.

phase I metabolism The first stage in the conversion of a foreign compound, such as a drug or toxin, into a form that can be eliminated by the body. Common reactions during this phase are oxidation, reduction, and hydrolysis; the resulting metabolites are chemically more reactive than the parent compound, enabling them to undergo the reactions of the second stage (see **phase II metabolism**).

phase II metabolism The second stage in adapting foreign compounds for elimination from the body (*compare* **phase I metabolism**). Phase II metabolism involves the addition of chemical groups (e.g. glycine or acetate), which usually makes the compound less toxic to body tissues and easier to excrete.

phellem See **cork**.

phelloderm See **cork cambium**.

phellogen See **cork cambium**.

phenetic Describing a system of *classification of organisms based on similarities and differences in as many observable characteristics as possible. A phenetic system does not aim to reflect evolutionary descent, although it may well do so. *Compare* **phylogenetic**.

phenology The study of the ways in which the timing and other aspects of periodic events, such as flowering in plants and breeding and migration in animals, are affected by climate and other environmental factors.

phenolphthalein A dye used as an acid-base indicator. It is colourless below pH 8 and red above pH 9.6. It is used in titrations involving weak acids and strong bases. It is also used as a laxative.

phenotype The observable characteristics of an organism. These are determined by its genes (*see* **genotype**), the dominance relationships between the *alleles, and by the interaction of the genes with the environment.

phenylalanine *See* **amino acid**.

phenylketonuria A genetic disorder in which there is disordered metabolism of the amino acid phenylalanine, leading to severe mental retardation of affected children. The disease is caused by the absence or deficiency of the enzyme phenylalanine hydroxylase, which results in the accumulation of phenylalanine in all body fluids. There are also high levels of the ketone phenylpyruvate in the urine, hence the name of the disease. The disease occurs in individuals who are homozygous for the defective recessive allele; both parents of such individuals are thus heterozygous carriers of the allele. The advent of *gene probes has greatly aided accurate diagnosis, of both phenylketonurics and carriers.

pheromone (ectohormone) A chemical substance emitted by an organism into the environment as a specific signal to another organism, usually of the same species. Pheromones play an important role in the social behaviour of certain animals, especially insects and mammals. They are used to attract mates, to mark trails, and to promote social cohesion and coordination in colonies (*see* **queen substance**). Pheromones are usually highly volatile organic acids or alcohols and can be effective at minute concentrations.

phloem (bast) A tissue that conducts food materials in vascular plants from regions where they are produced (notably the leaves) to regions, such as growing points, where they are needed. It consists of hollow tubes (sieve tubes) that run parallel to the long axis of the plant organ and are formed from elongated cells (*sieve elements) joined end to end and closely associated with *companion cells. The end walls of these cells are broken down to a greater or lesser extent to allow passage of materials. In young plants and in newly formed tissues of mature plants *primary phloem* is formed by the activity of the *apical meristem (*see* **protophloem; metaphloem**). In most plants *secondary phloem* is later differentiated by the vascular *cambium and this replaces the earlier formed phloem in older regions. *See also* **mass flow**. *Compare* **xylem**.

phloroglucinol A red dye (usually acidified with hydrochloric acid) that stains lignin in plant cells red.

phosphagen A compound found in animal tissues that provides a reserve of chemical energy in the form of high-energy phosphate bonds. The most common phosphagens are *creatine phosphate, occurring in vertebrate muscle and nerves, and arginine phosphate, found in most invertebrates. During tissue activity (e.g. in muscle contraction) phosphagens give up their phosphate groups, thereby generating *ATP from ADP. The phosphagens are then reformed when ATP is available.

phosphatase An enzyme that catalyses the removal of a phosphate group from an organic compound. *See* **alkaline phosphatase**.

phosphatide *See* phospholipid.

phosphatidylcholine *See* lecithin.

phosphodiester bond The covalent bond that links a phosphate group and a sugar group, by means of an oxygen bridge, in the sugar–phosphate backbone of a nucleic acid molecule. *See* **DNA**.

phosphoenolpyruvate (PEP) A three-carbon compound that is the substrate for carbon dioxide fixation during photosynthesis in C_4 plants (*see* **C_4 pathway**). PEP is also an intermediate in *glycolysis (being the immediate precursor of pyruvate) and the *glyoxylate cycle.

phosphoglyceric acid (PGA; 3-phosphoglycerate) *See* glycerate 3-phosphate.

phosphokinase *See* kinase.

Structure of the phospholipid lecithin (phosphatidylcholine), a phosphoglyceride

phospholipid (phosphatide) One of a group of lipids having both a phosphate group and one or more fatty acids. *Phosphoglycerides* are based on *glycerol; the three hydroxyl groups are esterified with two fatty acids and a phosphate group, which may itself be bound to one of a variety of simple organic groups (e.g. in *lecithin (phosphatidylcholine) it is choline; see formula). *Sphingolipids* are based on the alcohol sphingosine and contain only one fatty acid linked to an amino group. With their hydrophilic polar phosphate groups and long hydrophobic hydrocarbon 'tails', phospholipids readily form membrane-like structures in water. They are a major component of cell membranes (*see* **lipid bilayer**).

phosphorus Symbol P. A nonmetallic element that is a major *essential element for living organisms. It is an important constituent of tissues (especially bones and teeth) and of cells, being required for the formation of nucleic acids and energy-carrying molecules (e.g. ATP) and also involved in various metabolic reactions.

The phosphorus cycle

phosphorus cycle The cycling of phosphorus between the biotic and abiotic components of the environment (*see* **biogeochemical cycle**). Inorganic phosphates (PO_4^{3-}, HPO_4^{2-}, or $H_2PO_4^-$) are absorbed by plants from the soil and bodies of water and eventually pass into animals through food chains. Within living organisms phosphates are built up into nucleic acids and other organic molecules. When plants and animals die, phosphates are released and returned to the abiotic environment through the action of bacteria. On a geological time scale, phosphates in aquatic environments eventually become incorporated into and form part of rocks; through a gradual process of erosion, these phosphates are returned to the soil, seas, rivers, and lakes. Phosphorus-containing rocks are mined for the manufacture of fertilizers, which provide an additional supply of inorganic phosphate to the abiotic environment.

phosphorylation The introduction of a phosphate group into a biomolecule in a reaction that is normally controlled by a *phosphorylase* enzyme. Phosphate is able to combine easily with inert organic compounds, making them chemically active. The first stage in many biochemical reactions is phosphorylation. The conversion of AMP and ADP to *ATP occurs by phosphorylation reactions in two main metabolic pathways, *oxidative phosphorylation and *photophosphorylation. The formation of nucleotides also involves a phosphorylation reaction. The activity of many enzymes is controlled by phosphorylation: certain enzymes are activated when they are phosphorylated (*see* **kinase**), while others are deactivated. Phosphorylation of these enzymes is under the control of hormones and other messengers.

photic zone The upper layer of a sea or a lake, in which there is sufficient light for photosynthesis. The limit of the photic zone varies from less than a metre to more than 200 metres, depending on the turbidity of the water.

photoautotroph An autotrophic organism, such as a green plant or a photosynthetic bacterium, that synthesizes its organic materials from inorganic components using energy derived from the sun (solar energy) in the process of photosynthesis. *See* **autotrophic nutrition; phototroph.**

photochemical smog A noxious smog produced by the reaction of nitrogen oxides with hydrocarbons in the presence of ultraviolet light from the sun. The reaction is very complex and one of the products is ozone. *See also* **air pollution.**

photoheterotroph Any organism that uses energy derived from the sun to manufacture organic compounds from organic precursors in photosynthesis. For example, under certain conditions some purple *sulphur bacteria use organic acids (rather than hydrogen sulphide) as a source of hydrogen. *See* **phototroph.**

photolysis A chemical reaction produced by exposure to light or ultraviolet radiation. The photolysis of water, using energy from sunlight absorbed by chlorophyll, produces gaseous oxygen, electrons, and hydrogen ions and is a key reaction in *photosynthesis. *See* **photophosphorylation; photosystems I and II.**

photomicrography The use of photography to obtain a permanent record (a *photomicrograph*) of the image of an object as viewed through a microscope.

photonasty *See* **nastic movements.**

photoperiod *See* **photoperiodism.**

photoperiodism The response of an organism to changes in day length (*photoperiod*). Many plant responses are controlled by day length, the most notable being flowering in many species (*see* **florigen; day-neutral plant; long-day plant; short-day plant**). In plants the pigment *phytochrome is responsible for the regulation of photoperiodic responses (*see also* **dark period**). Activities in animals that are determined by photoperiod include breeding, *migration, and other seasonal events. *See also* **melatonin.**

photophosphorylation The formation of ATP from ADP and inorganic phosphate using light energy in *photosynthesis (compare **oxidative phosphorylation**). There are two pathways, noncyclic and cyclic photophosphorylation, which occur in the thylakoid membranes of the chloroplasts. In *noncyclic photophosphorylation* electrons derived from the *photolysis of water are raised to higher energy levels in *photosystems I and II and pass along an *electron transport chain of carrier molecules (*see* **ferredoxin; plastocyanin; plastoquinone**) to NADP reductase. This enzyme transfers electrons to $NADP^+$ to make NADPH, which provides reducing power for the dark reaction of photosynthesis. In *cyclic photophosphorylation* the electrons from photosystem I that are raised to a higher energy level are recycled through the electron carrier system back to photosystem I. Both pathways of electron flow cause H^+ ions to be pumped by a group of *cytochromes, the *cytochrome b_6–f complex*, across the thylakoid membrane. This creates a proton gradient that drives the phosphorylation of ADP to ATP by the enzyme ATP synthetase (*see* **chemiosmotic theory**).

photopic vision The type of vision that occurs when the cones in the eye are the principal receptors, i.e. when the level of illumination is high. Colours can be identified with photopic vision. *Compare* **scotopic vision**.

photoreceptor A sensory cell or group of cells that reacts to the presence of light. It usually contains a pigment that undergoes a chemical change when light is absorbed, thus stimulating a nerve. *See* **eye**.

photorespiration A metabolic pathway that occurs in plants in the presence of light, in which ribulose bisphosphate carboxylase, the enzyme involved in carbon dioxide fixation with *ribulose bisphosphate, accepts oxygen, in place of carbon dioxide, resulting in the formation of a two-carbon compound, glycolate, which is subsequently metabolized to carbon dioxide. When catalysing this reaction the enzyme is frequently referred to as ribulose bisphosphate carboxylase-oxygenase, or RUBISCO. Unlike respiration, there is no production of ATP. In C_3 plants (*see* **C_3 pathway**) photorespiration has the effect of reducing the rate of photosynthesis, as atmospheric oxygen can combine with RUBISCO. In C_4 plants (*see* **C_4 pathway**) the effect of photorespiration is negligible as the affinity of phosphoenolpyruvate carboxylase for carbon dioxide is extremely high. As oxygen is a competitive inhibitor of RUBISCO, photorespiration will increase as oxygen concentration increases or as carbon dioxide concentration decreases.

photosynthesis The chemical process by which green plants and other *phototrophs synthesize organic compounds from carbon dioxide and water in the presence of sunlight. In plants and most algae it occurs in the *chloroplasts and there are two principal reactions. In the *light-dependent* (or *light*) *reaction*, which requires the presence of light, energy from sunlight is absorbed by *photosynthetic pigments (chiefly the green pigment *chlorophyll) and used to bring about the *photolysis of water:

$$H_2O \rightarrow 2H^+ + 2e^- + \tfrac{1}{2}O_2$$

The electrons released by this reaction pass along a series of electron carrier molecules; as they do so they lose their energy, which is used to convert ADP to ATP in the process of *photophosphorylation. The electrons and protons produced by the photolysis of water are used to reduce NADP:

$$2H^+ + 2e^- + NADP^+ \rightarrow NADPH + H^+$$

The ATP and NADPH produced during the light reaction provide energy and reducing power, respectively, for the ensuing *light-independent* (or *dark*) *reaction*, which can take place in light or darkness. During this reaction carbon dioxide is reduced to carbohydrate in a metabolic pathway known as the *Calvin cycle. Photosynthesis can be summarized by the equation:

$$CO_2 + 2H_2O \rightarrow [CH_2O] + H_2O + O_2$$

Since virtually all other forms of life are directly or indirectly dependent on plants for food, photosynthesis is the basis for all life on earth. Furthermore virtually all the atmospheric oxygen has originated from oxygen released during photosynthesis.

Photosynthesis

photosynthetic pigments The plant pigments responsible for the capture of light energy during the light reaction of *photosynthesis. The green pigment *chlorophyll *a* is the *primary pigment*, absorbing blue and red light (*see* **photosystems I and II**). The *carotenoids and various other pigments are *accessory pigments*, absorbing light energy and passing this on to the chlorophyll *a* molecules.

photosystems I and II The two systems of *photosynthetic pigments in the thylakoid membranes of chloroplasts that are involved in the light reaction of photosynthesis. Each photosystem contains about 300 chlorophyll molecules that trap light energy, which is then passed to a *reaction centre*, comprising a chlorophyll *a* molecule, in each photosystem. In photosystem II a chlorophyll *a* molecule, known as P680, utilizes light of wavelength 680 nm; in photosystem I the chlorophyll *a* molecule, known as P700, absorbs light at a wavelength of 700 nm. Light energy is used in each reaction centre to raise electrons to higher energy levels to enable them to be taken up by electron acceptors. This causes P680 and P700 to become positively charged, or oxidized. The chlorophyll in photosystem II replaces its lost electrons with those that come from the *photolysis of water:

$$H_2O \rightarrow 2H^+ + 2e^- + \frac{1}{2}O_2$$

The oxygen produced is given off as a gas, and the H^+ ions, together with the electrons from photosystem I, reduce $NADP^+$ (*see* **photophosphorylation**).

phototaxis The movement of a cell (e.g. a gamete) or a unicellular organism in response to light. For example, certain algae (e.g. *Chlamydomonas*) can perceive light by means of a sensitive eyespot and move to regions of higher light concentration to enhance photosynthesis. *See* **taxis**.

phototroph Any organism that uses energy derived from the sun to manufacture organic compounds by photosynthesis. Most phototrophs are *photoautotrophs; a few bacteria are *photoheterotrophs.

phototropism (heliotropism) The growth of plant organs in response to light. Aerial shoots usually grow towards light, while some aerial roots grow away from light. The phototropic response is thought to be controlled by *auxin, a plant growth substance. *See* **tropism**.

phrenic nerve The nerve that forms a network in the diaphragm and is involved in the control of ventilation. *See* **inspiratory centre**.

pH scale A logarithmic scale for expressing the acidity or alkalinity of a solution. To a first approximation, the pH of a solution can be defined as $-\log_{10}c$, where c is the concentration of hydrogen ions in moles per cubic decimetre. A neutral solution at 25°C has a hydrogen-ion concentration of 10^{-7} mol dm^{-3}, so the pH is 7. A pH below 7 indicates an acid solution; one above 7 indicates an alkaline solution. pH stands for 'potential of hydrogen'. The scale was introduced by S. P. Sørensen in 1909.

phycobiliprotein Any one of a class of pigments in the cyanobacteria, red algae, and photosynthetic cryptophytes that act as *accessory pigments in photosynthesis, being attached to the *thylakoid membranes of the chloroplasts. Each comprises a protein bound to a *phycobilin*, a coloured tetrapyrrole prosthetic group (*see* **porphyrin**). The main phycobiliproteins are *phycocyanin and *phycoerythrin.

phycobiont The algal or bacterial component of a *lichen.

phycocyanin An accessory photosynthetic pigment occurring mainly in the cyanobacteria and red algae. It is a *phycobiliprotein, in which the pigmented prosthetic group is phycocyanobilin, which gives phycocyanin its blue colour.

phycoerythrin An accessory photosynthetic pigment occurring mainly in the red algae and cyanobacteria. It is a *phycobiliprotein, in which the pigmented prosthetic group is phycoerythrobilin, which gives phycoerythrin its red colour.

phycomycetes In older classification schemes, all primitive *fungi, many of which are found in water (e.g. the water moulds, which may be parasitic on fish) or in damp areas. Many are unicellular but those that form mycelia generally have hyphae lacking cross walls, which distinguishes them from the *Ascomycota and *Basidiomycota. They include the *Zygomycota.

phyletic *See* phylogenetic.

phyllotaxis (phyllotaxy) The arrangement of leaves on a plant stem. The leaves may be inserted in whorls or pairs at each node or singly up the stem. When arranged in pairs the two leaves arise on opposite sides of the stem and are usually at right angles to the leaf pairs above and below them. Single leaves may be inserted alternately or in a spiral pattern up the stem. Phyllotaxis generally results in the minimum of shading of leaves by those above them.

phylogenetic (phyletic) Describing a system of *classification of organisms that aims to show their evolutionary history. *Compare* **phenetic**.

phylogeny The evolutionary history of an organism or group of related organisms. *Compare* **ontogeny**.

phylum (*pl.* **phyla**) A category used in the *classification of organisms that consists of one or several similar or closely related classes. Examples of phyla are the Rhodophyta, Ascomycota, Bryophyta, Arthropoda, and Chordata. Phyla are grouped into kingdoms. In traditional plant classification schemes, the *division was used instead of the phylum.

physiological saline A liquid medium in which animal tissues may be kept alive for a few hours during experiments without pathological changes or distortion of the cells taking place. Such fluids are salt solutions that are isotonic with and have the same pH as the body fluids of the animal. A well-known example is *Ringer's solution*, formulated by the British physiologist S. Ringer (1835–1910), which is a mixture of sodium chloride, calcium chloride, sodium bicarbonate, and potassium chloride solutions.

physiological specialization The occurrence within a species of several forms that are identical in appearance but differ in physiology: these are termed *physiological races*. For example, many pathogenic fungi develop new physiological races in response to the strong selection pressure exerted when disease-resistant crop varieties are sown over large areas.

physiology The branch of biology concerned with the vital functions of plants and animals, such as nutrition, respiration, reproduction, and excretion.

physisorption *See* adsorption.

phyto- Prefix denoting plants. For example, phytopathology is the study of plant diseases.

phytochrome A protein-based plant pigment present in small quantities in many plant organs. It exists in two interconvertible forms: a physiologically active form, which forms when the plant is illuminated with red light or normal daylight; and an inactive form, formed when the plant is exposed to far-red light or darkness. The active form regulates many plant processes, such as seed germination and the initiation of flowering. *See also* **photoperiodism**.

phytogeography *See* plant geography.

phytohaemagglutinin (PHA) Any of various plant-derived compounds that induce changes in lymphocytes normally associated with antigen challenge. These changes include cell enlargement, increased RNA and DNA synthesis, and, finally, cell division. Response to PHAs is used to test for competence of cell-mediated *immunity, for example in patients suffering chronic virus infections.

phytohormone *See* growth substance.

phytoplankton The photosynthesizing *plankton, consisting chiefly of microscopic algae, such as diatoms, and dinoflagellates. Near the surface of the sea there may be many millions of such organisms per cubic metre. Members of the phytoplankton are of great importance as they form the basis of food for all other forms of aquatic life, being the primary *producers. *Compare* **zooplankton**.

pia mater The innermost of the three membranes (*meninges) that surround the brain and spinal cord of vertebrates. The pia mater lies immediately adjacent to the central nervous system, and the *choroid plexus, which secretes cerebrospinal fluid, is an extension of it.

pico- Symbol p. A prefix used in the metric system to denote 10^{-12}. For example, 10^{-12} farad = 1 picofarad (pF).

picornavirus One of a group of small RNA-containing viruses (*pico* = small; hence pico-RNA-virus) commonly present in the alimentary and respiratory tracts of vertebrates. They cause mild infections of these tracts but the group also includes the polioviruses, which attack the central nervous system causing poliomyelitis; and the causal agent of foot and mouth disease in cattle, sheep, and pigs.

pie chart A diagram in which percentages are shown as sectors of a circle. If x percent of the prey of a carnivore comprises species X, y percent species Y, and z percent species Z, a pie chart would show three sectors having central angles $3.6x°$, $3.6y°$, and $3.6z°$.

pigment A compound that gives colour to a tissue. Pigments perform a variety of functions. For example, *haemoglobin in vertebrate erythrocytes gives blood its characteristic red colour and enables the transport of oxygen throughout the body (*see* **respiratory pigment**). Other biological pigments include *chlorophyll, a photosynthetic pigment in plants that is responsible for their green coloration; and *melanin, a brown pigment in animals that provides protection from ultraviolet light and can be used in camouflaging colorations.

pileus The umbrella-shaped cap of certain fungi, such as mushrooms. Spores are produced from *gills or pores on the lower surface.

piliferous layer The part of the root epidermis that bears *root hairs. It extends over a region about 4–10 mm behind the root tip. Beyond this the piliferous layer is sloughed off to reveal the hypodermis.

pilomotor Describing nerve fibres or the muscles innervated by them that can cause the position of the body hair to be altered. The *arrector pili is a pilomotor muscle.

Piltdown man Fossil remains, purported to have been found by Charles Dawson at Piltdown, Sussex, in 1912, that were named *Eoanthropus dawsoni* and described as a representative of the true ancestors of modern man. The skull resembled that of a man but the jaw was apelike. In 1953 dating techniques showed the specimen to be a fraud.

pineal eye *See* **median eye**.

pineal gland An outgrowth of the *forebrain. In man its functions are obscure, but in other vertebrates it acts as an endocrine gland, secreting the hormone *melatonin.

pinna (auricle) The visible part of the *outer ear, present in some mammals. It is made of cartilage and its function is to channel sound waves into the external auditory meatus. In some species the pinna is movable and aids in detecting the direction from which a sound originates.

pinocytosis The process by which a living cell engulfs a minute droplet of liquid. It involves a technique similar to *phagocytosis. *See* **endocytosis**.

pipette A graduated tube used for transferring measured volumes of liquid.

Pisces A superclass of the *Gnathostomata (jawed chordates) comprising the fishes (*compare* **Tetrapoda**). The whole body of a fish is covered with a tough, usually scaly, skin (*see* **scales**), which extends over the eye and contains pigments and sometimes slime glands. The circulatory system is a single circuit with blood passing through two sets of capillaries, one at the gills and the other in the body tissues. Three *semicircular canals are present in the inner ear. The sense of smell is particularly well developed and pressure waves are detected by the lateral-line system (*see* **lateral-line canal**). Fossils of fish date back to the Ordovician period, 500–440 million years ago. There are two classes of modern fish: *Chondrichthyes (cartilaginous fishes) and *Osteichthyes (bony fishes).

pistil The female part of a flower, consisting either of a single *carpel (*simple pistil*) or a group of carpels (*compound pistil*).

pit 1. A depression or cavity in the secondary wall of a plant cell that facilitates the movement of substances between adjacent cells. The equivalent structures in primary cell walls are called *primary pit fields*, areas where plasmodesmata are concentrated and pit development usually occurs. A pit comprises a *pit membrane*, consisting of the middle lamella plus the primary wall; and a *pit cavity*, the depression in the secondary wall. Pits usually occur in pairs (called *pit pairs*) on either side of the middle lamella between two adjacent cells. **2. (coated pit)** *See* **endosome**.

pitfall trap A simple trap for small invertebrate animals consisting of a tin that is placed in the ground with its rim at ground level. The trap, which contains some kind of bait, can be covered by a tile suspended above ground level by stones so that rain does not enter the tin.

pith 1. (*or* **medulla**) The cylinder of *parenchyma tissue found in the centre of plant stems to the inside of the vascular tissue. It is light in weight and has been put to various commercial uses, notably the manufacture of pith helmets. **2. (*not in scientific usage*)** The white tissue below the rind of many citrus fruits. **3.** To destroy the central nervous system of an animal, especially a laboratory animal such as a frog, by severing its spinal cord.

Pithecanthropus *See* Homo.

pituitary gland (pituitary body; hypophysis) A pea-sized endocrine gland attached by a thin stalk to the *hypothalamus at the base of the brain. It consists of two lobes: the anterior and the posterior. The *anterior pituitary* (or *adenohypophysis*) secretes such hormones as *growth hormone, the *gonadotrophins, *prolactin, *thyroid-stimulating hormone, and *ACTH. Because these hormones regulate the growth and activity of several other endocrine glands, the anterior pituitary is often referred to as the *master endocrine gland*. Activity of the anterior pituitary itself is regulated by specific *releasing factors produced by the hypothalamus (*see also* **neuroendocrine system**). The *posterior pituitary* (or *neurohypophysis*) secretes the hormones *oxytocin and *antidiuretic hormone.

placenta 1. The organ in mammals and other viviparous animals by means of which the embryo is attached to the wall of the uterus. It is composed of embyronic and maternal tissues: extensions of the *chorion and *allantois grow into the uterine wall so that materials (e.g. oxygen, nutrients) can pass between the blood of the embryo and its mother (there is, however, no direct connection between the maternal and embryonic blood). The placenta is eventually expelled as part of the *afterbirth. **2.** A ridge of tissue on the ovary wall of flowering plants to which the ovules are attached. The arrangement of ovules on the placenta (*placentation*) is variable, depending on the number of carpels and whether they are free (*see* **apocarpy**) or fused (*see* **syncarpy**).

Placentalia *See* **Eutheria.**

placoid scale (denticle) *See* **scales.**

plagiotropism The tendency for a *tropism (growth response of a plant) to be orientated at an angle to the line of action of the stimulus concerned. For example, the growth of lateral branches and lateral roots is at an oblique angle to the stimulus of gravity (*plagiogeotropism*). *Compare* **orthotropism.**

planarians *See* **Turbellaria.**

plankton Minute *pelagic organisms that drift or float passively with the current in a sea or lake. Plankton includes many microscopic organisms, such as algae, protozoans, various animal larvae, and some worms. It forms an important food source for many other members of the aquatic community and is divided into *zooplankton and *phytoplankton. *Compare* **benthos; nekton; neuston.**

plant Any living organism of the kingdom Plantae. Plants are distinguished from other multicellular organisms by their life cycles, in which a haploid generation (*gametophyte) alternates with a diploid generation (*sporophyte). (*See also* **alternation of generations.**) Most plants manufacture carbohydrates by *photosynthesis, in which simple inorganic substances are built up into organic compounds. The radiant energy needed for this process is absorbed by *chlorophyll, a complex pigment not found in animals. Plants also differ from animals in the possession of *cell walls (usually composed of *cellulose). Plants are immobile, as there is no necessity to search for food, and they respond slowly to external stimuli. For a classification of the plant kingdom, see Appendix.

plant geography (phytogeography) The study of the distribution of world vegetation, with particular emphasis on the influence of the environmental factors that determine this distribution.

plantigrade Describing the gait of many mammals, including man, in which the whole lower surface of the foot is on the ground. *Compare* **digitigrade; unguligrade.**

planula The ciliated free-swimming larva of many coelenterates, consisting of a solid mass of cells. It eventually settles on a suitable surface and develops into a *polyp.

plaque 1. A thin layer of organic material covering all or part of the exposed surface of a tooth. It contains dissolved food (mostly sugar) and bacteria. The bacteria in plaque metabolize the sugar and produce acid, which eats into the surface of the enamel of the tooth and eventually causes tooth decay (*dental caries). **2.** A clear area in a bacterial culture grown on an agar plate due to *lysis of the bacteria by a bacteriophage.

plasma *See* **blood plasma.**

plasma cells Antibody-producing cells found in the epithelium of the lungs and gut and also in bone-forming tissue. They develop in the lymph nodes, spleen, and bone marrow when antigens stimulate lymphocytes to form the precursor cells that give rise to them (*see* **B cell**).

plasmagel The specialized outer gel-like *cytoplasm of living cells (such as *Amoeba*) that move by extruding part of the cell (known as a *pseudopodium) in the direction of motion. A reversible conversion of plasmagel to the more fluid *plasmasol is believed to be responsible for the continuous flow forward of cytoplasm necessary for forming a pseudopodium.

plasmalemma *See* **cell membrane.**

plasma membrane *See* **cell membrane.**

plasma protein Any of the protein constituents of blood plasma. They comprise *albumins, *globulins, and *fibrinogen.

plasmasol The specialized inner sol-like *cytoplasm of living cells that move by producing *pseudopodia. *Compare* **plasmagel.**

plasmid A structure in cells consisting of DNA that can exist and replicate independently of the chromosomes. In bacterial cells plasmids provide genetic instructions for certain cell activities (e.g. resistance to antibiotic drugs). They can be transferred from cell to cell in a bacterial colony (*see* **sex factor**). Bacterial plasmids are used to produce recombinant DNA for *gene cloning. Similar blocks of DNA that appear to move from one chromosome to another have been identified in the cells of a number of other species, including man. In organisms (other than man) that have been studied, it appears that they interfere with gene activity (*see* **transposon**). In man this has not been established, although they are almost identical to *retroviruses, which are known to affect gene action.

plasmin An enzyme, present in blood plasma, that breaks down a blood clot by destroying the fibrin threads of the clot and by inactivating factors involved in blood clotting, such as prothrombin and the *clotting factors. This occurs during *fibrinolysis. Plasmin is derived from an inactive precursor, *plasminogen.

plasminogen The inactive precursor of the enzyme *plasmin. Plasminogen is incorporated into blood clots and is converted to plasmin during *fibrinolysis.

plasmodesmata (*sing.* **plasmodesma**) Fine cytoplasmic strands that connect the *protoplasts of adjacent plant cells by passing through their cell walls. Plasmodesmata are cylindrical in shape (about 20–40 nm in diameter) and are lined by the plasma membrane of the two adjacent cells. The endoplasmic reticula of the two adjacent cells are connected by a narrower structure, the *desmotubule*, which runs through the centre of a plasmodesma.

plasmodium (*pl.* **plasmodia**) A multinucleate mass of protoplasm that comprises a stage in the life cycle of a plasmodial *slime mould. The plasmodium is the vegetative phase of the slime mould, often seen as a slimy layer on decaying wood; it is capable of slow amoeboid movement and feeds by engulfing particles of decaying organic matter.

plasmogamy Fusion of the cytoplasm of two (or more) cells. It precedes union of the nuclei in *fertilization and it occurs in *heterokaryosis.

plasmolysis The loss of water by *osmosis from a plant cell to the extent that the protoplasm shrinks away from the cell wall. This happens when the cell is placed in a solution that has a higher solute concentration than that of the cell sap, i.e. it has a lower (more negative) *water potential, since water always moves from an area of high water potential to an area of low water potential. *Compare* **turgor**.

plastid An *organelle within a plant cell, often occurring in large numbers. Apart from the nucleus, plastids are the largest solid inclusions in a plant cell. For convenience they are classified into those containing pigments (*chromoplasts) and those that are colourless (*leucoplasts), although changes from one to the other frequently occur. Plastids develop from *proplastids*, colourless bodies found in meristematic and immature cells; they also arise by division of existing plastids. *See also* **chloroplast**.

plastocyanin A blue copper-containing protein that is found in chloroplasts and acts as an electron *carrier molecule in the light reaction of *photosynthesis (*see* **photophosphorylation**). Plastocyanin consists of amino acid groups in association with a copper molecule which gives this compound a blue colour.

plastome Any genetic material that is contained in cytoplasmic organelles, such as mitochondria or chloroplasts.

plastoquinone A quinone, found in chloroplasts, that functions as one of the *carrier molecules of the electron transport chain in the light reaction of *photosynthesis. *See* **photophosphorylation**.

plastron *See* carapace.

plate count An estimate of the number of viable cells in a culture. A plate count of a bacterial culture is made by inoculating the culture plate with a dilute solution of the microorganisms and counting the number of cells that appear in the resultant culture.

platelet (**thrombocyte**) A minute disc-shaped cell fragment in mammalian blood. Platelets are formed as fragments of larger cells (*megakaryocytes*) found in red bone marrow; they have no nucleus. They play an important role in *blood clotting and release serotonin and other chemicals, which cause a chain of events leading to the formation of a plug at the site of the damage, thus preventing further blood loss. There are about 250 000 platelets per cubic millimetre of blood.

platelet-derived growth factor *See* growth factor.

plate tectonics The theory that the surface of the earth is made of lithospheric plates, which have moved throughout geological time resulting in the present-day positions of the continents. The theory explains the locations of mountain building as well as earthquakes and volcanoes. The rigid lithospheric plates consist of continental and oceanic crust together with the upper mantle, which lie above the weaker plastic asthenosphere. These plates move relative to each other across the earth. Six major plates (Eurasian, American, African, Pacific, Indian, and Antarctic) are recognized, together with a number of smaller ones. The plate margins coincide with zones of seismic and volcanic activity.

A *constructive* (or *divergent*) plate margin occurs when two plates move away from each other. It is marked by a mid-oceanic ridge where basaltic material wells up from the mantle to form new oceanic crust, in a process known as *sea-floor spreading*. The production of new crust at constructive plate margins is compensated for by the destruction of material along a *destructive* (or *convergent*) plate margin. Along these margins, which are also known as *subduction zones* and marked by an oceanic trench, one plate (usually oceanic) is forced to plunge down beneath the other (which may be continental or oceanic). The crust becomes partially melted and rises to form a chain of volcanoes in the upper plate parallel to the trench. When two continental plates collide the compression results in the formation of mountain chains. A third type of plate margin – the *transform plate margin* – occurs where two plates are slipping past each other.

Platyhelminthes A phylum of acoelomate invertebrates comprising the flatworms, characterized by a flattened unsegmented body. The simple nervous system shows some concentration of cells at the head end. The mouth leads to a simple branched gut without an anus. Flatworms are hermaphrodite but self-fertilization is unusual. Many species are parasitic. The phylum contains the classes *Turbellaria (planarians), *Trematoda (flukes), and *Cestoda (tapeworms).

pleiomorphism The existence of distinctly different forms during the life cycle of an individual, e.g. the caterpillar, pupa, and winged adult of a butterfly.

pleiotropic Describing an allele that has more than one effect in an organism. For example, the allele that causes the erythrocytes to have a distorted form in sickle-cell anaemia also causes these blood cells to rupture easily, thereby inducing anaemia.

Pleistocene The earlier epoch of the *Quaternary period (*compare* Holocene). It extended from the end of the Pliocene, about 2 million years ago, to the beginning of the Holocene, about 10 000 years ago. The Pleistocene is often known as the *Ice Age* as it was characterized by a series of glacials, in which ice margins advanced towards the equator, separated by interglacials when the ice retreated. *See also* **ice age**.

pleura (pleural membrane) The double membrane that lines the thoracic cavity and covers the exterior surface of the lungs. It is a *serous membrane forming a closed sac, with a small space (the *pleural cavity*) between the two layers.

plexus A compact branching network of nerves or blood vessels, such as the *brachial plexus* – a network of spinal nerves that supply branches to the forelimbs in vertebrates. *See also* **choroid plexus**.

Pliocene The fifth and final epoch of the *Tertiary period. Preceded by the Miocene and followed by the Pleistocene, it began about 7 million years ago and lasted for about 5 million years. Mammals similar to modern forms existed during the epoch and the australopithecines (*see Australopithecus*), early forerunners of man, appeared.

plumule **1.** *(in zoology)* A *down feather. **2.** *(in botany)* The part of a plant embryo that develops into the shoot system. It consists of the stem apex and first leaves. In seedlings showing *epigeal germination the plumule grows above the soil surface together with the cotyledons; in seeds showing *hypogeal germination, the plumule alone emerges. *Compare* **radicle**.

pneumatophore **1.** An aerial negatively geotropic root that acts as an organ of gaseous exchange. Pneumatophores are produced by *mangroves and other plants that grow in waterlogged oxygen-deficient soils. **2.** The gas-filled float of certain colonial cnidarians of the class Hydrozoa, such as *Physalis* (Portuguese man-of-war).

pod *See* **legume**.

podocyte A type of epithelial cell found in *Bowman's capsule. Podocytes possess major footlike processes, each supporting a series of minor processes. These minor processes are interwoven with those from other podocytes to form a number of slits, through which filtration can occur. These slits (approximately 0.1 μm wide) permit the passage of all plasma constituents but act as a barrier to blood cells.

poikilothermy The passive variation in the internal body temperature of an animal, which depends on the temperature of the environment. All animals except birds and mammals exhibit poikilothermy and are described as *ectotherms. Although unable to maintain a constant body temperature, they can respond to compensate for very low or very high temperatures. For example, the tissue composition (especially cell osmotic pressure) can change to regulate the blood flow to peripheral tissues (and thus increase heat loss or heat absorption), and the animals can actively seek sun or shade. Seasonal changes in metabolism are usually under hormonal control. In particularly hot climates, ectotherms may undergo *aestivation to escape the heat. *Compare* **homoiothermy**.

point mutation (gene mutation) A change in the nucleotide sequence of the DNA within a gene; a gene in which such a change has occurred is known as a *mutant gene or allele (*see* **mutation**). The DNA sequence can be altered in several ways; for example by *insertion, *substitution, *deletion, and *inversion. Point mutations result in a misreading of the genetic code during the translation phase of protein synthesis and usually change the order of amino acids making up a protein, which may or may not affect the function of that protein. *See also* **intergenic suppressor**. *Compare* **chromosome mutation**.

polar body *See* oogenesis.

polarity 1. The property of a cell, tissue, or organism of being structurally and/or functionally different at opposite ends of its long axis. For example, plants consist of roots, which grow in the direction of the force of gravity, and stems, which grow away from the gravitational force (*see* **geotropism**). **2.** The property of molecules of having an uneven distribution of electrons, so that one part has a positive charge and the other a negative charge. Such *polar molecules* include *water.

polar molecule *See* polarity.

polar nuclei Two haploid nuclei in the centre of the *embryo sac of flowering plants. These nuclei fuse with a male gamete nucleus to form a triploid endosperm nucleus, which subsequently divides to form the endosperm. *See also* **double fertilization**.

pollen The mass of grains containing the male gametes of seed plants, which are produced in large numbers in the *pollen sacs. The pollen grains of insect-pollinated plants may be spiny or pitted and are usually larger than those of wind-pollinated plants, which are usually smooth and light. The pollen grain represents the male *gametophyte generation; it contains two male nuclei: a *generative nucleus and a *tube nucleus. The wall of the mature pollen grain consists of the tough outer wall (*exine*) and the more delicate narrower *intine*. The latter gives rise to the *pollen tube. *See also* **pollination**.

pollen analysis *See* palynology.

pollen sac The structure in seed plants in which pollen is produced. In angiosperms there are usually four pollen sacs in each *anther; they contain the *microspore mother cells. In gymnosperms variable numbers of pollen sacs are borne on the microsporophylls that make up the male *cone.

pollen tube An outgrowth of a pollen grain, which transports the male gametes to the ovule. It will only grow if the pollen grain is compatible with the female tissue. In angiosperms, the pollen grain is deposited on the stigma and the pollen tube grows down through the style and into the ovule. In some conifers, e.g. *Pinus* (pines), the pollen tube penetrates the *nucellus but does not develop further until the following year, when the female part of the plant is mature. *See also* **embryo sac; fertilization; tube nucleus**.

pollex The innermost digit on the forelimb of a tetrapod vertebrate. It contains two phalanges (*see* **pentadactyl limb**) and in man and higher primates it is the thumb, which is opposable (i.e. capable of facing and touching the other digits) and gives the hand greater manipulating ability. In some mammals a pollex is absent. *Compare* **hallux**.

pollination The transfer of pollen from an anther (the male reproductive organ) to a stigma (the receptive part of the female reproductive organ), either of the same flower (*self-pollination*) or of a different flower of the same species (*cross-pollination*). Cross-pollination involves the action of a pollinating agent to effect transfer of the pollen (*see* **anemophily; entomophily; hydrophily**). *See also* **fertilization; incompatibility**.

pollutant Any substance, produced and released into the environment as a result of human activities, that has damaging effects on living organisms. Pollutants may be toxic substances (e.g. *pesticides) or natural constituents of the atmosphere (e.g. carbon dioxide) that are present in excessive amounts. *See* **pollution**.

pollution An undesirable change in the physical, chemical, or biological characteristics of the natural environment, brought about by man's activities. It may be harmful to human or nonhuman life. Pollution may affect the soil, rivers, seas, or the atmosphere (*see* **air pollution**). There are two main classes of *pollutants: those that are *biodegradable* (e.g. sewage), i.e. can be rendered harmless by natural processes and need therefore cause no permanent harm if adequately dispersed or treated; and those that are *nonbiodegradable* (e.g. heavy metals (such as lead) in industrial effluents (*see* **heavy-metal pollution**) and DDT and other chlorinated hydrocarbons used as *pesticides), which eventually accumulate in the environment and may be concentrated in food chains. Other forms of pollution in the environment include noise (e.g. from jet aircraft, traffic, and industrial processes) and thermal pollution (e.g. the release of excessive waste heat into lakes or rivers causing harm to wildlife). Recent pollution problems include the disposal of radioactive waste; *acid rain; *photochemical smog; increasing levels of human waste; high levels of carbon dioxide and other greenhouse gases in the atmosphere (*see* **greenhouse effect**); damage to the *ozone layer by nitrogen oxides, *chlorofluorocarbons (CFCs), and halons; and pollution of inland waters by agricultural *fertilizers and sewage effluent, causing eutrophication (*see* **eutrophic**). Attempts to contain or prevent pollution include strict regulations concerning factory emissions, the use of smokeless fuels, the banning of certain pesticides, the increasing use of lead-free petrol, restrictions on the use of chlorofluorocarbons, and the introduction, in some countries, of catalytic converters to cut pollutants in car exhausts.

polyacrylamide gel electrophoresis *See* PAGE.

Polychaeta A class of annelid worms in which each body segment has a pair of flattened fleshy lobes (*parapodia*) bearing numerous bristles (*chaetae). All polychaetes are aquatic and most of them are marine. They include the fanworms (*Sabella*), which construct tubes of sand, etc., in which they live; and the lugworms (*Arenicola*) and ragworms (*Nereis*), which burrow in sand or mud.

polycistronic Describing a type of messenger *RNA that can encode more than one polypeptide separately within the same RNA molecule. Bacterial messenger RNA is generally polycistronic. *Compare* **monocistronic**.

polyembryony 1. The formation of more than one embryo in a plant seed. Often one embryo develops from the fertilized egg cell, while the others have formed asexually from other tissues in the ovule. **2.** The formation of more than one embryo from a single animal zygote. *Identical twins are produced in this way.

polygene Any of a group of genes influencing a quantitative characteristic, e.g. height in man. *See* **multifactorial inheritance; polygenic characters**.

polygenic characters Phenotypic characters that are controlled by *polygenes. Polygenic characters display *continuous variation, each polygene having only a small influence on the particular character in question. *See also* **multifactorial inheritance**.

polymer A substance having large molecules consisting of repeated units (the monomers). There are a number of natural polymers, such as polysaccharides.

polymerase Any enzyme that catalyses the elongation of a polymeric molecule. *RNA polymerases* (types I to III) catalyse the synthesis of RNA using as a template either an existing DNA strand (*DNA-dependent RNA polymerase*) or an RNA strand. Type I is responsible for the synthesis of ribosomal RNA, type II for messenger RNA synthesis (*see* **transcription**), and type III for transfer RNA synthesis. *DNA polymerases* catalyse the elongation of a new DNA strand during *DNA replication, using an existing DNA strand as template. *RNA-directed DNA polymerase* is more usually known as *reverse transcriptase.

polymerase chain reaction (PCR) A technique used to replicate a fragment of DNA so as to produce many copies of a particular DNA sequence. PCR is commonly employed as an alternative to *gene cloning as a means of amplifying genetic material for *gene sequencing, and is also used to measure *gene expression. The two strands of the DNA are separated by heating and short sequences of a single DNA strand (*primers*) are added, together with a supply of free nucleotides and DNA *polymerase obtained from a bacterium that can withstand extreme heat. In a series of heating and cooling cycles, the DNA sequence doubles with each cycle and is thus rapidly amplified.

polymorphism The existence of three or more distinctly different forms within a plant or animal species. An example is the *caste system of social insects, in which there are the workers, drones, and queens. This is an *environmental polymorphism*, i.e. the differences are caused by environmental rather than genetic factors, in this case by the larvae receiving different types of food. There are also *heritable* or *genetic polymorphisms*, in which differences are the result of inherited characteristics. Two forms of genetic polymorphism exist: *transient polymorphism*, in which a particular form is in the process of spreading through a population, causing the relative proportion of phenotypes to alter; and *balanced polymorphism*, in which two or more forms coexist in a stable ratio within a population, each form possessing both advantageous and disadvantageous characteristics. An example of balanced polymorphism is the occurrence of *sickle-cell anaemia*, a genetic disease that principally affects Black populations of central Africa and is characterized by an abnormal form of the blood pigment haemoglobin (*haemoglobin S*) and sickle-shaped red blood cells. Three different types of individual occur in such populations: those who have two genes (*AA*) for normal haemoglobin and therefore do not suffer from the disease; those with one normal and one abnormal gene (*AS*), who suffer from a relatively mild form; and those with two abnormal genes (*SS*), who suffer a chronic and eventually fatal form of anaemia. Normally such a harmful gene would have been eliminated from the population by the process of natural selection, but it is maintained in this case because people with the mild form of the disease are resistant to a severe form of malaria endemic in central Africa.

See also restriction fragment length polymorphism.

polynucleotide *See* nucleotide.

polyol (polyhydric alcohol) Any compound that contains more than one alcohol group. The term is usually restricted to the sugar alcohols, which include such biologically important molecules as *myo-**inositol, *sorbitol* (common in the fruits of many plants), and *ribitol* (found in bacterial cell walls). Certain polyols act as *osmolytes in the kidney cells and play a role in osmoregulation.

polyp The sedentary stage in the life cycle of the *Cnidaria, consisting of a cylindrical body fixed at one end to a firm base and having a mouth surrounded by a ring of tentacles at the other. Some polyps (e.g. *Hydra*) are single; others (e.g. the corals and *Obelia*) form colonies. Polyps typically reproduce asexually by budding to form either new polyps or *medusae. The latter reproduce sexually giving rise to new polyps. Sea anemones are solitary polyps that reproduce sexually to form new polyps.

polypeptide A *peptide comprising ten or more amino acids. Polypeptides that constitute proteins usually contain 100–300 amino acids.

Shorter ones include certain antibiotics, e.g. gramicidin, and some hormones, e.g. *ACTH, which has 39 amino acids. The properties of a polypeptide are determined by the type and sequence of its constituent amino acids.

polypetalous Describing a flower in which the *corolla consists of separate petals. *Compare* **gamopetalous**.

polyphyletic Denoting any group of organisms the members of which have originated from several different ancestors. An example is the group including all insectivorous animals. Polyphyletic groups are not natural groups and do not have any place in natural classifications. *Compare* **monophyletic**.

polyphyodont Describing a type of dentition in which the teeth are continuously shed and replaced during the lifetime of the animal. Sharks and frogs have a polyphyodont dentition. *Compare* **diphyodont; monophyodont**.

polyploid Describing a nucleus that contains more than two sets of chromosomes (*see* **diploid**) or a cell or organism containing such nuclei. For example, *triploid plants have three sets of chromosomes and *tetraploid* plants have four. Polyploidy is far more common in plants than in animals; many crops, in particular, are polyploid (bread wheat, for example, is *hexaploid*, i.e. 6*n*). It can be induced chemically with *colchicine. *See also* **allopolyploid; autopolyploid**.

polyribosome (polysome) An aggregate of ribosomes in association with a single messenger RNA molecule during the *translation process of protein synthesis. In eukaryotes, polyribosomes are attached to the surface of the rough endoplasmic reticulum and the outer membrane of the nucleus; in bacteria they are found free in the cytoplasm.

polysaccharide Any of a group of carbohydrates comprising long chains of monosaccharide (simple sugar) molecules. *Homopolysaccharides* consist of only one type of monosaccharide; *heteropolysaccharides* contain two or more different types. Polysaccharides may have molecular weights of up to several million and are often highly branched. Some important examples are starch, glycogen, and cellulose.

polysepalous Describing a flower that possesses a calyx consisting of separate sepals. *Compare* **gamosepalous**.

polysome *See* **polyribosome**.

polyspermy The entry of several sperms into the egg during fertilization although only one sperm nucleus actually fuses with the egg nucleus. Polyspermy occurs in animals with yolky eggs (e.g. birds).

polysynaptic reflex A *reflex action that involves an electrical impulse being transferred from a sensory neurone to a motor neurone via at least one connecting neurone (*interneurone*) in the spinal cord. For example, stimulation of pain receptors in the skin initiates a *withdrawal reflex*, which involves several synapses with several motor neurones and results in the removal of the organism or part from the stimulus.

polyteny The condition of a chromosome, nucleus, or cell in which the DNA has repeatedly replicated, without subsequently separating. This results in a cablelike *giant chromosome*, consisting of parallel strands and often showing conspicuous transverse banding. Polyteny has been observed in the cells of certain insects of the order Diptera, notably in the salivary glands of *Drosophila*. Swellings (called *puffs*) have been observed in giant chromosomes at specific stages in the development of these insects. Puffs are associated with messenger RNA and are believed to be sites of gene transcription.

polythetic Describing a type of classification in which membership of a taxon is based on its constituent organisms sharing a large number of characteristics.

polytopic Describing a taxon that has arisen and exists in more than one geographical region.

polytypic Describing a species that exists in a variety of *forms or *subspecies that inhabit different geographical regions.

Polyzoa (Bryozoa) *See* Ectoprocta.

pome A type of fruit characteristic of apples and pears. The flesh of the fruit develops from the *receptacle of the flower, which completely encloses the fused carpels. After fertilization the carpels form the 'core' of the fruit, which contains the seeds. *See also* **pseudocarp**.

pons (pons Varolii) A thick tract of nerve fibres in the brain that links the medulla oblongata to the midbrain. Its function is to relay impulses between different parts of the brain. The pons is named after its discoverer, the Italian anatomist C. Varoli (1543–75).

population *(in ecology)* **1.** A group of individuals of the same species within a *community. The nature of a population is determined by such factors as density, *sex ratio, birth and death rates, emigration, and immigration. **2.** The total number of individuals of a given species or other class of organisms in a defined area, e.g. the population of rodents in Britain.

population dynamics The study of the fluctuations that occur in the numbers of individuals in animal and plant populations and the factors controlling these fluctuations. An important distinction is maintained between those factors that are dependent on population density and have a stabilizing effect (e.g. food supply) and those that are independent of population density (e.g. catastrophes, such as flooding).

population genetics The study of the distribution of inherited *variation among a group of organisms of the same species. The potential for change depends on the sum total of alleles that are available to the organisms (the *gene pool*), and estimates of changes in allele frequency in a population give an indication of its response to a changing condition.

Sigmoid growth curve · J-shaped growth curve

Population growth curves

population growth The increase in a population that occurs when the *birth rate is higher than the *death rate, or when immigration exceeds emigration, or when a combination of these factors is present. A growth curve, obtained by plotting population size against time, is typically S-shaped (sigmoid) or J-shaped (see graph). A sigmoid growth curve shows an initial phase of *exponential growth. The curve levels off when the environment has reached its *carrying capacity*, i.e. when food, space, and other conditions can support a given number of individuals without an increase in population numbers. A J-shaped growth curve shows an initial phase of exponential growth that ceases abruptly, with a sudden decrease in population numbers. This decrease may be caused by a number of factors, such as the end of the life cycle of the prey or any other factor contributing to *environmental resistance that may suddenly take effect. *See also* **bacterial growth curve**.

Porifera The phylum of marine and freshwater invertebrates that comprise the sponges, which live permanently attached to rocks or other surfaces. The body of a sponge is hollow and consists basically of an aggregation of cells between which there is little nervous coordination. The body is supported by an internal skeleton of spicules of chalk, silica, or fibrous protein (bath sponges have protein skeletons). Flagellated cells (*choanocytes*) cause water to flow in through openings (*ostia*) in the body wall and out through openings (*oscula*) at the top; food particles are filtered from the water by the choanocytes.

porphyrin Any of a group of organic pigments characterized by the possession of a cyclic group of four linked nitrogen-containing rings (a *tetrapyrrole* nucleus), the nitrogen atoms of which are often coordinated to metal ions. Porphyrins differ in the nature of their side-chain groups. They include the *chlorophylls, which contain magnesium; and *haem, which contains iron and forms the *prosthetic group of haemoglobin, myoglobin, and the cytochromes.

Generalized structure of a porphyrin

portal vein (**portal circulation; portal system**) Any vein that collects blood from one network of capillaries and transports it directly to a second capillary network in another region of the body, without returning to the heart. *See* **hepatic portal system**.

positive feedback *See* **feedback**.

postcaval vein *See* **vena cava**.

posterior 1. Designating the part of an animal that is to the rear, i.e. that follows when the animal is moving. In man and bipedal animals (e.g. kangaroos) the posterior surface is equivalent to the *dorsal surface. **2.** Designating the side of a flower or axillary bud that faces towards the flower stalk or main stem, respectively. *Compare* **anterior**.

postsynaptic membrane The membrane at the end of a neurone that receives an impulse at a *synapse.

potassium Symbol K. A soft silvery metallic element that is an *essential element for living organisms. The potassium ion, K^+, is the most abundant cation in plant tissues, being absorbed through the roots and being used in such processes as protein synthesis. In animals the passage of potassium and sodium ions across the nerve-cell membrane is responsible for the changes of electrical potential that accompany the transmission of impulses.

potassium–argon dating A *dating technique for certain rocks that depends on the decay of the radioisotope potassium–40 to argon–40, a process with a half-life of about 1.27×10^{10} years. It assumes that all the argon–40 formed in the potassium-bearing mineral accumulates within it and that all the argon present is formed by the decay of potassium–40. The mass of argon–40 and potassium–40 in the sample is estimated and the sample is then dated from the equation:

$$^{40}\text{Ar} = 0.1102\ ^{40}\text{K}(e^{\lambda t} - 1),$$

where λ is the decay constant and t is the time in years since the mineral cooled to about 300°C, when the ^{40}Ar became trapped in the crystal lattice. The method is effective for micas, feldspar, and some other minerals.

potentiation The synergistic interaction of two substances to produce an effect that is greater than the effect of either one of the substances in isolation. *See* **synergism**.

potometer An apparatus used to measure the rate of water loss from a shoot (*see* **transpiration**) under natural or artificial conditions.

poxvirus One of a group of DNA-containing viruses, often enclosed in an outer membrane, that typically produce skin lesions in vertebrates. They include those viruses causing smallpox (variola), cowpox (vaccinia), and myxomatosis. Some poxviruses produce tumours.

preadaptation An anatomical structure, physiological process, or behaviour pattern in an organism that is by chance highly suited to a new habitat to which the organism migrates or that improves the chance of the organism surviving following a change in environmental conditions. For example, the lungs that developed in certain fish were probably originally a buoyancy aid before these fish began to adapt to a new environment on dry land.

Precambrian The division of geological time from the formation of the earth, believed to be about 4600 million years ago, to the beginning of the Cambrian period, some 570 million years ago. The Precambrian thus represents most of geological time. Fossils are rare, although *stromatolites occur, and subsequent metamorphism of Precambrian rocks makes correlation of rocks and events extremely difficult. The Precambrian is divided by some authorities into the Archaean and Proterozoic eons. The largest areas of exposed Precambrian rocks are the shield areas, such as the Canadian (Laurentian) Shield and the Baltic Shield.

precaval vein *See* **vena cava**.

precipitin Any *antibody that combines with its specific soluble *antigen to form a precipitate. The term is sometimes applied to the precipitate itself. *See also* **agglutination**.

precocial species Any species of bird in which hatching occurs at a relatively late stage of development. The young of such species are capable of locomotion and leave the nest soon after hatching; they are described as *nidifugous. Compare* **altricial species**.

predation An interaction between two populations of animals in which one (the *predator*) hunts, captures, and kills the other (the *prey*) for food.

Predator–prey relationships form important links in many food chains. They are also important in regulating population sizes of both predator and prey, especially when the predator relies on a single prey species. The term predation is also used, more loosely, for any feeding relationship in which an organism feeds on any other living organism (animal or plant).

predator An animal that obtains its food by *predation. All predators are *carnivores, although not all carnivores are predators.

pregnancy See **gestation**.

premolar A broad ridged tooth in mammals that is situated behind the *canine teeth (when present) and in front of the *molars. Premolars are adapted for grinding and chewing food and are present in both the deciduous and *permanent teeth.

presbyopia A loss of accommodation that normally develops in human eyes over the age of 45–50 years. Vision of distant objects remains unchanged but accommodation of the eye to near objects is reduced as a result of loss of elasticity in the lens of the eye. The defect is corrected by reading glasses using weak converging lenses.

pressure potential Symbol Ψ_p. The component of *water potential due to the hydrostatic pressure that is exerted on water in a cell. In turgid plant cells it usually has a positive value as the entry of water causes the protoplast to push against the cell wall (see **turgor**). In xylem cells there is a negative pressure potential, or tension, as a result of transpiration.

presumptive Describing embryonic tissue that is not yet *determined but which will eventually develop into a certain kind of tissue by virtue of its position in the embryo.

presynaptic membrane The membrane of a neurone that releases neurotransmitter into the synaptic cleft between nerve cells (see **synapse**).

prey An animal that is a source of food for a predator. See **predation**.

prickle A hard sharp protective outgrowth, many of which may cover the surface of a plant. It contains cortical and vascular tissue and is not regarded as an epidermal outgrowth. Compare **spine; thorn**.

primary consumer See **consumer**.

primary growth The increase in size of shoots and roots of plants that results from the activity of the *apical (tip) meristems and subsequent expansion of the cells produced. The tissues thus produced are called primary tissues and the resultant plant parts constitute the primary plant body. Compare **secondary growth**.

primary producer See **producer**.

primary productivity The total amount of organic matter synthesized by the producers (e.g. green plants) of an ecosystem. See **productivity**.

primary structure *See* protein.

Primates An order of mammals that includes the monkeys, apes, and man. Primates evolved from arboreal insectivores 130 million years ago. They are characterized by thumbs and big toes that are opposable (i.e. capable of facing and touching the other digits), which permits manual dexterity, and forward-facing eyes allowing *binocular vision. The brain, particularly the cerebrum, is relatively large and well-developed, accounting for the intelligence and quick reactions of these mammals. The young are usually produced singly and undergo a long period of growth and development to the adult form.

primitive streak The longitudinal groove that develops in the *gastrula during the development of bird and mammal embryos. The cells in the primitive streak proliferate rapidly to form mesoderm cells, which migrate to the interior of the embryo.

primordium (*pl.* **primordia**) A group of cells that represents the initial stages in development of a plant organ. Root and shoot primordia are present in a young plant embryo while leaf primordia (or *leaf buttresses*) are seen as small bulges just below the shoot apex.

prion An abnormal form of a normal cell protein (PrP) found in the brain of mammals that is believed to be the agent responsible for the diseases scrapie in sheep, bovine spongiform encephalopathy (BSE) in cattle, and Creutzfeldt–Jacob disease in humans. Produced by mutation of the normal PrP gene, it is thought to interact with normal prion protein, causing it to accumulate in the brain and progressively damage and destroy brain cells. It may be transmitted to other individuals of the same or closely related species, by injection or possibly ingestion of infected tissue, but it is thought not to be transmissible between species that are not closely related.

Proboscidea The order of mammals that comprises the elephants. They are herbivorous, with a muscular trunk (*proboscis) used for drinking, bathing, and collecting food. The tusks are continuously growing upper incisors and the enormous ridged molar teeth are produced in sequence to replace worn teeth throughout life. The order, which evolved in the Eocene epoch, was formerly much larger and more widespread than it is today and included the extinct mammoths. There are only two species of modern elephants: the African and Indian species.

proboscis 1. The trunk of an elephant: a muscular and very flexible elongation of the nose, which has a finger-like extremity and is capable of picking up and moving objects, taking in water, collecting food, etc. **2.** The elongated mouthparts of certain invertebrates, such as the two-winged flies (Diptera).

procambium (provascular tissue) A plant tissue formed by the *apical meristems of shoots and roots. It consists of cells elongated parallel to the long axis of the plant. The procambium subsequently gives rise to the primary *vascular tissue.

procarcinogen The precursor of an active *carcinogen. The procarcinogen itself is not usually carcinogenic but is converted to the active carcinogen after it has been metabolized. For example, the drug diethylstilboestrol (a synthetic oestrogen no longer in clinical use) is metabolized to an epoxide intermediate, which can cause cervical cancer.

procaryote *See* prokaryote.

Proconsul A genus of extinct apes known from fossil remains, 50 million years old, found mostly in Kenya and assigned to the species *P. africanus*. A find made in 1948 was the first discovery of a skull of Miocene ape. It is believed that *Proconsul* was similar and possibly closely related to the genus *Dryopithecus*.

prodrug A drug that requires further metabolism in the liver before becoming biologically active. An example of a prodrug is the immunosuppressant azathioprine, which is metabolized first to cortisone and then to the pharmacologically active compound hydrocortisone.

producer An organism considered as a source of energy for those above it in a *food chain (i.e. at the next *trophic level). Green plants, which convert energy from sunlight into chemical energy, are *primary producers*; herbivores are *secondary producers*, as they utilize energy from green plants and supply energy for carnivores. *Compare* **consumer**.

productivity (production) *(in ecology)* The rate at which an organism, population, or community assimilates energy (*gross productivity*) or makes energy potentially available (as body tissue) to an animal that feeds on it (*net productivity*). The difference between these two rates is due to the rate at which energy is lost through excretion and respiration. Thus *gross primary productivity* is the rate at which plants (or other *producers) assimilate light energy, and *net primary productivity* is the rate at which energy is incorporated as plant tissue. It is measured in kilojoules per square metre per year ($kJ\,m^{-2}\,y^{-1}$). In terrestrial plants, much of the net productivity is not actually available to *consumers, e.g. tree roots are not eaten by herbivores. *See also* **energy flow**.

profundal Occurring in or designating the deep-water zone of an inland lake. Light intensity, oxygen concentration, and (during summer and autumn) temperature are markedly lower than in the surface layer. *Compare* **littoral; sublittoral**.

progenesis The maturation of gametes in an organism that is still otherwise at the juvenile stage of development (*see* **paedogenesis**). Progenesis can lead to *paedomorphosis.

progeny *See* offspring.

progesterone A hormone, produced primarily by the *corpus luteum of the ovary but also by the placenta, that prepares the inner lining of the uterus for implantation of a fertilized egg cell. If implantation fails, the corpus luteum degenerates and progesterone production ceases accordingly. If implantation occurs, the corpus luteum continues to secrete progesterone, under the influence of *luteinizing hormone and *prolactin, for several months of pregnancy, by which time the placenta has taken over this function. During pregnancy, progesterone maintains the constitution of the uterus and prevents further release of eggs from the ovary. Small amounts of progesterone are produced by the testes. *See also* **progestogen**.

progestogen One of a group of naturally occurring or synthetic hormones that maintain the normal course of pregnancy. The best known is *progesterone. In high doses progestogens inhibit secretion of *luteinizing hormone, thereby preventing ovulation, and alter the consistency of mucus in the vagina so that conception tends not to occur. They are therefore used as major constituents of *oral contraceptives.

proglottid (proglottis) *See* **Cestoda**.

prokaryote (procaryote) Any organism belonging to the kingdom Prokaryotae (or Monera), in which the genetic material is not enclosed in a cell nucleus. *Bacteria are prokaryotes. It is believed that eukaryotic cells (*see* **eukaryote**) probably evolved as symbiotic associations of prokaryotes. *See* **cell**.

prolactin (lactogenic hormone; luteotrophic hormone; luteotrophin) A hormone produced by the anterior pituitary gland. In mammals it stimulates the mammary glands to secrete milk (*see* **lactation**) and the corpus luteum of the ovary to secrete the hormone *progesterone. Secretion of prolactin is increased by suckling. In birds prolactin stimulates secretion of crop milk by the crop glands.

proline *See* **amino acid**.

prometaphase The period during the early part of *metaphase in which the chromosomes become attached to the spindle fibres and align themselves at right angles to the spindle poles in a plane through the centre of the cell.

promoter *(in protein synthesis)* The region of a DNA molecule that signals the start of *transcription. The promoter region contains a sequence of nucleotides (the *recognition site*) that induces binding between the DNA and the enzyme RNA *polymerase and a further sequence of nucleotides (the *initiation site*) that contains the first DNA nucleotide to be copied by the RNA polymerase. The tight binding between RNA polymerase and DNA at the recognition site destabilizes the DNA molecule, causing the double helix structure to unzip so that transcription can proceed.

pronation Rotation of the lower forearm so that the hand faces backwards or downwards with the radius and ulna crossed. *Compare* **supination**.

propagation 1. *(in botany)* See **vegetative propagation**. **2.** *(in neurophysiology)* The process whereby a nerve *impulse travels along the axon of a neurone. *Compare* **transmission**.

prophage The DNA of a temperate *bacteriophage following its incorporation into the host bacterium. The process of incorporation of the viral DNA is known as *lysogeny.

prophase The first stage of cell division, during which chromosomes contract and divide along their length (except for the centromeres) into chromatids. In *mitosis, the chromosomes remain separate from each other. In the first division of *meiosis, homologous chromosomes become paired (*see* **pairing**). By the end of first prophase the two chromosomes begin to move apart.

proplastid See **plastid**.

proprioceptor Any *receptor that is sensitive to movement, pressure, or stretching (*see* **stretch receptor**) within the body. Proprioceptors occurring in muscles, tendons, and ligaments are important for the coordination of muscular activity and the maintenance of balance and posture.

prop root Any of the modified roots that arise from the stem of certain plants and provide extra support. Such stems are usually tall and slender and the prop roots develop at successively higher levels as the stem elongates, as in the maize plant. *Buttress roots*, which develop at the base of the trunks of many tropical trees, are similar but tend to have a more flattened appearance. *Stilt roots* are stouter than prop roots. Those formed at the base of the mangrove tree provide firm anchorage in the soft mud of the swamps.

prosencephalon See **forebrain**.

prosoma The anterior section of the body (*see* **tagma**) of arachnids and other arthropods of the subphylum *Chelicerata, which consists of the fused head and thorax (cephalothorax) and bears the chelicerae and other appendages. *Compare* **opisthosoma**.

prostaglandin Any of a group of organic compounds derived from *essential fatty acids and causing a range of physiological effects in animals. Prostaglandins have been detected in most body tissues. They act at very low concentrations to cause the contraction of smooth muscle; natural and synthetic prostaglandins are used to induce abortion or labour in humans and domestic animals. Two prostaglandin derivatives have antagonistic effects on blood circulation: *thromboxane* A_2 causes blood clotting while *prostacyclin* causes blood vessels to dilate. Prostaglandins are also involved in inflammation, being released from affected tissues. *See also* **aspirin**.

prostate gland A gland in male mammals that surrounds and opens into the urethra where it leaves the bladder. During ejaculation it secretes a slightly alkaline fluid into the *semen that activates the sperms and prevents them from sticking together.

prosthetic group A tightly bound nonpeptide inorganic or organic component of a protein. Prosthetic groups may be lipids, carbohydrates, metal ions, phosphate groups, etc. Some *coenzymes are more correctly regarded as prosthetic groups.

protamine Any of a group of proteins of relatively low molecular weight found in association with the chromosomal *DNA of vertebrate male germ cells. They contain a single polypeptide chain comprising about 67% arginine. Protamines are thought to protect and support the chromosomes.

protandry 1. The condition in which the male reproductive organs (stamens) of a flower mature before the female ones (carpels), thereby ensuring that self-fertilization does not occur. Examples of protandrous flowers are ivy and rosebay willowherb. *Compare* **protogyny; homogamy.** *See also* **dichogamy. 2.** The condition in some hermaphrodite or colonial invertebrates in which the male gonads or individuals are sexually mature before the female ones. *Compare* **protogyny.**

protanopia *See* **colour blindness.**

protease (peptidase; proteinase; proteolytic enzyme) Any enzyme that catalyses the hydrolysis of proteins into smaller *peptide fractions and amino acids, a process known as *proteolysis.* Examples are *pepsin and *trypsin. Several proteases, acting sequentially, are normally required for the complete digestion of a protein to its constituent amino acids. *See also* **acid protease.**

protein Any of a large group of organic compounds found in all living organisms. Proteins comprise carbon, hydrogen, oxygen, and nitrogen and most also contain sulphur; molecular weights range from 6000 to several million. Protein molecules consist of one or several long chains (*polypeptides) of *amino acids linked in a characteristic sequence. This sequence is called the *primary structure* of the protein. These polypeptides may undergo coiling (*see* **alpha helix**) or pleating (*see* **beta sheet**), the nature and extent of which is described as the *secondary structure.* The three-dimensional shape of the coiled or pleated polypeptides is called the *tertiary structure,* the functional unit of which is a *domain. *Quaternary structure* specifies the structural relationship of the component polypeptides.

Proteins may be broadly classified into globular proteins and fibrous proteins. *Globular proteins* have compact rounded molecules and are usually water-soluble. Of prime importance are the *enzymes, proteins that catalyse biochemical reactions. Other globular proteins include the *antibodies, which combine with foreign substances in the body; the carrier proteins, such as *haemoglobin; the storage proteins (e.g. *casein in milk and *albumin in egg white), and certain hormones (e.g. *insulin). *Fibrous proteins* are generally insoluble in water and consist of long coiled strands or flat sheets, which confer strength and elasticity. In this category are *keratin and *collagen. Actin and myosin are the principal fibrous proteins of muscle, the interaction of which brings about muscle contraction. *Blood clotting involves the fibrous protein called fibrin.

When heated over 50°C or subjected to strong acids or alkalis, proteins lose their specific tertiary structure and may form insoluble coagulates (e.g. egg white). This usually inactivates their biological properties.

protein engineering The techniques used to alter the structure of proteins (especially enzymes) in order to improve their use to humans. This involves artificially modifying the DNA sequences that encode them so that, for example, new amino acids are inserted into existing proteins. Synthesized lengths of novel DNA can be used to produce new proteins by cells or other systems containing the necessary factors for *transcription and *translation. Alternatively, new proteins can be synthesized by *solid state synthesis*, in which polypeptide chains are assembled under the control of chemicals. One end of the chain is anchored to a solid support and the chemicals selectively determine which amino acids are added to the free end. The appropriate chemicals can be renewed during the process; when synthesized, the polypeptide is removed and purified. Protein engineering is used to synthesize enzymes (so-called 'designer enzymes') used in biotechnology. Further understanding of the three-dimensional tertiary structure of proteins is required for future development in protein engineering; this is being assisted by computer-aided modelling.

protein kinase An enzyme that catalyses the transfer of a phosphate group from ATP to an intracellular protein, thereby affecting the biological activity of the protein (*see* **kinase**). Protein kinases phosphorylate specific amino-acid residues of their target proteins, usually either serine, threonine, or tyrosine. They play an important role in increasing or decreasing enzyme activity and in transmitting signals from receptors on the cell surface. The activity of the protein kinases is itself controlled by cyclic AMP, calcium ions, or other intracellular chemicals. It can be reversed by the action of phosphatase enzymes in the cell.

protein sequencing The process of determining the amino-acid sequence of a protein or its component polypeptides. The technique most commonly used is Edman degradation (devised by Pehr Edman), in which the terminal amino-acid residues are removed sequentially and identified chromatographically. Each step is automated and the whole process can now be performed by a single machine – the sequenator. Large polypeptides must be cleaved into smaller peptides before sequencing.

The results of this chemical sequencing can often be compared with the amino-acid sequence deduced by *gene sequencing. The gene coding for the protein under investigation may be found by screening a *gene library, for example by *Western blotting. However, the base sequence of the gene gives only the amino-acid sequence of the nascent protein, i.e. before post-translational modification. The sequence of the functional protein can only be found by chemical analysis.

protein synthesis The process by which living cells manufacture proteins from their constituent amino acids, in accordance with the genetic information carried in the DNA of the chromosomes. This information is encoded in messenger *RNA, which is transcribed from DNA in the nucleus of the cell (*see* **genetic code; transcription**): the sequence of amino acids in a particular protein is determined by the sequence of nucleotides in messenger RNA. At the ribosomes the information carried by messenger RNA is translated into the sequence of amino acids of the protein in the process of *translation.

protein targeting The process whereby a newly synthesized protein is directed to its correct location within the cell. Protein targeting is determined by short sequences of amino acids in the protein (known as *signal sequences*), which direct it to the correct destination; for example to a mitochondrion or (in the case of a plant cell) a chloroplast. This movement can occur either during synthesis of the protein (cotranslational) or after synthesis is complete (post-translational).

proteoglycan One of a group of glycoproteins found in the matrix of *connective tissue, consisting of a polysaccharide chain (*see* **glycosaminoglycan**) bound to a small protein unit.

proteolysis The enzymic splitting of proteins. *See* **protease**.

proteolytic enzyme *See* **protease**.

prothallus A small flattened multicellular structure that represents the independent *gametophyte generation of clubmosses, horsetails, and ferns. In some of these plants a single prothallus bears both male and female sex organs. In others there are separate male and female prothalli.

prothoracic gland *See* **ecdysone**.

prothrombin (Factor II) One of the blood *clotting factors. It is the precursor of the enzyme thrombin, which catalyses the formation of the fibrin matrix of the blood clot. Prothrombin synthesis occurs in the liver and is dependent on adequate supplies of vitamin K. *See also* **blood clotting**.

Protista In older classifications, a kingdom into which all organisms of simple biological organization were classified. Originally proposed by E. Haeckel in 1866 to include the algae, bacteria, fungi, and protozoa, it was later restricted first to unicellular organisms, and then to protozoa, unicellular algae, and organisms then regarded as simple fungi. In modern classifications it has been replaced by the *Protoctista.

Protoctista A kingdom consisting of unicellular or simple multicellular organisms that possess nuclei and cannot be classified as animals, plants, or fungi. Protoctists include the *protozoa, *algae, *Dinoflagellata, *Oomycota, and *slime moulds. *See also* **Protista**.

protoderm A plant tissue formed by the *apical meristem of shoots and roots that subsequently gives rise to the epidermis.

protogyny **1.** The condition in which the female reproductive organs (carpels) of a flower mature before the male ones (stamens), thereby ensuring that self-fertilization does not occur. Examples of protogynous flowers are plantain and figwort. *Compare* **protandry; homogamy**. *See also* **dichogamy**. **2.** The condition in hermaphrodite or colonial invertebrates in which the female gonads or individuals are sexually mature before the male ones. *Compare* **protandry**.

protonema The first, usually filamentous, structure produced by a germinating moss or fern spore under dark conditions. The protonema of mosses bears buds that develop into the gametophyte plant; in ferns the protonema becomes the *prothallus.

protonephridium *See* **nephridium**.

protophloem The primary *phloem tissue that arises from the elongating regions of shoot or root apices. *Compare* **metaphloem**.

protoplasm The granular material comprising the living contents of a *cell, i.e. all the substances in a cell except large vacuoles and material recently ingested or to be excreted. It consists of a *nucleus embedded in a jelly-like *cytoplasm and includes the plasma (cell) membrane.

protoplast (energid) The living unit (protoplasm) of a cell, consisting of the nucleus and cytoplasm bounded by the cell membrane. Protoplasts of bacterial and plant cells can be prepared by removing the cell wall; they are used to study the processes involved in cell metabolism and reproduction.

Prototheria A subclass of mammals – the monotremes – that lay large yolky eggs. It contains only the duckbilled platypus and the spiny anteater. After hatching, the young feed on milk from simple mammary glands inside a maternal abdominal pouch. In the anteater the eggs are also incubated in this pouch, while the platypus builds an underground nest. Adult monotremes have no true teeth. Their skeleton resembles that of a reptile, and although they are warm-blooded the body temperature is somewhat variable. They are believed to have originated at least 150 million years ago.

protoxylem The primary *xylem tissue that is formed in the expanding region of roots and shoots before the shoot or root has completed the elongation process. In the secondary walls of these vessels the lignin is deposited in rings and spirals, which allows for further elongation. The protoxylem is succeeded by *metaxylem.

protozoa A group of unicellular or acellular, usually microscopic, organisms now classified in various phyla of the kingdom *Protoctista (*see* **Apicomplexa; Ciliophora; Rhizopoda; Zoomastigina**). They were formally regarded either as a phylum of simple animals or as members of the kingdom *Protista. They are very widely distributed in marine, freshwater, and moist terrestrial habitats; most protozoans are saprotrophs, but some are parasites, including the agents causing malaria (*Plasmodium*) and sleeping sickness (*Trypanosoma*), and a few contain chlorophyll and carry out photosynthesis, like plants. Protozoan cells may be flexible or rigid, with an outer *pellicle or protective *test*. In some (such as *Paramecium* and *Trypanosoma*) *cilia or *flagella are present for locomotion; others (such as *Amoeba*) have *pseudopodia for movement and food capture. *Contractile vacuoles occur in freshwater protozoans. Reproduction is usually asexual, by binary *fission, but some protozoans undergo a form of sexual reproduction (*see* **conjugation**).

provascular tissue *See* procambium.

proventriculus 1. The anterior part of the stomach of a bird, where digestive enzymes are secreted. Food passes from the proventriculus to the *gizzard. **2.** *See* **gastric mill**.

provirus The intermediate stage in the infection of a host cell by a *retrovirus in which the single RNA strand of the virus is converted into double-stranded DNA. This synthesis is performed by the enzyme *reverse transcriptase, which also digests the original viral RNA. The proviral DNA is then integrated into the host cell DNA and subsequently transcribed to form new RNA viruses. This integration also introduces retroviral oncogenes into the host, with the attendant risk of host-cell transformation to a cancer cell. A provirus, notably that of *HIV, can remain dormant for long periods before being transcribed.

proximal Denoting the part of an organ that is nearest to the organ's point of attachment. For example, the knuckles are at the proximal end of the fingers. *Compare* **distal**.

proximal convoluted tubule (first convoluted tubule) The section of a *nephron situated between Bowman's capsule and the loop of Henle in the vertebrate kidney. Reabsorption of salt, water, and glucose from the *glomerular filtrate occurs in this tubule; at the same time certain substances, including uric acid and drug metabolites, are actively transferred from the blood capillaries into the tubule. Both activities are facilitated by finger-like projections (*see* **brush border**) on the inner surface of the tubule, which increase its effective surface area.

pseudocarp (false fruit) A fruit that incorporates, in addition to the ovary wall, other parts of the flower, such as the *receptacle. For example, the fleshy part of the strawberry is formed from the receptacle and the 'pips' on the surface are the true fruits. *See also* **composite fruit; pome; sorosis; syconus**.

pseudocoelomate Describing any invertebrate animal whose body cavity is a *pseudocoel*, a cavity between the gut and the outer body wall derived from a persistent blastocoel (*see* **blastula**), rather than a true coelom. Pseudocoelomate animals include the Rotifera and Nematoda.

pseudoheart Any one of a series of contractile blood vessels in annelid worms that pump blood from the dorsal vessel to the ventral vessel.

pseudoparenchyma A tissue that superficially resembles plant parenchyma but is made up of an interwoven mass of hyphae (in fungi) or filaments (in algae). Examples of pseudoparenchymatous structures are the fruiting bodies (mushrooms, toadstools, etc.) of certain fungi and the thalli of certain red and brown algae.

pseudopodium (*pl.* **pseudopodia**) A temporary outgrowth of the cell of some protozoans (e.g. *Amoeba*), which serves as a feeding and locomotory organ. Pseudopodia may be blunt or threadlike, form a branching network, or be stiffened with an internal supporting rod. Phagocytic white blood cells also form pseudopodia to engulf invading bacteria. *See* **amoeboid movement**.

pseudopregnancy A state resembling pregnancy that may occur in some mammals (e.g. rabbits and rodents) in which many of the phenomena of pregnancy are present but there is no fetus developing in the uterus. It is caused by an extended dioestrus (*see* **oestrous cycle**) in the absence of fertilization.

Psilophyta A phylum of primitive *tracheophyte plants that contains two extant genera – *Psilotum* (whisk ferns) and *Tmesipteris* – as well as numerous extinct forms that flourished in the Devonian period. Psilophytes have rhizoids rather than roots; the stems, which show *dichotomous branching, may be naked or bear scalelike leaves.

Pteridophyta In traditional classification systems, a division of the plant kingdom that included ferns, horsetails, and clubmosses, i.e. the nonseed-bearing tracheophytes. These are now classified as separate phyla: *Filicinophyta (ferns), *Sphenophyta (horsetails), *Lycopodophyta (clubmosses), and *Psilophyta.

Pteridospermales (**seed ferns**) An extinct order of gymnosperms that flourished in the Carboniferous period. They possessed characteristics of both the ferns and the seed plants in reproducing by means of seeds and yet retaining fernlike leaves. Their internal anatomy combined both fern and seed-plant characteristics.

pterodactyls *See* **Pterosauria**.

Pteropsida In older classifications, a subdivision of tracheophytes that contained the ferns and seed plants, or a class of the *Pteridophyta containing only the ferns.

Pterosauria An extinct order of flying reptiles – the pterodactyls – that lived in the Jurassic and Cretaceous periods (180–70 million years ago).

Pterodactyls had beaked jaws and an elongated fourth finger that supported a membranous wing. They had long jointed tails, no feathers, and could probably only fly by soaring.

Pterygota A subclass of insects that typically possess wings, although some are secondarily wingless. It comprises the *exopterygotes and the *endopterygotes. *Compare* **Apterygota**.

ptyalin An enzyme that digests carbohydrates (*see* **amylase**). It is present in mammalian *saliva and is responsible for the initial stages of starch digestion.

puberty *See* **adolescence**.

pubis One of the three bones that make up each half of the *pelvic girdle. It is the most anterior of the three pelvic bones. In mammals and many reptiles the pubes are united at a slightly movable joint, the *pubic symphysis*. *See also* **ilium; ischium**.

puff *(in genetics) See* **polyteny**.

pullulan A water-soluble polysaccharide composed of glucose units that are polymerized in such a way as to make it viscous and impermeable to oxygen. Pullulan is used in adhesives, food packaging, and moulded articles. It is derived from the fungus *Aureobasidium pullulans*.

pulmonary Of or relating to the lungs.

pulmonary artery The artery that conveys deoxygenated blood from the right ventricle of the heart to the lungs, where it receives oxygen.

pulmonary circulation The part of the circulatory system of birds and mammals that transports deoxygenated blood from the right side of the heart to the lungs and returns oxygenated blood to the left side of the heart. *Compare* **systemic circulation**. *See* **double circulation**.

pulmonary valve *See* **semilunar valve**.

pulmonary vein The vein that conveys oxygenated blood from the lungs to the left atrium of the heart.

pulp cavity The central region of a tooth, which is connected by a narrow channel at the tip of the root with the surrounding tissues. The pulp cavity contains the *pulp* – connective tissue in which blood vessels and nerve fibres are embedded, and it is lined with *odontoblasts, which produce the *dentine.

pulse *(in physiology)* A series of waves of dilation that pass along the arteries, caused by pressure of blood pumped from the heart through contractions of the left ventricle. In man it can be felt easily where arteries pass close to the skin surface, e.g. at the wrist.

pulvinus A group of cells at the base of a leaf or leaflet in certain plants that, by rapidly losing water, brings about changes in the position of the leaves. In the sensitive plant (*Mimosa pudica*), the pulvinus is responsible for the folding of the leaves that occurs at nightfall or when the plant is touched or injured.

punctuated equilibrium A theory proposing that plant and animal species usually arise very quickly in terms of geological time (in less than 100 000 years) and seldom through a process of gradual change. It thus questions the traditional Darwinian theory of evolution, citing as evidence the discontinuities observed in the fossil records of certain animal groups (e.g. the ammonites).

Punnett square A representation of all possible genotypes produced in the *F_2 generation of a Mendelian genetic cross; it was devised by R. C. Punnett. All possible male gametes are listed horizontally against a grid and all possible female gametes are listed vertically against the grid. The various combinations of these gametes are then written into the grid. See illustration at *dihybrid cross.

pupa The third stage of development in the life cycle of *endopterygote insects. During the pupal stage locomotion and feeding cease and *metamorphosis from the larva to the adult form takes place. There are three types of pupa. The commonest is the *exarate* or free pupa, in which the wings and other appendages are visible and movable. In the *obtect* type the wings are stuck to the body and immovable, as in the *chrysalis* of a butterfly or moth; and in the *co-arctate* type an exarate pupa develops within a hard barrel-shaped *puparium*, as in the housefly and other Diptera.

pupil *See* iris.

pupillary reflex An alteration in the size of the pupil in response to a change in the light intensity. An increase in light intensity causes both pupils to contract (*see* **iris**) even if only one retina is directly stimulated. This is known as a *consensual reflex. Conversely, a decrease in light intensity causes dilation of the pupils.

pure line A population of plants or animals all having a particular feature that has been retained unchanged through many generations. The organisms are *homozygous and are said to 'breed true' for the feature concerned.

Purine

purine An organic nitrogenous base (see formula), sparingly soluble in water, that gives rise to a group of biologically important derivatives, notably *adenine and *guanine, which occur in *nucleotides and nucleic acids (DNA and RNA).

Purkyne cell (Purkinje cell) A type of nerve cell large numbers of which are found in the cortex of the *cerebellum. Purkyne cells are *multipolar neurones having large bodies from which arise several dendrons with highly branched *dendrites, which form a flat fan extending towards the surface of the cerebellum. They are named after the Czech physiologist J. E. Purkyne (1787–1869).

Purkyne fibres (Purkinje fibres) Modified fibres in the mammalian heart that originate in the *bundle of His and spread out in a network over the ventricles. Action potentials generated in the sinoatrial node (the *pacemaker of the heart) are conducted extremely rapidly through the ventricles, due to the extensive branching of the Purkyne fibres, causing both ventricles to contract almost simultaneously.

putrefaction The microbial decomposition of organic matter, especially the anaerobic breakdown of proteinaceous material with the production of foul-smelling amines.

Pycnogonida *See* **Chelicerata**.

pyknotic Describing a nucleus of a damaged cell that has decreased in volume and become darker due to some degree of condensation of the nuclear *chromatin. *See also* **karyorrhexis**.

pyloric sphincter *See* **sphincter; stomach**.

pyramidal cell A type of neurone found in the cerebral cortex of the brain. Pyramidal cells have a pyramid-shaped cell body and its dendrites extend from both the body and the axon.

pyramid of biomass A diagrammatic representation of the amount of organic material (*see* **biomass**), measured in grams of dry mass per square metre ($g\,m^{-2}$), found in a particular habitat at ascending *trophic levels of a *food chain. Biomass decreases at each ascending level of the food chain. A pyramid of biomass is a more accurate representation of the flow of energy through a food chain than a *pyramid of numbers, but seasonal variations in the rate of turnover of the organisms at a particular level may result in higher or lower values for the amount of biomass sampled at a particular time than the average amount over the whole year. The best representation of energy flow in a food chain is a *pyramid of energy.

pyramid of energy A diagrammatic representation of the amount of energy, measured in kilojoules per square metre per year ($kJ\,m^{-2}\,yr^{-1}$), available at ascending *trophic levels of a *food chain in a particular habitat. A pyramid of energy is the most accurate representation of the *energy flow through a food chain as it indicates how much energy is lost at each trophic level (through respiration, etc.). *See also* **bomb calorimeter**. *Compare* **pyramid of biomass; pyramid of numbers**.

Pyramid of energy

pyramid of numbers A diagramatic representation of the numbers of animals found in an area at ascending *trophic levels of a *food chain (see illustration). Because only a small proportion of the energy taken in by an organism is converted to tissue and is thus available to consumers at the next trophic level, the number of organisms that can be supported at each level is generally much less than the number at the level that supplies its food (i.e. the level below). *See also* **pyramid of biomass; pyramid of energy**.

Pyramid of numbers for a woodland food chain

pyranose A *sugar having a six-membered ring containing five carbon atoms and one oxygen atom.

pyrenocarp *See* **drupe.**

pyrenoid A spherical protein body found in the *chloroplasts of many algae. Pyrenoids are associated with the storage of starch: layers of starch are often found around them.

pyrethrum An insecticide (*see* **pesticide**) prepared from the flowers of

Chrysanthemum cinerariifolium. It penetrates the insect's cuticle and is fast-acting, nontoxic to other animals and to plants, and readily biodegradable.

pyridoxine *See* **vitamin B complex**.

Pyrimidine

pyrimidine An organic nitrogenous base (see formula), sparingly soluble in water, that gives rise to a group of biologically important derivatives, notably *uracil, *thymine, and *cytosine, which occur in *nucleotides and nucleic acids (DNA and RNA).

HC ———— CH
HC   CH
      N
      H

Pyrrole

pyrrole An organic nitrogen-containing compound (see formula) that forms part of the structure of *porphyrins.

pyruvic acid (2-oxopropanoic acid) A colourless liquid organic acid, $CH_3COCOOH$. Pyruvate is an important intermediate compound in metabolism, being produced during *glycolysis and converted to acetyl coenzyme A, required for the *Krebs cycle. Under anaerobic conditions pyruvate is converted to lactate or ethanol.

quadrat An ecological sampling unit consisting of a small square area of ground within which all species of interest are noted or measurements taken. Quadrats may be spaced over a larger area to form an overall view when a total survey would be impracticable, or they may be used to sample along a *transect.

quadrate A paired bone in the upper jaw of bony fishes, amphibians, reptiles, and birds that articulates with the lower jawbone. It is absent in mammals, being reduced to a small bone (the incus) in the middle ear (*see* **ear ossicles**).

quadriceps A group of *extensor muscles that cover the anterior surface and sides of the mammalian femur. These muscles join at the base and are connected to the tibia by a single tendon. *Muscle spindles in the quadriceps are responsible for the knee-jerk reflex (*see* **stretch reflex**).

qualitative variation *See* discontinuous variation.

quantitative structure–activity relationship (QSAR) A statistical algorithm that quantitatively defines the relationship between the chemical structure of a drug and its effect on an organism. QSAR studies are often used to predict the activity or toxicity of new drugs. Similar methods can be used to predict the metabolism of new drugs (*quantitative structure–metabolism relationships*).

quantitative structure–metabolism relationship (QSMR) *See* quantitative structure–activity relationship.

quantitative variation *See* continuous variation.

quarantine A period of isolation imposed on an animal that moves from an area where particular diseases are prevalent to an area where those diseases are not prevalent. In the UK domestic animals entering the country are quarantined for the necessary incubation period (*see* **infection**) in order to prevent the spread of rabies.

Quaternary The second period of the Cenozoic era, which began about 2 million years ago, following the *Tertiary period, and includes the present. It is subdivided into two epochs – the *Pleistocene and *Holocene. The beginning of the Quaternary is usually based on the onset of a worldwide cooling. During the period four principal glacial phases occurred in Europe and North America, in which ice advanced towards the equator, separated by interglacials during which conditions became warmer and the ice sheets and glaciers retreated. The last glacial ended about 10 000 years ago. Man became the dominant terrestrial species during the Quaternary. Among the fauna adapted to the cold conditions were the mammoth and the woolly rhinoceros.

quaternary structure *See* protein.

queen substance A *pheromone, *trans*-9-keto-2-decenoic acid, that is secreted by the mandibular salivary glands of a queen honeybee and inhibits the development of ovaries in the worker bees in the colony. The queen substance also affects the behaviour of the workers, preventing them from maintaining special brood cells for larvae that develop into new queens. As the queen ages, the secretion of queen substance diminishes: the workers then construct brood cells for future queens, which they feed exclusively with *royal jelly.

quinone Any of various compounds derived from benzene and containing C=O groups in an unsaturated ring. An example is *plastoquinone.

race 1. *(in biology)* A category used in the *classification of organisms that consists of a group of individuals within a species that are geographically, ecologically, physiologically, or chromosomally distinct from other members of the species. The term is frequently used in the same sense as *subspecies. *Physiological races*, for example, are identical in appearance but differ in function. They include strains of fungi adapted to infect different varieties of the same crop species. **2.** *(in anthropology)* A distinct human type possessing several characteristics that are genetically inherited. The major races of man are Mongoloid, Caucasian, Negroid, and Australoid.

raceme A type of *racemose inflorescence in which the main flower stalk is elongated and bears stalked flowers. An example is the lupin. *See also* **panicle**.

racemose inflorescence (indefinite inflorescence) A type of flowering shoot (*see* **inflorescence**) in which the growing region at the tip of the flower stalk continues to produce new flower buds during growth. As a result, the youngest flowers are at the top and the oldest flowers are at the base of the stalk. In a flattened inflorescence, the youngest flowers are in the centre and the oldest flowers are on the outside. Types of racemose inflorescence include the *capitulum, *catkin, *corymb, *raceme, *spadix, *spike, and *umbel (see illustration). *Compare* **cymose inflorescence**.

rachis (rhachis) **1.** The main axis of a compound leaf or an inflorescence. **2.** The shaft of a *feather. **3.** The backbone.

rad *See* **radiation units**.

radial symmetry The arrangement of parts in an organ or organism such that cutting through the centre of the structure in any direction produces two halves that are mirror images of each other. The stems and roots of plants usually show radial symmetry, while all animals belonging to the Cnidaria (e.g. jellyfish) and Echinodermata (e.g. starfish) are radially symmetrical – and typically sessile – in their adult form. The term *actinomorphy* is used to describe radial symmetry in flowers (e.g. a buttercup flower). *Compare* **bilateral symmetry**.

radiation 1. Energy travelling in the form of electromagnetic waves or photons. *See* **electromagnetic spectrum**. **2.** A stream of particles, especially alpha- or beta-particles from a radioactive source or neutrons from a nuclear reactor. *See also* **background radiation**.

radiation damage Harmful changes that occur to inanimate materials and living organisms as a result of exposure to energetic electrons, nucleons, fission fragments, or high-energy electromagnetic radiation. In organisms this can cause changes to cells that alter their genetic structure,

spike
(e.g. plantain)

raceme
(e.g. lupin)

corymb
(e.g. candytuft)

umbel
(e.g. hogweed)

capitulum
(e.g. daisy)

1 = oldest flower

Types of racemose inflorescences

interfere with their division, or kill them. In humans, these changes can lead to *radiation sickness*, *radiation burns* (from large doses of radiation), or to long-term damage of several kinds, the most serious of which result in various forms of cancer (especially leukaemia).

radiation units Units of measurement used to express the activity of a radionuclide and the *dose of ionizing radiation. The units *curie*, *roentgen*, *rad*, and *rem* are not coherent with SI units but their temporary use with SI units has been approved while the derived SI units *becquerel*, *gray*, and *sievert* become familiar.

The becquerel (Bq), the SI unit of activity, is the activity of a radionuclide decaying at a rate, on average, of one spontaneous nuclear transition per second. Thus 1 Bq = 1 s^{-1}. The former unit, the curie (Ci), is equal to 3.7×10^{10} Bq. The curie was originally chosen to approximate the activity of 1 gram of radium−226.

The gray (Gy), the SI unit of absorbed dose, is the absorbed dose when the energy per unit mass imparted to matter by ionizing radiation is 1 joule per kilogram. The former unit, the rad (rd), is equal to 10^{-2} Gy.

The sievert (Sv), the SI unit of dose equivalent, is the dose equivalent when the absorbed dose of ionizing radiation multiplied by the stipulated dimensionless factors is $1 \, J \, kg^{-1}$. As different types of radiation cause different effects in biological tissue a weighted absorbed dose, called the *dose equivalent*, is used in which the absorbed dose is modified by multiplying it by dimensionless factors stipulated by the International Commission on Radiological Protection. The former unit of dose equivalent, the rem (originally an acronym for *r*oentgen *e*quivalent *m*an), is equal to 10^{-2} Sv.

In SI units, exposure to ionizing radiation is expressed in coulombs per kilogram, the quantity of X- or gamma-radiation that produces ion pairs carrying 1 coulomb of charge of either sign in 1 kilogram of pure dry air. The former unit, the roentgen (R), is equal to $2.58 \times 10^{-4} \, C \, kg^{-1}$.

radicle The part of a plant embryo that develops into the root system. The tip of the radicle is protected by a root cap and points towards the micropyle. On germination it breaks through the testa and grows down into the soil. *Compare* **plumule**.

radioactive age The age of an archaeological or geological specimen as determined by a process that depends on a radioactive decay. *See* **carbon dating; fission-track dating; potassium–argon dating; rubidium–strontium dating; uranium–lead dating**.

radioactive tracing *See* **labelling**.

radioactive waste (nuclear waste) Any solid, liquid, or gaseous waste material that contains radionuclides (radioactive atomic nuclei). These wastes are produced in the mining and processing of radioactive ores, the normal running of nuclear power stations and other reactors, the manufacture of nuclear weapons, and in hospitals and research laboratories. Because high-level radioactive wastes can be extremely dangerous to all living matter and because they may contain radionuclides having half-lives of many thousands of years, their disposal has to be controlled with great stringency.

High-level waste (e.g. spent nuclear fuel) requires to be cooled artificially and is therefore stored for several decades by its producers before it can be disposed of. Intermediate-level waste (e.g. processing plant sludge and reactor components) is solidified, mixed with concrete, packed in steel drums, and stored in special sites at power stations before being buried in concrete chambers in deep mines or below the seabed. Low-level waste (e.g. solids or liquids lightly contaminated by radioactive substances) is disposed of in steel drums in special sites in concrete-lined trenches. In the UK, a company (Nirex Ltd) was set up by the nuclear industry and the government in 1988 to handle the disposal of nuclear waste. There are also nuclear reprocessing plants at Dounreay and Sellafield (Thermal Oxide

Reprocessing Plant). Other countries have similar arrangements. Since 1983 no wastes have been disposed of in steel drums cast in concrete in the Atlantic deeps.

radioactivity The spontaneous disintegration of certain atomic nuclei accompanied by the emission of alpha-particles (helium nuclei), beta-particles (electrons or positrons), or gamma radiation (short-wavelength electromagnetic waves). *Natural radioactivity* is the result of the spontaneous disintegration of naturally occurring radioisotopes. The rate of disintegration is uninfluenced by chemical changes or any normal changes in their environment. However, radioactivity can be induced in many nuclides by bombarding them with neutrons or other particles. *See also* **radiation units**.

radiobiology The branch of biology concerned with the effects of radioactive substances on living organisms and the use of radioactive tracers to study metabolic processes (*see* **labelling**).

radiocarbon dating *See* **carbon dating**.

radiography The process or technique of producing images of an opaque object on photographic film or on a fluorescent screen by means of radiation (either particles or electromagnetic waves of short wavelength, such as X-rays and gamma-rays). The photograph produced is called a *radiograph*. The process is widely used in diagnostic *radiology, using X-rays. *See also* **autoradiography**.

radioimmunoassay (RIA) A sensitive quantitative method for detecting trace amounts of a biomolecule, based on its capacity to displace a radioactively labelled form of the molecule from combination with its antibody. *See also* **immunoassay**.

radioisotope (radioactive isotope) An isotope of an element that is radioactive. *See* **labelling**.

radiology The study and use of X-rays, radioactive materials, and other ionizing radiations for medical purposes, especially for diagnosis (*diagnostic radiology*) and the treatment of cancer and allied diseases (*therapeutic radiology* or *radiotherapy*).

radiometric dating (radioactive dating) *See* **dating techniques**; **radioactive age**.

radio-opaque (radiopaque) Describing a medium that is opaque to X-rays and gamma rays. Examples are barium salts, used in diagnostic radiology of the digestive tract.

radiotherapy *See* **radiology**.

radius *(in anatomy)* The smaller of the two bones in the lower section of the forelimb of a tetrapod vertebrate (*compare* **ulna**). The radius articulates with some carpal bones and the ulna at the wrist and with the *humerus at the elbow. This sophisticated articulation of the radius enables man (and some other animals) to twist the forearm (*see* **pronation; supination**).

radula A tonguelike organ of molluscs, consisting of a horny strip whose surface is studded with rows of horny teeth for rasping food. In some species it is modified for scraping or boring.

rainforest Any major terrestrial *biome in which trees form the dominant plants and annual rainfall is high (over 200 cm). Tropical rainforest is restricted to equatorial regions, such as the Amazon basin, central west Africa, and SE Asia. It is dominated by broadleaved evergreens and shows a very rich species diversity (*see* **biodiversity**). The leafy crowns of the trees typically form three layers of canopy, since the trees grow to different heights, which prevents much sunlight from reaching ground level. This limits the number of herbaceous plants and small shrubs that grow on the forest floor, but *epiphytes, vines, and creepers are abundant. The average temperature is about 27°C, which – together with the high humidity – encourages rapid decomposition of leaf litter, releasing minerals that replace those leached from the soil by the heavy rain. If the forest canopy is removed, the soil is destroyed rapidly due to leaching by rain. The soil of a rainforest, known as *latosol*, is acidic and typically red, due to the oxidation of iron oxide (Fe_2O_3) in the topsoil. Rainforests are thought to contain many undiscovered plant species that could be of benefit in the fields of medicine and biotechnology. The continued destruction of rainforest in many parts of the world, particularly in South America (*see* **deforestation**), will not only result in the loss of these and other species but also contributes to the *greenhouse effect.

r.a.m. *See* **relative atomic mass**.

Ramapithecus A genus of extinct primates that lived between 8 and 14 million years ago. Fossil remains of ramapithecines have been found in India and Pakistan, the Near East, and East Africa. Jaw fragments suggest that they chewed from side to side and had fairly short muzzles, both of which are humanoid features. They were possibly the earliest representatives of the hominid line after this split from the pongids (apes) during the Miocene. *See also* **Dryopithecus; Australopithecus**.

ramus communicans (*pl.* **rami communicantes**) A structure containing nerve fibres that runs from a spinal nerve to a *ganglion of the sympathetic nervous system (*see* **autonomic nervous system**). The sympathetic ganglia form a series down each side of the vertebral column; each is connected to the ventral root of a spinal nerve by a pair of rami communicantes.

rank (category) The position or status of a *taxon in a *classification hierarchy. Examples of ranks are the family, genus, and species.

raphe **1.** A ridge on a seed marking the line of fusion between an anatropous (inverted) ovule and the *funicle. **2.** A groove on the valve of certain diatoms. **3.** Any other groove, ridge, or suture line marking the line of fusion between two originally separate structures.

rapid eye movement (REM) *See* **sleep**.

rate-limiting step The slowest step in a metabolic pathway or series of chemical reactions, which determines the overall rate of the other reactions in the pathway. In an enzymatic reaction, the rate-limiting step is generally the stage that requires the greatest *activation energy or the transition state of highest *free energy.

Ratitae (Palaeognathae) A group comprising the flightless birds, including the ostrich, kiwi, and emu. They have long legs, heavy bones, small wings, a flat breastbone, and curly feathers. These birds are thought to have descended from a variety of flying birds and are not representatives of a single homologous group.

ray **1.** *(in optics)* A narrow beam of radiation. **2.** *(in botany)* *See* **medullary ray**.

reabsorption *(in excretion)* *See* **selective reabsorption**.

reactant *See* **chemical reaction**.

reaction *See* **chemical reaction**.

reaction time (latent period) The period of time between the detection of a stimulus at a sensory receptor and the performance of the appropriate response by the effector organ. This delay is caused by the time taken for the impulse to travel across the synapses of adjacent neurones. The reaction time for a *reflex response, involving only a single linking synapse, is very short.

recapitulation The theory that some stages of evolution are repeated in the development of an individual organism, i.e. that *phylogeny is repeated in *ontogeny. Now largely discredited, the theory was proposed by the German zoologist Ernst Haeckel (1834–1919).

Recent *See* **Holocene**.

receptacle **1.** (thalamus *or* torus) The tip of a flower stalk, which bears the petals, sepals, stamens, and carpels. The way the receptacle develops determines the position of the flower parts. It can be dilated and dome-shaped, saucer-shaped, or hollow and enclosing the gynoecium (*see* **epigyny; hypogyny; perigyny**). In some plants it may become part of the fruit (*see* **pseudocarp**). **2.** A swollen part of the thallus of some algae, e.g. *Fucus*, that bears the conceptacles in which the sex organs are situated.

receptor *(in physiology)* A cell or group of cells specialized to detect a particular stimulus and to initiate the transmission of impulses via the sensory nerves. The eyes, ears, nose, skin, and other sense organs all contain specific receptors responding to external stimuli (*see* **exteroceptor**); other receptors are sensitive to changes within the body (*see* **interoceptor**). *See* *also* **baroreceptor; chemoreceptor; mechanoreceptor; osmoreceptor; proprioceptor.**

receptor site An area of a cell membrane, consisting of a specially adapted membrane protein, that can bind with a specific hormone, neurotransmitter, drug, or other chemical, thereby initiating a change within the cell. For example, the binding of protein hormones to receptor sites of target cells initiates the synthesis of a *second messenger inside the cell.

recessive The *allele that is not expressed in the *phenotype when two different alleles are present in the cells of an organism. The aspect of a characteristic controlled by a recessive allele only appears when two such alleles are present, i.e. in the *double recessive* condition. *Compare* **dominant**.

recipient An individual who receives tissues or organs of the body from another (the *donor).

reciprocal cross A *cross reversing the roles of males and females to confirm the results obtained from an earlier cross. For example, if the pollen (male) from tall plants is transferred to the stigmas (female) of dwarf plants in one cross, the reciprocal cross would use the pollen of dwarf plants to pollinate the stigmas of tall plants.

recognition site *See* **promoter**.

recombinant DNA DNA that contains genes from different sources that have been combined by the techniques of *genetic engineering rather than by breeding experiments. Genetic engineering is therefore also known as *recombinant DNA technology*.

recombination The rearrangement of genes that occurs when reproductive cells (gametes) are formed (*see* **crossing over**). Recombination results in offspring that have a combination of characteristics different from that of their parents. Recombination can also be induced artificially by *genetic engineering techniques.

recon The smallest unit of DNA in which a mutation can occur. A recon consists of a single pair of nucleotides.

recruitment The activation of extra motor neurones in order to bring about an increased response to a stimulus that is present at an even intensity.

rectal gland A salt-secreting gland found in the cloaca of cartilaginous fishes.

rectum The portion of the *alimentary canal between the *colon and the *anus. Its main function is the storage of *faeces prior to elimination.

recycling 1. The recovery and processing of materials after they have been used, which enables them to be reused. For example, used paper, cans, and glass can be broken down into their constituents, which form the raw materials for the manufacture of new products. **2.** The continual movement of essential elements between the biotic (living) and abiotic (nonliving) components of the environment. *See* **carbon cycle; nitrogen cycle; oxygen cycle; phosphorus cycle; sulphur cycle**.

red algae *See* **Rhodophyta**.

red blood cell *See* **erythrocyte**.

redox *See* **oxidation–reduction**.

reducing sugar A monosaccharide or disaccharide sugar that can donate electrons to other molecules and can therefore act as a reducing agent. The possession of a free ketone (–CO–) or aldehyde (–CHO) group enables most monosaccharides and disaccharides to act as reducing sugars. Reducing sugars can be detected by *Benedict's test. *Compare* **nonreducing sugar**.

reduction *See* **oxidation–reduction**.

reduction division *See* **meiosis**.

reflex An automatic and innate response to a particular stimulus. A reflex response is extremely rapid. This is because it is mediated by a simple nervous circuit called a *reflex arc*, which at its simplest involves only a receptor linked to a sensory neurone, which synapses with a motor neurone (supplying the effector) in the spinal cord. Such reflexes are known as *monosynaptic spinal reflexes*; an example is the *stretch reflex. Other spinal reflexes involve more than one synapse (*see* **polysynaptic reflex**); an example is the *withdrawal reflex* of the hand from a painful stimulus (such as fire). *Cranial reflexes* are mediated by pathways in the cranial nerves and brain; examples are the blinking and swallowing reflexes. *See also* **conditioning**.

reflex action An automatic movement produced in response to a stimulus (*see* **reflex**).

reflex arc *See* **reflex**.

reforestation The replanting of trees on areas of land where forests have been cleared by felling or burning (*see* **deforestation**) or by natural means. Reforestation is particularly important in countries, such as Brazil, where large areas of forest have been destroyed by deforestation.

refractory period The period after the transmission of an impulse in a nerve or muscle in which the membrane of the axon or muscle fibre regains its ability to transmit impulses (*see* **action potential**). This period lasts approximately 3 milliseconds and is divided into an *absolute refractory period*, during which a second impulse may not be generated; and a *relative refractory period*, during which it is possible to generate an impulse only if there is an abnormally strong stimulus.

regeneration The growth of new tissues or organs to replace those lost or damaged by injury. Many plants can regenerate a complete plant from a shoot segment or a single leaf, this being the basis of many horticultural propagation methods (*see* **cutting**). The capacity for regeneration in animals is less marked. Some planarians and sponges can regenerate whole organisms from small pieces, and crustaceans (e.g. crabs), echinoderms (e.g. brittlestars), and some reptiles and amphibians can grow new limbs or tails (*see* **autotomy**), but in mammals regeneration is largely restricted to wound healing.

regma A dry fruit that is characteristic of the geranium family. It is similar to the *carcerulus but breaks up into one-seeded parts, each of which splits open to release a seed.

regulation *(in embryology)* The processes involved in animal embryonic development that counteract any abnormalities that may arise in the different developmental stages. Regulative (or regulation) embryos or eggs can compensate for the removal of sections of the embryo or egg at an early stage of development so that subsequent development is not affected; in such embryos the direction of development of the cells is not *determined until cleavage is well advanced. Regulation also includes *twinning*: the formation of two embryos from the cleavage of a single embryo (*see* **twins**).

regulator Any organism that can maintain a constant *internal environment largely independently of the external environment. This is generally achieved by homeostatic mechanisms (*see* **homeostasis**). Regulators tend to occupy habitats in which environmental conditions are variable.

regulator gene *See* operon.

regulatory enzyme Any enzyme that is involved in controlling the different metabolic pathways in the cell by switching them on or off. Regulatory enzymes exist in active and inactive forms; they include *allosteric enzymes and those enzymes whose activity is controlled by *kinases.

regulatory protein (gene-regulatory protein) Any protein that influences the regions of a DNA molecule that are transcribed by RNA polymerase during the process of *transcription. These proteins, which include *transcription factors, therefore help control the synthesis of proteins in cells.

reinforcement *(in animal behaviour)* Increasing (or decreasing) the frequency of a particular behaviour through *conditioning, by arranging for some biologically important event (the *reinforcer*) always to follow another event. In instrumental conditioning an *appetitive reinforcer*, or *reward* (e.g. food), given after a response made by the animal, increases that response; an *aversive reinforcer*, or *punishment* (e.g. an electric shock) decreases the response.

Reissner's membrane The membrane in the *cochlea of the inner ear that separates the *vestibular canal from the cochlear duct. It is named after the German anatomist Ernst Reissner (1824–78).

relative atomic mass (atomic weight; r.a.m.) Symbol A_r. The ratio of the average mass per atom of the naturally occurring form of an element to 1/12 of the mass of a carbon–12 atom.

relative density (r.d.) The ratio of the density of a substance to the density of some reference substance. For liquids or solids it is the ratio of the density (usually at 20°C) to the density of water (at its maximum density). This quantity was formerly called *specific gravity*.

relative growth rate A measurement of the *productivity of a plant, defined as the increase in dry mass per unit of plant mass over a specified period of time.

relative molecular mass (molecular weight) Symbol M_r. The ratio of the average mass per molecule of the naturally occurring form of an element or compound to 1/12 of the mass of a carbon–12 atom. It is equal to the sum of the relative atomic masses of all the atoms that comprise a molecule.

relative refractory period *See* **refractory period**.

relaxin A mammalian hormone produced by the corpus luteum and placenta during the terminal stages of pregnancy. It relaxes the pubic symphysis and dilates the cervix of the uterus, thereby aiding parturition.

releaser *See* **sign stimulus**.

releasing factor (releasing hormone) A hormone that is produced by the hypothalamus and stimulates the release of a hormone from the anterior *pituitary gland into the bloodstream. Each hormone has a specific releasing factor; for example, thyrotrophin-releasing hormone stimulates the release of *thyroid-stimulating hormone.

rem *See* **radiation units**.

renal Of or relating to the *kidney. For example, the *renal artery* and *renal vein* convey blood towards and away from the kidney, respectively.

renal capsule *See* **Bowman's capsule**.

renal tubule Any of the sections of a *nephron of the vertebrate kidney that are concerned with reabsorption of water and solutes from the *glomerular filtrate. *See also* **proximal convoluted tubule; loop of Henle; distal convoluted tubule**.

renaturation The reconstruction of a protein or nucleic acid that has been *denatured such that the molecule resumes its original function. Some proteins can be renatured by reversing the conditions (of temperature, pH, etc.) that brought about denaturation.

renin A proteolytic enzyme (*see* **protease**) that is involved in the formation of the hormone *angiotensin, which raises blood pressure. Renin is secreted into the blood by cells of the kidney glomeruli under the control of the sympathetic nervous system; its release also occurs in response to a fall in blood-sodium levels and to falling blood pressure. Renin catalyses cleavage of circulating α_2-globulin to produce angiotensin I, precursor of the active hormone.

rennin (chymase) An enzyme secreted by cells lining the stomach in mammals that is responsible for clotting milk. It acts on a soluble milk protein (*caseinogen*), which it converts to the insoluble form *casein. This ensures that milk remains in the stomach long enough to be acted on by protein-digesting enzymes.

repetitive DNA Sequences of nucleotides that occur repeatedly in chromosomal DNA. Sequences range in length from tens to many thousands of base pairs and they can be repeated thousands of times in the genome. Repetitive DNA includes certain repeated sequences concentrated in the region of the centromere; known as *satellite DNA, it is not usually transcribed. Much repetitive DNA does not seem to have a function (*see* **selfish DNA**) but some is transcribed.

replacing bone *See* cartilage bone.

replicon A DNA sequence that is replicated as a unit from a single initiation site (origin of replication). The genome of a bacterium or a virus comprises a single replicon; eukaryotes contain a number of replicons on each chromosome.

repolarization The restoration of the *resting potential in neurones or muscle fibres following the passage of a nerve impulse. Repolarization is brought about by diffusion of potassium ions out of the neurone and by active elimination of sodium ions (*see* **sodium pump**).

repressor A protein that can prevent the expression of a gene. *See* operon.

reproduction The production of new individuals more or less similar in form to the parent organisms. This may be achieved by a number of means (*see* **sexual reproduction; asexual reproduction**) and serves to perpetuate or increase a species.

reproductive system The organs that are involved in the process of *sexual reproduction in an organism. The reproductive system of a flowering plant is found in the *flower and consists of the stamens (male organs) and carpels (female organs). In mammals the reproductive system consists of the testes, epididymis, sperm duct, and penis in the male and the ovaries, fallopian tubes, and uterus in the female (see illustration).

Reptilia The class that contains the first entirely terrestrial vertebrates, which can live in dry terrestrial habitats as their skin is covered by a layer of horny scales, preventing water loss. They breathe atmospheric oxygen by means of lungs assisted by respiratory movements principally involving the

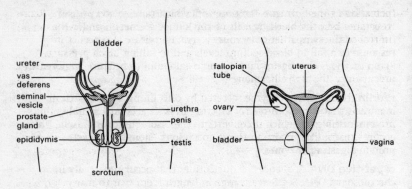

Human male reproductive system   *Human female reproductive system*

ribs (there is no diaphragm). Reptiles are cold-blooded (*see* **poikilothermy**) but behavioural patterns make it possible for them to maintain a fairly even body temperature throughout the day. Fertilization is internal and the majority of reptiles lay eggs on land. These eggs have a porous shell to provide protection from desiccation and allow gas exchange. In some reptiles the eggs are retained within the body of the mother until the young are ready to hatch, thereby greatly reducing juvenile mortality (*see* **ovoviviparity**).

The class includes the modern crocodiles, lizards and snakes (*see* **Squamata**), and tortoises and turtles, as well as many extinct forms, such as the *dinosaurs and *Pterosauria.

residual volume The amount of air remaining in the lungs after maximum expiration, which cannot be expelled from the lungs voluntarily. An average human has a residual volume of about 1 litre. *See also* **vital capacity**.

resin A naturally occurring acidic polymer secreted by many trees (especially conifers) into ducts or canals. Resins are found either as brittle glassy substances or dissolved in essential oils. Their functions are probably similar to those of gums and mucilages, i.e. protective.

resistance **1.** *(in microbiology)* The degree to which pathogenic microorganisms remain unaffected by antibiotics and other drugs. Genes for antibiotic resistance are often carried on *transposons, which can spread across species barriers. **2.** *(in ecology)* **a.** The degree to which a *pest can withstand the effects of a pesticide. It depends on the selection and spread within a pest population of genes that confer the ability to destroy, or minimize the effects of, a pesticide. **b.** *See* **environmental resistance**. **3.** *(in immunology)* The degree of *immunity to infection that an animal possesses.

resistance response A long-term response to any stimulus that potentially threatens the wellbeing of an organism. An example of a resistance response is the release of *ACTH (adrenocorticotrophic hormone) in response to stress. This hormone stimulates the production of mineralocorticoid hormones (e.g. *aldosterone), which promote the excretion of hydrogen ions produced in excess as a result of the increased metabolic activity resulting from the *alarm response. Resistance responses diminish as normal conditions resume.

resolving power A measure of the ability of an optical instrument to form separable images of close objects or to separate close wavelengths of radiation. The resolving power of a microscope is usually taken as the minimum distance between two points that can be separated; the smaller the resolving power, the better the resolution.

respiration The metabolic process in animals and plants in which organic substances are broken down to simpler products with the release of energy, which is incorporated into special energy-carrying molecules (*see* **ATP**) and subsequently used for other metabolic processes. In most plants and animals respiration requires oxygen, and carbon dioxide is an end product. The exchange of oxygen and carbon dioxide between the body tissues and the environment is called *external respiration* (*see* **ventilation**). In many animals the exchange of gases takes place at *respiratory organs (e.g. *lungs in air-breathing vertebrates) and is assisted by *respiratory movements (e.g. breathing). In plants oxygen enters through pores on the plant surface and diffuses through the tissues via intercellular spaces or dissolved in tissue fluids.

  Respiration at the cellular level is known as *internal* (*cellular* or *tissue*) *respiration* and can be divided into two stages. In the first, *glycolysis, glucose is broken down to pyruvate. This does not require oxygen and is a form of *anaerobic respiration. In the second stage, the *Krebs cycle, pyruvate is broken down by a cyclic series of reactions to carbon dioxide and water. This is the main energy-yielding stage and requires oxygen. The processes of glycolysis and the Krebs cycle are common to all plants and animals that respire aerobically (*see* **aerobic respiration**).

respiratory chain *See* **electron transport chain.**

respiratory movement The muscular movement that enables the passage of air to and from the lungs or other *respiratory organs of an animal. The mechanism of the movement varies with the species. In insects abdominal muscles relax and contract rhythmically to encourage the flow of air through the *tracheae. In amphibians air is drawn into the lungs by a pumping action of the muscles in the floor of the mouth. *Breathing* in mammals involves the muscle of the *diaphragm and the *intercostal muscles between the ribs. Contraction of these muscles lowers the diaphragm and raises the ribs, so that the lungs expand and air is drawn in (*see* **inspiration**). Relaxation has the opposite effect and forces air out during *expiration.

respiratory organ Any animal organ across which exchange of carbon dioxide and oxygen takes place. The surface membranes of such organs are always moist, thin, and well supplied with blood. Examples are the *lungs of air-breathing vertebrates, the *gills of fish, the *tracheae of insects, and the *lung books of arachnids.

respiratory pigment 1. A coloured compound that is capable of reversibly binding with oxygen at high oxygen concentrations and releasing it at low oxygen concentrations. Most such pigments are present in the blood (*blood pigments*), transporting oxygen within the circulatory system from the *respiratory organs to the tissues of the body; an exception is *myoglobin, which occurs in the muscles. In vertebrates the respiratory pigment is *haemoglobin, contained in the erythrocytes (red blood cells). *See also* **chlorocruorin; haemocyanin; haemerythrin**. **2.** Any of the proteins involved in cellular respiration as components of an *electron transport chain. They include the *cytochromes.

respiratory quotient (RQ) The ratio of the volume of carbon dioxide produced by an organism during respiration to the volume of oxygen consumed. The RQ is usually about 0.8.

respirometer Any device that measures an organism's oxygen uptake. Simple respirometers consist of a chamber (in which the organism is placed) connected to a *manometer. Carbon dioxide is chemically removed from the chamber so that only oxygen uptake is measured. Human oxygen consumption is generally measured by a device known as a *spirometer*, which can also be used to measure depth and frequency of breathing.

response The physiological, muscular, or behavioural activity that can be elicited by a *stimulus.

resting potential The difference in electrical potential that exists across the membrane of a nerve cell that is not in the process of transmitting a nerve impulse. The resting potential is maintained by means of the *sodium pump. *Compare* **action potential**.

restriction enzyme (restriction endonuclease) A type of enzyme that can cleave molecules of foreign DNA at a particular site. Restriction enzymes are produced by many bacteria and protect the cell by cleaving (and therefore destroying) the DNA of invading viruses. The bacterial cell is protected from attack by its own restriction enzymes by modifying the bases of its DNA during replication. Restriction enzymes are widely used in the techniques of genetic engineering (*see* **gene cloning; gene library; gene sequencing; genetic fingerprinting; restriction mapping**).

restriction fragment length polymorphism (RFLP) The occurrence of different cleavage sites for *restriction enzymes in the DNA of different individuals of the same species. Cleavage of DNA from different individuals with restriction enzymes thus produces differing sets of restriction fragments. The deletion of existing restriction sites or the creation of new ones is the result of random base changes in the noncoding stretches of DNA (*introns) between genes. RFLPs have provided geneticists with a powerful set of genetic markers for mapping the genome (*see* **restriction mapping**) and for identifying particular genes (*see* **gene tracking**).

restriction mapping A technique for determining the sites at which a length of DNA (e.g. from a chromosome) is cleaved by *restriction enzymes. By cleaving the DNA with various such enzymes, both individually and in combination, and analysing the resultant number and size of fragments by electrophoresis, a *restriction map*, indicating the order of restriction sites in the original DNA, can be deduced. Restriction mapping is routinely applied to organisms undergoing genetic investigation. Gene deletions or rearrangements that alter the restriction sites can be detected as changes in the pattern of fragments obtained. This may be used, for instance, to diagnose certain genetic abnormalities in the fetus. The fragments are separated by gel electrophoresis and identified using specific *gene probes, as in the *Southern blotting technique. The absence of a certain fragment in a fetal DNA digest can be diagnostic of a pathological change in the fetal gene containing the corresponding restriction site.

restriction point *See* **cell cycle**.

reticular formation *See* **brainstem**.

reticuloendothelial system The network of *macrophages in the body's tissues, which is concentrated in the bone marrow, liver, spleen, and lymph nodes.

reticulum The second of four chambers that form the stomach of ruminants. *See* **Ruminantia**.

retina The light-sensitive membrane that lines the interior of the eye. The retina consists of two layers. The outer layer (*pigment epithelium*) is pigmented, which prevents the back reflection of light and consequent decrease in visual acuity. The inner layer contains nerve cells, blood vessels, and two types of light-sensitive cells (*rods and *cones). Light passing through the lens stimulates individual rods and cones, which generates nerve impulses that are transmitted through *bipolar* and *ganglion cells* to the optic nerve, and hence to the brain, where the visual image is formed. Information can also be transferred horizontally within the retina via a network of *horizontal* and *amacrine cells* (see illustration).

retinal (retinene) *See* **rhodopsin**; **vitamin A**.

retinol *See* **vitamin A**.

retrovirus An RNA-containing virus that converts its RNA into DNA by means of the enzyme *reverse transcriptase; this enables it to become integrated into its host's DNA. Retroviruses can cause cancer in animals: they contain *oncogenes (cancer-causing genes), which are activated when the virus enters its host cell and starts to replicate. *See also* **provirus**.

reverse transcriptase An enzyme, occurring in *retroviruses, that catalyses the formation of double-stranded DNA using the single RNA

pigment epithelium

rod

cone

horizontal cell

bipolar cell

amacrine cell

ganglion cell

to optic nerve

Structure of the retina

strand of the viral genome as template. This enables the viral genome to be inserted into the host's DNA and replicated by the host. Reverse transcriptase is thus an RNA-directed DNA *polymerase. The enzyme is used in genetic engineering for producing *complementary DNA from messenger RNA.

RFLP *See* **restriction fragment length polymorphism.**

R_F value *(in chromatography)* The distance travelled by the solvent front divided by the distance travelled by a given component. For a given system at a known temperature, it is a characteristic of the component and can be used to identify components. For example, the photosynthetic pigments of an organism and the metabolites of a drug excreted in the urine can be identified by their R_F values in paper or thin-layer chromatography.

rhabdom *See* compound eye.

rhachis *See* rachis.

rhesus factor (Rh factor) An *antigen whose presence or absence on the surface of red blood cells forms the basis of the rhesus *blood group system. (The factor was first recognized in rhesus monkeys.) Most people possess the Rh factor, i.e. they are rhesus positive (Rh+). People who lack the factor are Rh−. If Rh+ blood is given to an Rh− patient, the latter develops anti-Rh antibodies. Subsequent transfusion of Rh+ blood results in *agglutination, with serious consequences. Similarly, an Rh− pregnant woman carrying an Rh+ fetus may develop anti-Rh antibodies in her blood; these will react with the blood of a subsequent Rh+ fetus, causing anaemia in the newborn baby.

rhizoid One of a group of delicate and often colourless hairlike outgrowths found in certain algae and the gametophyte generation of bryophytes and ferns. They anchor the plant to the substrate and absorb water and mineral salts.

rhizome A horizontal underground stem. It enables the plant to survive from one growing season to the next and in some species it also serves to propagate the plant vegetatively. It may be thin and wiry, as in couch grass, or fleshy and swollen, as in *Iris*. Compact upright underground stems, as in rhubarb, strawberry, and primrose, are often called *rootstocks*.

Rhizopoda A phylum of the Protoctista that contains the amoebas. They are characterized by the possession of *pseudopodia, which are used for locomotion and engulfing food particles. Rhizopods are found in freshwater and marine habitats and the soil. They reproduce by binary *fission. *See also* Amoeba; protozoa.

Rhodophyta (red algae) A phylum of *algae that are often pink or red in colour due to the presence of the pigments *phycocyanin and *phycoerythrin. Members of the Rhodophyta may be unicellular or multicellular; the latter form branched flattened thalli or filaments. They are commonly found along the coasts of tropical areas. Sexual reproduction is by means of carpospores (*see* **carpogonium**).

rhodopsin (visual purple) The light-sensitive pigment found in the *rods of the vertebrate retina. It consists of a protein component, *opsin*, linked to a nonprotein *chromophore, *retinal* (or *retinene*), a derivative of *vitamin A. Light falling on the rod is absorbed by the retinal, which changes its form and separates from the opsin component; this initiates the transmission of a nerve impulse to the brain. The great sensitivity of rhodopsin allows vision in dim light (night vision). *See also* **dark adaptation**.

rhombencephalon *See* hindbrain.

rhytidome *See* bark.

rib One of a series of slender curved bones that form a cage to enclose, support, and protect the heart and lungs (*see* **thorax**). Ribs occur in pairs, articulating with the *thoracic vertebrae of the spinal column at the back and (in reptiles, birds, and mammals) with the *sternum (breastbone) in front. Movements of the rib cage, controlled by *intercostal muscles between the ribs, are important in breathing (*see* **respiratory movement**).

riboflavin *See* **vitamin B complex**.

ribonuclease *See* **RNAase**.

ribonucleic acid *See* **RNA**.

ribonucleoprotein (RNP) Any complex of protein and RNA that forms during the synthesis of RNA in eukaryotes; the protein is involved in the packaging and condensation of the RNA. Certain RNPs are restricted to the nucleus whereas others are found in both the nucleus and the cytoplasm. The most common RNP occurring in the nucleus is *heterogeneous nuclear RNP (hnRNP)*, which consists of protein bound to the primary transcript of DNA (*see* **transcription**). It may be associated with *small nuclear RNP (snRNP)*, which is involved in the removal of intron sequences from the primary transcript to form messenger RNA, which eventually leaves the nucleus (*see* **gene splicing**).

ribose A *monosaccharide, $C_5H_{10}O_5$, rarely occurring free in nature but important as a component of *RNA (ribonucleic acid). Its derivative *deoxyribose*, $C_5H_{10}O_4$, is equally important as a constituent of *DNA (deoxyribonucleic acid), which carries the genetic code in chromosomes.

ribosomal RNA *See* **ribosome; RNA**.

ribosome A small spherical body within a living cell that is the site of *protein synthesis. Ribosomes consist of a type of RNA (called *ribosomal RNA*) and protein. Usually there are many ribosomes in a cell, either attached to the *endoplasmic reticulum or free in the cytoplasm. During protein synthesis they are associated with messenger RNA as *polyribosomes in the process of *translation.

ribozyme Any RNA molecule that can catalyse changes to its own molecular structure. Self-splicing introns (*see* **gene splicing**) are examples of ribozymes.

ribulose A ketopentose sugar (*see* **monosaccharide**), $C_5H_{11}O_5$, that is involved in carbon dioxide fixation in photosynthesis as a component of *ribulose bisphosphate.

ribulose bisphosphate (RuBP) A five-carbon sugar that is combined with carbon dioxide to form a six-carbon intermediate in the first stage of the dark reaction of *photosynthesis (*see* **Calvin cycle**). The enzyme that mediates the carboxylation of ribulose bisphosphate, *ribulose bisphosphate carboxylase*, is also involved in *photorespiration.

rickets A childhood condition caused by decalcification of bone, resulting in deformed bones. Rickets is associated with chronic deficiency of *vitamin D or calcium and with disorders that cause poor phosphate reabsorption from the kidney *nephrons.

rickettsia A very small rod-shaped nonmotile microorganism that is now classified as a bacterium (being included in the phylum Omnibacteria), although – like a virus – it is totally parasitic, being unable to reproduce outside the cells of its host. Rickettsias can infect such arthropods as ticks, fleas, lice, and mites, through which they can be transmitted to vertebrates, including man. The group includes the causal agents of trench fever, Rocky Mountain spotted fever, and forms of typhus.

rigor mortis The stiffening of the body of an animal after death, due to a temporary rigidity of the muscles. This condition arises because ATP, which is no longer synthesized after death, is required to break down the bridges that form between actin and myosin filaments in muscle tissue during contraction.

Ringer's solution *See* **physiological saline**.

ring species Two apparently distinct species that are connected by a series of intermediate geographical and structural subspecies between which interbreeding can occur. This phenomenon is an example of evolution in action and in such cases it is often difficult to define a boundary between the species.

ritualization An evolutionary process in which the form or context of an action is altered because it comes to play a role in social communication. For example, many *courtship and greeting ceremonies in animals include ritual food presentation (though the quantities of food may be negligible), derived from the action of feeding the young.

RNA (ribonucleic acid) A complex organic compound (a nucleic acid) in living cells that is concerned with *protein synthesis. In some viruses, RNA is also the hereditary material. Most RNA is synthesized in the nucleus and then distributed to various parts of the cytoplasm. An RNA molecule consists of a long chain of *nucleotides in which the sugar is *ribose and the bases are adenine, cytosine, guanine, and uracil (see illustration; *compare* **DNA**). *Messenger RNA (mRNA)* is responsible for carrying the *genetic code transcribed from DNA to specialized sites within the cell (known as *ribosomes), where the information is translated into protein composition (*see* **transcription; translation**). *Ribosomal RNA (rRNA)* is present in ribosomes; it is single-stranded but helical regions are formed by *base pairing within the strand. *Transfer RNA (tRNA, soluble RNA, sRNA)* is involved in the assembly of amino acids in a protein chain being synthesized at a ribosome. Each tRNA is specific for an amino acid and bears a triplet of bases complementary with a triplet on mRNA (*see* **anticodon; elongation**).

RNAase (RNase; ribonuclease) Any enzyme that catalyses the cleavage of nucleotides in RNA. Each RNAase has a specificity for a different cleavage site. For example, RNAase A is a digestive enzyme secreted by the pancreas

Detail of molecular structure of sugar–phosphate backbone. Each ribose unit is attached to a phosphate group and a base, forming a nucleotide

Single-stranded structure of RNA

The four bases of RNA

Molecular structure of RNA

that hydrolyses phosphodiester bonds in the nucleotide chain. Other RNAases are active at the cellular level, for instance in modifying transfer RNA and ribosomal RNA after transcription.

RNA polymerase *See* **polymerase**.

RNP *See* **ribonucleoprotein**.

rod A type of light-sensitive receptor cell present in the *retinas of vertebrates. Rods contain the pigment *rhodopsin and are essential for vision in dim light. They are not evenly distributed on the retina, being absent in the *fovea and occupying all of the retinal margin *See also* **dark adaptation**. *Compare* **cone**.

Rodentia An order of mammals characterized by a single pair of long curved incisors in each jaw. These teeth are specialized for gnawing: they continue growing throughout life and have enamel only on the front so that they wear to a chisel-shaped cutting edge. Rodents often breed throughout the year and produce large numbers of quickly maturing young. The order includes the squirrels, beavers, rats, mice, and porcupines.

roentgen The former unit of dose equivalent (*see* **radiation units**). It is named after the discoverer of X-rays, W. K. Roentgen (1845–1923).

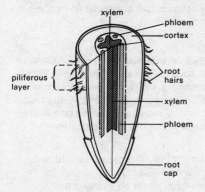

Section through the tip of a plant root

root 1. *(in botany)* The part of a vascular plant that grows beneath the soil surface in response to gravity and water. It anchors the plant in the soil and absorbs water and mineral salts. Unlike the stem, it never produces leaves, buds, or flowers and never contains chlorophyll. The *radicle (embryonic root) may give rise either to a *tap root system* with a single main *tap root* from which *lateral roots* develop, or a *fibrous root system*, with many roots of equal size. The *apical meristem at the root tip gives rise to a protective sheath, the *root cap, and to the primary tissues of the root. The vascular tissues usually form a central core (see illustration). This distinguishes roots from stems, in which the vascular tissue often forms a ring. A short distance behind the root tip *root hairs* develop from the epidermis and greatly increase the surface area for absorption of water and minerals. Beyond this, lateral roots develop.

Roots may be modified in various ways. Some are swollen with food to survive the winter, as in the carrot. Certain plants, such as orchids, have absorptive aerial roots; others, such as ivy, have short clasping roots for climbing. The roots of leguminous plants, such as beans and peas, contain *root nodules, which have an important role in nitrogen fixation. Other modifications include *prop roots, stilt roots, and buttress roots, which support the plant; and *pneumatophores.

2. *(in dentistry)* The portion of a *tooth that is not covered with enamel

and is embedded in a socket in the jawbone. Incisors, canines, and premolars have single roots; molars normally have several roots.
3. *(in anatomy)* The point of origin of a nerve in the central nervous system. There are two roots for every *spinal nerve (*see* **dorsal root; ventral root**).

root cap (calyptra) A cone-shaped structure that covers the root tip and develops as a result of cell division by a meristem at the root apex (*see* **calyptrogen**). It protects the root tip as it grows between the soil particles. The cells are constantly worn away by friction and are replaced by the meristem.

root hair *See* root.

root nodule A swelling on the roots of certain plants, especially those of the family Leguminosae, that contains bacteria (notably *Rhizobium*) capable of fixing atmospheric nitrogen into ammonia, which is subsequently converted to nitrates and amino acids (*see* **nitrogen fixation**). Plants that possess root nodules increase soil fertility by increasing the nitrate content of the soil. The practice of *crop rotation will normally include the cultivation of a leguminous species.

root pressure The pressure that forces water, absorbed from the soil, to move through the roots and up the stem of a plant. This pressure can be demonstrated by cutting a stem, from which water will exude. A *manometer can be attached to a plant stem to measure the root pressure. Root pressure is believed to be due to both the osmosis of water, from the soil into the root cells, and the active pumping of salts into the *xylem tissue, which maintains a concentration gradient along which the water will move. *See also* **transpiration**.

rootstock *See* rhizome.

Rotifera A phylum of microscopic (0.04–2.00 mm) *pseudocoelomate aquatic animals characterized by a crown of cilia at the head end. These are used in locomotion and in some species for feeding: the crown resembles a rotating wheel when the cilia are beating. Rotifers possess jaws and are covered with a layer of chitin (the *lorica*). There is no circulatory system and gas exchange occurs across the body surface. Some rotifers reproduce by *parthenogenesis.

roughage *See* (dietary) fibre.

rough endoplasmic reticulum (rough ER) *See* endoplasmic reticulum.

round window (fenestra rotunda) A membrane-covered opening between the middle ear and the inner ear (*see* **ear**), situated below the *oval window. Pressure waves transmitted through the perilymph in the *cochlea are released into the middle ear through the round window, which bulges out into the middle-ear cavity.

roundworms *See* Nematoda.

royal jelly A highly nutritious food substance that is secreted from the hypopharyngeal and mandibular glands of worker honeybees and fed to the newly hatched larvae. For the first three days after hatching, larvae that will develop into worker bees are fed on royal jelly, after which their diet changes to pollen and nectar. However, larvae that will develop into queen bees continue to be fed exclusively on royal jelly throughout the larval stage.

r selection A type of selection that favours organisms with a high rate of reproduction (*r value*). Organisms that are r selected (*r strategists*) are able to colonize a habitat rapidly, utilizing the food and other resources before other organisms are established and begin to compete. The r strategists tend to be relatively small organisms with short life spans (e.g. bacteria) and often live in temporary or unstable environments; characteristically their survival depends on their ability to produce large numbers of offspring rather than on their ability to compete. *Compare* **K selection**. *See* **survival curve**.

rubidium–strontium dating A method of dating geological specimens based on the decay of the radioisotope rubidium–87 into the stable isotope strontium–87. Natural rubidium contains 27.85% of rubidium–87, which has a half-life of 4.7×10^{11} years. The ratio $^{87}\text{Rb}/^{87}\text{Sr}$ in a specimen gives an estimate of its age (up to several thousand million years).

Section of the stomach of a ruminant

rumen The first of four chambers that form the stomach of ruminants. *See* **Ruminantia**.

Ruminantia A suborder of hooved mammals (*see* **Artiodactyla**) comprising the sheep, cattle, goats, deer, and antelopes. They are characterized by a four-chambered stomach (see illustration). In the first chamber, the *rumen*, swallowed food is digested by *cellulase, an enzyme secreted by symbiotic anaerobic microorganisms that live in the rumen. Some products of digestion are absorbed in the rumen; the remaining partly digested food is regurgitated and chewed to a pulp – the process known as 'chewing the cud'. This food mass is then swallowed and enters the *reticulum*, where further absorption of digested food may occur; then to the *omasum*, where water is removed; and finally to the *abomasum*, in which further digestion takes place.

runner A stem that grows horizontally along the soil surface and gives rise to new plants either from axillary or terminal buds. Runners are seen in the creeping buttercup and the strawberry. *Offsets*, e.g. those of the houseleek, are short runners.

rusts A group of parasitic fungi of the phylum *Basidiomycota. Many of these species attack the leaves and stems of cereal crops: characteristic rust-coloured streaks of spores appear on infected plants. The life cycles of some rusts may be complex; many form a number of different types of spore and some require two different host plants. *Compare* **smuts**.

saccharide *See* **sugar**.

Saccharomyces An industrially important genus of yeasts. *S. cerevisiae*, of which there are at least 1000 strains, is used in baking (*see* **baker's yeast**), brewing, and wine making; it is also used in the production of *single-cell protein and ergosterol, a steroid. The other main yeast used in the production of beer is *S. uvarum* (or *carlsbergensis*); it is distinguished from *S. cerevisiae* by its ability to ferment the disaccharide melibose using α-galactosidase, an enzyme not produced by *S. cerevisiae*.

saccharose *See* **sucrose**.

sacculus (saccule) A chamber of the *inner ear from which the *cochlea arises in reptiles, birds, and mammals. It bears patches of sensory epithelium concerned with balance (*see* **macula**).

sacral vertebrae The vertebrae that lie between the lumbar and the caudal vertebrae in the *vertebral column. The function of the sacral vertebrae is to articulate securely with the *pelvic girdle, and they are usually fused to form a single bone (the *sacrum*) to provide a firm support. The number of sacral vertebrae varies from animal to animal. Amphibians have a single sacral vertebra, reptiles have two, and mammals have three or more.

safranin A stain used in optical microscopy that colours lignified tissues, cutinized tissues, and nuclei red and chloroplasts pink. It is used mainly for plant tissues, in conjunction with a green or blue counterstain.

sagittal Describing a section through an organism that is bisected in the longitudinal plane.

salicylic acid (1-hydroxybenzoic acid) A naturally occurring carboxylic acid, HOC_6H_4COOH, found in certain plants. It is used in making aspirin and in the foodstuffs and dyestuffs industries.

saline Describing a chemical compound that is a salt, or a solution containing a salt. *See also* **physiological saline**.

salinization An increase in the salt content of the soil (or of a body of fresh water), which can lead to the retardation of plant growth and eventually renders the soil infertile. This problem is particularly acute in hot areas, where water readily evaporates from the soil, and it may also occur as a consequence of *irrigation.

saliva A watery fluid secreted by the *salivary glands in the mouth. Production of saliva is stimulated by the presence of food in the mouth and also by the smell or thought of food. Saliva contains mucin, which lubricates food and eases its passage into the oesophagus, and in some animals salivary *amylase (or ptyalin), which begins the digestion of starch. The saliva of insects is rich in digestive enzymes, and that of bloodsucking animals contains an anticoagulant.

salivary glands Glands in many terrestrial animals that secrete *saliva into the mouth. In man there are three pairs: the sublingual, submandibular, and the submaxillary glands. The salivary gland cells of some insect larvae produce giant chromosomes (*see* **polyteny**), which are widely used in the study of genetics and protein synthesis.

Salmonella A genus of rod-shaped Gram-negative bacteria that inhabit the intestine and cause disease (*salmonellosis*) in humans and animals. They are aerobic or facultatively anaerobic, and most are motile. Salmonellae can exist for long periods outside their host, and may be found, for example, in sewage and surface water. Humans may become infected by consuming contaminated water or food, especially animal products, such as eggs, meat, and milk, or vegetables that have been fertilized with contaminated manure. The bacteria can also be transmitted from human or animal carriers by unhygienic food preparation. Various species of *Salmonella* cause gastroenteritis and septicaemia; typhoid fever and paratyphoid fever are caused by *S. typhi* and *S. paratyphi*, respectively.

salt A compound formed by reaction of an acid with a base, in which the hydrogen of the acid has been replaced by metal or other positive ions. Typically, salts are crystalline ionic compounds, such as Na^+Cl^- (*sodium chloride).

samara A dry single-seeded indehiscent fruit in which the fruit wall hardens and extends to form a long membranous winglike structure that aids dispersal. Examples are ash and elm fruits. The sycamore fruit is a double samara and technically a *schizocarp. *See also* **achene**.

sampling The selection of small groups of entities to represent a large number of entities in statistics. In *random sampling* each individual of a population has an equal chance of being selected as part of the sample. In *stratified random sampling*, the population is divided into strata, each of which is randomly sampled and the samples from the different strata are pooled. In *systematic sampling*, individuals are chosen at fixed intervals; for example, every tenth animal in a population. In *sampling with replacement*, each individual chosen is replaced before the next selection is made.

sap 1. The sugary fluid that is found in the phloem tissue of plants. Sap is the medium in which carbohydrates, produced in photosynthesis, and other organic molecules are transported and stored in plants. **2. (cell sap)** The fluid that is contained in the *vacuoles of plant cells. It is a solution of organic and inorganic compounds, including sugars, amino acids, salts, pigments, and waste products.

saprotroph (saprobe; saprobiont) Any organism that feeds by absorbing dead organic matter. Most saprotrophs are bacteria and fungi. Saprotrophs are important in *food chains as they bring about decay and release nutrients for plant growth. *Compare* **parasitism**.

sapwood (alburnum) The outer wood of a tree trunk or branch. It consists of living *xylem cells, which both conduct water and provide structural support. *Compare* **heartwood**.

sarcolemma The contractile membrane that surrounds a muscle fibre. *See also* **transverse tubules**.

Structure of a sarcomere

sarcomere Any of the functional units that make up the myofibrils of *voluntary muscle. Each sarcomere is bounded by two membranes (*Z lines), which provide the points of attachment of *actin filaments; another membrane (the *M band* or *line*) is the point of attachment of the *myosin filaments. The sarcomere is divided into various bands reflecting the arrangement of the filaments (see illustration). During muscle contraction the actin and myosin filaments slide over each other (*see* **sliding filament theory**) and the length of the sarcomere shortens: the Z lines are drawn closer together and the I and H bands become narrower.

sarcoplasm The cytoplasm of a muscle fibre. Sarcoplasm contains chemicals that are required for muscle contraction, including glycogen, ATP, and phosphocreatine. In addition the sarcoplasm of active muscles tends to be rich in mitochondria.

sarcoplasmic reticulum The specialized endoplasmic reticulum in a muscle fibre, which forms a network of membrane-lined cavities surrounding the contractile myofibrils that run through the fibre. The release of calcium ions from the sarcoplasmic reticulum into the cytosol following stimulation from an action potential (*see* **transverse tubules**) causes subsequent contraction of the myofibrils (*see* **sarcomere**).

satellite DNA A portion of the DNA in plant and animal cells that consists of repeating sequences of nucleotides (*see* **repetitive DNA**). Some 30% of human DNA consists of such satellite regions. They do not appear to form part of any *genes but their function is unclear. The term is sometimes applied to DNA other than that found in the cell nucleus, in particular *mitochondrial DNA, which differs in its density.

saturated Denoting a compound consisting of molecules that have only single bonds (i.e. no double or triple bonds). Saturated compounds can undergo substitution reactions but not addition reactions. *See* **fatty acid**. *Compare* **unsaturated**.

savanna *See* **grassland**.

scala Any one of three fluid-filled canals of the *cochlea in the inner ear: the *scala media* (cochlear duct), *scala tympani* (tympanic canal), and *scala vestibuli* (vestibular canal).

scales The small bony or horny plates forming the body covering of fish and reptiles. The wings of some insects, notably the Lepidoptera (butterflies and moths), are covered with tiny scales that are modified cuticular hairs.

In fish there are three types of scales. *Placoid scales* (*denticles*), characteristic of cartilaginous fish, are small and toothlike, with a projecting spine and a flattened base embedded in the skin. They are made of *dentine, have a pulp cavity, and the spine is covered with a layer of enamel. Teeth are probably modified placoid scales. *Cosmoid scales*, characteristic of lungfish and coelacanths, have an outer layer of hard *cosmin* (similar to dentine) covered by modified enamel (*ganoine*) and inner layers of bone. The scale grows by adding to the inner layer only. In modern lungfish the scales are reduced to large bony plates. *Ganoid scales* are characteristic of primitive ray-finned fishes, such as sturgeons. They are similar to cosmoid scales but have a much thicker layer of ganoine and grow by the addition of material all round. The scales of modern teleost fish are reduced to thin bony plates.

In reptiles there are two types of scales: horny epidermal *corneoscutes* sometimes fused with underlying bony dermal *osteoscutes*.

scanning electron microscope *See* **electron microscope**.

scapula (shoulder blade) The largest of the bones that make up each half of the *pectoral (shoulder) girdle. It is a flat triangular bone, providing anchorage for the muscles of the forelimb and an articulation for the *humerus at the *glenoid cavity. It is joined to the *clavicle (collar bone) in front.

scavenger An animal that feeds on dead organic matter. Scavengers (such as hyenas) may feed on animals killed by predators or they may be *detritivores.

Schiff's reagent A reagent used for testing for aldehydes and ketones; it consists of a solution of fuchsin dye that has been decolorized by sulphur dioxide. Aliphatic aldehydes restore the pink immediately, whereas aromatic ketones have no effect on the reagent. Aromatic aldehydes and aliphatic ketones restore the colour slowly. It is named after the German chemist Hugo Schiff (1834–1915).

schizocarp A dry indehiscent fruit formed from carpels that develop into separate one-seeded fragments called *mericarps*, which may be dehiscent, as in the *regma, or indehiscent, as in the *cremocarp and *carcerulus.

schizogeny The localized separation of plant cells to form a cavity (surrounded by the intact cells) in which secretions accumulate. Examples are the resin canals of some conifers and the oil ducts of caraway and aniseed fruits. *Compare* **lysigeny**.

Schwann cell A cell that forms the *myelin sheath of a nerve fibre. Each cell is responsible for a single length of the fibre (called an *internode*); adjacent internodes are separated by small gaps (*nodes of Ranvier*). During its development the cell wraps itself around the fibre, so the sheath consists of concentric layers of Schwann cell membrane. These cells are named after the German anatomist T. Schwann (1810–82).

scintillation counter A type of particle or radiation counter that makes use of the flash of light (scintillation) emitted by an excited atom falling back to its ground state after having been excited by a passing photon or particle. The scintillating medium is usually either solid or liquid and is used in connection with a photomultiplier, which produces a pulse of current for each scintillation. The pulses are counted to enable the radioactivity of the source to be calculated. The distribution of a radiolabelled compound, such as a drug, in an organism can be determined in this way by testing tissue samples from different organs after drug administration.

scion *See* **graft**.

sclera *See* **sclerotic**.

sclereid A type of *sclerenchyma cell that is shorter than a *fibre; its lignified walls typically contain branched *pits. Sclereids are abundant in seed coats, nut shells, and in pear fruits.

sclerenchyma A plant tissue whose cell walls have become impregnated with lignin. Due to the added strength that this confers, sclerenchyma plays an important role in support; it is found in the stems and also in the midribs of leaves. The cell walls contain *pits, enabling the exchange of substances between adjacent cells. Mature sclerenchyma cells are dead, since the lignin makes the cell wall impermeable to water and gases. Sclerenchyma cells take the form of *fibres or *sclereids. *Compare* **collenchyma; parenchyma**.

scleroprotein Any of a group of proteins found in the exoskeletons of some invertebrates, notably insects. Scleroproteins are formed by conversion of the relatively soft elastic larval protein by a natural tanning process (*sclerotization*) involving orthoquinones. These are secreted and form cross linkages between polypeptides of the proteins, producing a hard rigid covering.

sclerotic (sclera) The tough external layer of the vertebrate eye. At the front of the eye, the sclera is modified to form the *cornea.

scolex *See* **Cestoda**.

scorpions *See* **Arachnida**.

scotopic vision The type of vision that occurs when the rods in the eye are the principal receptors, i.e. when the level of illumination is low. With scotopic vision colours cannot be identified. *Compare* **photopic vision**.

SCP *See* **single-cell protein**.

scrotum The sac of skin and tissue that contains and supports the *testes in most mammals. It is situated outside the body cavity and allows sperm to develop at the optimum temperature, which is slightly lower than body temperature.

scurvy A disease caused by deficiency of *vitamin C, which results in poor collagen formation. Symptoms include anaemia, skin discoloration, and tooth loss. Scurvy was a common disease among sailors in the 16th–18th centuries, when no fresh food was available on long sea voyages.

scutellum The tissue in a grass seed that lies between the embryo and the endosperm. It is the modified cotyledon of grasses, being specialized for the digestion and absorption of the endosperm.

Scyphozoa *See* **Cnidaria**.

seasonal isolation *See* **isolating mechanism**.

seaweeds Large multicellular *algae living in the sea or in the intertidal zone. They are commonly species of the *Chlorophyta, *Phaeophyta, and *Rhodophyta.

sebaceous gland A small gland occurring in mammalian *skin. Its duct opens into a hair follicle, through which it discharges *sebum onto the skin surface.

sebum The substance secreted by *sebaceous glands onto the surface of the *skin. It is a fatty mildly antiseptic material that protects, lubricates, and waterproofs the skin and hair and helps prevent desiccation.

second Symbol s. The SI unit of time equal to the duration of 9 192 631 770 periods of the radiation corresponding to the transition between two hyperfine levels of the ground state of the caesium–133 atom.

secondary consumer *See* **consumer**.

secondary growth (secondary thickening) The increase in thickness of plant shoots and roots through the activities of the vascular *cambium and *cork cambium. It is seen in most dicotyledons and gymnosperms but not in monocotyledons. The tissues produced by secondary growth are called *secondary tissues* and the resultant plant or plant part is the *secondary plant body*. *Compare* **primary growth**.

secondary sexual characteristics External features of a sexually mature animal that, although not directly involved in copulation, are significant in reproductive behaviour. The development of such features is controlled by sex hormones (androgens or oestrogens); they may be seasonal (e.g. the antlers of male deer or the body colour of male sticklebacks) or permanent (e.g. breasts in women or facial hair in men). In humans they develop during *adolescence.

secondary structure *See* protein.

secondary thickening *See* secondary growth.

second convoluted tubule *See* distal convoluted tubule.

second messenger A chemical within a cell that is responsible for initiating the response to a signal from a chemical messenger (such as a hormone, neurotransmitter, or growth factor) that cannot enter the target cell itself, for example because it is not lipid-soluble and is therefore unable to cross the plasma membrane. A common second messenger is *cyclic AMP; the signal for its formation within the cell is transmitted from hormone receptors on the cell surface by *G-protein. Inositol 1,4,5-trisphosphate (*see* **inositol**) is another example of a second messenger.

secretin A hormone produced by the anterior part of the small intestine (the *duodenum and *jejunum) in response to the presence of hydrochloric acid from the stomach. It causes the pancreas to secrete alkaline pancreatic juice and stimulates bile production in the liver. Secretin, whose function was first demonstrated in 1902, was the first substance to be described as a hormone.

secretion 1. The manufacture and discharge of specific substances into the external medium by cells in living organisms. (The substance secreted is also called the secretion.) Secretory cells are often specialized and organized in groups to form *glands. The substances produced may be released directly into the blood (*endocrine secretion*; *see* **endocrine gland**) or through a duct (*exocrine secretion*; *see* **exocrine gland**). Secretions can be classified according to the manner of their discharge. *Merocrine (eccrine) secretion* occurs without the secretory cells sustaining any permanent change; in *apocrine secretion* the cells release a secretory vesicle incorporating part of the secretory cell membrane; and *holocrine secretion* involves the disruption of the entire cell to release its accumulated secretory vesicles. **2.** The process by which a substance is pumped out of a cell against a concentration gradient. Secretion has an important role in adjusting the composition of urine as it passes through the *nephrons of the kidney.

seed The structure in angiosperms and gymnosperms that develops from the ovule after fertilization. Occasionally seeds may develop without fertilization taking place (*see* **apomixis**). The seed contains the *embryo and nutritive tissue, either as *endosperm or food stored in the *cotyledons. Angiosperm seeds are contained within a *fruit that develops from the ovary wall. Gymnosperm seeds lack an enclosing fruit and are thus termed *naked*. The seed is covered by a protective layer, the *testa. During development of the testa the seed dries out and enters a resting phase (dormancy) until conditions are suitable for germination.

Annual plants survive the winter or dry season as seeds. The evolution of the seed habit enabled plants to colonize the land, since seed plants do not depend on water for fertilization (unlike the lower plants).

seed coat *See* testa.

seed ferns *See* Pteridospermales.

seed leaf *See* cotyledon.

seed plant Any plant that produces seeds. Most seed plants belong to the phyla *Angiospermophyta (flowering plants) or *Coniferophyta (conifers).

segmentation 1. *See* metameric segmentation. **2.** *See* cleavage.

segregation The separation of pairs of *alleles during the formation of reproductive cells so that they contain one allele only of each pair. Segregation is the result of the separation of *homologous chromosomes during *meiosis. *See also* **Mendel's laws**.

Selachii The major subclass of the Chondrichthyes (cartilaginous fishes), containing the sharks, rays, skates, and similar but extinct forms. Their sharp teeth develop from the toothlike placoid *scales (denticles) and are rapidly replaced as they wear out.

selection The process by which one or more factors acting on a population produce differential mortality and favour the transmission of specific characteristics to subsequent generations. *See* **artificial selection; directional selection; disruptive selection; natural selection; sexual selection; stabilizing selection**.

selection coefficient Symbol s. A value used to calculate the rate of change of a particular phenotype within a population. The selection coefficient indicates the comparative *fitness of genotypes within a population. For example, a selection coefficient of 100 indicates that 1 out of 100 individuals for a given genotype will fail to reproduce and pass on their genes to the subsequent generation.

selection pressure The extent to which organisms possessing a particular characteristic are either eliminated or favoured by environmental demands. It indicates the degree of intensity of *natural selection.

selective breeding *See* breeding.

selective reabsorption The absorption of some of the components of the *glomerular filtrate back into the blood as the filtrate flows through the *nephrons of the kidney. Glucose, amino acids, and salts can be reabsorbed against a concentration gradient and their transport across the nephron into the capillaries requires energy (*see* **active transport**). Other components, such as ammonia and urea, are secreted rather than absorbed (*see* **secretion**), while certain ions, including potassium, can be both secreted and absorbed by the tubules according to the overall ionic balance throughout the body.

self-fertilization *See* fertilization.

selfish DNA Regions of DNA that apparently have no function (it is also known as 'junk' DNA) and exist between those regions of DNA that represent the genes. It has the ability to integrate and transpose within the genome, and in some cases consists of repeated DNA sequences (*see* **repetitive DNA**). This DNA is said to be 'selfish' as it exists only to pass copies of itself from one generation to another; it does so by using the organism in which it is contained as a survival machine. This is known as the *selfish DNA theory*. The greatest amounts of selfish DNA are found in vertebrates and higher plants. The presence of selfish DNA may be due to an unrecognizable function that it performs or because the cell has no way of halting its increase in the genome.

self-pollination *See* pollination.

self-sterility The condition found in many hermaphrodite organisms in which male and female reproductive cells produced by the same individual will not fuse to form a zygote, or if they do, the zygote is unable to develop into an embryo. In plants this is usually termed *self-incompatibility* (*see* **incompatibility**).

Seliwanoff's test A biochemical test to identify the presence of ketonic sugars, such as fructose, in solution. It was devised by the Russian chemist F. F. Seliwanoff. A few drops of the reagent, consisting of resorcinol crystals dissolved in equal amounts of water and hydrochloric acid, are heated with the test solution and the formation of a red precipitate indicates a positive result.

semen A slightly alkaline fluid (pH 7.2–7.6) containing sperm and various secretions that is produced by a male mammal during copulation and is introduced into the body of the female by *ejaculation. Spermatozoa are produced by the *testes and the secretions by the *prostate gland, *seminal vesicles, and *Cowper's glands. Semen also contains enzymes that activate the sperm after ejaculation.

semicircular canals The sense organ in vertebrates that is concerned with the maintenance of physical equilibrium (sense of balance). It occurs in the *inner ear and consists of three looped canals set at right angles to each other and attached to the *utriculus. The canals contain a fluid (*endolymph*) that flows in response to movements of the head and body. A swelling (*ampulla) at one attachment point of each canal contains sensory cells that respond to movement of the endolymph in any of the three planes. These sensory cells initiate nervous impulses to the brain.

semiconservative replication The generally accepted method of *DNA replication, in which the two strands of the DNA helix separate and free nucleotides pair with the exposed bases on the single chains to form two new DNA molecules, each containing one original and one newly synthesized strand of DNA. *Compare* **conservative replication; dispersive replication**.

semilunar valve Either of two *valves in the heart, found in the pulmonary artery (*pulmonary valve*) and in the aorta (*aortic valve*), that prevent the backflow of blood into the right and left ventricles from the pulmonary artery and the aorta, respectively, thus maintaining blood flow in a single direction. Each semilunar valve consists of three pockets, which close by filling with blood when the force of contraction that expels the blood from the heart diminishes.

seminal receptacle *See* **spermatheca**.

seminal vesicle 1. A pouch or sac in many male invertebrates and lower vertebrates that is used for storing sperm. **2.** One of a pair of glands in male mammals that secrete a liquid component of *semen into the vas deferens. This secretion is alkaline, which neutralizes the acidic conditions in the female genital tract, and contains fructose, used by the sperm as a source of energy.

seminiferous tubules *See* **testis**.

semipermeable membrane *See* **partially permeable membrane**.

senescence The changes that occur in an organism (or a part of an organism) between maturity and death, i.e. ageing. Characteristically there is a deterioration in functioning as the cells become less efficient in maintaining and replacing vital cell components. In animals this results in a decline in physical ability and, in man, there is also often a reduction in mental ability. Not all the parts of the body necessarily become senescent at the same time or age at the same rate. For example, in deciduous trees the shedding of senescent leaves in the autumn is a normal physiological process.

sensation The raw data detected by the *senses. For example, red is a colour sensation. *Compare* **perception**.

sense organ A part of the body of an animal that contains or consists of a concentration of *receptors that are sensitive to specific stimuli (e.g. sound, light, pressure, heat). Stimulation of these receptors initiates the transmission of nervous impulses to the brain, where sensory information is analysed and interpreted. Examples of sense organs are the *eye, *ear, *nose, and *taste bud.

senses The faculties that enable animals to perceive information about their external environment or about the state of their bodies in relation to this environment (*see* **sense organ; vision; hearing; balance; olfaction; taste; touch**). Specific *receptors are sensitive to pain, temperature, chemicals, etc.

sensitivity (irritability) One of the fundamental properties of all organisms: the capacity to detect, interpret, and respond to changes in the environment (e.g. the stimuli of light, touch, chemicals, etc.). Multicellular animals have specialized *sense organs and *effector organs for this purpose; in unicellular organisms, which lack a nervous system, the reception of, and response to, a stimulus occur in the same cell.

sensitization 1. *(of a cell)* The alteration of the integrity of a cell membrane resulting from the reaction of specific *antibodies with *antigens on the surface of the cell. In the presence of *complement, the cell ruptures. **2.** *(of an individual)* Initial exposure to a specific antigen such that re-exposure to the same antigen causes a severe immune response (*see* **anaphylaxis**).

sensory cell *See* **receptor**.

sensory neurone A nerve cell (*see* **neurone**) that transmits information about changes in the internal and external environment to the central nervous system. Sensory neurones are of two types. *Somatic sensory neurones* occur in peripheral nerves in the skin, skeletal muscle, joints, and bones. *Visceral sensory neurones* are located in sympathetic and parasympathetic nerves in the heart, lungs, and other organs.

sepal One of the parts of a flower making up the *calyx. Sepals are considered to be modified leaves with a simpler structure. They are usually green and often hairy but in some plants, e.g. monk's hood, they may be brightly coloured.

sepsis Infection of the soft tissues or blood by pathogenic microorganisms, arising usually after these have entered the body through a skin wound. It results in destruction of the tissues by the pathogens or their *toxins.

septum (*pl.* **septa**) Any dividing wall in a plant or animal. Examples are the septa that separate the chambers of the heart.

sere A complete *succession of plant communities, which results in the climax community. A sere is composed of a series of different plant communities that change with time. These communities are known as *seral stages* or *seral communities*.

serine *See* **amino acid**.

serology The laboratory study of blood serum and its constituents, particularly *antibodies and *complement, which play a part in the *immune response.

serotonin (5-hydroxytryptamine) A compound (see formula), derived from the amino acid tryptophan, that affects the diameter of blood vessels and also functions as a *neurotransmitter. In the brain it is thought to influence mood: drugs, such as *lysergic acid diethylamide (LSD), that alter serotonin levels have hallucinogenic effects.

Serotonin

serous membrane (serosa) A tissue consisting of a layer of *mesothelium attached to a surface by a thin layer of connective tissue. Serous membrane lines body cavities that do not open to the exterior; the *peritoneum, *pleura, and serous *pericardium are examples.

Sertoli cells (sustentacular cells) Cells that line the seminiferous tubules in the *testis. Named after Italian histologist Enrico Sertoli (1842–1910), Sertoli cells protect the *spermatids and convey nutrients to both the developing and mature spermatozoa. They also produce a hormone, *inhibin*, which can inhibit *follicle-stimulating hormone and thereby regulate production of spermatozoa.

serum *See* **blood serum**.

sessile 1. Describing animals that live permanently attached to a surface, i.e. sedentary animals. Many marine animals, e.g. sea anemones and limpets, are sessile. **2.** Describing any organ that does not possess a stalk where one might be expected. For example, the leaves of the oak (*Quercus robur*) are attached directly to the twigs.

seta (*pl.* **setae**) **1.** A bristle or hair in many invertebrates. Setae are produced by the epidermis and consist either of a hollow projection of cuticle containing all or part of an epidermal cell (as in insects) or are composed of chitin (as in the *chaetae of annelid worms). **2.** *See* **sporogonium**.

sewage Waste matter from industrial and domestic sources that is dissolved or suspended in water. Raw (untreated) sewage is a pollutant. It has a high content of organic matter (notably faeces and nitrogenous waste) and therefore provides a rich source of food for many decomposers (bacteria, fungi) and *detritivores, some of which are pathogenic to humans. The release of raw (untreated) sewage into a river causes eutrophication (*see* **eutrophic**); there is a sudden increase in the *biochemical oxygen demand (BOD), as the organisms that feed on sewage proliferate and use up the available dissolved oxygen in the river. Oxygen-sensitive organisms, such as fish, will die. Certain organisms can proliferate in particular concentrations of sewage, depending on their tolerance, and can be used as markers of the extent to which a river is polluted by sewage. For example, *Tubifex* worms are able to tolerate high concentrations of sewage.

Sewage can be treated before release. This involves a number of stages, including filtration, sedimentation, and microbial degradation (notably by *methanogens). When most of the solid waste has been removed, the remaining liquid (*effluent*) is discharged into rivers, etc. During sedimentation, particulate organic matter accumulates at the bottom of large tanks. This material, known as *sludge*, is periodically removed, further decomposed by microorganisms (*see* **activated sludge process**), and then sold as fertilizer or dumped.

Sewall Wright effect *See* **genetic drift**. It is named after US statistician Sewall Wright (1889–1988).

sex chromosome A chromosome that operates in the sex-determining mechanism of a species. In most animals there are two kinds of sex chromosomes: the *X chromosome*, similar in size to the other chromosomes, and the smaller *Y chromosome*. In man and many animals a combination of two X chromosomes gives a female while one X and one Y chromosome gives a male. Sex chromosomes carry genes governing the development of sex organs and secondary sexual characteristics. They also carry other genes unrelated to sex (*see* **sex linkage**). *See also* **sex determination**.

sex determination The method by which the distinction between males and females is established in a species. It is usually under genetic control. Equal numbers of males and females are produced when sex is determined by *sex chromosomes or by a contrasting pair of alleles. In some species (e.g. bees) females develop from fertilized eggs and males from unfertilized eggs. This does not produce equal numbers.

sex factor A *plasmid found in the cytoplasm of certain bacteria that is responsible for initiating *conjugation and gene transfer. The sex factor in *Escherichia coli* is known as the *F factor*; it can be transferred from a donor cell (F^+) to a recipient (F^-) by conjugation. Occasionally the F factor may become integrated with the chromosome of an F^+ bacterium to form a *high-frequency recombinant* (*Hfr*) cell. This enables the part of the host chromosome nearest to the F factor to be transferred to the recipient by conjugation.

sex hormones Steroid hormones that control sexual development. The most important are the *androgens and *oestrogens.

sex linkage The tendency for certain inherited characteristics to occur far more frequently in one sex than the other. For example, red–green colour blindness and *haemophilia affect men more often than women. This is because the genes governing normal colour vision and blood clotting occur on the X *sex chromosome. Women have two X chromosomes. If one carries an abnormal allele it is likely that its effects will be masked by a normal allele on the other X chromosome. However, men only have one X chromosome and any abnormal alleles therefore will not be masked. *See also* **carrier**.

sex ratio The ratio of the number of females to the number of males in a *population. Because the mortality rates in the two sexes may be different, the sex ratios in different age classes may differ.

sexual cycle *See* **menstrual cycle; oestrous cycle**.

sexual intercourse (coitus; copulation; mating) The process by which spermatozoa from a male are deposited in the body of a female during *sexual reproduction. In mammals the penis of the male becomes erect and stiff as its tissues become filled with blood, enabling it to be inserted into the vagina of the female. Thrusting movements of the penis result in *ejaculation, in which *semen, containing spermatozoa, is deposited in the vagina.

sexually transmitted disease (STD) Any disease that is passed from one individual to another during sexual intercourse or other types of sexual activity. These diseases have been traditionally referred to as *venereal diseases*. They include gonorrhoea, caused by the bacterium *Neisseria gonorrhoeae*; syphilis, due to infection by the bacterium *Treponema pallidum*; genital herpes, which is caused by a herpesvirus; and *AIDS, resulting from infection with *HIV, a retrovirus. The transmission of sexually transmitted diseases can be reduced by limiting the number of sexual partners and by the use of condoms (*see* **birth control**), which reduces the risk of contact with body fluids that harbour the microorganisms that cause these diseases.

sexual reproduction A form of reproduction that involves the fusion of two reproductive cells (*gametes) in the process of *fertilization. Normally, especially in animals, it requires two parents, one male and the other female. However, most plants bear both male and female reproductive organs and self-fertilization may occur, as it does in hermaphrodite animals. Gametes are formed by *meiosis, a special kind of cell division in the parent reproductive organs that both reassorts the genetic material and halves the chromosome number. Meiosis thus ensures genetic variability in the gametes and therefore in the offspring resulting from their subsequent fusion. Sexual reproduction, unlike *asexual reproduction, therefore generates variability within a species. However, it depends on there being reliable means of bringing together male and female gametes, and many elaborate mechanisms have evolved to ensure this.

sexual selection The means by which it is assumed that certain *secondary sexual characteristics, particularly of male animals, have evolved. Females presumably choose to mate with the male that gives the best courtship display and therefore has the brightest coloration, etc.: these features would be inherited by its male offspring and would thus tend to become exaggerated down the generations.

shoot The aerial part of a vascular plant. It develops from the *plumule and consists of a stem supporting leaves, buds, and flowers.

short-day plant A plant in which flowering can be induced or enhanced by short days, usually of less than 12 hours of daylight. Examples are strawberry and chrysanthemum. *See* **photoperiodism**. *Compare* **day-neutral plant; long-day plant**.

short-sightedness *See* **myopia**.

shoulder girdle *See* **pectoral girdle**.

shunt vessel A blood vessel that links an artery directly to a vein, allowing the blood to bypass the capillaries in certain areas. Shunt vessels can control blood flow by constriction and dilation. In *endotherms the shunt vessels dilate in response to cold, thereby cutting off the blood flow to the extremities and preventing heat loss.

siblings Individuals that have both parents in common.

sickle-cell anaemia *See* **polymorphism**.

sieve element A type of plant cell occurring within the *phloem. Sieve elements combine to form a series of tubes (*sieve tubes*) connecting the leaves, shoots, and roots in a fine network. Food materials are transported from one element to another via perforations termed *sieve areas* or *sieve plates*. This is an active (energy-requiring) process but the mechanism is uncertain. Sieve elements contain little cytoplasm and no nucleus. It is believed that their metabolic activities may be controlled by *companion cells in angiosperms and by albuminous cells in gymnosperms.

sievert The SI unit of dose equivalent (*see* **radiation units**). It is named after the Swedish physicist Rolf Sievert (1896–1966).

sieve tube A tube within the *phloem tissue of a plant, composed of joined *sieve elements.

sigmoid growth curve *See* **population growth**.

sign stimulus (releaser) The essential feature of a stimulus, which is necessary to elicit a response. For example, a red belly (characteristic of courting male sticklebacks) is the sign stimulus necessary to provoke an attack from a rival male; even a very crude model fish is attacked if it has a red undersurface. Similarly, territorial fighting in male robins is triggered by the sign stimulus of a red breast; a bunch of red feathers is enough to induce territorial behaviour. *See also* **stimulus filtering**.

silent mutation An alteration in the genetic code that has no apparent effect on the phenotype of an organism.

silicula A type of *capsule formed from a bicarpellary ovary. It is longitudinally flattened and divided lengthwise into two cavities (*loculi*). It is broader than a *siliqua. Examples include the fruits of *Alyssum* and candytuft.

siliqua A type of *capsule formed from a bicarpellary ovary. It resembles a *silicula but is longer than it is broad; an example is the fruit of the wallflower. *See also* **lomentum**.

Silurian A geological period of the Palaeozoic era following the Ordovician period and extending until the beginning of the Devonian period. It began about 440 million years ago and lasted for about 45 million years. The Silurian was named by Roderick Murchison (1792–1871) after an ancient British tribe that inhabited South Wales, where he observed rocks of this period. The majority of Silurian life was marine but during the later part of the period primitive plants began to make their appearance on land. Trilobites and graptolites became less common, brachiopods were numerous and varied, crinoids became common for the first time, and corals also increased. The only known vertebrates during the Silurian were primitive fish; the first jawed fish appeared later in the period. The Caledonian orogeny (mountain-building period) reached its peak towards the end of the Silurian.

single-cell protein (SCP) Protein produced by microorganisms, such as bacteria, yeasts, and unicellular algae, that is extracted for use as a component of human and animal foods.

single circulation The type of circulatory system that occurs in fishes, in which the blood passes only once through the heart in each complete circuit of the body *Compare* **double circulation**.

sinoatrial node *See* pacemaker.

sinus A saclike cavity or organ in an animal, such as the *sinus venosus.

sinusoid A tiny blood vessel or blood-filled space in an organ. Sinusoids replace capillaries in certain organs, notably the liver; they allow more direct contact between the blood and the tissue it is supplying.

sinus venosus A thin-walled chamber of the heart of vertebrate embryos that conveys deoxygenated blood to the single atrium. It is retained in adult fishes but in higher vertebrates it becomes incorporated into the left atrium.

siphonaceous (siphonous) Describing algae and other protoctists in which the thallus is coenocytic (*see* **coenocyte**) and often takes the form of a tubular structure.

Siphonaptera An order of secondarily wingless insects comprising the fleas. The body of a flea is laterally compressed and bears numerous backward-directed spines. Fleas live as blood-sucking ectoparasites of mammals and birds, having mouthparts adapted to piercing their host, injecting saliva to prevent clotting, and sucking up the blood. The long bristly legs can transmit energy stored in the elastic body wall to leap relatively long distances (over 300 mm horizontally). Apart from causing irritation, fleas can transmit disease organisms, most notably bubonic plague bacteria, which can be carried from rats to humans by the rat flea

(*Xenopsylla cheopsis*). The whitish wormlike legless larvae feed on organic matter. After two moults the larva spins a cocoon and undergoes metamorphosis into the adult.

Siphunculata (Anoplura) An order of secondarily wingless insects comprising the sucking lice: blood-sucking ectoparasites of mammals, with piercing and sucking mouthparts forming a snoutlike proboscis. They constitute an irritating pest to humans and domestic animals and can transmit diseases, including typhoid. The human louse (*Pediculus humanus*) exists in two forms: the head louse (*P. humanus capitis*) and the body louse (*P. humanus corporis*).

Site of Special Scientific Interest *See* SSSI.

SI units Système International d'Unités: the international system of units now recommended for all scientific purposes. A coherent and rationalized system of units derived from the *m.k.s. units* (a metric system based on the metre, kilogram, and second), SI units have now replaced *c.g.s. units and *Imperial units. The system has seven *base units* and two *dimensionless units* (formerly called *supplementary units*), all other units being derived from these nine units. There are 18 derived units with special names. Each unit has an agreed symbol (a capital letter or an initial capital letter if it is named after a scientist, otherwise the symbol consists of one or two lower-case letters). Decimal multiples of the units are indicated by a set of prefixes; when possible a prefix representing 10 raised to a power that is a multiple of three should be used. See Appendix.

skeletal muscle *See* voluntary muscle.

skeleton The structure in an animal that provides mechanical support for the body, protection for internal organs, and a framework for anchoring the muscles. The skeleton may be external (*see* **exoskeleton**) or internal (*see* **endoskeleton**). Both types require *joints to allow *locomotion. The skeleton of higher vertebrates consists of a system of *bones (*see* **appendicular skeleton; axial skeleton**). Soft-bodied animals have a *hydrostatic skeleton.

skin The outer layer of the body of a vertebrate (see illustration). It is composed of two layers, the *epidermis and *dermis, with a complex nervous and blood supply. The skin may bear a variety of specialized structures, including *hair, *scales, and *feathers. This skin has an important role in protecting the body from mechanical injury, water loss, and the entry of harmful agents (e.g. disease-causing bacteria). It is also a sense organ, containing receptors sensitive to pain, temperature, and pressure (*see* **Meissner's corpuscle; Pacinian corpuscle**). In warm-blooded animals it helps regulate body temperature by means of hair, fur, or feathers and *sweat glands.

skull The skeleton of the head. In mammals it consists of a *cranium enclosing the brain and the bones of the face and jaw. All the joints between the individual bones of the skull are immovable (*see* **suture**)

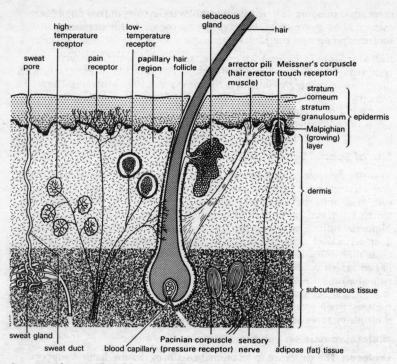

Structure of mammalian skin

except for the joint between the mandible (lower jaw) and the rest of the skull. There is a large opening (*foramen magnum*) at the base of the skull through which the spinal cord passes from the brain.

sleep A readily reversible state of reduced awareness and metabolic activity that occurs periodically in many animals. Usually accompanied by physical relaxation, the onset of sleep in humans and other mammals is marked by a change in the electrical activity of the brain, which is recorded by an *electroencephalogram as waves of low frequency and high amplitude (*slow-wave sleep*). This is interspersed by short bouts of high-frequency low-amplitude waves (similar to wave patterns produced when awake) associated with restlessness, dreaming, and rapid eye movement (REM); this is called *REM* (or *paradoxical*) *sleep* and is often accompanied by an increased pulse rate and dilation of the pupils. Several regions of the brain are involved in sleep, especially the reticular formation of the *brainstem.

sleep movements *See* **nyctinasty**.

sliding filament theory A proposed mechanism of muscle contraction

in which the *actin and *myosin filaments of striated muscle slide over
each other to shorten the length of the muscle fibres (*see* **sarcomere**).
Myosin-binding sites on the actin filaments are exposed when calcium ions
bind to *troponin molecules in these filaments. This allows bridges to form
between actin and myosin, which requires ATP as an energy source.
Hydrolysis of ATP in the heads of the myosin molecules causes the heads to
change shape and bind to the actin filaments. The release of ADP from the
myosin heads causes a further change in shape and generates mechanical
energy that causes the actin and myosin filaments to slide over one
another (see illustration).

Sliding filament theory of muscle contraction

slime moulds Small simple organisms widely distributed in damp
habitats on land. They exist either as free cells (*myxamoebas*) or as
multinucleate aggregates of cells (*see* **plasmodium**), depending on the stage
of the life cycle. In the *plasmodial* (or *true*) *slime moulds* the nuclei of the
plasmodium are not separated by cell membranes; in *cellular slime moulds*
the myxamoebas retain their cell membranes after aggregating to form a
'pseudoplasmodia'. Cell walls are generally absent. Slime moulds show
*amoeboid movement and feed by ingesting small particles of food. They
reproduce by means of spores. Slime moulds were formerly classified as
fungi. They are now usually classified as two phyla, Myxomycota
(plasmodial slime moulds) and Acrasiomycota (cellular slime moulds), of the
kingdom Protoctista.

slow virus Any of a group of quasi-viral or subviral agents formerly
thought to be responsible for the diseases scrapie in sheep and bovine
spongiform encephalopathy (BSE) in cattle. These diseases are now generally
regarded as being caused by abnormal *prion proteins.

slow-wave sleep *See* sleep.

sludge *See* sewage.

small intestine The portion of the *alimentary canal between the stomach and the large intestine. It is subdivided into the *duodenum, *jejunum, and *ileum. It plays an essential role in the final digestion and absorption of food.

smell *See* olfaction.

smooth endoplasmic reticulum (smooth ER) *See* endoplasmic reticulum.

smooth muscle *See* involuntary muscle.

smuts A group of parasitic fungi of the phylum *Basidiomycota. Many of these species attack the ears of cereal crops, replacing the grain by a mass of dark spores. *Compare* rusts.

snakes *See* Squamata.

social behaviour Any behaviour exhibited by a group of animals that interact with each other. Social behaviour ranges from moving as a herd in order to minimize the effects of predators to performing designated roles in highly organized societies. For example, within a colony of bees specific tasks, including tending the larvae, foraging for food, and controlling the temperature within the colony by wing fanning, are performed by different individuals.

sodium Symbol Na. A soft silvery element that is a major *essential element required by animals. It is important in maintaining the *acid–base balance and in controlling the volume of extracellular fluid and functions in the transmission of nerve impulses (*see* **sodium pump**).

sodium chloride (common salt) A colourless crystalline solid, NaCl, that is soluble in water. Sodium chloride has a key role in biological systems in maintaining electrolyte balances. It is used as a food preservative (*see* **food preservation**).

sodium fluoride A crystalline compound, NaF. It is highly toxic but in very dilute solution (less than 1 part per million) it is used in the *fluoridation of water for the prevention of tooth decay on account of its ability to replace OH groups with F groups in the material of dental enamel.

sodium pump A mechanism by which sodium ions are transported out of a neurone across the cell membrane. The process requires energy in the form of ATP, being a form of *active transport. It maintains the different concentration of sodium ions on either side of the neurone membrane necessary for establishing the *resting potential of the neurone.

softwood *See* wood.

soil The layer of unconsolidated particles derived from weathered rock, organic material (*humus), water, and air that forms the upper surface over much of the earth and supports plant growth. The formation of soil depends on the parent material (i.e. the original material from which the soil is derived), the climate and topography of the area, the organisms present in the soil, and the time over which the soil has been developing. Soils are often classified in terms of their structure and texture. The structure of a soil is the way in which the individual soil particles are bound together to form aggregates or peds. The structure types include platy, blocky, granular, and crumbs. The texture of a soil denotes the proportion of the various particle sizes that it contains. The four main texture classes are sand, silt, *clay, and *loam, of which loams are generally the best agricultural soils as they contain a mixture of all particle sizes. A number of distinct horizontal layers can often be distinguished in a vertical section (profile) of soil – these are known as *soil horizons*. Four basic horizons are common to most soils: an uppermost A horizon (or *topsoil*) containing the organic matter; an underlying B horizon (or *subsoil*), which contains little organic material and is strongly leached; a C horizon consisting of weathered rock; and a D horizon comprising the bedrock.

soil erosion The removal and thinning of the soil layer due to climatic and physical processes, such as high rainfall, which is greatly accelerated by certain human activities, such as *deforestation. Soil erosion can lead to a loss of agricultural land and if unchecked, eventually results in *desertification.

sol A *colloid in which small solid particles are dispersed in a liquid continuous phase.

solenocytes *See* flame cells.

solute The substance dissolved in a solvent in forming a *solution.

solute potential Symbol Ψ_s. The component of *water potential that is due to the presence of solute molecules. It always has a negative value as solutes lower the water potential of the system.

solution A homogeneous mixture of a liquid (the *solvent) with a gas or solid (the *solute*). In a solution, the molecules of the solute are discrete and mixed with the molecules of solvent. There is usually some interaction between the solvent and solute molecules. For example, when sodium chloride dissolves in water the sodium ions attract polar water molecules, with the negative oxygen ion pointing towards the positive Na^+ ion. This is called *hydration*.

solvent A liquid that dissolves another substance or substances to form a *solution. *Polar solvents* are compounds, such as water, in which there is some separation of charge in the chemical bonds. These solvents are capable of dissolving ionic compounds or covalent compounds that ionize. *Nonpolar solvents* are compounds, such as benzene, that do not dissolve ionic compounds but will dissolve nonpolar covalent compounds.

somatic **1.** Relating to all the cells of an animal or plant other than the

reproductive cells. Thus a somatic *mutation is one that is not heritable.
2. Relating to organs and tissues of the body other than the gut and its associated structures. The term is applied especially to voluntary muscles, the sense organs, and the nervous system. *Compare* **visceral**.

somatic cell hybridization *See* **cell fusion**.

somatic sensory neurone *See* **sensory neurone**.

somatostatin (growth hormone inhibiting hormone; GHIH) A hormone, secreted by the hypothalamus, that inhibits the release of *growth hormone from the anterior pituitary gland. The secretion of somatostatin is stimulated by various factors, including very high blood glucose levels, which result from the effect that growth hormone has on glucose metabolism.

somatotrophin *See* **growth hormone**.

somite Any of a series of segmented blocks of tissue in animal embryos that develop from the mesoderm. In vertebrate embryos they lie on the dorsal side of the notochord and give rise to the vertebral column, ribs, dermis, and striated muscle. In invertebrates showing *metameric segmentation the somites give rise to the body segments.

sonicator A device used to break down cells using ultrasonic waves. Ultrasonic waves are transmitted through a metal rod, which is placed in a suspension of the cells. The vibrations cause the cell membranes to rupture and the cellular contents are released into the surrounding medium.

soredium (*pl.* **soredia**) A small body, consisting of a core of algal cells surrounded by fungal hyphae, that functions as a structure of vegetative propagation in *lichens. Soredia, which range from 25 to 100 μm in diameter, are released like spores and are dispersed in air currents. *Compare* **isidium**.

sorosis A type of *composite fruit formed from an entire inflorescence spike. Mulberry and pineapple fruits are examples.

sorus 1. Any of the spore-producing structures on the undersurface of a fern frond, visible as rows of small brown dots. **2.** A reproductive area on the thallus of some algae, e.g. *Laminaria*. **3.** Any of various spore-producing structures in certain fungi.

Southern blotting A chromatographic technique for isolating and identifying specific fragments of DNA, such as the fragments formed as a result of DNA cleavage by *restriction enzymes. The mixture of fragments is subjected to electrophoresis through an agarose gel, followed by denaturation to form single-stranded fragments. These are transferred, or 'blotted', onto a nitrocellulose filter where they are immobilized in their relative positions. Specific *gene probes labelled with a radioisotope are then added. These hybridize with any complementary fragments on the filter, which are subsequently revealed by autoradiography. The technique was devised by E. M. Southern. A similar technique for detecting RNA fragments is called *Northern blotting*, by analogy. *See also* **Western blotting**.

spadix A flowering shoot (a type of *spike) with a large fleshy floral axis bearing small, usually unisexual, flowers. It is protected by a large petal-like bract, the *spathe*, and is characteristic of plants of the family Araceae (e.g. calla lily).

spathe *See* spadix.

special creation The belief, in accordance with the Book of Genesis, that every species was individually created by God in the form in which it exists today and is not capable of undergoing any change. It was the generally accepted explanation of the origin of life until the advent of *Darwinism. The idea has recently enjoyed a revival, especially among members of the fundamentalist movement in the USA, partly because there still remain problems that cannot be explained entirely by Darwinian theory. However, special creation is contradicted by fossil evidence and genetic studies, and the pseudoscientific arguments of *creation science* cannot stand up to logical examination.

specialization 1. Increasing *adaptation of an organism to a particular environment. **2.** *See* **physiological specialization**.

speciation The development of one or more species from an existing species. It occurs when *sympatric or *allopatric populations diverge so much from the parent population that interbreeding can no longer occur between them. *See also* **parapatric speciation**.

species A category used in the *classification of organisms that consists of a group of similar individuals that can usually breed among themselves and produce fertile offspring. Similar or related species are grouped into a genus. Within a species groups of individuals may become reproductively isolated because of geographical or behavioural factors. Such populations may, because of different selection pressures, develop different characteristics from the main population and so form a distinct *subspecies. *See also* **binomial nomenclature**.

species diversity *See* biodiversity.

spectrin A protein that forms part of the fibrous network underlying the plasma membrane of erythrocytes.

spectroscopy The study of methods of producing and analysing *spectra using spectroscopes and other instruments. Spectroscopy is employed in a wide range of biological research areas, such as biochemistry and toxicology, for the identification of metabolites and other compounds of biological significance. *See* **mass spectroscopy**.

spectrum (*pl.* spectra) A range of electromagnetic energies arrayed in order of increasing or decreasing wavelength or frequency. The *emission spectrum* of a body or substance is the characteristic range of radiations it emits when it is heated, bombarded by electrons or ions, or absorbs photons. The *absorption spectrum* of a substance is produced by examining, through the substance and through a spectroscope, a continuous spectrum of radiation. The energies removed from the continuous spectrum by the absorbing medium show up as black lines or bands; with a substance capable of emitting a spectrum these are in exactly the same positions in the spectrum as the emission lines and bands would occur in the emission spectrum.

Emission and absorption spectra may show a *continuous spectrum*, a *line spectrum*, or a *band spectrum*. A continuous spectrum contains an unbroken sequence of frequencies over a relatively wide range; it is produced by incandescent solids, liquids, and compressed gases. Line spectra are discontinuous lines produced by excited atoms and ions as they fall back to a lower energy level. Band spectra (closely grouped bands of lines) are characteristic of molecular gases or chemical compounds. Absorption spectra of chlorophylls and other photosynthetic pigments are important in the study of photosynthesis. *See* **action spectrum**.

sperm 1. A single *spermatozoon. **2.** Spermatozoa, collectively.

spermatheca (seminal receptacle) A sac or receptacle in some female or hermaphrodite animals (e.g. earthworms) in which sperm from the mate is stored until the eggs are ready to be fertilized.

spermatid A nonmotile cell, produced during *spermatogenesis, that subsequently differentiates into a mature spermatozoon. Four spermatids are formed after two meiotic divisions of a primary *spermatocyte and therefore contain the haploid number of chromosomes.

spermatium (*pl.* **spermatia**) A type of nonmotile male sex cell that is produced by the red algae (Rhodophyta) and by certain fungi (e.g. the *rusts).

spermatocyte A diploid cell in the testis that divides by meiosis to give rise to four *spermatids (*see* **spermatogenesis**). A *primary spermatocyte* develops from a *spermatogonium. Two *secondary spermatocytes* result from the first meiotic division of a primary spermatocyte; each of these produces two spermatids after the second meiotic division.

spermatogenesis The series of cell divisions in the testis that results in the production of spermatozoa. Within the seminiferous tubules of the testis germ cells grow and divide by mitosis to produce *spermatogonia. These divide by mitosis to produce *spermatocytes, which divide by meiosis to produce *spermatids. The spermatids, which thus have half the number of chromosomes of the original germ cells, then develop into spermatozoa.

spermatogonium (*pl.* **spermatogonia**) Any of the diploid cells in the walls of the seminiferous tubules of the testis that give rise to the primary *spermatocytes. *See also* **spermatogenesis**.

Spermatophyta In traditional classifications, a division of the plant kingdom containing plants that reproduce by means of *seeds. In modern systems seed plants are grouped into separate phyla, the most important of which are the *Angiospermophyta and *Coniferophyta.

spermatozoid *See* antherozoid.

spermatozoon (sperm) The mature mobile reproductive cell (*see* **gamete**) of male animals, which is produced by the testis (*see* **spermatogenesis**). It consists of a head section containing a *haploid nucleus and an *acrosome*, which allows the sperm to penetrate the egg at fertilization; a middle section containing *mitochondria to provide the energy for movement; and a tail section, which lashes to drive the sperm forward.

sperm competition Competition between sperm from different males to reach and fertilize the egg cell of a single female. Sperm competition can occur among rodents in which a male mates a number of times with the same female, with a rest period between successive matings during which the sperm journeys towards the egg. If a second male mates with the female during a rest period its own sperm may disrupt the movement of sperm from the first male and succeed in fertilizing the egg cell. Certain animals in which sperm competition is possible have evolved features to minimize this interference. For example, in moths and butterflies the male cements the opening of the female genitalia after mating, thereby preventing further matings with other males. An ingenious mechanism operates in the fly *Johannseniella nitida*, in which the female eats a copulating male except for his genitalia, which remain in the body of the female and prevent further mating.

Sphenophyta A phylum of *tracheophyte plants, the only living members of which are the horsetails (*Equisetum*). Horsetails have a perennial creeping rhizome supporting erect jointed stems bearing whorls of thin leaves. Spores are produced by terminal conelike structures. The group has a fossil record extending back to the Palaeozoic with its greatest development in the Carboniferous period, when giant tree forms were the dominant vegetation with the *Lycopodophyta.

spherosome A small spherical organelle found in the cytoplasm of plant cells. Up to 1 mm in diameter and bounded by a single membrane, it synthesizes and stores lipids.

sphincter A specialized muscle encircling an opening or orifice. Contraction of the sphincter tends to close the orifice. Examples are the *anal sphincter* (round the opening of the anus) and the *pyloric sphincter* (at the lower opening of the *stomach).

sphingolipid *See* phospholipid.

spiders *See* Arachnida.

spike A type of *racemose inflorescence in which stalkless flowers arise from an undivided floral axis, as in plantain and *Orchis*. In the family Gramineae (sedges and grasses) the flowers are grouped in clusters called *spikelets* (see illustration), which may be arranged to form a compound spike (as in wheat). *See* **glume; lemma; palea.**

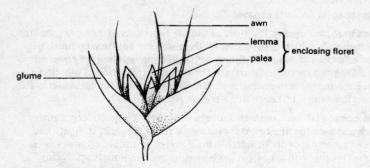

Structure of a spikelet

spinal column *See* **vertebral column.**

spinal cord The part of the vertebrate central nervous system that is posterior to the brain and enclosed within the *vertebral column. It consists of a hollow core of *grey matter (H-shaped in cross section) surrounded by an outer layer of *white matter; the central cavity contains *cerebrospinal fluid. The white matter contains numerous longitudinal nerve fibres organized into distinct tracts: *ascending tracts* consist of sensory neurones, conducting impulses towards the brain; *descending tracts* consist of motor neurones, transmitting impulses from the brain. Paired *spinal nerves arise from the spinal cord.

Transverse section of the spinal cord

spinal nerves Pairs of nerves that arise from the *spinal cord (*compare* **cranial nerves**). In man there are 31 pairs (one from each of the vertebrae). Each nerve arises from a *dorsal root and a *ventral root and contains both motor and sensory fibres (i.e. they are mixed nerves). The spinal nerves form an important part of the *peripheral nervous system.

spinal reflex *See* reflex.

spindle 1. A structure formed from fine fibres of protein in the cytoplasm of cells at the beginning of the *metaphase stage of cell division (*see* meiosis; mitosis). Two poles become established at diametrically opposite points close to the nucleus. Fibres radiating from each pole towards the opposite one diverge to their greatest extent midway between. This widest part of the spindle is the *equator*. Chromosomes are attached by their centromeres to the spindle fibres at the equator. **2.** *See* **muscle spindle**.

spindle attachment *See* centromere.

spine 1. *See* **vertebral column. 2.** A hard pointed protective structure on a plant that is formed through modification of a leaf, part of a leaf, or a stipule. The edge of the holly leaf is drawn out into spines, but in cacti the whole leaf is modified as a spine. *Compare* **prickle; thorn**.

spinneret A small tubular appendage from which silk is produced in spiders and some insects. Spiders have four to six spinnerets on the hind part of the abdomen, into which numerous silk glands open. The silk is secreted as a fluid and hardens on contact with the air. Various types of silk are produced depending on its use (e.g. for webs, egg cocoons, etc.). The spinnerets that produce the cocoons of insects are not homologous with those of spiders. For example, the spinneret of the silkworm is in the pharynx and the silk is produced by modified salivary glands.

spinocerebellar tracts Pathways of neurones that transfer sensory information from the spinal cord to the cerebellum of the brain.

spiracle 1. A small paired opening that occurs on each side of the head in cartilaginous fish. It is the reduced first *gill slit, its small size resulting from adaptations of the skeleton for the firm attachment of the jaws. In modern teleosts (bony fish) the spiracle is closed up. In tetrapods the first gill slit develops into the middle ear cavity. **2.** Any of the external openings of the *tracheae along the side of the body of an insect.

spirillum Any rigid spiral-shaped bacterium. Generally, spirilla are Gram-negative (*see* **Gram's stain**), aerobic, and highly motile, bearing one or more tufts of flagella. They occur in soil and water, feeding on organic matter.

spirochaete Any nonrigid corkscrew-shaped bacterium that moves by means of muscular flexions of the cell. Most spirochaetes are Gram-negative (*see* **Gram's stain**), anaerobic, and feed on dead organic matter. They are particularly common in sewage-polluted waters. Some, however, cause disease in man; *Treponema*, the agent of syphilis, is an example.

spirometer *See* respirometer.

spleen A vertebrate organ, lying behind the stomach, that is basically a collection of *lymphoid tissue. Its functions include producing lymphocytes and destroying foreign particles. It acts as a reservoir for erythrocytes and can regulate the number in circulation. It is also the site for the breakdown of worn-out erythrocytes and it stores the iron they contain.

splicing *See* gene splicing.

sponges *See* Porifera.

spongy bone *See* bone.

spongy mesophyll *See* mesophyll.

spontaneous generation The discredited belief that living organisms can somehow be produced by nonliving matter. For example, it was once thought that microorganisms arose by the process of decay and even that vermin spontaneously developed from household rubbish. Controlled experiments using sterilized media by Pasteur and others finally disproved these notions. *Compare* **biogenesis**. *See also* **biopoiesis**.

sporangiophore The simple or branched stalk that bears one or more sporangia.

sporangium (*pl.* **sporangia**) A reproductive structure in plants that produces asexual spores. A *megasporangium* produces megaspores, which give rise to the female gametophyte; in seed plants it is represented by the *ovule. A *microsporangium* produces microspores, which give rise to the male gametophyte; it is represented in seed plants by the *pollen sac. *See also* **sporophyll**.

spore A reproductive cell that can develop into an individual without first fusing with another reproductive cell (*compare* **gamete**). Spores are produced by plants, fungi, bacteria, and some protozoa. A spore may develop into an organism resembling the parent or into another stage in the life cycle, either immediately or after a period of dormancy. In plants showing *alternation of generations, spores are formed by the *sporophyte generation and give rise to the *gametophyte generation. In ferns, the rows of brown reproductive structures on the undersurface of the fronds are spore-producing bodies.

spore mother cell (**sporocyte**) A diploid cell that gives rise to four haploid spores by meiosis. *See also* **megaspore mother cell**; **microspore mother cell**.

sporocyte *See* spore mother cell.

sporogonium The *sporophyte generation in bryophytes. It is made up of an absorptive *foot*, a stalk (*seta*), and a spore-producing *capsule*. It may be completely or partially dependent on the *gametophyte.

sporophore (fructification) The aerial spore-producing part of certain fungi; for example, the stalk and cap of a mushroom.

sporophyll A leaf that bears *sporangia (spore-producing structures). In ferns the sporophylls are the normal foliage leaves, but in other plants the sporophylls are modified and arise in specialized structures such as the strobilus (cone) of clubmosses and horsetails and gymnosperms and the flower of angiosperms. Most plants produce spores of two different sizes (small *microspores* and large *megaspores*). The sporophylls bearing these are called *microsporophylls* and *megasporophylls* respectively.

sporophyte The generation in the life cycle of a plant that produces spores. The sporophyte is *diploid but its spores are *haploid. It is either completely or partially dependent on the *gametophyte generation in bryophytes but is the dominant plant in the life cycle of clubmosses, horsetails, ferns, and seed plants. *See also* **alternation of generations**.

Sporozoa *See* **Apicomplexa**.

sport An individual organism that shows the effects of a mutation. The term is usually applied to atypical forms of plants in horticulture.

squalene An intermediate compound formed in the synthesis of cholesterol; it is a hydrocarbon containing 30 carbon atoms. The immediate oxidation of squalene to squalene 2,3-epoxide is the last common step in the synthesis of *sterols in animals, plants, and fungi.

Squamata An order of reptiles comprising the lizards and snakes. They appeared at the end of the Triassic period, about 170 million years ago, and have invaded a wide variety of habitats. Most lizards have four legs and a long tail, eardrums, and movable eyelids. Snakes are limbless reptiles that lack eardrums; the eyes are covered by transparent immovable eyelids and the articulation of the jaws is very loose, enabling a wide gape to facilitate swallowing prey whole.

squamous epithelium A type of *epithelium that is made up of flattened cells and therefore presents a small distance over which substances have to pass. This type of epithelium is found lining the alveoli in the lungs and Bowman's capsule in the kidney.

SSSI (Site of Special Scientific Interest) The legal designation for an area of land in England, Scotland, or Wales that has been identified by English Nature, Scottish National Heritage, or the Countryside Council for Wales as being of special interest because of its flora, fauna, or geological or physiographical features. Such sites are protected from development activities and funds are available for their conservation and management. There are over 6000 SSSIs in Britain; similar sites in Northern Ireland are designated *Areas of Special Scientific Interest (ASSIs)*.

stabilizing selection *Natural selection that acts to maintain the constancy of a species over successive generations. It involves selection against the extremes of the range of phenotypes for a particular characteristic. For example, babies whose birth weight is substantially below or above the average of 3.6 kg have a greater mortality than babies of average birth weight. *Compare* **directional selection; disruptive selection**.

staining A technique in which cells or thin sections of biological tissue that are normally transparent are immersed in one or more coloured dyes (*stains*) to make them more clearly visible through a microscope. Staining heightens the contrast between the various cell or tissue components. Stains are usually organic salts with a positive and negative ion. If the colour comes from the negative ion (organic anion), the stain is described as *acidic*, e.g. *eosin. If the colour comes from the positive ion (organic cation), the stain is described as *basic*, e.g. *haematoxylin. *Neutral stains* have a coloured cation and a coloured anion; an example is *Leishman's stain. Cell constituents are described as being *acidophilic* if they are stained with acidic dyes, *basophilic* if receptive to basic dyes, and *neutrophilic* if receptive to neutral dyes. *Vital stains* are used to colour the constituents of living cells without harming them (*see* **vital staining**); *nonvital stains* are used for dead tissue.

Counterstaining involves the use of two or more stains in succession, each of which colours different cell or tissue constituents. *Temporary staining* is used for immediate microscopical observation of material, but the colour soon fades and the tissue is subsequently damaged. *Permanent staining* does not distort the cells and is used for tissue that is to be preserved for a considerable period of time.

Electron stains, used in the preparation of material for electron microscopy, are described as *electron-dense* as they interfere with the transmission of electrons. Examples are lead citrate, phosphotungstic acid (PTA), and uranyl acetate (UA).

stamen One of the male reproductive parts of a flower. It consists of an upper fertile part (the *anther) on a thin sterile stalk (the *filament*).

staminode A sterile stamen.

standard deviation A measure of the dispersion of data in statistics. For a set of values $a_1, a_2, a_3, \dots a_n$, the mean m is given by $(a_1 + a_2 + \dots + a_n)/n$. The *deviation* of each value is absolute value of the difference from the mean $|m - a_1|$, etc. The standard deviation is the square root of the mean of the squares of these values, i.e.

$$\sqrt{[(|m - a_1|^2 + \dots |m - a_n|^2)/n]}$$

standard error of the mean (SEM) The *standard deviation divided by the square root of the number of samples. Whereas the standard deviation of a population provides a measure of the scatter of the data, the SEM is a measure of the accuracy of the calculated population *mean.

standing biomass *See* **standing crop**.

standing crop The total amount of living material in a specified population at a particular time, expressed as *biomass (*standing biomass*) or its equivalent in terms of energy. The standing crop may vary at different times of the year; for example, in a population of deciduous trees between summer and winter.

stapes (stirrup) The third of the three *ear ossicles of the mammalian middle ear.

Staphylococcus A genus of spherical nonmotile Gram-positive bacteria that occur widely as saprotrophs or parasites. The cells occur in grapelike clusters. Many species inhabit the skin and mucous membranes, and some cause disease in humans and animals. *S. aureus* infection can lead to boils and abscesses in humans; this species also produces *toxins that irritate the gastrointestinal tract and result in staphylococcal food poisoning.

starch A *polysaccharide consisting of various proportions of two glucose polymers, *amylose and *amylopectin. It occurs widely in plants, especially in roots, tubers, seeds, and fruits, as a carbohydrate energy store. Starch is therefore a major energy source for animals. When digested it ultimately yields glucose. Starch granules are insoluble in cold water but disrupt if heated to form a gelatinous solution.

start codon (initiation codon) The triplet of nucleotides on a messenger *RNA molecule (*see* **codon**) at which the process of *translation is initiated. In eukaryotes the start codon is AUG (*see* **genetic code**), which codes for the amino acid methionine; in bacteria the start codon can be either AUG, coding for N-formyl methionine, or GUG, coding for valine. *Compare* **stop codon**.

stationary phase **1.** *See* **bacterial growth curve**. **2.** *See* **chromatography**.

statoblast A chitinous bud that is produced asexually by members of the *Ectoprocta. The statoblast stores food and is able to remain dormant for long periods, withstanding drought and extreme temperatures.

statocyst (otocyst) A balancing organ found in many invertebrates. It consists of a fluid-filled sac lined with sensory hairs and contains granules of calcium carbonate, sand, etc. (*statoliths*). As the animal moves the statoliths stimulate different hairs, giving a sense of the position of the body or part of it. The *semicircular canals in the ears of vertebrates act on the same principle and have a similar function.

statocyte Any plant cell that contains *statoliths.

statolith **1.** *See* **statocyst**. **2.** Any starch grain or other inclusion in plant cells that acts as a sensor to gravity. Starch statoliths are found in cells at the root tips and in the tissues close to the vascular bundles in shoots; under the influence of gravity they sink to the bottom of the cell. Their mechanism of action in triggering the transport of growth substances across the cell membrane is not understood.

stearic acid (octadecanoic acid) A solid saturated *fatty acid, $CH_3(CH_2)_{16}COOH$, that occurs widely (as *glycerides) in animal and vegetable fats.

stele The vascular tissue (i.e. *xylem and *phloem) of *tracheophyte plants, together with the endodermis and pericycle (when present). The arrangement of stelar tissues is very variable. In roots the stele often forms a solid core, which better enables the root to withstand tension and compression. In stems it is often a hollow cylinder separating the cortex and pith. This arrangement makes the stem more resistant to bending stresses. Monocotyledons and dicotyledons can usually be distinguished by the pattern of their stelar tissue. In monocotyledons the vascular bundles are scattered throughout the stem whereas in dicotyledons (and gymnosperms) they are arranged in a circle around the pith.

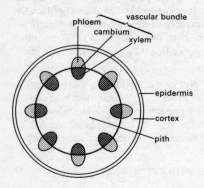

Transverse section through a herbaceous stem

stem The part of a plant that usually grows vertically upwards towards the light and supports the leaves, buds, and reproductive structures (see illustration). The leaves develop at the *nodes and side or branch stems develop from buds at the nodes. The stems of certain species are modified as bulbs, corms, rhizomes, and tubers. Some species have twining stems; others have horizontal stems, such as *runners. Another modification is the *cladode. Erect stems may be cylindrical or angular; they may be covered with hairs, prickles, or spines and many exhibit secondary growth and become woody (*see* **growth ring**). In addition to its supportive function, the stem contains *vascular tissue that conducts food, water, and mineral salts between the roots and leaves. It may also contain chloroplasts and carry out photosynthesis.

stem cell A cell that is not differentiated itself but can undergo unlimited division to form other cells, which either remain as stem cells or differentiate to form specialized cells. For example, stem cells in the bone

marrow divide to produce daughter cells that differentiate into various
types of immune cell (e.g. monocytes, lymphocytes, mast cells). *See also*
haemopoietic tissue.

steno- A prefix denoting narrowness. For example, *stenohaline* aquatic
organisms can tolerate only small variations in the salinity of the water.
Compare **eury-**.

stereoisomerism The existence of chemical compounds (*stereoisomers*)
that have the same molecular formulae and functional groups, but differ
in the arrangement of groups in space. Optical isomerism is one form of
this (*see* **optical activity**).

sterigma (*pl.* **sterigmata**) A small spore-bearing stalk found in certain
fungi. In the Basidiomycota, sterigmata develop as outgrowths of a
basidium, each bearing a basidiospore at its tip.

sterile 1. *(of organisms)* Unable to produce offspring. *See also* **hybrid;
incompatibility; self-sterility; sterilization. 2.** *(of objects, food, etc.)* Free
from contaminating microorganisms. *See* **sterilization**.

sterilization 1. The process of destroying microorganisms that
contaminate food, wounds, surgical instruments, etc. Common methods
of sterilization include heat treatment (*see* **autoclave; pasteurization**)
and the use of *disinfectants and *antiseptics. **2.** The operation of
making an animal or human incapable of producing offspring. Men
are usually sterilized by tying and then cutting the *vas deferens
(*vasectomy*); in women the operation often involves permanently
blocking the fallopian tubes by means of clips (*tubal occlusion*). *See also* **birth
control**.

sternum (breastbone) 1. A shield-shaped or rod-shaped bone in terrestrial
vertebrates, on the ventral side of the thorax, that articulates with the
*clavicle (collar bone) of the pectoral girdle and with most of the ribs. It is
absent in fish, and in birds it bears a *keel. **2.** The ventral portion of each
segment of the exoskeleton of arthropods.

steroid nucleus

cholesterol (a sterol)

testosterone (an androgen)

Steroid structure

steroid Any of a group of lipids derived from a saturated compound
called cyclopentanoperhydrophenanthrene, which has a nucleus of four

rings (see formulae). Some of the most important steroid derivatives are the steroid alcohols, or *sterols. Other steroids include the *bile acids, which aid digestion of fats in the intestine; the sex hormones (*androgens and *oestrogens); and the *corticosteroid hormones, produced by the adrenal cortex. *Vitamin D is also based on the steroid structure.

sterol Any of a group of *steroid-based alcohols having a hydrocarbon side-chain of 8–10 carbon atoms. Sterols exist either as free sterols or as esters of fatty acids. Animal sterols (*zoosterols*) include *cholesterol and lanosterol. The major plant sterol (*phytosterol*) is beta-sitosterol, while fungal sterols (*mycosterols*) include *ergosterol.

stigma 1. The glandular sticky surface at the tip of a carpel of a flower, which receives the pollen. In insect-pollinated plants the stigmas are held within the flower, whereas in wind-pollinated species they hang outside it. **2.** *See* **eyespot**.

stilt root *See* **prop root**.

stimulus Any change in the external or internal environment of an organism that provokes a physiological or behavioural response in the organism. In an animal specific *receptors are sensitive to stimuli.

stimulus filtering The process of separating useful sensory information from the many thousands of stimuli present in the external environment, so that only potentially useful information is sent to the brain. Many sensory organs are adapted so that they receive stimuli only within a certain range. For example, the human eye only detects colours in the visible region of the spectrum and not in the ultraviolet or infrared regions. Another method of stimulus filtering involves *sign stimuli, which are more important than other stimuli in eliciting a response and, once detected, may lead to other stimuli being ignored.

stipe 1. The stalk that forms the lower portion of the fruiting body of certain fungi, such as mushrooms, and supports the umbrella-shaped cap. **2.** The stalk between the holdfast and blade (*lamina*) of certain brown algae, notably kelps.

stipule An outgrowth from the petiole or leaf base of certain plants. Those of the garden pea are leaflike photosynthetic organs. The stipules of the lime tree are scalelike and protect the winter buds, whereas those of the false acacia (*Robinia*) are modified as spines.

stock *See* **graft**.

stolon A long aerial side stem that gives rise to a new daughter plant when the bud at its apex touches the soil. Plants that multiply in this way include blackberry and currant bushes. Gardeners often pin down stolons to the soil to aid the propagation of such plants. This process is termed *layering*.

stoma (*pl.* **stomata**) A pore, large numbers of which are present in the epidermis of leaves (especially on the undersurface) and young shoots. Stomata function in gas exchange between the plant and the atmosphere. Each stoma is bordered by two semicircular *guard cells* (specialized epidermal cells), whose movements (due to changes in water content) control the size of the aperture. The term stoma is also used to mean both the pore and its associated guard cells.

stomach The portion of the vertebrate *alimentary canal between the oesophagus and the small intestine. It is a muscular organ, capable of dramatic changes in size and shape, in which ingested food is stored and undergoes preliminary digestion. Cells lining the stomach produce *gastric juice, which is thoroughly mixed with the food by muscular contractions of the stomach. The resultant acidic partly digested food mass (*chyme) is discharged into the *duodenum through the pyloric *sphincter for final digestion and absorption. Some herbivorous animals (*see* **Ruminantia**) have multichambered stomachs from which food is regurgitated, rechewed, and swallowed again.

stomium A region of thin-walled cells in certain spore-producing structures that ruptures to release the spores. For example, in the sporangium of the fern *Dryopteris* the stomium ruptures when the annulus dries out.

stop codon (**termination codon**) The triplet of nucleotides on a messenger *RNA molecule (*see* **codon**) at which the process of *translation ends. It cannot code for an amino acid as there is no *anticodon that complements it: this effectively stops the formation of a polypeptide chain at that point. The three stop codons are UGA, UAA, and UAG (*see* **genetic code**). *Compare* **start codon**.

storage compound *See* **food reserves**.

stratification 1. The arrangement of the components of an entity in layers (*strata*). Stratification is a feature of sedimentary rocks and *soils. It is also seen in *stratified epithelium, and thermal stratification can occur in some lakes (*see* **thermocline**). **2.** The practice of placing certain seeds between layers of peat or sand and then exposing them to low temperatures for a period, which is required before they will germinate. *See* **vernalization**.

stratified epithelium *Epithelium that is made up of a number of layers of cells and is present in areas of the body that are subject to wear and tear. Stratified epithelium containing *keratin forms the outer layer of the skin; stratified epithelial cells that do not contain keratin are found in moist areas, such as the mouth and vagina.

stratosphere The layer of the earth's atmosphere that lies above the troposphere and extends to about 50 km above the earth's surface. The temperature within the stratosphere remains fairly constant but can rise in the upper regions of this layer due to absorption of ultraviolet radiation by ozone. *See* **ozone layer**.

stratum corneum The layers of dead keratinized cells that form the outermost layers of mammalian *epidermis. The stratum corneum provides a water-resistant barrier between the external environment and the living cells of the *skin.

Streptococcus A genus of spherical Gram-positive bacteria occurring widely in nature, typically as chains or pairs of cells. Many are saprotrophic and exist as usually harmless commensals inhabiting the skin, mucous membranes, and intestine of humans and animals. Others are parasites, some of which cause diseases, including scarlet fever (*S. pyogenes*; group A streptococci), endocarditis (*S. viridans*), and pneumonia (*S. pneumoniae*).

streptomycin *See* **Actinobacteria; antibiotics**.

stretch receptor A specialized group of cells found in muscles (*see* **muscle spindle**) or tendons that is sensitive to mechanical stress. When muscles or tendons are stretched, the strain is recorded by the stretch receptors.

stretch reflex (myotactic reflex) The *reflex initiated when a muscle is stretched; an example is the *knee-jerk reflex*. Stretching of a muscle causes impulses to be generated in the *muscle spindles. These impulses are transmitted by sensory neurones to the spinal cord, where the sensory neurones synapse with motor neurones; these initiate contraction of the same muscle so that it returns to its original length. Since the reflex action involves the transmission of impulses across only one set of synapses, the response is rapid and described as *monosynaptic*.

striated muscle *See* **voluntary muscle**.

stridulation The production of sounds by insects rubbing one part of the body against another. The parts of the body involved vary from species to species. Stridulation is typical of the Orthoptera (grasshoppers, crickets, cicadas), in which the purpose of the sounds is usually to bring the sexes together, although they are also used in territorial behaviour, warning, etc.

strobilus 1. A type of *composite fruit that is formed from a complete inflorescence. It produces *achenes enclosed in bracts and when mature becomes cone-shaped. The hop fruit is an example. **2.** *See* **cone**.

stroke volume The volume of blood ejected at each stroke of the heart during *systole. A typical stroke volume for a normal person at rest is 70 ml; this increases to a maximum of 120 ml during strenuous exercise. *See* **cardiac output**.

stroma Tissue that forms the framework of an organ; for example, the tissue of the ovary that surrounds the reproductive cells, or the gel-like matrix of *chloroplasts that surrounds the grana.

stromatolite A rocky cushion-like mass formed by the unchecked growth of millions of lime-secreting blue-green bacteria (*see* **Cyanobacteria**). Stromatolites are found only in areas where other organisms that would normally keep down the bacterial numbers cannot survive, such as extremely salty bays. It is thought that such formations were abundant 2000 million years ago, when blue-green bacteria were the most advanced form of life on earth, and that the white rings of fossilized microorganisms found in rocks of this age are the remains of stromatolites.

strontium Symbol Sr. A soft yellowish metallic element. The isotope strontium–90 is present in radioactive fallout (half-life 28 years), and can be metabolized with calcium so that it collects in bone.

structural gene *See* operon.

strychnine A colourless poisonous crystalline alkaloid found in certain plants.

style The stalk of a carpel, between the stigma and the ovary. In many plants it is elongated to aid pollination.

subarachnoid space The space between the *arachnoid membrane and the *pia mater, two of the membranes (*meninges) that surround the brain and spinal cord. It is filled with *cerebrospinal fluid.

subclavian artery A paired artery that passes beneath the collar bone (clavicle) and branches to supply blood to the arm. The left subclavian artery arises from the aorta; the right from the innominate artery.

subcutaneous tissue The tissue that lies immediately beneath the *dermis (*see* **skin**). It is made up of loose fibrous *connective tissue, muscle, and fat (*see* **adipose tissue**), which in some animals (e.g. whales and hibernating mammals) forms an insulating layer or an important food store.

suberin A mixture of waxy substances, similar to *cutin, present in the thickened cell walls of many trees and shrubs, particularly in corky tissues. The deposition of suberin (*suberization*) provides a protective water-impermeable layer. *See also* **Casparian strip**.

sublittoral 1. Designating or occurring in the shallow-water zone of a sea, over the continental shelf and below the low tide mark. **2.** Designating or occurring in the zone of a lake below the littoral zone, to a depth of 6–10 metres.

submucosa A layer of *areolar connective tissue that lies beneath the *mucosa, for example in the stomach and intestine.

subsoil *See* soil.

subspecies A group of individuals within a *species that breed more freely among themselves than with other members of the species and resemble each other in more characteristics. Reproductive isolation of a subspecies may become so extreme that a new species is formed (*see* speciation). Subspecies are sometimes given a third Latin name, e.g. the mountain gorilla, *Gorilla gorilla beringei* (*see also* **binomial nomenclature**).

substitution *(in genetics)* A *point mutation in which one base pair in the DNA sequence is replaced by another. There are two types of substitution mutation: *transition* mutations, in which a pyrimidine base (i.e. thymine or cytosine) is replaced by another pyrimidine base or a purine base (adenine or guanine) is replaced by another purine base; and *transversion* mutations, in which a pyrimidine base is replaced by a purine base, or vice versa. Most substitutions tend to alter one amino acid in a protein chain, which may or may not affect the functioning of that protein. Sickle-cell anaemia is an example of a substitution mutation in which thymine is replaced by adenine in the triplet coding for the sixth amino acid in the β-chain of haemoglobin.

substrate **1.** *(in biochemistry)* The substance upon which an *enzyme acts in biochemical reactions. **2.** *(in biology)* The material on which a sedentary organism (such as a barnacle or a plant) lives or grows. The substrate may provide nutrients for the organism or it may simply act as a support.

subtilisin Any of a group of protein-digesting enzymes (*see* **protease**) produced by *Bacillus subtilis* and related species. Subtilisins have broad specificities and work over a wide range of pH values. *Subtilisin Carlsberg* is an alkaline protease, derived from *Bacillus licheniformis*, that is used widely in detergents. Roughly 30–40% of the peptide bonds in casein, a milk protein, are hydrolysed by subtilisin Carlsberg.

succession *(in ecology)* The sequence of communities that develops in an area from the initial stages of colonization until a stable mature *climax community* is achieved. Many factors, including climate and changes brought about by the colonizing organisms, influence the nature of a succession; for example, after many years shrubs produce soil deep enough to support trees, which then shade out the shrubs. *See also* **sere**.

succinate A salt of succinic (butanedioic) acid, $HOOC(CH_2)_2COOH$, a four-carbon fatty acid. Succinate occurs in living organisms as an intermediate in metabolism, especially in the *Krebs cycle.

succulent A plant that conserves water by storing it in fleshy leaves or stems. Succulents are found either in dry regions or in areas where there is sufficient water but it is not easily obtained, as in salt marshes. Such plants are often modified to reduce water loss by transpiration. For example, the leaves of cacti are reduced to spines. *See also* **crassulacean acid metabolism; xerophyte**.

succus entericus A mixture of enzymes produced by glands within the small intestine. These enzymes are responsible for most of the digestive processes that occur in the alimentary canal and include *proteases, *lipases, and *amylases.

sucker (turion) A shoot that arises from an underground root or stem and grows at the expense of the parent plant. Suckers can be dug up with a portion of root attached and used to propagate a plant. If, however, a plant is grafted onto a different rootstock, as many roses are, any suckers will be of the wild rootstock, rather than the ornamental scion, and must be removed.

sucrase A carbohydrate-digesting enzyme, produced in the brush border of the small intestine, that breaks down the disaccharide sucrose into the monosaccharides glucose and fructose.

sucrose (cane sugar; beet sugar; saccharose) A sugar comprising one molecule of glucose linked to a fructose molecule. It occurs widely in plants and is particularly abundant in sugar cane and sugar beet (15–20%), from which it is extracted and refined for table sugar. If heated to 200°C, sucrose becomes caramel.

sugar (saccharide) Any of a group of water-soluble *carbohydrates of relatively low molecular weight and typically having a sweet taste. The simple sugars are called *monosaccharides. More complex sugars comprise between two and ten monosaccharides linked together: *disaccharides contain two, trisaccharides three, and so on. The name is often used to refer specifically to *sucrose (cane or beet sugar).

sulpha drugs *See* sulphonamides.

sulphonamides Organic compounds containing the group $-SO_2.NH_2$. The sulphonamides are amides of sulphonic acids. Many have antibacterial action and are also known as *sulpha drugs*, including sulphadiazine, $NH_2C_6H_4SO_2NHC_4H_3N_2$, sulphathiazole, $NH_2C_6H_4SO_2NHC_5H_2NS$, and several others. They act by preventing bacteria from reproducing and are used to treat a variety of bacterial infections, especially of the gut and urinary system.

sulphur Symbol S. A yellow nonmetallic element that is an *essential element in living organisms, occurring in the amino acids cysteine and methionine and therefore in many proteins. It is also a constituent of various cell metabolites, e.g. *coenzyme A. Sulphur is absorbed by plants from the soil as the sulphate ion (SO_4^{2-}). *See* **sulphur cycle**.

sulphur bacteria Any of various groups of unrelated bacteria that utilize sulphur, sulphide, or sulphate in their metabolism. The anaerobic photoautotrophic *green sulphur bacteria* and *purple sulphur bacteria* can use hydrogen sulphide as a source of electrons in photosynthesis, producing sulphur as a by-product:

$$2H_2S + CO_2 \rightarrow H_2O + CH_2O + 2S$$

The chemoautotrophic *sulphur-oxidizing bacteria*, such as *Thiobacillus*, derive energy from the oxidation of sulphur or its compounds (e.g. sulphide), producing sulphate (SO_4^{2-}). The anaerobic heterotrophic *sulphate-reducing bacteria*, such as *Desulfovibrio*, require sulphate for respiration, deriving energy from its reduction to sulphur or hydrogen sulphide. The aerobic *filamentous sulphur bacteria* (e.g. *Beggiatoa*), which belong to the phylum Myxobacteria (gliding bacteria), can grow by oxidizing sulphides to sulphates.

sulphur bridge *See* **disulphide bridge**.

The sulphur cycle

sulphur cycle The cycling of sulphur between the biotic (living) and abiotic (nonliving) components of the environment (*see* **biogeochemical cycle**). Most of the sulphur in the abiotic environment is found in rocks, although a small amount is present in the atmosphere as sulphur dioxide (SO_2), produced by combustion of fossil fuels. Sulphate (SO_4^{2-}), derived from the weathering and oxidation of rocks, is taken up by plants and incorporated into sulphur-containing proteins. In this form sulphur is passed along food chains to animals. Decomposition of dead organic matter and faeces by anaerobic sulphate-reducing bacteria returns sulphur to the abiotic environment in the form of hydrogen sulphide (H_2S). Hydrogen sulphide can be converted back to sulphate or to elemental sulphur by the action of different groups of photosynthetic and sulphide-oxidizing bacteria (*see* **sulphur bacteria**). Elemental sulphur becomes incorporated into rocks.

sulphur dioxide (**sulphur(IV) oxide**) A colourless liquid or pungent gas, SO_2, formed by sulphur burning in air. Sulphur dioxide is produced by burning fossil fuels in power stations; it dissolves in water to give a mixture of sulphuric and sulphurous acids. *See also* **acid rain; air pollution**.

summation 1. *(in neurophysiology)* The combined effect of the changes in electric potential elicited in one or more postsynaptic membranes by the transmission of impulses at *synapses that is sufficient to trigger an action potential in the postsynaptic neurone. Summation occurs when one or a few postsynaptic potentials alone are insufficient to elicit a response in the postsynaptic neurone; it may consist of the effect of two or more potentials evoked simultaneously at different synapses on the same neurone (*spatial summation*) or in rapid succession at the same synapse (*temporal summation*). **2.** *See* **synergism**.

superior Describing a structure that is positioned above or higher than another structure in the body. For example, in flowering plants the ovary is described as superior when located above the other organs of the flower (*see* **hypogyny**). *Compare* **inferior**.

supernormal stimulus 1. *(in animal behaviour)* A stimulus that produces a more vigorous response than the normal stimulus eliciting that particular response. For example, a female herring gull will brood a giant egg in preference to its own eggs, which are smaller. A supernormal stimulus is an exaggerated *sign stimulus. **2.** *(in neurophysiology)* A stimulus that is more intense than a normal stimulus and is capable of inducing a response in a nerve fibre during the relative *refractory period.

superoxide dismutase (SOD) A widely distributed enzyme that removes the superoxide radical ($O_2\cdot$), with the formation of molecular oxygen and hydrogen peroxide, in the following reaction:

$$O_2\cdot + O_2\cdot + 2H^+ \rightarrow O_2 + H_2O_2$$

The hydrogen peroxide formed is removed by the action of *catalase. The superoxide anion, which damages tissues, is a *free radical formed by the partial reduction of molecular oxygen; it is produced during the metabolic breakdown of various toxins (including drugs and chemical poisons) and as part of the immune response to the presence of bacteria and virus-infected cells.

supination Rotation of the lower forearm so that the hand faces forwards or upwards with the radius and ulna parallel. *Compare* **pronation**.

supplementary units *See* **SI units**.

suprarenal glands *See* **adrenal glands**.

surface tension Symbol γ. The property of a liquid that makes it behave as if its surface is enclosed in an elastic skin. The property results from intermolecular forces: a molecule in the interior of a liquid experiences a force of attraction from other molecules equally from all sides, whereas a molecule at the surface is only attracted by molecules below it in the liquid. The surface tension is defined as the force acting over the surface per unit length of surface perpendicular to the force. It is measured in newtons per metre. It can equally be defined as the energy required to increase the surface area by one square metre, i.e. it can be measured in joules per metre squared (which is equivalent to $N m^{-1}$).

The surface tension of water is very strong, due to the intermolecular hydrogen bonding, and is responsible for the formation of drops, bubbles, and *meniscuses* (the curved surfaces of columns of liquid), as well as the rise of water in a capillary tube (*capillarity*), the absorption of liquids by porous substances, and the ability of liquids to wet a surface. Capillarity is very important in plants as it is largely responsible for the transport of water, against gravity, within the plant.

surfactant (surface active agent) A substance, such as a *detergent, added to a liquid to increase its spreading or wetting properties by reducing its *surface tension.

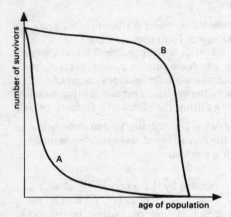

Survival curves

survival curve A graph that shows the relationship between the survival and age of a population of a particular species. Survival can be represented as the percentage of individuals of the original population that have survived after a specified time. Some types of organism, notably those that produce many offspring (e.g. bacteria), have a low survival rate in the early stages of development (line A on graph; *see* **r selection**). Other organisms, which have a low rate of reproduction, tend to be long-lived and their population numbers are maintained over a long period (line B on graph; *see* **K selection**). This type of survival curve is typical of a human population in a developed country.

suspension culture A type of culture in which intact cells are maintained in suspension in the culture medium so that they are distributed evenly within it. *Compare* **monolayer culture**.

suspensor The chain of cells that anchors a plant embryo in the surrounding gametophyte tissue. In flowering plants the suspensor attaches the embryo to the embryo sac and extends to push the embryo into the endosperm.

suspensory ligaments *See* lens.

suture The line marking the junction of two body structures. Examples are the immovable joints between the bones of the skull and, in plants, the seam along the edge of a pea or bean pod.

SV40 A DNA-containing virus belonging to the *papovavirus group that is much used in cancer research for its ability to induce *transformations in cells. It was originally isolated from monkeys – hence its name, which is an abbreviation of *Simian Virus 40*.

swallowing *See* deglutition.

sweat The salty fluid secreted by the *sweat glands onto the surface of the skin. Excess body heat is used to evaporate sweat, thereby resulting in cooling of the skin surface. Small amounts of urea are excreted in sweat.

sweat gland A small gland in mammalian skin that secretes *sweat. The distribution of sweat glands on the body surface varies between species: they occur over most of the body surface in man and higher primates but have a more restricted distribution in other mammals.

swim bladder (air bladder; gas bladder) An air-filled sac lying above the alimentary canal in bony fish that regulates the buoyancy of the animal. Air enters or leaves the bladder either via a pneumatic duct opening into the oesophagus or stomach or via capillary blood vessels, so that the specific gravity of the fish always matches the depth at which it is swimming. This makes the fish weightless, so less energy is required for locomotion. In lungfish it also has a respiratory function. The lungs of tetrapods are homologous with the swim bladder, which has developed its hydrostatic function by specialization.

syconus A type of *composite fruit formed from a hollow fleshy inflorescence stalk inside which tiny flowers develop. Small *drupes, the 'pips', are produced by the female flowers. An example is the fig.

symbiont An organism that is a partner in a symbiotic relationship (*see* symbiosis).

symbiosis An interaction between individuals of different species (*symbionts*). The term symbiosis is usually restricted to interactions in which both species benefit (*see* **cooperation; mutualism**), but it may be used for other close associations, such as *commensalism. Many symbioses are obligatory (i.e. the participants cannot survive without the interaction); for example, a lichen is an obligatory symbiotic relationship between an alga or a blue-green bacterium and a fungus.

symmetry Regularity in the arrangement of the parts of an organism. *See* **bilateral symmetry; radial symmetry**.

sympathetic nervous system Part of the *autononomic nervous system. Its nerve endings release noradrenaline or adrenaline as a neurotransmitter and its actions tend to antagonize those of the *parasympathetic nervous system, thus achieving a balance in the organs they serve. For example, the sympathetic nervous system decreases salivary gland secretion, increases heart rate, and constricts blood vessels, while the parasympathetic nervous system has opposite effects.

sympathetic tone The condition of a muscle when the *tone is maintained predominantly by impulses from the *sympathetic nervous system.

sympatric Describing groups of similar organisms that, although in close proximity and theoretically capable of interbreeding, do not interbreed because of differences in behaviour, flowering time, etc. *See* **isolating mechanism.** *Compare* **allopatric.**

symphysis A *joint that is only slightly movable; examples are the joints between the vertebrae of the vertebral column and that between the two pubic bones in the pelvic girdle. The bones at a symphysis articulate by means of smooth layers of cartilage and strong fibres.

symplast The system of *protoplasts in plants, which are interconnected by *plasmodesmata. This effectively forms a continuous system of cytoplasm bounded by the plasma membranes of the cells. The movement of water through the symplast is known as the *symplast pathway.* It is one of the routes by which water travels across the root cortex from the root hairs and is the only means by which water crosses the *endodermis. *Compare* **apoplast.**

sympodium The composite primary axis of growth in such plants as lime and horse chestnut. After each season's growth the shoot tip of the main stem stops growing (sometimes terminating in a flower spike); growth is continued by the tip of one or more of the lateral buds. *Compare* **monopodium.**

symport A membrane protein that effects the transport of a solute across a membrane but only when it also transports another type of solute in the same direction. For example, the protein that carries the sugar lactose across membranes in bacteria functions as a hydrogen-ion symport. *Compare* **antiport; uniport.**

synapse The junction between two adjacent neurones (nerve cells), i.e. between the axon ending of one (the *presynaptic neurone*) and the dendrites of the next (the *postsynaptic neurone*). The swollen tip of the axon of the presynaptic neurone, called the *synaptic knob,* contains vesicles of *neurotransmitter substance. At a synapse, the membranes of the two cells (the *pre-* and *postsynaptic membranes*) are in close contact, with only a minute gap (the *synaptic cleft*) between them. A nerve *impulse is transmitted across the synapse by the release from the presynaptic membrane of neurotransmitter, which diffuses across the synaptic cleft to the postsynaptic membrane. This triggers the propagation of the impulse from the dendrite along the length of the postsynaptic neurone. Most neurones have more than one synapse. *See also* **excitatory postsynaptic potential**.

- end of axon of presynaptic neurone
- mitochondria
- synaptic knob
- synaptic vesicles containing neurotransmitter
- dendrite of postsynaptic neurone
- postsynaptic membrane
- synaptic cleft
- presynaptic membrane

Structure of a synapse

synapsis *See* **pairing**.

synaptic cleft *See* **synapse**.

synaptic knob *See* **synapse**.

syncarpy The condition in which the female reproductive organs (*carpels) of a flower are joined to each other. It occurs, for example, in the primrose. *Compare* **apocarpy**.

syncytium (*pl.* **syncytia**) A group of animal cells in which cytoplasmic continuity is maintained. For example, the cells of striated muscle form a syncytium. In some syncytia the cells remain discrete but are joined together by cytoplasmic bridges. For example, developing male germ cells in the testes of mammals are linked throughout the process of differentiation until an individual sperm is released into the lumen of the seminiferous tubule (*see* **spermatogenesis**).

synecology The study of ecology at the level of the *community. A synecological study aims to investigate the relationships between different species that form a community and their interactions with the surrounding environment. Synecology involves both *biotic and *abiotic factors. *Compare* **autecology**.

synergids The two nuclei in the *embryo sac of flowering plants that are closely associated with the oosphere, or egg cell, to form the *egg apparatus*.

synergism (summation) **1.** The phenomenon in which the combined action of two substances (e.g. drugs or hormones) produces a greater effect than would be expected from adding the individual effects of each substance. *See also* **potentiation**. **2.** The combined action of one muscle (the *synergist*) with another (the *agonist*) in producing movement. *Compare* **antagonism**.

syngamy *See* **fertilization**.

synovial membrane The membrane that lines the ligament surrounding a freely movable joint (such as that at the hip or elbow). It secretes a fluid (*synovial fluid*) that lubricates the layers of cartilage forming the articulating surfaces of the joint.

synthesis The formation of chemical compounds from more simple compounds. *See* **biosynthesis**.

syphilis *See* **sexually transmitted disease**.

syrinx. The sound-producing organ of a bird, situated at the lower end of the trachea where it splits into the bronchi. It has a complex structure with a number of vibrating membranes.

systematics The study of the diversity of organisms and their natural relationships. It is sometimes used as a synonym for *taxonomy. The term *biosystematics* (or *experimental taxonomy*) describes the experimental study of diversity, especially at the species level. Biosystematic methods include breeding experiments, field work, biochemical work (known as *chemosystematics*), and cytotaxonomy.

Système International d'Unités *See* **SI units**.

systemic arch A paired blood vessel in the embryos of tetrapods that carries blood from the aorta to the trunk and hind limbs. It is derived from the fourth *aortic arch. Adult amphibians and reptiles retain both arches, while birds and mammals only have one.

systemic circulation The part of the circulatory system of birds and mammals that transports oxygenated blood from the left ventricle of the heart to the tissues in the body and returns deoxygenated blood from the tissues to the right atrium of the heart. *Compare* **pulmonary circulation**. *See* **double circulation**.

T

2,4,5-T 2,4,5-trichlorophenoxyacetic acid (2,4,5-trichlorophenoxyethanoic acid): a synthetic *auxin frequently used as a herbicide and defoliant. There are now moves to restrict its use, as it tends to become contaminated with the toxic chemical dioxin.

tachycardia An increase in heart rate. The heart rate is controlled by the opposing actions of the sympathetic nervous system, which accelerates the heartbeat; and the parasympathetic nervous system, which slows it down. The heart rate is normally approximately 70 beats per minute. During exercise there is an increase in sympathetic activity and tachycardia occurs: the heart rate increases to up to 110 beats per minute. Immediately after exercise, at rest, the heart rate decreases to around 60 beats per minute (*bradycardia*), due to an increase in parasympathetic activity.

tactic movement *See* taxis.

tagma (*pl.* **tagmata**) A section of the body of an arthropod that is formed by the fusion of mesodermal *somites and has a distinct function and structure. The basic tagmata are the head, thorax, and abdomen, but the form of the tagmata (known as *tagmosis*) varies between arthropod groups, each group having its characteristic tagmosis. For example, many crustaceans have a *cephalothorax and abdomen, while arachnids have a *prosoma and *opisthosoma.

tagmosis *See* tagma.

taiga A terrestrial *biome consisting mainly of evergreen coniferous forests (mainly pine, fir, and spruce), which occurs across subarctic North America and Eurasia. In certain parts, such as northeastern Siberia, deciduous conifers and broadleaved trees, such as larch and birch, are dominant. Over most of the taiga the ground is permanently frozen within about one metre of the surface, which prevents water from filtering down to deeper levels in the soil. This means that bogs may form in depressions. For at least six months of the year temperatures are below freezing but there is a short growing season lasting 3–5 months. The soil in taiga areas is acidic and infertile. *Compare* **tundra**.

tannin One of a group of complex organic chemicals commonly found in leaves, unripe fruits, and the bark of trees. Their function is uncertain though the unpleasant taste may discourage grazing animals. Some tannins have commercial uses, notably in the production of leather and ink.

tapetum A reflecting layer, containing crystals of guanine, in the *choroid of the eye of many nocturnal vertebrates. It reflects light back onto the retina, thus improving vision and causing the eyes to shine in the dark.

tapeworms *See* **Cestoda.**

tap root *See* **root.**

tarsal (tarsal bone) One of the bones that form the ankle (*see* **tarsus**) in terrestrial vertebrates.

tarsus The ankle (or corresponding part of the hindlimb) in terrestrial vertebrates, consisting of a number of small bones (*tarsals*). The number of tarsal bones varies with the species: humans, for example, have seven.

tartrazine A *food additive (E102) that gives foods a yellow colour. Tartrazine can cause a toxic response in the immune system.

taste 1. The sense that enables the flavour of different substances to be distinguished (*see* **taste bud**). **2.** The flavour of a substance.

taste bud A small sense organ in most vertebrates, specialized for the detection of taste. In terrestrial animals taste buds are concentrated on the upper surface of the *tongue. They contain *gustatory receptors*, which are sensitive to four types of taste: sweet, salt, bitter, or sour. The taste bud transmits information about a particular type of taste to the brain via nerve fibres. The four types of taste bud show distinct distribution patterns on the surface of the human tongue.

In fishes, taste buds are distributed over the entire surface of the body and provide information about the surrounding water.

taxis (taxic response; tactic movement) The movement of a cell (e.g. a gamete) or a microorganism in response to an external stimulus. Certain microorganisms have a light-sensitive region that enables them to move towards or away from high light intensities (positive and negative *phototaxis respectively). Many bacteria move in response to chemical stimuli (*chemotaxis*); a specific example is *aerotaxis*, in which atmospheric oxygen is the stimulus. Taxic responses are restricted to cells that possess cilia, flagella, or some other means of locomotion. The term is usually not applied to the movements of higher animals. *See also* **geotaxis.** *Compare* **kinesis; tropism**.

taxon Any named taxonomic group of any *rank in the hierarchical *classification of organisms. Thus the taxa Papilionidae, Lepidoptera, Insecta, and Arthropoda are named examples of a family, order, class, and phylum, respectively.

taxonomy The study of the theory, practice, and rules of *classification of living and extinct organisms. The naming, description, and classification of a given organism draws on evidence from a number of fields. *Classical taxonomy* is based on morphology and anatomy. *Biochemical taxonomy* studies similarities in the structure of certain proteins and nucleic acids. *Cytotaxonomy* compares the size, shape, and number of chromosomes of different organisms. *Numerical taxonomy* uses mathematical procedures to assess similarities and differences and establish taxonomic groups. *See also* **systematics.**

TCA cycle *See* **Krebs cycle**.

T cell (T lymphocyte) Any of a population of *lymphocytes that are the principal agents of cell-mediated *immunity. T cells are derived from the bone marrow but migrate to the thymus to mature (hence *T* cell). Subpopulations of T cells play different roles in the immune response and can be characterized by their surface antigens (*see* **CD**). *Helper T cells* recognize foreign antigens provided these are presented by cells (such as macrophages and B lymphocytes) bearing class II *histocompatibility antigens. The helper T cell carries *T-cell receptors*, which recognize the class II antigens on the antigen-presenting cell. Interleukin-1 released by inducer cells stimulates helper T cells to release interleukin-2, which in turn triggers the release of other cytokines (*see* **interleukin**). Consequently there is a proliferation of B lymphocytes and the generation of *effector T cells*, i.e. cytotoxic T cells, delayed hypersensitivity T cells, and suppressor T cells.

Cytotoxic T cells recognize foreign antigen on the surface of virus-infected cells and destroy the cell by releasing cytolytic proteins. *Suppressor T cells* are important in regulating the activity of other lymphocytes and are crucial in maintaining tolerance to self tissues. *Delayed hypersensitivity T cells* mediate delayed hypersensitivity by releasing various *lymphokines.

tectorial membrane *See* **organ of Corti**.

teeth *See* **deciduous teeth; permanent teeth; dentition; tooth**.

Teleostei The major superorder of the *Osteichthyes (bony fish), containing about 20 000 species. Teleosts have colonized an extensive variety of habitats and show great diversity of form. The group includes the eel, seahorse, plaice, and salmon. They have been the dominant fish since the Cretaceous period (about 70 million years ago).

telocentric *See* **centromere**.

telomere The end of a chromosome, which consists of repeated sequences of DNA that perform the function of ensuring that each cycle of *DNA replication has been completed. Each time a cell divides some sequences of the telomere are lost; eventually (after 60–100 divisions in an average cell) the cell dies. Replication of telomeres is directed by *telomerase*, an enzyme consisting of RNA and protein that is inactive in normal cells. Its presence in tumours is linked to the uncontrolled multiplication of cancer cells.

telophase A stage of cell division. In *mitosis the chromatids that separated from each other at *anaphase collect at the poles of the spindle. A nuclear membrane forms around each group, producing two daughter nuclei with the same number and kind of chromosomes as the original cell nucleus. In the first telophase of *meiosis, complete chromosomes from the pairs that separated at first anaphase form the daughter nuclei. The number of chromosomes in these nuclei is therefore half the number in the original one. In the second telophase, daughter nuclei are formed from chromatids (as in mitosis).

temperature inversion An abnormal increase in air temperature that occurs in the troposphere, the lowest level of the earth's atmosphere. This can lead to pollutants becoming trapped in the troposphere (*see* **air pollution**).

template Any molecule that acts as a pattern for the synthesis of a new molecule. For example, the two nucleotide chains of a DNA molecule can separate and each acts as a template for the synthesis of the missing chain (*see* **DNA replication**).

tendon A thick strand or sheet of tissue that attaches a muscle to a bone. Tendons consist of *collagen fibres and are therefore inelastic: they ensure that the force exerted by muscular contraction is transmitted to the relevant part of the body to be moved.

tendril A slender branched or unbranched structure found in many climbing plants. It may be a modified stem, leaf, leaflet, or petiole. Tendrils respond to contact with solid objects by twining around them (*see* **haptotropism**). The cells that touch the object lose water and decrease in volume in comparison to the outer cells, thus causing the tendril to curve.

tentacle Any of the soft flexible appendages in aquatic invertebrate animals that are used principally for feeding. Water flows over the tentacles, which are able to capture food and direct it to the oral aperture. Tentacles are possessed by many cnidarians, some echinoderms (including sea cucumbers), and by cephalopod molluscs, in which the tentacles bear rows of suckers.

tera- Symbol T. A prefix used in the metric system to denote one million million times. For example, 10^{12} volts = 1 teravolt (TV).

teratogen Any environmental factor that acts on a fetus to cause congenital abnormality. Examples include ionizing radiation (e.g. X-rays), nutritional deficiencies, drugs (e.g. thalidomide), toxic chemicals, and virus infections (e.g. rubella).

tergum The protective cuticle found on the dorsal surface of body segments in arthropods.

termination codon *See* **stop codon**.

terpenes A group of unsaturated hydrocarbons present in plants (*see* **essential oil**). Terpenes consist of isoprene units, $CH_2:C(CH_3)CH:CH_2$. Monoterpenes have two units, $C_{10}H_{16}$, sesquiterpenes three units, $C_{15}H_{24}$, diterpenes four units, $C_{20}H_{32}$, etc.

territory A fixed area that an animal or group of animals defends against intrusion from others of its species by various types of *territorial behaviour*. Outside the territory (which may contain food sources, hiding places, and nesting sites) others are not threatened. Many mammals indicate their territory boundaries with scent markings, while birds sing territorial songs that repel would-be intruders. Animals in neighbouring territories normally respect each other's boundaries, which reduces overt *aggression. Some animals are territorial only at certain times of the year, usually the breeding season (*see* **courtship**).

Tertiary The older geological period of the Cenozoic era (*compare* **Quaternary**). It began about 65 million years ago, following the Cretaceous period, and extended to the beginning of the Quaternary, about 2 million years ago. It is subdivided into the *Palaeocene, *Eocene, *Oligocene, *Miocene, and *Pliocene epochs in ascending order. The Tertiary period was characterized by the rise of the modern mammals and the development of shrubs, grasses, and other flowering plants.

tertiary consumer *See* **consumer**.

tertiary structure *See* **protein**.

testa (seed coat) The lignified or fibrous protective covering of a seed that develops from the integuments of the ovule after fertilization. *See also* **hilum; micropyle**.

test cross A mating (cross) made to identify hidden *recessive alleles in an individual of unknown genotype. This individual is crossed with one that is *homozygous for the allele being investigated (i.e. a homozygous recessive). The homozygous recessive individual may be the parent of the individual being investigated (*see* **back cross**).

testicle *See* **testis**.

testis (testicle) The reproductive organ in male animals in which spermatozoa are produced. In vertebrates there are two testes; as well as sperm, they produce steroid hormones (*see* **androgen**). In most animals the testes are within the body cavity but in mammals, although they develop within the body near the kidneys, they come to hang outside the body cavity in a *scrotum. Most of the vertebrate testis is made up of a mass of *seminiferous tubules*, lined with *Sertoli cells, in which the sperms develop (*see* **spermatogenesis**). It is connected to the outside by means of the *vas deferens. *See* **reproductive system**.

testosterone The principal male sex hormone. *See* **androgen**.

tetanus 1. A powerful sustained contraction of a voluntary muscle resulting from the summation of a series of rapid muscular contractions (*twitches*) that are induced by repeated stimulation of the muscle. **2.** A disease caused by the bacterium *Clostridium tetani*, which generally enters the body via an open wound. A toxin produced by the bacterium irritates the nerves, which induce muscle spasms that begin in the jaw muscles (giving the disease the informal name of *lockjaw*).

tetrad A group of four *haploid cells formed at the end of the second division of *meiosis.

tetraploid Describing a nucleus, cell, or organism that has four times ($4n$) the haploid number (n) of chromosomes. *See also* **polyploid**.

Tetrapoda A superclass of jawed chordates (*Gnathostomata) comprising all vertebrate animals with four limbs. Tetrapods include amphibians, reptiles, birds, and mammals. The skeleton of the limbs of all tetrapods is based on the same five-digit pattern (*see* **pentadactyl limb**).

tetraspore Any one of four spores produced by meiosis in an independent diploid sporophyte (*tetrasporophyte*) in the life cycle of certain red algae. The tetraspores are produced within a *tetrasporangium* and germinate to produce the sexual haploid generation. The tetrasporophyte develops from a *carpospore.

thalamus 1. *(in anatomy)* Part of the vertebrate *forebrain that lies above the hypothalamus. It relays sensory information to the cerebral cortex and is also concerned with the translation of impulses into conscious sensations. **2.** *(in botany)* *See* **receptacle**.

thalassaemia A blood disease caused by a *substitution mutation in the DNA sequence that controls the synthesis of the α- or β-chains of *haemoglobin. Thalassaemia results in anaemia and growth retardation.

Thallophyta A former division of the plant kingdom containing relatively simple plants, i.e. those with no leaves, stems, or roots. It included the algae, bacteria, fungi, and lichens.

thallus A relatively undifferentiated vegetative body with no true roots, stems, leaves, or vascular system. It is found in the algae, fungi, and bryophytes and in the gametophyte generation of clubmosses, horsetails, and ferns.

theca *See* **capsule**.

therapeutic half-life *(in pharmacology)* The time taken for half the dose of a drug to be excreted: used to calculate the most effective and nontoxic dosing intervals. It can be determined by administering a therapeutic dose of the drug labelled with a radioisotope (*see* **labelling**) and measuring the time for half of it to be excreted in the urine.

therapeutic index (therapeutic window) The therapeutic usefulness of a drug, expressed as a ratio of the maximum nontoxic dose and the minimum effective dose of the drug. The therapeutic index is commonly used as a measure of the range of the dose considered to be safe. However, due to the variation in individual response to a given drug, the therapeutic index cannot be calculated accurately.

thermal denaturation *See* **denaturation**.

thermocline A steep temperature gradient that exists in the middle zone (the *metalimnion*) of a lake and gives rise to thermally induced vertical stratification of the water. The metalimnion lies between the relatively warm *epilimnion* above and the cold *hypolimnion* below. The thermocline may represent a temperature change of 1°C for every incremental depth of 1 metre of water. It may be short-lived, especially in shallow lakes where wind action can mix the water from different levels. However, it can exist for most of the summer period in temperate lakes and sometimes nearly all year in tropical lakes. A thermocline can speed up the process of eutrophication by preventing the diffusion of oxygen from the epilimnion to the hypolimnion (*see* **eutrophic**).

thermogenesis The production of heat, for example in the tissues of hibernating animals in order to maintain their body temperature. Heat is generated by preventing synthesis of ATP in the mitochondria, which results in an increase in oxygen consumption and heat production.

thermography A medical technique that makes use of the infrared radiation from the human skin to detect an area of elevated skin temperature that could be associated with an underlying cancer. The heat radiated from the body varies according to the local blood flow, thus an area of poor circulation produces less radiation. A tumour, on the other hand, has an abnormally increased blood supply and is revealed on the *thermogram* (or *thermograph*) as a 'hot spot'. The technique is used particularly in mammography, the examination of the infrared radiation emitted by human breasts in order to detect breast cancer.

thermoluminescence Luminescence produced in a solid when its temperature is raised. It arises when free charge carriers, trapped in a solid as a result of exposure to ionizing radiation, unite and emit photons of light. The process is made use of in *thermoluminescent dating*, which assumes that the number of charge carriers trapped in a sample of pottery is related to the length of time that has elapsed since the pottery was fired. By comparing the luminescence produced by heating a piece of pottery of unknown age with the luminescence produced by heating similar materials of known age, a fairly accurate estimate of the age of an object can be made.

thermoluminescent dating *See* thermoluminescence.

thermonasty *See* nastic movements.

thermoreceptor A group of specialized cells (*see* **receptor**) in the skin that is sensitive to temperature.

thiamin(e) *See* vitamin B complex.

thigmotropism (haptotropism) The growth of an aerial plant organ in response to localized physical contact. For example, when a tendril of sweet pea touches a supporting structure, it curves in the direction of the support and coils around it. *See* **tropism**.

thin-layer chromatography A technique for the analysis of liquid mixtures using *chromatography. The stationary phase is a thin layer of an absorbing solid (e.g. alumina) prepared by spreading a slurry of the solid on a plate (usually glass) and drying it in an oven. A spot of the mixture to be analysed is placed near one edge and the plate is stood upright in a solvent. The solvent rises through the layer by capillary action carrying the components up the plate at different rates (depending on the extent to which they are absorbed by the solid). After a given time, the plate is dried and the location of spots noted. It is possible to identify constituents of the mixture by the distance moved in a given time. The technique needs careful control of the thickness of the layer and of the temperature.

thoracic cavity The space within the *thorax, which in vertebrates contains the heart, lungs, and rib cage.

thoracic duct The main collecting vessel of the *lymphatic system, running longitudinally in front of the backbone. The thoracic duct drains its lymph into the superior vena cava.

thoracic vertebrae The *vertebrae of the upper back, which articulate with the *ribs. They lie between the *cervical vertebrae and the *lumbar vertebrae and are distinguished by a number of articulating facets for attachment of the ribs. In man there are 12 thoracic vertebrae.

thorax The anterior region of the body trunk of animals. In vertebrates it contains the heart and lungs within the rib cage. It is particularly well-defined in mammals, being separated from the *abdomen by the *diaphragm. In insects the thorax is divided into an anterior *prothorax*, a middle *mesothorax*, and a posterior *metathorax*, each of which bears a pair of legs; the hindmost two segments also both carry a pair of wings. In other arthropods, especially crustaceans and arachnids, the thorax is fused with the head to form a *cephalothorax*.

thorn A hard side stem with a sharp point at the tip, replacing the growing point. In some plants the development of thorns and subsequent suppression of the growing points may be a response to dry conditions. Examples are the thorns of gorse and hawthorn. *Compare* **prickle; spine**.

thread cell (**nematoblast; cnidoblast**) A specialized cell found only in the ectoderm of the *Cnidaria. It contains a *nematocyst*, a fluid-filled sac within which lies a long hollow coiled thread. When a small sensory projection (*cnidocil*) on the surface of the thread cell is touched, e.g. by prey, the thread is shot out and adheres to the prey, coils round it, or injects poison into it. Numerous thread cells on the tentacles of jellyfish produce their sting.

threat display *See* **agonistic behaviour**.

threonine *See* **amino acid**.

threshold (*in physiology*) The minimum intensity of a stimulus that is necessary to initiate a response.

thrombin An enzyme that catalyses the conversion of fibrinogen to fibrin. *See* **blood clotting; prothrombin**.

thrombocyte *See* **platelet**.

thromboplastin A lipoprotein, released from damaged tissues and platelets at the site of a wound, that – together with various *clotting factors and calcium ions – converts *prothrombin to the enzyme thrombin. *See* **blood clotting**.

thrombosis The obstruction of a blood vessel by a mass of blood cells and fibrin (*thrombus*), which can result from excessive *blood clotting.

thylakoid Any of the flattened saclike membranous structures that are stacked on top of one another to form the grana (*see* **granum**) of plant *chloroplasts. Chlorophyll and other photosynthetic pigments are situated in the thylakoid membranes, which are the site for the light reaction of *photosynthesis (*see also* **photophosphorylation**). The photolysis of water (*see* **photosystems I and II**) occurs in the space between the thylakoid membranes.

thymidine A nucleoside consisting of one thymine molecule linked to a D-ribose sugar molecule.

thymine A *pyrimidine derivative and one of the major component bases of *nucleotides and the nucleic acid *DNA.

thymus An organ, present only in vertebrates, that is concerned with development of *lymphoid tissue and hence the antibody-producing white blood cells (*see* **T cell**) and the *immune response. In mammals it is a bilobed organ in the region of the lower neck, above and in front of the heart. It shrinks in size after sexual maturity and is therefore believed to function only during early life.

thyrocalcitonin *See* **calcitonin**.

thyroglobulin (TGB) A glycoprotein, made in the thyroid gland, that consists of about 5000 amino acids, some of which are tyrosine residues. TGB is the precursor of the thyroid hormones, thyroxine and triiodothyronine. Iodine binds to the tyrosine residues in thyroglobulin, which is then hydrolysed into *iodotyrosines* that combine to form triiodothyronine (T_3) or thyroxine (tetraiodothyronine or T_4).

Thyroxine

thyroid gland A bilobed endocrine gland in vertebrates, situated in the base of the neck. It secretes two iodine-containing *thyroid hormones*, thyroxine (T_4; see formula) and triiodothyronine (T_3), which are formed in the gland from *thyroglobulin; they control the rate of all metabolic processes in the body and influence physical development and activity of the nervous system. Growth and activity of the thyroid is controlled by *thyroid-stimulating hormone, secreted by the anterior *pituitary gland. The thyroid gland also contains *C cells*, which secrete *calcitonin.

thyroid-stimulating hormone (TSH; thyrotrophin) A hormone, secreted by the anterior pituitary gland, that controls the synthesis and secretion of the two thyroid hormones, thyroxine and triiodothyronine, in the *thyroid gland. The secretion of thyroid-stimulating hormone is controlled by *thyrotrophin-releasing hormone (TRH)* from the hypothalamus. The release of TRH depends on many factors, including the levels of TSH, glucose, and thyroxine in the blood and the rate of metabolism in the body.

thyrotrophin-releasing hormone (TRH) *See* thyroid-stimulating hormone.

thyroxine (T$_4$) The principal hormone of the *thyroid gland. *See also* thyroglobin.

tibia 1. The larger of the two bones of the lower hindlimb of terrestrial vertebrates (*compare* **fibula**). It articulates with the *femur at the knee and with the *tarsus at the ankle. The tibia is the major load-bearing bone of the lower leg. **2.** The fourth segment of an insect's leg, which is attached to the femur.

tidal volume The volume of air taken in or expelled by an animal breathing normally at rest during each cycle of *ventilation. The average human has a tidal volume of approximately 500 cm^3.

tight junction (zonula occludens) The region between the plasma membranes of two adjacent cells that are so closely positioned that there is no intercellular space between them. This type of junction fuses cells together and provides a selective barrier to the diffusion of substances between cells. *Compare* **desmosome; gap junction**.

tiller A shoot that develops at the base of a plant stem from an axillary bud. Tillers are often produced in response to injury of the main stem, as occurs when a tree is lopped.

time-lapse photography A form of photography used to record a slow process, such as plant growth. A series of single exposures of the object is made on film at predetermined regular intervals. The film produced can then be projected at normal ciné speeds and the process appears to be taking place at an extremely high rate.

tissue A collection of similar cells organized to carry out one or more particular functions. For example, in animals nervous tissue is specialized to perceive and transmit stimuli. An organ, such as a lung or kidney, contains many different types of tissues.

tissue culture The growth of the tissues of living organisms outside the body in a suitable culture medium. Culture (or nutrient) media contain a mixture of nutrients either in solid form (e.g. in *agar) or in liquid form (e.g. in *physiological saline). Tissue culture has proved to be invaluable for gaining information about factors that control the growth and differentiation of cells. Culture of plant tissues has resulted in the regeneration of complete plants, enabling commercial propagation (e.g. of orchids) and – through culture of meristem tissues – the production of virus-free crop plants. *See also* **explantation**.

tissue fluid The fluid, consisting of water, ions, and dissolved gases and food substances, that is formed when blood is ultrafiltered (*see* **ultrafiltration**) from the capillaries into the intercellular spaces. The pressure in the arterial capillaries causes most components of the blood to pass across the capillary walls; blood cells and most of the plasma proteins are retained in the capillaries. The tissue fluid surrounds the body cells, facilitating the exchange of nutrients and waste materials. At the venous end of the capillaries, the tissue fluid is drawn into the capillaries by *osmosis.

tissue typing The process of identifying the human leucocyte antigens (*see* **HLA system**) in both the recipient and the potential donor before organ transplantation. If the donor and recipient tissue types do not match closely, the transplanted organ will be rejected.

titre 1. The number of infectious virus particles present in a suspension. **2.** A measure of the amount of *antibody present in a sample of serum, given by the highest dilution of the sample that results in the formation of visible clumps with the appropriate antigen (*see* **agglutination**).

toads *See* Amphibia.

tobacco mosaic virus (TMV) A rigid rod-shaped RNA-containing virus that causes distortion and blistering of leaves in a wide range of plants, especially the tobacco plant. It is transmitted by insects when they feed on plant tissue. TMV was the first virus to be discovered.

tocopherol *See* **vitamin E**.

tolerance 1. The ability of an organism to withstand extreme variations in environmental conditions, such as drought. **2.** The build-up of *resistance to drugs or other chemicals (such as pesticides), which occurs after prolonged use or application. Increasingly large doses of the chemical are required to produce the desired effect in the organism.
3. (immunological tolerance) Failure of the immune response in an animal exposed to a potential antigen. It commonly results from previous exposure to the antigen during fetal development, before the cells of the immune system are fully functional.

Tollens reagent A reagent used in testing for aldehydes, named after German chemist B. C. G. Tollens (1841–1918). It is made by adding sodium hydroxide to silver nitrate to give silver(I) oxide, which is dissolved in aqueous ammonia (giving the complex ion $[Ag(NH_3)_2]^+$). The sample is warmed with the reagent in a test tube. Aldehydes reduce the complex Ag^+

ion to metallic silver, forming a bright silver mirror on the inside of the tube (hence the name *silver-mirror test*). Ketones give a negative result.

tomography The use of X-rays to photograph a selected plane of a human body with other planes eliminated. The *CT* (*computerized tomography*) *scanner* is a ring-shaped X-ray machine that rotates through 180° around the horizontal patient, making numerous X-ray measurements every few degrees. The vast amount of information acquired is built into a three-dimensional image of the tissues under examination by the scanner's own computer. The patient is exposed to a dose of X-rays only some 20% of that used in a normal diagnostic X-ray.

tone (tonus) The state of sustained tension in muscles that is necessary for the maintenance of posture. In a tonic muscle contraction, only a certain proportion of the muscle fibres are contracting at any given time; the rest are relaxed and recovering for subsequent contractions. The fibres involved in tone contract more slowly than the fast fibres used for rapid responses by the same muscle. The proportions of slow and fast fibres depend on the function of the muscle.

tongue A muscular organ of vertebrates that in most species is attached to the floor of the mouth. It plays an important role in manipulating food during chewing and swallowing and in terrestrial species it bears numerous *taste buds on its upper surface. In some advanced vertebrates the tongue is used in the articulation of sounds, particularly in human speech.

tonoplast (vacuole membrane) The single membrane that bounds the *vacuole of plant cells.

tonsil A mass of *lymphoid tissue, several of which are situated at the back of the mouth and throat in higher vertebrates. In humans there are the *palatine tonsils* at the back of the mouth, *lingual tonsils* below the tongue, and *pharyngeal tonsils* (or *adenoids*) in the pharynx. They are concerned with the production of *lymphocytes and therefore with defence against infection.

tonus *See* tone.

tooth Any of the hard structures in vertebrates that are used principally for biting and chewing food but also for attack, grooming, and other functions. In fish and amphibians the teeth occur all over the palate, but in higher vertebrates they are concentrated on the jaws. They evolved in cartilaginous fish as modified placoid *scales, and this is reflected in their structure: a body of bony *dentine with a central *pulp cavity and an outer covering of *enamel on the exposed surface (*crown*). The portion of the tooth embedded in the jawbone is the *root* (see illustration).

In mammals there are four different types of teeth, specialized for different functions (*see* **canine; incisor; molar; premolar**). Their number varies with the species (*see* **dental formula**). *See also* **deciduous teeth; permanent teeth**.

Section through an incisor tooth

top carnivore *See* **consumer**.

topsoil *See* **soil**.

tornaria A ciliated larva produced by some acorn worms (*see* **Hemichordata**).

torus *See* **receptacle**.

totipotent 1. Describing differentiated plant cells that, when isolated, have the ability to develop into an entire new plant if provided with the suitable growing medium. **2.** Describing embryonic cells at a stage before their fate is irreversibly determined, when they have the ability to develop into any differentiated cell given the appropriate stimulation.

touch The sense that enables the texture of objects and substances to be perceived. Touch receptors occur in the *skin, being concentrated in the tips of the finger in humans (*see* **Meissner's corpuscle**).

toxicology The science of the study of poisons. Originally developed by Paracelsus (1493–1541), toxicology is concerned with the investigation of the deleterious effects of all foreign substances (*xenobiotics) on living organisms.

toxin A poison produced by a living organism, especially a bacterium. An *endotoxin* is released only when the bacterial cell dies or disintegrates. An *exotoxin* is secreted by a bacterial cell into the surrounding medium. In the body a toxin acts as an *antigen, producing an *immune response.

trace element *See* **essential element**.

trace fossil *See* **fossil**.

trachea 1. (*or* **windpipe**) The tube in air-breathing vertebrates that conducts air from the throat to the *bronchi. It is strengthened with incomplete rings of cartilage. **2.** An air channel in insects and most other terrestrial arthropods. Tracheae occur as ingrowths of the body wall. They open to the exterior by *spiracles* and branch into finer channels (*tracheoles*) that terminate in the tissues (*see also* **air sac**). Pumping movements of the abdominal muscles cause air to be drawn into and out of the tracheae.

tracheid A type of cell occurring within the *xylem of conifers, ferns, and related plants. Tracheids are elongated and their walls are usually extensively thickened by deposits of lignin. Water flows from one tracheid to another through unthickened regions (pits) in the cell walls. *Compare* **vessel element**.

tracheole *See* **trachea**.

tracheophyte Any plant that has elaborate tissues, including *vascular tissue; a conspicuous *sporophyte generation; and complex leaves with waterproof cuticles. Tracheophytes include plants of the phyla *Psilophyta, *Lycopodophyta, *Sphenophyta, *Filicinophyta, *Coniferophyta, and *Angiospermophyta. In traditional classification systems these were regarded as classes of the division Tracheophyta.

tracing (radioactive tracing) *See* **labelling**.

transaminase An enzyme that catalyses the transfer of an amino group from one molecule to another. Transaminases play an important role in the synthesis of amino acids (*see* **transamination**).

transamination A biochemical reaction in amino acid metabolism in which an amine group is transferred from an amino acid to a keto acid to form a new amino acid and keto acid. The coenzyme required for this reaction is pyridoxal phosphate.

transcriptase *See* **reverse transcriptase**.

transcription The process in living cells in which the genetic information of *DNA is transferred to a molecule of messenger *RNA (mRNA) as the first step in *protein synthesis (*see also* **genetic code**). Transcription takes place in the cell nucleus or nuclear region and is regulated by *transcription factors. It involves the action of RNA *polymerase enzymes in assembling the nucleotides necessary to form a complementary strand of mRNA from the DNA template (*see also* **promoter**), and (in eukaryote cells) the subsequent removal of the noncoding sequences from this primary transcript (*see* **gene splicing**) to form a functional mRNA molecule. *See also* **reverse transcriptase**. *Compare* **translation**.

transcription factor Any of a group of proteins that can increase or decrease the binding of RNA polymerases to the DNA molecule during the process of *transcription. This is achieved by the ability of the transcription factors to bind to the DNA molecule (*see* **DNA-binding proteins**). Transcription factors contain *finger domains, which are often in repeated sequences called *multifinger loops*.

transduction **1.** The transfer of genetic material from one bacterial cell to another by means of a *bacteriophage. **2.** The conversion of stimuli

detected in the *receptor cells into electric impulses, which are transported
by the nervous system.

transect A straight line across an expanse of ground along which
ecological measurements are taken, continuously or at regular intervals.
Thus an ecologist wishing to study the numbers and types of organisms at
different distances above the low-tide line might sample at five-metre
intervals along a number of transects perpendicular to the shore.

transferase Any of a class of enzymes that catalyse the transfer of a
group of atoms from one molecule to another.

transferrin A plasma protein (a β-*globulin) that can bind to iron and
transport it to the liver for storage (*see* **ferritin**) or to the cells of bone
marrow, where it is used for the formation of haemoglobin. One
transferrin molecule can bind to two iron ions.

transfer RNA *See* RNA.

transformation 1. A permanent heritable change in a cell, particularly a
bacterial cell, that occurs as a result of its acquiring foreign DNA.
Nonvirulent bacterial cells can be transformed into virulent forms if
cultured in a medium containing killed virulent bacteria. **2.** The
conversion of a normal cell into a malignant cell (*see* **cancer**), which can be
brought about by the action of *carcinogens or *oncogenic viruses.

transgenic Describing an organism whose genome incorporates and
expresses genes from another species. Transgenic individuals are created by
genetic engineering, using suitable *vectors to insert the desired foreign
gene into the fertilized egg or early embryo of the host. For example, the
gene for rat growth hormone can be inserted into fertilized mouse eggs to
produce mice with cells that produce rat growth hormone. Transgenic
organisms offer considerable commercial potential; for example, transgenic
farm animals may be engineered to produce therapeutic proteins in their
milk or extra growth hormone for improved meat production. Similarly,
plants may have disease resistance conferred upon them by transgenic
techniques.

transient polymorphism *See* polymorphism.

transition *(in genetics)* *See* substitution.

transition zone *See* hypocotyl.

translation The process in living cells in which the genetic information
encoded in messenger *RNA (mRNA) in the form of a sequence of
nucleotide triplets (*codons) is translated into a sequence of amino acids in
a polypeptide chain during *protein synthesis (see illustration). Translation
takes place on *ribosomes in the cell cytoplasm (*see* **initiation factor**). The
ribosomes move along the mRNA 'reading' each codon in turn. Molecules of
transfer RNA (tRNA), each bearing a particular amino acid, are brought to
their correct positions along the mRNA molecule: base pairing occurs

The stages of translation in protein synthesis

between the bases of the codons and the complementary base triplets of tRNA (*see* **anticodon**). In this way amino acids are assembled in the correct sequence to form the polypeptide chain (*see* **elongation**).

translocation 1. *(in botany)* The movement of minerals and chemical compounds within a plant. There are two main processes. The first is the uptake of soluble minerals from the soil and their passage upwards from the roots to various organs by means of the water-conducting vessels (*xylem). The second is the transfer of organic compounds, synthesized by the leaves, both upwards and downwards to various organs, particularly the growing points. This movement occurs within the *phloem tubes. *See also* **mass flow. 2. *(in genetics)*** A type of *chromosome mutation in which a section of one chromosome is broken off and becomes attached to another chromosome, resulting in a loss of genetic information from the first chromosome.

transmission 1. *(in neurophysiology)* The one-way transfer of a nerve *impulse from one neurone to another across a *synapse. *See also* **neurotransmitter.** *Compare* **propagation. 2. *(in medicine)*** The spread of an *infection from person to person. This can occur in various ways, such as close contact with an infected person, including sexual contact (*see* **sexually transmitted disease**); contact with a *vector or a *carrier of the disease; consuming food or drink contaminated with the infecting microorganism; and breathing in contaminated droplets of moisture, produced by coughing and sneezing.

transmission electron microscope *See* **electron microscope**.

transmitter *See* **neurotransmitter**.

transpiration The loss of water vapour by plants to the atmosphere. It occurs mainly from the leaves through pores (stomata) whose primary function is gas exchange. The water is replaced by a continuous column of water (and dissolved nutrients) moving upwards from the roots within the *xylem vessels. The flow of this column of water is known as the *transpiration stream*, which is maintained by *root pressure and a combination of cohesive and adhesive forces in the xylem vessels according to the *cohesion–tension theory* (*see* **cohesion**). *See also* **potometer**.

transplantation *See* **graft**.

transposon A mobile genetic element that can become integrated at many different sites along a chromosome. The simplest types of transposon are known as *insertion sequences*; these are well known in *E. coli* and other bacteria and comprise just one or two genes. Their presence can cause deletions, inversions, and other major rearrangements of adjacent genes. Certain bacterial *plasmids can also behave as transposons. Complex, or composite, transposons carry other genes, such as antibiotic resistance genes. Comparable elements in eukaryotes are called *transposable elements*. Maize, for example, has several families of such elements (also known as *controlling elements*). These can undergo transposition, causing chromatid breaks and modifying the action of neighbouring genes.

transverse tubules (T tubules) A series of infoldings of the plasma membrane (sarcolemma) of muscle fibres that envelop each myofibril at the Z line (*see* **sarcomere**). The T tubules transfer action potentials from the sarcolemma to the *sarcoplasmic reticulum, which releases calcium ions into the cytosol causing contraction of the muscle fibres.

transversion *(in genetics) See* **substitution**.

Trematoda A class of parasitic flatworms (*see* **Platyhelminthes**) comprising the flukes, such as *Fasciola* (liver fluke). Flukes have suckers and hooks to anchor themselves to the host and their body surface is covered by a protective cuticle. The whole life cycle may either occur within one host or require one or more intermediate hosts to transmit the infective eggs or larvae (*see* **cercaria; miracidium**). *Fasciola hepatica*, for example, undergoes larval development in a land snail (the intermediate host) and infects sheep (the primary host) when contaminated grass containing the larvae is swallowed.

Triassic The earliest period of the Mesozoic era. It began about 225 million years ago, following the Permian, the last period of the Palaeozoic era, and extended until about 190 million years ago when it was succeeded by the Jurassic. It was named, by F. von Alberti in 1834, after the sequence of three divisions of strata that he studied in central Germany – Bunter, Muschelkalk, and Keuper. The Triassic rocks are frequently difficult to distinguish from the underlying Permian strata and the term *New Red Sandstone* is often applied to rocks of the Permo-Triassic. During the period marine animals diversified: molluscs were the dominant invertebrates – ammonites were abundant and bivalves replaced the declining brachiopods. Reptiles were the dominant vertebrates and included turtles, phytosaurs, dinosaurs, and the marine ichthyosaurs.

tribe A category used in the *classification of plants and animals that consists of several similar or closely related genera within a family. For example the Bambuseae, Oryzeae, Paniceae, and Aveneae are tribes of grasses.

tricarboxylic acid cycle *See* **Krebs cycle**.

triceps A muscle that runs parallel to the humerus and works antagonistically with the *biceps, causing the arm to extend. *See* **antagonism; voluntary muscle**.

trichogyne *See* **carpogonium**.

trichome A hairlike projection from a plant epidermal cell. Examples include root hairs and the stinging hairs of nettle leaves.

trichromatic theory *See* **colour vision**.

tricuspid valve (right atrioventricular valve) A valve, consisting of three flaps, situated between the right atrium and the right ventricle of the mammalian heart. When the right ventricle contracts, forcing blood into the pulmonary artery, the tricuspid valve closes the aperture to the atrium, thereby preventing any backflow of blood. The valve reopens to allow blood to flow from the atrium into the ventricle. *Compare* **bicuspid valve**.

triglyceride (triacylglycerol) An ester of glycerol (propane-1,2,3-triol) in which all three hydroxyl groups are esterified with a fatty acid. Triglycerides are the major constituent of fats and oils and provide a concentrated food energy store in living organisms as well as cooking fats and oils, margarines, etc. Their physical and chemical properties depend on the nature of their constituent fatty acids. In *simple triglycerides* all three fatty acids are identical; in *mixed triglycerides* two or three different fatty acids are present.

triiodothyronine (T_3) A hormone secreted by the *thyroid gland. *See also* **thyroglobulin**.

trilobite An extinct marine arthropod belonging to the class Trilobita (some 4000 species), fossils of which are found in deposits dating from the Precambrian to the Permian period (590–280 million years ago). Trilobites were typically small (1–7 cm long); the oval flattened body comprised a head (covered by a semicircular dorsal shield) and a thorax and abdomen, which were protected by overlapping dorsal plates with a raised central part and flattened lateral portions, presenting a three-lobed appearance. The head bore a pair of antenna-like appendages and a pair of compound eyes; nearly all body segments bore a pair of Y-shaped (biramous) appendages – one branch for locomotion and the other fringed for respiratory exchange. Trilobites were bottom-dwelling scavengers.

trimethylamine A nitrogenous compound, $(CH_3)_3N$, that acts as an *osmolyte in the cells of the kidney. This compound can also be converted to *trimethylamine oxide*, which is excreted by many marine fish and gives them their characteristic odour.

triose A sugar molecule that contains three carbon atoms. *See* **monosaccharide**.

triplet code *See* **codon; genetic code**.

triploblastic Describing an animal having a body composed of three embryonic cell layers: the *ectoderm, *mesoderm, and *endoderm. Most multicellular animals are triploblastic; the coelenterates, which are *diploblastic, are an exception.

triploid Describing a nucleus, cell, or organism that has three times $(3n)$ the haploid number (n) of chromosomes (*see also* **polyploid**). Triploid organisms are normally sterile as their lack of *homologous chromosomes prevents pairing during meiosis. This can be useful to plant breeders, for example in banana cultivation: sterile triploid bananas can be propagated asexually and will not contain any seeds.

trisomy The condition of a nucleus, cell, or organism in which one of the pairs of homologous chromosomes has gained an additional chromosome, resulting in a chromosome number of $2n + 1$ (*see* **aneuploid**). Trisomy is the cause of a number of human genetic abnormalities, including *Down's syndrome; Patau's syndrome, in which there is an extra chromosome 13 (*trisomy 13*); and Edwards' syndrome, in which there is an extra chromosome 18 (*trisomy 18*).

tritanopia *See* **colour blindness**.

tritiated compound *See* **labelling**.

trochanter **1.** Any of several bony knobs on the femur of vertebrates to which muscles are attached. **2.** The second segment of an insect's leg, between the *coxa and the *femur.

trochophore The pelagic planktonic larva of polychaete worms, some molluscs, and certain other invertebrates. It is top-shaped and usually has two bands of cilia encircling the body.

trophic level The position that an organism occupies in a *food chain. For example, green plants (which obtain their energy directly from sunlight) are the primary *producers; herbivores are primary *consumers (and secondary producers). A carnivore that eats only herbivores is a secondary consumer and a tertiary producer. Many animals feed at several different trophic levels.

trophoblast The layer of epithelial cells that forms the outer wall of a blastocyst (see **blastula**). It embeds the blastocyst into the wall of the uterus (see **implantation**) and contributes to the *chorion.

tropism The directional growth of a plant organ in response to an external stimulus, such as light, touch, or gravity. Growth towards the stimulus is a *positive tropism*; growth away from the stimulus is a *negative tropism*. See also **geotropism; hydrotropism; orthotropism; phototropism; plagiotropism; thigmotropism**. Compare **nastic movements; taxis**.

tropomyosin A protein found in the *actin filaments in muscles. The molecule consists of two elongated strands that run along the length of the filament. When the muscle is at rest, the tropomyosin molecule covers the binding site of the actin molecule, where interaction with myosin occurs. On contraction of the muscle, the tropomyosin is displaced by another protein, *troponin, allowing the interaction of actin with myosin. See **sliding filament theory**.

troponin A complex of three polypeptide chains that are found at regular intervals along the length of an *actin filament. During muscle contraction, troponin binds to calcium ions, displacing *tropomyosin and exposing the binding site on the actin filament. This allows the interaction of actin and myosin to occur. See **sliding filament theory**.

troposphere The lowest level of the earth's atmosphere, extending from the earth's surface to a height of about 10 km (its thickness varies from 7 km at the poles to 28 km at the equator). Within the troposphere temperature falls with increasing height, although *temperature inversions can occur.

trypsin An enzyme that digests proteins (see **endopeptidase; protease**). It is secreted in an inactive form (*trypsinogen*) by the pancreas into the duodenum. There, trypsinogen is acted on by an enzyme (*enterokinase*) produced in the duodenum to yield trypsin. The active enzyme plays an important role in the digestion of proteins in the anterior portion of the small intestine. It also activates other proteases in the pancreatic juice (see **carboxypeptidase; chymotrypsin**).

trypsinogen See **trypsin**.

tryptophan See **amino acid**.

T tubules See **transverse tubules**.

tube feet See **Echinodermata**.

tube nucleus One of the two nuclei produced when the haploid nucleus of a *pollen grain divides by mitosis (compare **generative nucleus**). The tube nucleus is thought to control growth of the *pollen tube.

tuber A swollen underground stem or root in certain plants. It enables the plant to survive the winter or dry season and is also a means of propagation. A *stem tuber*, such as the potato, forms at the end of an underground stem. Each tuber represents several nodes and internodes. The following season several new plants develop from the terminal and axillary buds (eyes). *Root tubers*, such as those of the dahlia, are modified food-storing adventitious roots and may also give rise to new plants.

tubicolous Describing invertebrate animals that live in tubes that they have themselves constructed. For example, the polychaete fanworms (*Sabella*) build tubes of sand particles.

tubulin A protein of which the *microtubules of cells are formed.

Tullgren funnel A device used to remove and collect small animals, such as insects, from a sample of soil or leaf litter. The sample is placed on a coarse sieve fixed across the wide end of a funnel and a 100-watt light bulb, in a metal reflector, is placed about 25 cm above the funnel. The heat from the bulb dries and warms the sample, causing the animals to move downwards and fall through the sieve into the funnel, which directs them into a collecting dish or tube below. The dish can contain water or alcohol to prevent the animals from escaping.

tumour *See* neoplasm.

tundra A terrestrial *biome characterized by a lack of trees and a permanently frozen subsoil. Tundra lies to the north of the *taiga in North America and Eurasia; the vegetation is dominated by grasses, sedges, lichens, mosses, heathers, and low shrubs. The growing season, which occurs during the warmest part of the year when the average daily mean temperature is about 10°C, lasts only 2–4 months, during which the topsoil thaws to a depth of 30 cm, allowing roots to penetrate it. However, below this level the soil is permenently frozen (*permafrost*); water cannot filter through the soil and may lie in surface depressions during the growing season. *Compare* taiga.

Tunicata (Urochordata) A subphylum of marine nonvertebrate chordates in which the body is typically enclosed in a protective *tunic* of a cellulose-like material and the notochord and dorsal hollow nerve cord are evident only in the free-swimming tadpole-like larval stage. The pharynx is specialized for filter feeding (*see* endostyle). There are three classes: the sessile Ascidiacea (sea squirts), the pelagic Thaliacea (salps), and the pelagic Larvacea (in which the larval form is retained into adulthood).

Turbellaria A class of free-living flatworms (*see* **Platyhelminthes**) comprising the planarians, which occur in wet soils, fresh water, and marine environments. Their undersurface is covered with cilia, used for gliding over stones and weeds. Planarians can also swim by means of undulations of the body.

turgor The condition in a plant cell when its *vacuole is distended with water, pushing the protoplast against the cell wall. In this condition the force causing water to enter the cell by *osmosis is balanced by the hydrostatic pressure (*pressure potential) that has built up in the cell sap due to the pressure exerted by the cell wall on the protoplast (*see also* **water potential**). Turgidity assists in maintaining the rigidity of plants; a decrease in turgidity leads to *wilting. *Compare* **plasmolysis**.

turion 1. A winter bud, covered with scale leaves and mucilage, that is produced by certain aquatic plants, such as frogbit. Turions become detached and remain dormant on the pond or lake bottom during the winter before developing into new plants the following season. **2.** *See* **sucker**.

Turner's syndrome A genetic disorder of women caused by the absence of the second *sex chromosome (such women are XO, rather than the normal XX). It is characterized by a lack of ovaries and menstrual cycle. Affected women are sterile and lack secondary sexual characteristics, although the external genitalia are present. The syndrome is named after the US endocrinologist H. H. Turner (1892–1970), who first described it.

turnover number *(in biochemistry)* The number of molecules of substrate that are converted into product per minute by one molecule of enzyme. It is used to express the speed of action of an enzyme. *Catalase has an extremely high turnover number (6×10^6).

twins Two individuals born to the same mother at the same time. Twins can develop from the same egg (*see* **identical twins**) or from two separately fertilized eggs (*see* **fraternal twins**).

tylose A balloon-like extension of a parenchyma cell that protrudes into the lumen of a neighbouring xylem vessel or tracheid through a *pit in the cell wall. Tyloses form most commonly in older woody tissue, possibly in response to injury; they may eventually block the vessels and thus help prevent the spread of fungi and other pathogens within the plant. Tyloses may become filled with tannins, gums, pigments, etc., giving heartwood its dark colour, and their walls can remain thin or become lignified.

tympanic cavity *See* **middle ear**.

tympanum (tympanic membrane; eardrum) The membrane that separates the *outer ear from the *middle ear. It vibrates in response to sound waves and transmits these vibrations via the *ear ossicles of the middle ear to the site of hearing (the *cochlea of the *inner ear). In amphibians and some reptiles there is no external ear and the tympanum is exposed at the skin surface.

type specimen The specimen used for naming and describing a *species or subspecies. If this is the original specimen collected by the author who named the species it is termed a *holotype*. The type specimen is not necessarily the most characteristic representative of the species. The term *type* is also applied to any taxon selected as being representative of the rank to which it belongs. For example, the genus *Solanum* (potato) is said to be the type genus of the family Solanaceae.

tyrosine *See* **amino acid**.

U

ubiquinone *See* **coenzyme Q**.

ulna The larger of the two bones in the forearm of vertebrates (*compare* **radius**). It articulates with the outer carpals at the wrist and with the humerus at the elbow.

ultracentrifuge A high-speed centrifuge used to measure the rate of sedimentation of colloidal particles or to separate macromolecules, such as proteins or nucleic acids, from solutions. Ultracentrifuges are electrically driven and are capable of speeds up to 60 000 rpm.

ultrafiltration The process in which hydrostatic pressure causes water and small dissolved molecules and ions to move across a membrane against a *concentration gradient. Ultrafiltration is responsible for the formation of *tissue fluid and *glomerular filtrate from blood. In both these processes the ultrafiltered fluid has the same composition as the plasma except that it does not contain blood cells or large protein molecules.

ultramicroscope A form of microscope that reveals the presence of particles that cannot be seen with a normal optical microscope. Colloidal particles, smoke particles, etc., are suspended in a liquid or gas in a cell with a black background and illuminated by an intense cone of light that enters the cell from the side and has its apex in the field of view. The particles then produce diffraction-ring systems, appearing as bright specks on the dark background.

ultramicrotome *See* **microtome**.

ultrasonics The study and use of pressure waves that have a frequency in excess of 20 000 Hz and are therefore inaudible to the human ear. Ultrasound is used in medical diagnosis, particularly in conditions such as pregnancy, in which X-rays could have a harmful effect.

ultrastructure The submicroscopic, almost molecular, structure of living cells, which is revealed by the use of an electron microscope.

ultraviolet microscope A *microscope that has quartz lenses and slides and uses *ultraviolet radiation as the illumination. The use of shorter wavelengths than the visible range enables the instrument to resolve smaller objects and to provide greater magnification than the normal optical microscope. The final image is either photographed or made visible on a fluorescent screen by means of an image converter.

ultraviolet radiation (UV) Electromagnetic radiation having wavelengths between that of violet light and long X-rays, i.e. between 400 nanometres and 4 nm. In the range 400–300 nm the radiation is known as the *near ultraviolet*. In the range 300–200 nm it is known as the *far* *ultraviolet*. Below 200 nm it is known as the *extreme ultraviolet* or the *vacuum ultraviolet*, as absorption by the oxygen in the air makes the use of evacuated apparatus essential. The sun is a strong emitter of UV radiation but only the near UV reaches the surface of the earth as the *ozone layer of the atmosphere absorbs all wavelengths below 290 nm. Ultraviolet radiation is classified in three ranges according to its effect on the skin: UV-A (320–400 nm), UV-B (290–320 nm), and UV-C (230–290 nm). The longest-wavelength range, UV-A, is not harmful in normal doses and is used clinically in the treatment of certain skin complaints, such as psoriasis. It is also used to induce *vitamin D formation in patients that are allergic to vitamin D preparations. UV-B causes reddening of the skin followed by pigmentation (tanning). Excessive exposure can cause severe blistering. UV-C, with the shortest wavelengths, is particularly damaging. It has been claimed that short-wavelength ultraviolet radiation causes skin cancer and that the risk of contracting this has been increased by the depletion of the ozone layer.

Most UV radiation for practical use is produced by various types of mercury-vapour lamps. Ordinary glass absorbs UV radiation and therefore lenses and prisms for use in the UV are made from quartz.

umbel A type of *racemose inflorescence in which stalked flowers arise from the same point on the flower axis, resembling the spokes of an umbrella. An involucre (cluster) of bracts may occur at the point where the stalks emerge. This arrangement is characteristic of the family Umbelliferae (e.g. carrot, hogweed, parsley, parsnip), in which the inflorescence is usually a compound umbel.

umbilical cord The cord that connects the embryo to the *placenta in mammals. It contains a vein and two arteries that carry blood between the embryo and placenta. It is severed after birth to free the newly born animal from the placenta, and shrivels to leave a scar, the navel, on the animal.

undernourishment *See* malnutrition.

ungulate A herbivorous mammal with hoofed feet (*see* **unguligrade**). Ungulates are grouped into two orders: *Artiodactyla and *Perissodactyla.

unguligrade Describing the gait of ungulates (e.g. horses and cows), in which only the tips of the digits (i.e. the hooves) are on the ground and the rest of the foot is off the ground. *Compare* **digitigrade**; **plantigrade**.

unicellular Describing tissues, organs, or organisms consisting of a single cell. For example, the reproductive organs of some algae and fungi are unicellular. Unicellular organisms include bacteria, protozoans, and certain algae. *Compare* **acellular**; **multicellular**.

unipolar neurone A *neurone that has one main process, the axon, extending from its cell body. Unipolar neurones include many sensory neurones and many vertebrate motor neurones and interneurones. *Compare* **bipolar neurone**; **multipolar neurone**.

uniport A membrane protein that transports molecules across a membrane in one direction only. *Compare* **antiport; symport**.

Uniramia A subphylum or superclass of the *Arthropoda that contains the classes *Chilopoda (centipedes), *Diplopoda (millipedes), Pauropoda and Symphyla (centipede-like animals), and *Insecta (insects). Members are characterized by *uniramous appendages.

uniramous appendage A type of appendage that is characteristic of insects and other members of the subphylum *Uniramia. It consists of an unbranched series of segments (*see* **coxa; trochanter; femur**). *Compare* **biramous appendage**.

unisexual Describing animals or plants with either male or female reproductive organs but not both. Most of the more advanced animals are unisexual but plants are often *hermaphrodite. Flowers that contain either stamens or carpels but not both are also described as unisexual. *See also* **monoecious; dioecious**.

unit A specified measure of a physical quantity, such as length, mass, time, etc., specified multiples of which are used to express magnitudes of that physical quantity. For scientific purposes previous systems of units have now been replaced by *SI units.

unsaturated Denoting a compound having double or triple bonds in its molecules. Unsaturated compounds can undergo addition reactions as well as substitution. *See* **fatty acid**. *Compare* **saturated**.

uracil A *pyrimidine derivative and one of the major component bases of *nucleotides and the nucleic acid *RNA.

uranium–lead dating A group of methods of *dating certain rocks that depends on the decay of the radioisotopes uranium–238 to lead–206 (half-life 4.5×10^9 years) or the decay of uranium–235 to lead–207 (half-life 7.1×10^8 years). One form of uranium–lead dating depends on measuring the ratio of the amount of helium trapped in the rock to the amount of uranium present (since the decay $^{238}U \rightarrow {}^{206}Pb$ releases eight alpha-particles). Another method of calculating the age of the rocks is to measure the ratio of radiogenic lead (^{206}Pb, ^{207}Pb, and ^{208}Pb) present to nonradiogenic lead (^{204}Pb). These methods give reliable results for ages of the order 10^7–10^9 years.

urea (carbamide) A white crystalline water-soluble solid, $CO(NH_2)_2$. Urea is the major end product of nitrogen excretion in mammals, being synthesized by the *urea cycle. Urea is synthesized industrially from ammonia and carbon dioxide for use in urea–formaldehyde resins and pharmaceuticals, as a source of nonprotein nitrogen for ruminant livestock, and as a nitrogen fertilizer.

urea cycle (ornithine cycle) The series of biochemical reactions that converts ammonia to *urea during the excretion of metabolic nitrogen. Urea formation occurs in mammals and, to a lesser extent, in some other animals. The liver converts ammonia to the much less toxic urea, which is excreted in solution in *urine.

ureotelic Describing animals that excrete nitrogenous waste in the form of *urea. Most terrestrial vertebrates are ureotelic, converting ammonium ions formed during the breakdown of amino acids into urea. *See* **urea cycle**. *Compare* **ammonotelic; uricotelic**.

ureter The duct in vertebrates that conveys urine from the *kidney to the *bladder.

urethra The duct in mammals that conveys urine from the *bladder to be discharged to the outside of the body. In males the urethra passes through the penis and is joined by the *vas deferens; it therefore also serves as a channel for sperm.

uric acid The end product of purine breakdown in most mammals, birds, terrestrial reptiles, and insects and also (except in mammals; *see* **urea**) the major form in which metabolic nitrogen is excreted. Being fairly insoluble, uric acid can be expelled in solid form, which conserves valuable water in arid environments. The accumulation of uric acid in the synovial fluid of joints causes gout.

uricotelic Describing animals that excrete nitrogenous waste in the form of *uric acid. Uricotelic animals include birds and reptiles. *Compare* **ammonotelic; ureotelic**.

uridine A nucleoside consisting of one uracil molecule linked to a D-ribose sugar molecule. The derived nucleotide uridine diphosphate (UDP) is important in carbohydrate metabolism.

urinary system The collection of organs and tissues that perform *osmoregulation and *excretion. The mammalian urinary system consists of two *kidneys each linked to the bladder by a ureter.

urine The aqueous fluid formed by the excretory organs of animals for the removal of metabolic waste products. In higher animals, urine is produced by the *kidneys, stored in the *bladder, and excreted through the *urethra or *cloaca. Apart from water, the major constituents of urine are one or more of the end products of nitrogen metabolism – ammonia, urea, uric acid, and creatinine. It may also contain various inorganic ions, the pigments urochrome and urobilin, amino acids, and purines. Precise composition depends on many factors, especially the habitat of a particular species: aquatic animals produce copious volumes; terrestrial animals need to conserve water and produce much less (about 1.0–1.5 litres per day in humans).

uriniferous tubule *See* **nephron**.

Urochordata *See* **Tunicata**.

uterus (womb) The organ of female mammals in which the embryo develops. Paired in most mammals but single in humans, it is situated between the bladder and rectum and is connected to the *fallopian tubes and to the *vagina. The lining (*see* **endometrium**) shows cyclical changes (*see* **menstrual cycle; oestrous cycle**) associated with egg production and provides a thick spongy layer in which the fertilized egg becomes embedded. The outer wall of the uterus (*myometrium*) is thick and muscular; by contracting, it forces the fully grown fetus through the vagina to the outside.

utriculus (utricle) A chamber of the *inner ear from which the *semicircular canals arise. It bears patches of sensory epithelium concerned with detecting changes in the direction and speed of movement (*see* **macula**).

UV *See* **ultraviolet radiation**.

V

vaccination *See* **immunization**.

vaccine A liquid preparation of treated disease-producing microorganisms or their products used to stimulate an *immune response in the body and so confer resistance to the disease (*see* **immunization**). Vaccines are administered orally or by injection (*inoculation*). They take the form of dead viruses or bacteria that can still act as antigens, live but weakened microorganisms (*see* **attenuation**), specially treated *toxins, or antigenic extracts of the microorganism.

vacuole A space within the cytoplasm of a living *cell that is filled with air, water or other liquid, *cell sap, or food particles. In plant cells there is usually one large vacuole bounded by a single-layered membrane (*tonoplast* or *vacuole membrane*); animal cells usually have several small vacuoles. *See also* **contractile vacuole**.

vacuole membrane (tonoplast) *See* **vacuole**.

vagal tone The effect produced on the heart when only the parasympathetic nerve fibres (which are carried in the *vagus nerve) are controlling the heart rate. The parasympathetic nerve fibres slow the heart rate from approximately 70 beats per minute to 60 beats per minute.

vagina The tube leading from the uterus to the outside. Sperm are deposited in the vagina during copulation and the fully developed fetus is born through it. In a number of mammals the vagina may be sealed when the animal is not sexually receptive and only open during oestrus. Its lining produces mucus, which prevents friction and the entry of infective organisms.

vagus nerve The tenth *cranial nerve: a paired nerve that supplies branches to many major internal organs. It carries motor nerve fibres to the heart, lungs, and viscera and sensory fibres from the viscera.

valine *See* **amino acid**.

valve 1. (*in anatomy*) Any of various structures for restricting the flow of a fluid through an aperture or along a tube to one direction. Valves in the heart (*see* **bicuspid valve; semilunar valve; tricuspid valve**), veins, and lymphatic vessels consist of two or three flaps of tissue (*cusps*) fastened to the walls. The cusps are flattened to the walls to allow the normal passage of blood or lymph, but a reverse flow causes them to block the vessel or aperture, so preventing further backflow. **2. (*in botany*) a.** Any of the parts that make up a capsule or other dry fruit that sheds its seeds. **b.** One of the two halves of the cell wall of a diatom.

variation The differences between individuals of a plant or animal species. Variation may be the result of environmental conditions; for example, water supply and light intensity affect the height and leaf size of a plant. Differences of this kind, acquired during the lifetime of an individual, are not transmitted to succeeding generations since the genes are not affected. *Genetic variation*, due to differences in genetic constitution, is inherited (*see* **continuous variation; discontinuous variation**). The most important sources of genetic variation are *mutation and *recombination (*see also* **crossing over**). It is also increased by *outbreeding. Wide genetic variation improves the ability of a species to survive in a changing environment, since the chances that some individuals will tolerate a particular change are increased. Such individuals will survive and transmit the advantageous genes to their offspring.

variegation The occurrence of differently coloured patches, spots, or streaks in plant leaves, petals, or other parts, due to absence of pigment or different combinations of pigment in the affected area of the part. Variegation may be brought about by infection, for example *tobacco mosaic virus infection, or by genetic differences between the cells of the variegated part.

variety A category used in the *classification of plants and animals below the *species level. A variety consists of a group of individuals that differ distinctly from but can interbreed with other varieties of the same species. The characteristics of a variety are genetically inherited. Examples of varieties include breeds of domestic animals and the *races of man. *See also* **cultivar**. *Compare* **subspecies**.

varve dating (geochronology) An absolute *dating technique using thin sedimentary layers of clays called *varves*. The varves, which are particularly common in Scandinavia, have alternate light and dark bands corresponding to winter and summer deposition. Most of them are found in the Pleistocene series, where the edges of varve deposits can be correlated with the annual retreat of the ice sheet, although some varve formation is taking place in the present day. By counting varves it is possible to establish an absolute time scale for fossils up to about 20 000 years ago.

vasa recta Thin-walled blood vessels that branch from the efferent arterioles leaving each *glomerulus in the vertebrate kidney (*see* **nephron**). The vasa recta form U-shaped loops adjacent to the *loop of Henle and eventually drain into the renal vein.

vascular bundle (fascicle) A long continuous strand of conducting (vascular) tissue in tracheophyte plants that extends from the roots through the stem and into the leaves. It consists of *xylem and *phloem, which are separated by a *cambium in plants that undergo secondary thickening. *See* **vascular tissue**.

vascular cambium *See* **cambium**.

vascular plants All plants possessing organized *vascular tissue. *See* **tracheophyte**.

vascular system 1. A specialized network of vessels for the circulation of fluids throughout the body tissues of an animal. All animals, apart from simple invertebrate groups, possess a *blood vascular system*, which enables the passage of respiratory gases, nutrients, excretory products, and other metabolites into and out of the cells. In vertebrates it consists of a muscular *heart, which pumps blood through major blood vessels (*arteries) into increasingly finer branches until in the *capillaries it is in intimate contact with tissues. It then returns to the heart via another network of vessels (the *veins). This *circulation also enables a stable *internal environment for tissue function (*see* **homeostasis**), the transmission of chemical messengers (*hormones) around the body, and a means of defending the body against pathogens and damage via the immune system. A *water vascular system* is characteristic of the *Echinodermata.
2. The system of *vascular tissue in plants.

vascular tissue (vascular system) The tissue that conducts water and nutrients through the plant body in higher plants (*tracheophytes). It consists of *xylem and *phloem. Since the xylem and phloem tissues are always in close proximity to each other, distinct regions of vascular tissue can be identified (*see* **vascular bundle**). The possession of vascular tissue has enabled the higher plants to attain a considerable size and dominate most terrestrial habitats.

vas deferens One of a pair of ducts carrying sperm from the testis (or *epididymis) to the outside, in mammals through the *urethra.

vas efferens Any of various small ducts carrying sperm. In reptiles, birds, and mammals they convey sperm from the seminiferous tubules of the testis to the *epididymis; in invertebrates they carry sperm from the testis to the vas deferens.

vasoactive intestinal peptide *See* VIP.

vasoconstriction The reduction in the internal diameter of blood vessels, especially arterioles or capillaries. The constriction of arterioles is mediated by the action of nerves on the smooth muscle fibres of the arteriole walls and results in an increase in blood pressure.

vasodilation (vasodilatation) The increase in the internal diameter of blood vessels, especially arterioles or capillaries. The vasodilation of arterioles is mediated by the action of nerves on the smooth muscle fibres of the arteriole walls and results in a decrease in blood pressure.

vasomotor centre The region of the *medulla oblongata of the brain that mediates the diameter of the blood vessels (via the action of the *vasomotor nerves) and hence controls blood pressure (*see* vasoconstriction; vasodilation).

vasomotor nerves The nerves of the *autonomic nervous system that control the diameter of blood vessels. *Vasoconstrictor nerves* decrease the diameter (*see* **vasoconstriction**); *vasodilator nerves* increase it (*see* **vasodilation**).

vasopressin *See* antidiuretic hormone.

vector **1.** An animal, usually an insect, that passively transmits disease-causing microorganisms from one animal or plant to another or from an animal to man. *Compare* **carrier**. **2. (cloning vector)** A vehicle used in *gene cloning to insert a foreign DNA fragment into the genome of a host cell. For bacterial hosts three different types of vector are used: *bacteriophages, *plasmids, and their hybrid derivatives, *cosmids. The foreign DNA is spliced into the vector using specific *restriction enzymes and *ligases to cleave the vector DNA and join the foreign DNA to the two ends created (*insertional vectors*). In some phage vectors, part of the viral genome is enzymically removed and replaced with the foreign DNA (*replacement vectors*). *Retroviruses can be effective vectors for introducing recombinant DNA into mammalian cells. In plants, derivatives of the tumour-inducing (Ti) plasmid of the crown gall bacterium, *Agrobacterium tumefaciens*, are used as vectors.

vegetative propagation (vegetative reproduction) **1.** A form of *asexual reproduction in plants whereby new individuals develop from specialized multicellular structures (e.g. *tubers, *bulbs) that become detached from the parent plant. Examples are the production of strawberry plants from *runners and of gladioli from daughter *corms. Artificial methods of vegetative propagation include grafting (*see* **graft**), *budding, and making *cuttings. **2.** Asexual reproduction in animals, e.g. budding in *Hydra*.

vein **1.** A blood vessel that carries blood towards the heart. Most veins carry deoxygenated blood (the *pulmonary vein is an exception). The largest veins are fed by smaller ones, which are formed by the merger of *venules. Veins have thin walls and a relatively large internal diameter. *Valves within the veins ensure that the flow of blood is always towards the heart. *Compare* **artery**. **2.** A vascular bundle in a leaf (*see* **venation**). **3.** Any of the tubes of chitin that strengthen an insect's wing.

velamen A whitish spongy sheath of dead empty cells that surrounds the aerial roots of epiphytic plants, such as certain orchids. It absorbs any surface water on the roots.

velum *See* annulus.

vena cava Either of the two large veins that carry deoxygenated blood into the right atrium of the heart. The *precaval vein* (*anterior* or *superior vena cava*) receives blood from the head and forelimbs; the *postcaval vein* (*posterior* or *inferior vena cava*) drains blood from the trunk and hindlimbs.

venation **1.** The arrangement of veins (vascular bundles) in a leaf. The leaves of dicotyledons have a central main vein (midrib) with side branches that themselves further subdivide to form a network (*net* or *reticulate venation*). The leaves of monocotyledons have parallel veins (*parallel* *venation*). **2.** The arrangement of the veins in an insect's wing, which is often important in classification.

venter The swollen base of an *archegonium, in which the egg cell (oosphere) develops.

ventilation The process by which a continuous exchange of gases is maintained across respiratory surfaces. Often called external *respiration, this is achieved by *respiratory movements; in air-breathing vertebrates it is movement of air into and out of the lungs (*see also* **air sac; expiration; inspiration; trachea; ventilation centre**). The *ventilation rate* (or *respiration rate*) of an animal is the volume of air breathed per minute, i.e. *tidal volume × number of breaths per minute. It can be measured with the aid of a *respirometer.

ventilation centre The group of neurones in the *medulla oblongata of the brain that controls the process of *ventilation. The partial pressure of carbon dioxide in the blood and the pH of the blood are monitored by chemoreceptors in the arteries. These include the *carotid bodies in the carotid arteries and the *aortic bodies* in the wall of the aorta close to the heart. The ventilation centre responds to an increase in the amount of carbon dioxide in the blood by increasing the rate of breathing. Within the ventilation centre are subcentres that control inspiration (*see* **inspiratory centre**) and expiration (*expiratory centre*).

ventral Describing the surface of a plant or animal that is nearest or next to the ground or other support, i.e. the lower surface. In bipedal animals, such as man, it is the forward-directed (*anterior) surface. *Compare* **dorsal**.

ventral aorta The artery in vertebrate embryos that carries blood from the ventricle of the heart to the *aortic arches. In adult fish it branches into afferent branchial arteries supplying the gills; in adult tetrapods it is represented by the ascending part of the *aorta. *Compare* **dorsal aorta**.

ventral root The part of a *spinal nerve that leaves the spinal cord on the ventral side and contains motor fibres. *Compare* **dorsal root**. *See* **spinal cord**.

ventricle 1. A chamber of the *heart that receives blood from an *atrium and pumps it into the arterial system. Amphibians and fish have a single ventricle, but mammals, birds, and reptiles have two, pumping deoxygenated blood to the lungs and oxygenated blood to the rest of the body, respectively. **2.** Any of the four linked fluid-filled cavities in the brain of vertebrates. One of these cavities is in the *medulla oblongata, two are in the cerebral hemispheres (*see* **cerebrum**), and the fourth is in the posterior part of the *forebrain. The ventricles contain cerebrospinal fluid filtered from the blood by the *choroid plexus.

venule A small blood vessel that receives blood from the capillaries and transports it to a vein.

vermiform appendix *See* **appendix**.

vernalization The application of a cold treatment to germinating seeds or seedlings to ensure flowering. Many plants, including winter cereals, will not flower unless subjected to a period of chilling early in their development. Winter cereals are therefore sown in the autumn for flowering the following year. However, if germinating seeds are artificially vernalized they can be sown in the spring for flowering the same year.

vertebra Any of the bones that make up the *vertebral column. In mammals each vertebra typically consists of a main body, or *centrum*, from which arises a *neural arch* through which the spinal cord passes, and *transverse processes* projecting from the side. There are five groups of vertebrae, specialized for various functions and varying in number with the species. In man, for example, there are 7 *cervical vertebrae, 12 *thoracic vertebrae, 5 *lumbar vertebrae, 5 fused *sacral vertebrae, and 5 fused *caudal vertebrae (forming the *coccyx).

vertebral column (backbone; spinal column; spine) A flexible bony column in vertebrates that extends down the long axis of the body and provides the main skeletal support. It also encloses and protects the *spinal cord and provides attachment for the muscles of the back. The vertebral column consists of a series of bones (*see* **vertebra**) separated by discs of cartilage (*intervertebral discs). It articulates with the skull by means of the *atlas vertebra, with the ribs at the *thoracic vertebrae, and with the pelvic girdle at the sacrum (*see* **sacral vertebrae**).

vertebrate Any one of a large group of animals comprising all those members of the phylum *Chordata that have backbones (*see* **vertebral column**). Vertebrates include the fishes, amphibians, reptiles, birds, and mammals.

vesicle A small, usually fluid-filled, membrane-bound sac within the cytoplasm of a living cell. Vesicles form part of the *Golgi apparatus.

vessel **1.** *(in botany)* A tube within the *xylem composed of joined *vessel elements. Vessels facilitate the efficient movement of water from the roots to the shoots and leaves of a plant. **2.** *(in zoology)* Any of various tubular structures through which substances are transported, especially a blood vessel or a lymphatic vessel.

vessel element A type of cell occurring within the *xylem of flowering plants, many of which, end to end, form water-conducting *vessels*. Vessel elements are frequently very broad and have side walls thickened by deposits of lignin over most of the surface area. However, the end walls are broken down to provide connections with the cells both above and below them. *Compare* **tracheid**.

vestibular apparatus The part of the inner ear that is responsible for balance. The vestibular apparatus is continuous with the cochlea. It consists of the three *semicircular canals, which detect movements of the head (*see* **ampulla**), and the *utriculus and *sacculus, which detect the position of the head (*see* **macula**). *See* **ear**.

vestibular canal A canal in the *cochlea of the inner ear that connects with the *oval window. It contains *perilymph, through which pressure waves are transmitted from the oval window via *Reissner's membrane to the endolymph within the cochlear duct. *See also* **organ of Corti**.

vestibule A chamber that leads to a body cavity or that links one cavity to another. For example, a vestibule leads from the vulva into the *vagina in the mammalian reproductive system.

vestigial organ Any part of an organism that has diminished in size during its evolution because the function it served decreased in importance or became totally unnecessary. Examples are the human appendix and the wings of the ostrich.

viable count A measure of the number of living cells within a culture.

vibrio Any comma-shaped bacterium. Generally, vibrios are Gram-negative (*see* **Gram's stain**), motile, and aerobic. They are widely distributed in soil and water and while most feed on dead organic matter some are parasitic, e.g. *Vibrio cholerae*, the causal agent of cholera.

villus (*pl.* **villi**) A microscopic outgrowth from the surface of some tissues and organs, which serves to increase the surface area of the organ. Numerous villi line the interior of the small intestine. Their shape may vary from finger-like (in the *duodenum) to spadelike (in the *ileum). Intestinal villi are specialized for the absorption of soluble food material: each contains blood vessels and a lymph vessel (*see* **lacteal**).

Chorionic villi occur on the chorion of the mammalian placenta, where they increase the surface area for the exchange of materials between the fetal and maternal blood.

VIP (vasoactive intestinal peptide) A peptide hormone secreted by endocrine cells of the upper part of the small intestine in response to the entry of partially digested food from the stomach. VIP, along with *secretin, stimulates the pancreas to produce a thin watery secretion containing bicarbonate. This raises the pH in the intestine in preparation for secretion of pancreatic enzymes. VIP also occurs as a neuropeptide in central nervous tissue.

virion *See* **virus**.

virology The scientific study of *viruses. *See* **microbiology**.

virulence The disease-producing ability of a microorganism. *See also* **pathogen**.

virus A particle that is too small to be seen with a light microscope or to be trapped by filters but is capable of independent metabolism and reproduction within a living cell. Outside its host cell a virus is completely inert. A mature virus (a *virion*) ranges in size from 20 to 400 nm in diameter. It consists of a core of nucleic acid (DNA or RNA) surrounded by a protein coat (*see* **capsid**). Some (the *enveloped viruses*) bear an outer envelope consisting of proteins and lipids. Inside its host cell the virus initiates the synthesis of viral proteins and undergoes replication. The new virions are released when the host cell disintegrates. Viruses are parasites of animals, plants, and some bacteria (*see* **bacteriophage**). Viral diseases of animals include the common cold, influenza, AIDS, herpes, hepatitis, polio, and rabies (*see* **adenovirus; arbovirus; herpesvirus; HIV; myxovirus; papovavirus; picornavirus; poxvirus**); some viruses are also implicated in the development of cancer (*see* **retrovirus**). Plant viral diseases include various forms of yellowing and blistering of leaves and stems (*see* **tobacco mosaic virus**). Antibiotics are ineffective against viral diseases but *vaccines (if available) provide good protection.

visceral Relating to the internal organs (the *viscera*) that lie in the coelomic cavities of animals, i.e. in the thoracic and abdominal cavities of mammals. *Compare* **somatic**.

visceral sensory neurone *See* **sensory neurone**.

vision The sense that enables perception of objects in the environment by means of the *eyes.

visual acuity Sharpness of vision: the ability of the eye to distinguish between objects that lie close together. This hinges on the ability of the eye to focus incoming light to form a sharp image on the retina. Visual acuity depends on the *cone cells, which are most densely packed in the *fovea, close to the centre of the retina, and are therefore in the optimum position to receive focused light. In addition, each cone cell synapses with a single bipolar cell in the *retina and is thus able to send a separate signal, via the optic nerve fibres, to the brain.

visual cortex (striate cortex) The region of the *cerebral cortex of the brain where sensory information from the eyes is interpreted.

visual purple *See* **rhodopsin**.

vital capacity The total amount of air that can be exhaled after maximum inspiration. The vital capacity of an average human is about 4.5 litres; in trained male athletes it can be 6 litres or more. However, some air always remains in the lungs (*see* **residual volume**).

vital staining A technique in which a harmless dye is used to stain living tissue for microscopical observation. The stain may be injected into a living animal and the stained tissue removed and examined (*intravital staining*) or the living tissue may be removed directly and subsequently stained (*supravital staining*). Microscopic organisms, such as protozoa, may be completely immersed in the dye solution. Vital stains include trypan blue, vital red, and Janus green, the latter being especially suitable for observing mitochondria.

vitamin One of a number of organic compounds required by living organisms in relatively small amounts to maintain normal health. There are some 14 generally recognized major vitamins: the water-soluble *vitamin B complex (containing 9) and *vitamin C and the fat-soluble *vitamin A, *vitamin D, *vitamin E, and *vitamin K. Most B vitamins and vitamin C occur in plants, animals, and microorganisms; they function typically as *coenzymes. Vitamins A, D, E, and K occur only in animals, especially vertebrates, and perform a variety of metabolic roles. Animals are unable to manufacture many vitamins themselves and must have adequate amounts in the diet. Foods may contain vitamin precursors (called *provitamins*) that are chemically changed to the actual vitamin on entering the body. Many vitamins are destroyed by light and heat, e.g. during cooking.

vitamin A (retinol) A fat-soluble vitamin that cannot be synthesized by mammals and other vertebrates and must be provided in the diet. Green plants contain precursors of the vitamin, notably carotenes, that are converted to vitamin A in the intestinal wall and liver. The aldehyde derivative of vitamin A, *retinal*, is a constituent of the visual pigment *rhodopsin. Deficiency affects the eyes, causing night blindness, *xerophthalmia* (dryness and thickening of the cornea), and eventually total blindness. The role of vitamin A in other aspects of metabolism is less clear; it may be involved in controlling *ATP production and the growth of epithelial cells.

vitamin B complex A group of water-soluble vitamins that characteristically serve as components of *coenzymes. Plants and many microorganisms can manufacture B vitamins but dietary sources are essential for most animals. Heat and light tend to destroy B vitamins.

Vitamin B_1 (thiamin(e)) is a precursor of the coenzyme thiamine pyrophosphate, which functions in carbohydrate metabolism. Deficiency leads to *beriberi in humans and to polyneuritis in birds. Good sources include brewer's yeast, wheatgerm, beans, peas, and green vegetables.

Vitamin B_2 (riboflavin) occurs in green vegetables, yeast, liver, and milk. It is a constituent of the coenzymes *FAD and FMN, which have an important role in the metabolism of all major nutrients as well as in the oxidative phosphorylation reactions of the *electron transport chain. Deficiency of B_2 causes inflammation of the tongue and lips and mouth sores.

Vitamin B_6 (pyridoxine) is widely distributed in cereal grains, yeast, liver, milk, etc. It is a constituent of a coenzyme (pyridoxal phosphate) involved in amino acid metabolism. Deficiency causes retarded growth, dermatitis, convulsions, and other symptoms.

Vitamin B_{12} (cobalamin) is manufactured only by microorganisms and natural sources are entirely of animal origin. Liver is especially rich in it. One form of B_{12} functions as a coenzyme in a number of reactions, including the oxidation of fatty acids and the synthesis of DNA. It also works in conjunction with *folic acid (another B vitamin) in the synthesis of the amino acid methionine and it is required for normal production of red blood cells. Vitamin B_{12} can only be absorbed from the gut in the presence of a glycoprotein called *intrinsic factor*; lack of this factor or deficiency of B_{12} results in pernicious anaemia.

Other vitamins in the B complex include *nicotinic acid, *pantothenic acid, *biotin, and *lipoic acid. *See also* **choline; inositol**.

vitamin C (ascorbic acid) A colourless crystalline water-soluble vitamin found especially in citrus fruits and green vegetables. Most organisms synthesize it from glucose but man and other primates and various other species must obtain it from their diet. It is required for the maintenance of healthy connective tissue; deficiency leads to *scurvy. Vitamin C is readily destroyed by heat and light.

vitamin D A fat-soluble vitamin occurring in the form of two steroid derivatives: *vitamin D_2 (ergocalciferol*, or *calciferol*), found in yeast; and *vitamin D_3 (cholecalciferol*), which occurs in animals. Vitamin D_2 is formed from a steroid by the action of ultraviolet light and D_3 is produced by the action of sunlight on a cholesterol derivative in the skin. Fish-liver oils are the major dietary source. The active form of vitamin D is manufactured in response to the secretion of *parathyroid hormone, which occurs when blood calcium levels are low. It causes increased uptake of calcium from the gut, which increases the supply of calcium for bone synthesis. Vitamin D deficiency causes *rickets in growing animals and osteomalacia in mature animals. Both conditions are characterized by weak deformed bones.

vitamin E (tocopherol) A fat-soluble vitamin consisting of several closely related compounds, deficiency of which leads to a range of disorders in different species, including muscular dystrophy, liver damage, and infertility. Good sources are cereal grains and green vegetables. Vitamin E prevents the oxidation of unsaturated fatty acids in cell membranes, so maintaining their structure (*see* **antioxidants**).

vitamin K A fat-soluble vitamin consisting of several related compounds that act as coenzymes in the synthesis of several proteins (including prothrombin) necessary for blood clotting. Deficiency of vitamin K, which leads to extensive bleeding, is rare because a form of the vitamin is manufactured by intestinal bacteria. Green vegetables and egg yolk are good sources.

vitelline membrane *See* **egg membrane**.

vitreous humour The colourless jelly that fills the space between the lens and the retina of the vertebrate eye.

viviparity 1. *(in zoology)* A form of reproduction in animals in which the developing embryo obtains its nourishment directly from the mother via a *placenta or by other means. Viviparity occurs in some insects and other arthropods, in certain fishes, amphibians, and reptiles, and in the majority of mammals. *Compare* **oviparity; ovoviviparity. 2. *(in botany)* a.** A form of *asexual reproduction in certain plants, such as the onion, in which the flower develops into a budlike structure that forms a new plant when detached from the parent. **b.** The development of young plants on the inflorescence of the parent plant, as seen in certain grasses and the spider plant.

vocal cords A pair of elastic membranes that project into the *larynx in air-breathing vertebrates. Vocal sounds are produced when expelled air passing through the larynx vibrates the cords. The pitch of the sound produced depends on the tension of the cords, which is controlled by muscles and cartilages in the larynx.

voltage clamp An experimental apparatus for measuring the flow of ions through membrane *channels in cells. As the flow of ions amounts to a flow of current this can easily be measured. Typically a cell is placed in solution with three electrodes, two intracellular and one extracellular. One of the intracellular electrodes is used to measure voltage across the membrane relative to the extracellular electrode; the second intracellular electrode passes current into the cell. When the membrane becomes either less negative (depolarization) or more negative (hyperpolarization) these effects can be detected by the electrodes.

voluntary Controlled by conscious thought. *See* **voluntary muscle**. *Compare* **involuntary**.

Structure and action of a voluntary muscle

voluntary muscle (skeletal, striped, *or* striated muscle) Muscle that is under the control of the will and is generally attached to the skeleton. An individual muscle consists of bundles of long *muscle fibres*, each bounded by a *sarcolemma and containing *sarcoplasm, *sarcoplasmic reticulum, and many nuclei. The whole muscle is covered with a strong connective tissue sheath (*epimysium*) and attached at each end to a bone by inextensible *tendons. Running through each fibre are smaller fibres (*myofibrils*) having alternate light and dark bands, which contain protein filaments responsible for the muscle's contractile ability and give the muscle its typical striped appearance under the microscope. The functional unit of a myofibril is the *sarcomere. See illustration.

The end of the muscle that is attached to a nonmoving bone is called the *origin* of the muscle; the end attached to a moving bone is the *insertion*. As a muscle contracts it becomes shorter and fatter, moving one bone closer to the other. Since a muscle cannot expand, another muscle (the *extensor*) is required to move the bone in the opposite direction and stretch the first muscle (known as the *flexor*). The flexor and extensor are described as *antagonistic muscles*. See illustration.

volutin Polyphosphate granules that are used as food reserves by certain bacteria, including the cyanobacteria.

vulva The female external genitalia, comprising in women two pairs of fleshy folds of tissue, the labia (*see* **labium**); the *clitoris; and the vaginal opening.

waggle dance *See* **dance of the bees**.

Wallace's line An imaginary line that runs between the Indonesian islands of Bali and Lombok and represents the separation of the Australian and Oriental faunas. It was proposed by the naturalist A. R. Wallace (1823–1913), who had noted that the mammals in Southeast Asia are different from and more advanced than their Australian counterparts. He suggested this was because the Australian continent had split away from Asia before the better adapted placental mammals evolved in Asia. Hence the isolated Australian marsupials and monotremes were able to thrive while those in Asia were driven to extinction by competition from placental mammals. *See also* **zoogeography**.

warfarin 3-(alpha-acetonylbenzyl)-4-hydroxycoumarin: a synthetic *anticoagulant used both therapeutically in clinical medicine and, in lethal doses, as a rodenticide (*see* **pesticide**).

warm-blooded animal *See* **endotherm**.

warning coloration (aposematic coloration) The conspicuous markings of an animal that make it easily recognizable and warn would-be predators that it is a poisonous, foul-tasting, or dangerous species. For example, the yellow-and-black striped abdomen of the wasp warns of its sting. *See also* **mimicry**.

waste product Any product of metabolism that is not required for further metabolic processes and is therefore excreted from the body. Common products include *nitrogenous wastes (such as *urea and ammonia), carbon dioxide, and *bile.

water A colourless liquid, H_2O. In the gas phase water consists of single H_2O molecules in which the H–O–H angle is 105°. The structure of liquid water is still controversial; hydrogen bonding of the type $H_2O...H–O–H$ imposes a high degree of structure (*see* **hydrogen bond**) and current models supported by X-ray scattering studies have short-range ordered regions, which are constantly disintegrating and re-forming. This ordering of the liquid state is sufficient to make the density of water at about 0°C higher than that of the relatively open-structured ice; the maximum density occurs at 3.98°C. This accounts for the well-known phenomenon of ice floating on water and the contraction of water below ice, a fact of enormous biological significance for all aquatic organisms. Water is a powerful *solvent for both polar and ionic compounds; molecules or ions in solution are frequently strongly hydrated. Pure liquid water is very weakly dissociated into H_3O^+ (hydroxonium) and OH^- (hydroxyl) ions by self-ionization and consequently any compound that increases the concentration of the positive ion, H_3O^+, is acidic and compounds increasing the concentration of the negative ion, OH^-, are basic (*see* **acid**). The phenomena of ion transport in water and the division of materials into *hydrophilic* (water loving) and *hydrophobic* (water hating) substances are central features of almost all biological chemistry.

water cycle *See* **hydrological cycle**.

water potential Symbol Ψ. The difference between the chemical potential of the water in a biological system and the chemical potential of pure water at the same temperature and pressure. It is manifested as a force acting on water molecules in a solution separated from pure water by a membrane that is permeable to water molecules only and can be expressed as the sum of the *solute potential and the *pressure potential:

$$\Psi = \Psi_s + \Psi_p$$

Water potential is measured in kilopascals (kPa). The water potential of pure water is zero; aqueous solutions of increasing concentration have increasingly negative values. Water tends to move from areas of high (less negative) water potential to areas of low (more negative) water potential. *Osmosis in plants is now described in terms of water potential.

water vascular system *See* **Echinodermata**.

Watson–Crick model The double-stranded twisted ladder-like molecular structure of *DNA as determined by the US biochemist James Watson (1928–   ) and the British biochemist Francis Crick (1916–   ) at Cambridge, England, in 1953. It is commonly known as the *double helix*.

wax *(in biology)* Esters of fatty acids, usually having a protective function. Examples are the beeswax forming part of a honeycomb and the wax coating on some leaves, fruits, and seed coats, which acts as a protective water-impermeable layer supplementing the functions of the cuticle. The seeds of a few plants contain wax as a food reserve.

weed *See* **pest**.

Weismannism The theory of the *continuity of the germ plasm* published by August Weismann (1834–1914) in 1886. It proposes that the contents of the reproductive cells (sperms and ova) are passed on unchanged from one generation to the next, unaffected by any changes undergone by the rest of the body. It thus rules out any possibility of the inheritance of acquired characteristics, and has become fundamental to neo-Darwinian theory.

Western blotting (protein blotting) An *immunoassay for determining very small amounts of a particular protein in tissue samples or cells. The sample is subjected to electrophoresis on SDS-polyacrylamide gel to separate constituent proteins. The resultant protein bands are then 'blotted' onto a polymer sheet. A radiolabelled antibody specific for the target protein is added; this binds to the protein, which can then be detected by autoradiography. A variation of this technique is used to screen bacterial colonies containing cDNA clones in order to isolate those colonies expressing a particular protein. The name is derived by analogy to that of *Southern blotting.

whalebone (baleen) Transverse horny plates hanging down from the upper jaw on each side of the mouth of the toothless whales (see **Cetacea**), forming a sieve. Water, containing plankton on which the whale feeds, enters the open mouth and is then expelled with the mouth slightly closed, so that food is retained on the baleen plates.

whales See Cetacea.

whey See curd.

white blood cell See leucocyte.

white matter Part of the tissue that makes up the central nervous system of vertebrates. It consists chiefly of nerve fibres enclosed in whitish *myelin sheaths. Compare grey matter.

wild type Describing the form of an *allele possessed by most members of a population in their natural environment. Wild-type alleles are usually *dominant.

wilting The condition that arises in plants when more water is lost by evaporation than is absorbed from the soil. This causes the cells to lose their *turgor and the plant structure droops. Plants can normally recover from wilting if water is added to the soil, but permanent wilting and possible death can result if the plant does not have access to water for a long period of time. In certain plants wilting is important as a mechanism to avoid overheating: when the leaves droop they are taken out of direct contact with the sun's rays. When the sun sets the plant can begin to transpire at the normal rate and the cells of the leaves regain their turgor.

windpipe See trachea.

wind pollination See anemophily.

wing See flight.

withdrawal reflex See polysynaptic reflex; reflex.

womb See uterus.

wood The hard structural and water-conducting tissue that is found in many perennial plants and forms the bulk of trees and shrubs. It is composed of secondary *xylem and associated cells, such as fibres. The wood of angiosperms is termed *hardwood*, e.g. oak and mahogany, and that of gymnosperms *softwood*, e.g. pine and fir. New wood is added to the outside of the old wood each growing season by divisions of the vascular cambium (see **growth ring**). Only the outermost new wood (*sapwood) functions in water conduction; the inner wood (*heartwood) provides only structural support.

wort See brewing.

xanthophyll A member of a class of oxygen-containing *carotenoid pigments, which provide the characteristic yellow and brown colours of autumn leaves.

Xanthophyta A phylum of mostly freshwater organisms of the kingdom Protoctista, traditionally known as yellow-green algae, that possess carotenoid pigments (including xanthins), which are responsible for their colour, in addition to chlorophylls. Xanthophytes occur in a variety of forms – unicellular, colonial, filamentous, and siphonaceous; motile cells have two unequal-sized flagella. Storage products are oil and the polysaccharide *chrysolaminarin*.

X chromosome *See* **sex chromosome**.

xenobiotic Any substance foreign to living systems. Xenobiotics include drugs, pesticides, and carcinogens. *Detoxification of such substances occurs mainly in the liver.

xeric Denoting conditions characterized by an inadequate supply of water. Xeric conditions exist in arid habitats, extremely cold habitats, and in salt marshes. Certain plants are adapted to live in such conditions. *See* **halophyte; xerophyte**.

xeromorphic Describing the structural modifications of certain plants (*xerophytes) that enable them to reduce water loss, particularly from their leaves and stems.

xerophthalmia *See* **vitamin A**.

xerophyte A plant that is adapted to live in conditions in which there is either a scarcity of water in the soil, or the atmosphere is dry enough to provoke excessive transpiration, or both. Xerophytes have special structural (*xeromorphic*) and functional modifications, including swollen water-storing stems or leaves (*see* **succulent**) and specialized leaves that may be hairy, rolled, or reduced to spines or have a thick cuticle to lower the rate of transpiration. Examples of xerophytes are desert cacti and many species growing on sand dunes and exposed moorlands. Some *halophytes have xeromorphic features. *Compare* **mesophyte; hydrophyte**.

X-ray crystallography The use of *X-ray diffraction to determine the structure of crystals or molecules, such as nucleic acids. The technique involves directing a beam of X-rays at a crystalline sample and recording the diffracted X-rays on a photographic plate. The diffraction pattern consists of a pattern of spots on the plate, and the crystal structure can be worked out from the positions and intensities of the diffraction spots. X-rays are diffracted by the electrons in the molecules and if molecular crystals of a compound are used, the electron density distribution in the molecule can be determined.

X-rays Electromagnetic radiation of shorter wavelength than ultraviolet radiation and longer wavelength than gamma radiation. The range of wavelengths is 10^{-11} m to 10^{-9} m. X-rays can pass through many forms of matter and they are therefore used medically and industrially to examine internal structures. X-rays are produced for these purposes by an X-ray tube.

xylem A tissue that transports water and dissolved mineral nutrients in vascular plants. In flowering plants it consists of hollow *vessels* that are formed from cells (*vessel elements) joined end to end. The end walls of the vessel elements are perforated to allow the passage of water. In less advanced vascular plants, such as conifers and ferns, the constituent cells of the xylem are called *tracheids. In young plants and at the shoot and root tips of older plants *primary xylem* is formed by the apical meristems (*see* **protoxylem; metaxylem**). In plants showing secondary growth this xylem is replaced in most of the plant by *secondary xylem*, formed by the vascular *cambium. The walls of the xylem cells are thickened with lignin, the extent of this thickening being greatest in secondary xylem. Xylem contributes greatly to the mechanical strength of the plant: *wood is mostly made up of secondary xylem. *See also* **fibre**. *Compare* **phloem**.

xylenes *See* **dimethylbenzenes**.

Y chromosome *See* sex chromosome.

yeasts A group of unicellular fungi of the class Hemiascomycetae and phylum *Ascomycota. They occur as single cells or as groups or chains of cells; yeasts reproduce asexually by *budding and sexually by producing ascospores. Yeasts of the genus *Saccharomyces ferment sugars and are used in the baking and brewing industries (*see* **baker's yeast**).

yellow body *See* corpus luteum.

yolk The food stored in an egg for the use of the embryo. It can consist mainly of protein (*protein yolk*) or of phospholipids and fats (*fatty yolk*). The eggs of oviparous animals (e.g. birds) contain a relatively large yolk.

yolk sac One of the protective membranes surrounding the embryos of birds, reptiles, and mammals (*see* **extraembryonic membranes**). The embryo derives nourishment from the yolk sac via a system of blood vessels. In birds and reptiles the yolk sac encloses the yolk; in most mammals a fluid replaces the yolk.

Y organ Either of a pair of glands in the head of certain crustaceans that secrete a hormone resembling *ecdysone and are believed to play a role in the moulting process.

zeatin A naturally occurring *cytokinin that was first identified in 1963 in the grains of the maize plant (*Zea mais*).

zinc Symbol Zn. A blue-white metallic element that is a trace element (*see* **essential element**) required by living organisms. It functions as the prosthetic group of a number of enzymes.

Z line (Z band) The thin membrane that separates adjacent *sarcomeres in the fibres of striated muscle. It is visible under an electron microscope as a thin dark line in the middle of a lighter band. The centre point of each thin (actin) filament is anchored to the Z line and exerts a pull on the membrane when the fibre contracts.

zona pellucida A layer of *glycoprotein that surrounds the plasma membrane of a mammalian egg cell. It develops as a jelly coat around the primary oocyte and is surrounded by the *granulosa cells.

zonation The distribution of the different species of a community into separate zones, which are created by variations in the environment. A clear example of zonation occurs on a rocky shore, where different species of seaweed (*Fucus*) occupy different zones, according to their ability to withstand desiccation. For example, the species found in the splash zone, which is never completely submerged in water, is better adapted to exposure than those found in zones lower down the shore, where they are submerged for longer periods. Animals, particularly stationary species, such as barnacles, also exhibit zonation on a rocky shore; as with the seaweeds, this may depend on the ability of different species to withstand desiccation. Competition between species may also contribute to zonation.

zone fossil *See* **index fossil**.

zoogeography The study of the geographical distributions of animals. The earth can be divided into several zoogeographical regions separated by natural barriers, such as oceans, deserts, and mountain ranges. The characteristics of the fauna of each region are believed to depend particularly on the process of *continental drift and the stage of evolution reached when the various land masses became isolated. For example Australia, which has been isolated since Cretaceous times, has the most primitive native mammalian fauna, consisting solely of marsupials and monotremes. *See also* **Wallace's line**.

zooid An individual member of a colony of invertebrate animals, especially an individual of the phylum *Ectoprocta.

zoology The scientific study of animals, including their anatomy, physiology, biochemistry, genetics, ecology, evolution, and behaviour.

Zoomastigina A phylum of the Protoctista containing heterotrophic *protozoa that can be parasitic or free-living. They possess one or more *flagella for locomotion. The phylum includes *Trypanosoma* (the parasite causing sleeping sickness) and *Naegleria*.

zoonosis (*pl.* **zoonoses**) An infectious disease of nonhuman vertebrates that can be transmitted to humans. Rabies and anthrax are well-known examples, and certain midges and the tsetse fly act as carriers for a variety of nematode-worm zoonoses.

zooplankton The animal component of *plankton. All major animal phyla are represented in zooplankton, as adults, larvae, or eggs; some are just visible to the naked eye but most cannot be seen without magnification. Near the surface of the sea there may be many thousands of such animals per cubic metre.

zoosporangium *See* zoospore.

zoospore A spore that possesses one or more flagella and is therefore motile. Released from a sporangium (called a *zoosporangium*), zoospores are produced by many algae and certain other protoctists, such as the potato blight (*Phytophthora infestans*) and other members of the *Oomycota.

zwitterion (**ampholyte ion**) An ion that has a positive and negative charge on the same group of atoms. Zwitterions can be formed from compounds that contain both acid groups and basic groups in their molecules. For example, the amino acid glycine has the formula $H_2N.CH_2.COOH$. However, under neutral conditions, it exists in the different form of the zwitterion $^+H_3N.CH_2.COO^-$, which can be regarded as having been produced by an internal neutralization reaction (transfer of a proton from the carboxyl group to the amino group). Glycine therefore has some properties characteristic of ionic compounds, e.g. a high melting point and solubility in water. In acid solutions, the positive ion $^+H_3NCH_2COOH$ is formed. In basic solutions, the negative ion $H_2NCH_2COO^-$ predominates. The name comes from the German *zwei*, two.

zygomorphy *See* **bilateral symmetry**.

Zygomycota A phylum of saprotrophic or parasitic fungi that includes the bread mould (*Mucor*). Their hyphae lack cross walls and they can reproduce asexually by sporangiospores formed within a *sporangium or sexually by means of *zygospores.

zygospore A zygote with a thick resistant wall, formed by some algae and fungi (*see* **Zygomycota**). It results from the fusion of two gametes, neither of which is retained by the parent in any specialized sex organ (such as an oogonium). It enters a resting phase before germination. *Compare* **oospore**.

zygote A fertilized female *gamete: the product of the fusion of the nucleus of the ovum or ovule with the nucleus of the sperm or pollen grain. *See* **fertilization**.

zymase An enzyme in yeast cells that plays an essential role in *fermentation processes.

zymogen Any inactive enzyme precursor that, following secretion, is chemically altered to the active form of the enzyme. For example, the protein-digesting enzyme *trypsin is secreted by the pancreas as the zymogen trypsinogen. This is changed in the small intestine by the action of another enzyme, enterokinase, to the active form.

Appendix 1. SI units

TABLE 1.1 Base and dimensionless SI units

Physical quantity	Name	Symbol
length	metre	m
mass	kilogram	kg
time	second	s
electric current	ampere	A
thermodynamic temperature	kelvin	K
luminous intensity	candela	cd
amount of substance	mole	mol
*plane angle	radian	rad
*solid angle	steradian	sr

*dimensionless units

TABLE 1.2 Derived SI units with special names

Physical quantity	Name of SI unit	Symbol of SI unit
frequency	hertz	Hz
energy	joule	J
force	newton	N
power	watt	W
pressure	pascal	Pa
electric charge	coulomb	C
electric potential difference	volt	V
electric resistance	ohm	Ω
electric conductance	siemens	S
electric capacitance	farad	F
magnetic flux	weber	Wb
inductance	henry	H
magnetic flux density (magnetic induction)	tesla	T
luminous flux	lumen	lm
illuminance	lux	lx
absorbed dose	gray	Gy
activity	becquerel	Bq
dose equivalent	sievert	Sv

TABLE 1.3 **Decimal multiples and submultiples to be used with SI units**

Submultiple	Prefix	Symbol	Multiple	Prefix	Symbol
10^{-1}	deci	d	10	deca	da
10^{-2}	centi	c	10^2	hecto	h
10^{-3}	milli	m	10^3	kilo	k
10^{-6}	micro	μ	10^6	mega	M
10^{-9}	nano	n	10^9	giga	G
10^{-12}	pico	p	10^{12}	tera	T
10^{-15}	femto	f	10^{15}	peta	P
10^{-18}	atto	a	10^{18}	exa	E
10^{-21}	zepto	z	10^{21}	zetta	Z
10^{-24}	yocto	y	10^{24}	yotta	Y

TABLE 1.4 **Conversion of units to SI units**

From	To	Multiply by
in	m	2.54×10^{-2}
ft	m	0.3048
sq. in	m^2	6.4516×10^{-4}
sq. ft	m^2	9.2903×10^{-2}
cu. in	m^3	1.63871×10^{-5}
cu. ft	m^3	2.83168×10^{-2}
l(itre)	m^3	10^{-3}
gal(lon)	l(itre)	4.546 09
miles/hr	$m\ s^{-1}$	0.477 04
km/hr	$m\ s^{-1}$	0.277 78
lb	kg	0.453 592
g cm^{-3}	kg m^{-3}	10^3
lb/in^3	kg m^{-3}	$2.767\ 99 \times 10^4$
dyne	N	10^{-5}
poundal	N	0.138 255
lbf	N	4.448 22
mmHg	Pa	133.322
atmosphere	Pa	$1.013\ 25 \times 10^5$
hp	W	745.7
erg	J	10^{-7}
eV	J	$1.602\ 10 \times 10^{-19}$
kW h	J	3.6×10^6
cal	J	4.1868

Appendix 2. Classification of the animal kingdom

*Only major phyla and classes are shown

Appendix 3. Classification of the plant kingdom

*Extinct and mostly extinct groups are excluded

Appendix 4. Geological time scale

Era	Period	Epoch	millions of years ago
Cenozoic	Quaternary	Holocene	
		Pleistocene	1.8
	Tertiary	Pliocene	
		Miocene	
		Oligocene	
		Eocene	
		Palaeocene	65
Mesozoic	Cretaceous		
			136
	Jurassic		
			190
	Triassic		225
Palaeozoic	Permian		
			280
	Carboniferous		
			345
	Devonian		
			395
	Silurian		
			440
	Ordovician		
			500
	Cambrian		
			570
Precambrian	Precambrian		
			4600